AF905550

Conjugated Polymer Synthesis and Materials Applications

Conjugated Polymer Synthesis and Materials Applications

Edited by Yoshiki Chujo

WILEY-VCH

The Editor

Yoshiki Chujo
Department of Polymer Chemistry
Graduate School of Engineering
Kyoto University
Kyoto 615-8510
Japan

Cover Design: Wiley
Cover Image: Courtesy of Yoshiki Chujo

Typesetting: Lumina Datamatics, Madurai, India

Library of Congress Card No.: applied for

British Library Cataloguing-in-Publication Data
A catalogue record for this book is available from the British Library.

Bibliographic information published by the Deutsche Nationalbibliothek
The Deutsche Nationalbibliothek lists this publication in the Deutsche Nationalbibliografie; detailed bibliographic data are available on the Internet at <http://dnb.d-nb.de>.

Print ISBN: 9783527354436
ePDF ISBN: 9783527849444
ePub ISBN: 9783527849437
oBook ISBN: 9783527849451

© 2026 WILEY-VCH GmbH, Boschstraße 12, 69469 Weinheim, Germany

All rights reserved, including those of translation into other languages, text and data mining and training of artificial intelligence technologies or similar technologies. No part of this publication may be reproduced, stored in a retrieval system, or transmitted, in any form or by any means, electronic, mechanical, photocopying, recording or otherwise, without written permission from the publishers.

Registered names, trademarks, etc. used in this book, even when not specifically marked as such, are not to be considered unprotected by law.

The manufacturer's authorized representative according to the EU General Product Safety Regulation is WILEY-VCH GmbH, Boschstr. 12, 69469 Weinheim, Germany, e-mail: Product_Safety@wiley.com.

In view of ongoing research, equipment modifications, changes in governmental regulations, and the constant flow of information relating to the use of experimental reagents, equipment, and devices, the reader is urged to review and evaluate the information provided in the package insert or instructions for each chemical, piece of equipment, reagent, or device for, among other things, any changes in the instructions or indication of usage and for added warnings and precautions. While the publisher and the authors have used their best efforts in preparing this work, including a review of the content of the work, neither the publisher nor the authors make any representations or warranties with respect to the accuracy or completeness of the contents of this work and specifically disclaim all warranties, including without limitation any implied warranties of merchantability or fitness for a particular purpose. No warranty may be created or extended by sales representatives, written sales materials or promotional statements for this work. The fact that an organization, website, or product is referred to in this work as a citation and/or potential source of further information does not mean that the publisher and authors endorse the information or services the organization, website, or product may provide or recommendations it may make. This work is sold with the understanding that the publisher is not engaged in rendering professional services. The advice and strategies contained herein may not be suitable for your situation. You should consult with a specialist where appropriate. Further, readers should be aware that websites listed in this work may have changed or disappeared between when this work was written and when it is read. Neither the publisher nor authors shall be liable for any loss of profit or any other commercial damages, including but not limited to special, incidental, consequential, or other damages.

Printed and bound by CPI Group (UK) Ltd, Croydon, CR0 4YY

C9783527354436_051225

Contents

About the Editor

Yoshiki Chujo

Yoshiki Chujo completed his PhD at Kyoto University in 1980. After working at Nagoya University, Virginia Tech in the United States, and Kyoto University, he has been Professor at Kyoto University during 1994–2018. He is now a Professor Emeritus of Kyoto University. His research interests focus on polymer synthesis and organic–inorganic hybrid materials. He has received the Award for Encouragement of Research in Polymer Science, the Society of Polymer Science, Japan (SPSJ), the Award of the SPSJ, the Scientific Award of the Chemical Society of Japan (CSJ), the Award of the CSJ, the BCSJ Award from the CSJ, the Fellow of the Royal Society of Chemistry, the SPSJ Award for Outstanding Achievement in Polymer Science and Technology, the Honorary Member of the SPSJ (the past President of the SPSJ), and the Medal with Purple Ribbon from Emperor.

Preface

In 2010, I have edited the book "Conjugated Polymer Synthesis – Methods and Reactions" published from WILEY-VCH. I am very happy to hear that this is one of the well-received books in polymer chemistry. Since then, new methodologies have been developed, and the materials from such conjugated polymers are all set to be very important in various fields of materials applications. As a result, I feel that now is the good timing to publish new and updated topics in these fields.

Looking around us, the functional polymeric materials are used everywhere, and it is no exaggeration to say that we are inundated with functional polymers in the present day, from everyday items such as kitchen utensils, stationery, home appliances, and automobiles to computers, artificial organs, and space shuttles, as well as the clothes we wear today. In other words, the functional polymers are indispensable to our lives. It should be pointed out here that conjugated polymers are one of the most important typical functional polymers, having high potential in various fields of applications.

This book "Conjugated Polymer Synthesis and Materials Applications" aims to summarize the fundamentals and major developments in the topics of synthesis of new conjugated polymers, novel methodologies for the preparation of inorganic-elements containing conjugated polymers, and recent applications of these conjugated polymers in electrical, optical, and biomedical fields.

In each of the chapters, all of them written by internationally acclaimed experts, the book covers the whole spectrum of the synthesis of new conjugated polymers from fundamentals to materials applications. I hope that the readers will enjoy this new chemistry and methodology developed throughout the book. Finally, as an editor of this book, I would like to express my sincere gratitude to all authors for their excellent chapter contributions. I also thank the corresponding editorial staff members of WILEY-VCH, especially Dr. Sakeena Quraishi, Sarah Higginbotham, and Vishali Chandra Mohan for their helpful and professional efforts to make it possible to publish this excellent book.

May 2025 Yoshiki Chujo

1

Transition Metal-catalyzed Polycondensation

Takaki Kanbara[1], *and Junpei Kuwabara*[2]

[1]*Institute of Pure and Applied Sciences, University of Tsukuba, Tsukuba, Ibaraki, Japan*
[2]*Tsukuba Research Center for Energy Materials Science (TREMS), Institute of Pure and Applied Sciences, University of Tsukuba, Tsukuba, Ibaraki, Japan*

1.1 Introduction

Transition metal-catalyzed homo- and cross-coupling reactions have emerged as versatile synthetic methodologies that are widely employed for the synthesis of a variety of organic compounds [1–9]. Cross-coupling reactions of various organometallic (Mg, Zn, Sn, B, and Si) reagents and aryl halides are suitable for bond formation at sp^2 and sp carbons. Profs. Negishi, Suzuki, and Heck were awarded the Nobel Prize in Chemistry in 2010 for their pioneering work on Pd-catalyzed cross-coupling reactions in organic synthesis [10–13].

Scheme 1.1 shows the general mechanism of cross-coupling reactions. Ni- and Pd-based complexes are commonly employed as transition metal catalysts. In general, cross-coupling reactions involve three steps: an oxidative addition reaction across the carbon–halogen bond as an electrophile, transmetalation with the main group in the organometallic reagents acting as nucleophiles, and reductive elimination, resulting in the formation of the carbon–carbon bonds and regeneration of the active catalyst. Various organometallic reagents, including Grignard reagents (Kumada–Tamao–Corriu), zinc (Negishi), stannane (Migita–Kosugi–Stille), boron (Suzuki–Miyaura), silane (Hiyama), copper (Sonogashira), lithium, and mercury compounds, can be used [1–13].

In terms of polymer synthesis, various π-conjugated polymers have been designed and developed by the transition metal-catalyzed homo- and cross-coupling polycondensation over the last few decades [14–17]. This chapter describes a fundamental overview of the transition metal-catalyzed homo- and cross-coupling polymerization reactions used in the synthesis of various π-conjugated polymers. The utilization of transition metal-catalyzed C—H bond functionalization and carbon–nitrogen bond formations in synthesizing π-conjugated polymers is also involved.

Conjugated Polymer Synthesis and Materials Applications, First Edition. Edited by Yoshiki Chujo.
© 2026 Wiley-VCH GmbH. All rights reserved, including rights for text and data mining and training of artificial intelligence technologies or similar technologies. Published 2026 by Wiley-VCH GmbH.

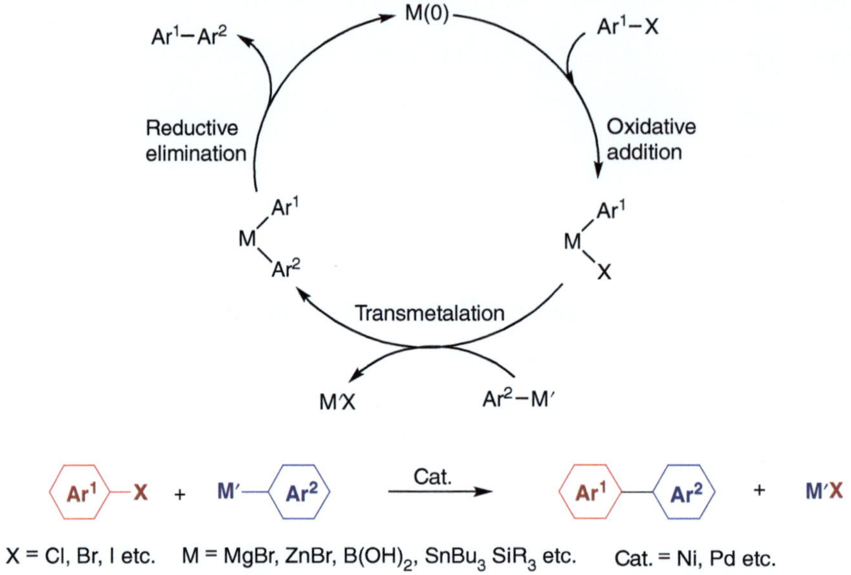

Scheme 1.1 General mechanism of metal-catalyzed cross-coupling reaction. Ar^1 and Ar^2 denote aromatic units, M and M′ represent an organometallic moiety, and X represents halogen or pseudo-halogen.

1.2 Ni-catalyzed Polycondensation

The utilization of transition metal-catalyzed cross-coupling reactions to π-conjugated polymers was first reported by Yamamoto et al. in the synthesis of poly(*p*-phenylene), which was prepared by Ni-catalyzed Grignard coupling of 1,4-dibromobenzene [14, 18, 19]. One of the C—Br bonds reacts with Mg to form the Grignard reagent and undergoes further coupling reactions based on the Kumada–Tamao–Corriu cross-coupling reaction (Scheme 1.2a). Polythiophene (PTh) and poly(3-alkylthiophene) s (PRThs) were synthesized using the same protocol (Scheme 1.2b) [20].

While the reactions are simple and provide several π-conjugated polymers, the regioregularity cannot be controlled using the conventional protocol (Scheme 1.3). The synthesis of regioregular PRThs involving head-to-tail (HT) repeating units can be realized by the Ni-catalyzed polycondensation of regio-controlled organometallic monomers (Scheme 1.4). The reaction of 2,5-dibromo-3-alkylthiophene with metallic Mg or Zn was shown to give the organometallic species at the 5-position. Further addition of the Ni catalyst initiated polymerization to give PRThs with high HT regioregularity (HT-P3RTh, Schemes 1.4a, b) [21–23]. The active monomers for regioregular PRThs were also generated by deprotonation of 2-bromo-3-alkylthiophene with Knochel–Hauser base, 2,2,6,6-tetramethyl piperidinyl magnesium chloride lithium chloride complex (TMPMgCl·LiCl) (Scheme 1.4c) [23]. Regioregular PRThs have been of particular interest because regioregular polymers exhibit remarkable physical properties such as higher crystallinity and electrical conductivity in the solid state than regio-random PRThs [14, 24]. It should be noted

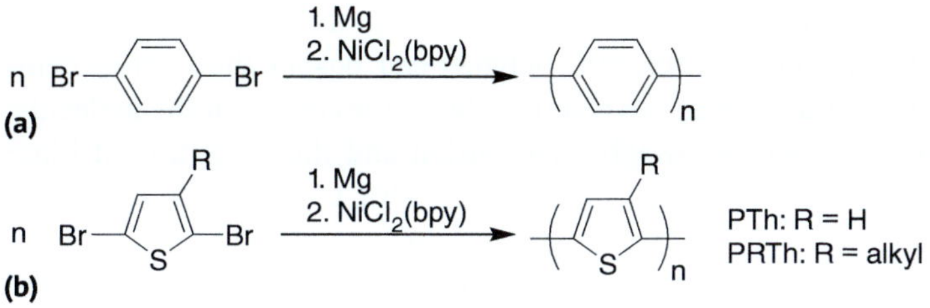

Scheme 1.2 Ni-catalyzed Grignard coupling polycondensation of (a) 1,4-dibromobenzene and (b) 2,5-dibromo-3-alkylthiophene.

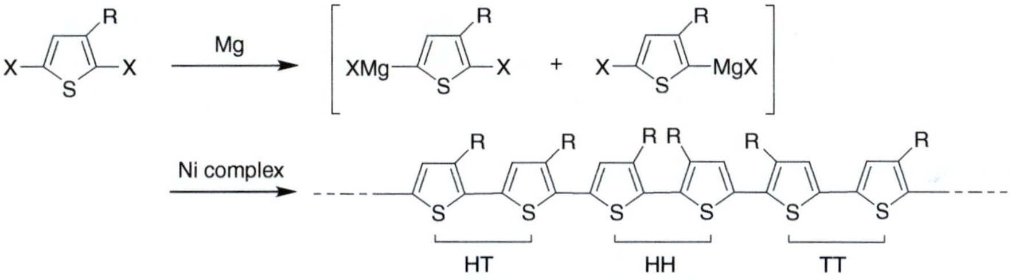

Scheme 1.3 Synthesis of regioirregular PRTh.

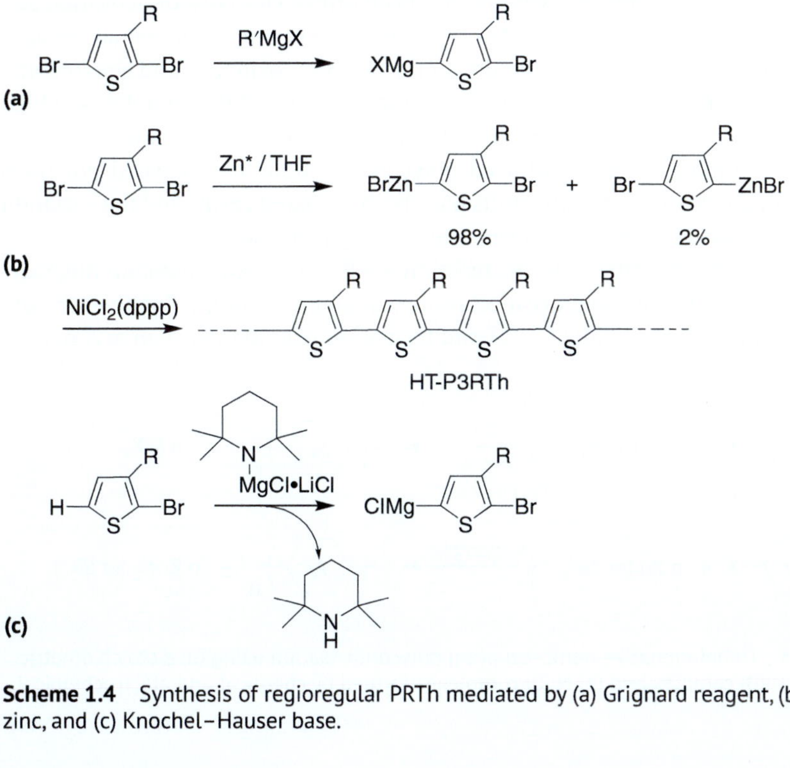

Scheme 1.4 Synthesis of regioregular PRTh mediated by (a) Grignard reagent, (b) Rieke zinc, and (c) Knochel–Hauser base.

that the polymerization reaction in Scheme 1.4 could proceed in a chain-growth manner, in which propagation occurs at the polymer end, despite the reaction being based on a polycondensation. Thus, the reaction allows the control of the molecular weight with a narrow molecular weight distribution and the formation of block copolymers by successive monomer additions [23, 25–30]. Chapter 2 will provide a detailed description of the chain-growth polycondensation.

Dehalogenative homo-coupling polycondensation of dihaloaromatic monomers using zerovalent Ni complexes ($Ni(0)L_m$) such as $Ni(PPh_3)_4$ and $Ni(cod)_2$ (cod = 1,5-cyclooctadiene), has also contributed to design various π-conjugated polymers (Scheme 1.5a) [14, 18]. The protocol proceeds under mild conditions and enables the polycondensation of dihaloaromatic monomers, such as 2,5-dibromopyridine and 1,4-dichloroanthracene, which has not been achieved by other polycondensation methods. While the reaction generally requires a stoichiometric amount of Ni(0) complex, the $Ni(0)L_m$ complexes formed in situ by chemical (e.g. by Zn, NaH, and hydrazine hydrate) or electrochemical reduction of Ni(II) compounds are available for polycondensation [31–35] thus facilitating the following catalytic reactions (Scheme 1.5b).

1.3 Pd-catalyzed Polycondensation

Pd complexes are known as useful catalysts for various C—C coupling reactions, particularly the Pd-catalyzed Migita–Kosugi–Stille, Suzuki–Miyaura, Sonogashira, and Mizoroki–Heck coupling reactions have been utilized for polycondensation giving various π-conjugated polymers [14–17]. This is due to some inherent advantages such as increased diversity and tenability of the catalysts, oxidative and aqueous stability, and relatively facile isolation and structural analysis of the complexes, which aid their mechanistic and methodological developments. With the development of these cross-coupling reactions utilized polycondensation, π-conjugated polymers consisting of alternating aromatic units have been prepared easily, and the variation of the π-conjugated polymers has been dramatically increased.

The Migita–Kosugi–Stille coupling reaction is a Pd-catalyzed cross-coupling reaction between aryl halides and organostannic compounds and the Suzuki–Miyaura coupling reaction is a versatile C—C bond formation reaction between aryl halides

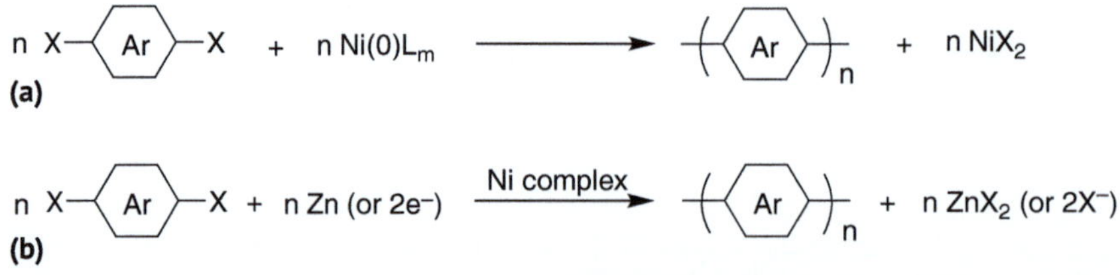

Scheme 1.5 Dehalogenative homo-coupling polycondensation using (a) a stoichiometric amount of Ni(0) complex and (b) Ni(0) complexes formed by chemical and electrochemical reduction.

and organoboronic acids. One important difference between the cross-coupling reactions is that the Suzuki–Miyaura coupling reaction requires the activation of the organoboronic acid with a base. This activation facilitated transmetalation (Scheme 1.6). The major advantage of these cross-coupling reactions is that they can tolerate various functional groups and proceed under mild conditions. The Suzuki–Miyaura coupling reaction can even be conducted in aqueous media. The feature of the coupling reactions offers the opportunity to design a variety of π-conjugated polymers with functional moieties manifesting many highly desirable properties (Scheme 1.7a, b) [14–17, 36, 37]. In addition, these cross-coupling polymerization protocols also proceed in a chain-growth polymerization manner from the initiator unit derived from the Pd catalyst [25–30]; the details are to be described in Chapter 2.

The Pd-catalyzed cross-coupling of aryl halides with terminal alkynes employing co-catalytic Cu(I) halides and an amine base is referred as the Sonogashira coupling reaction [9]. The polycondensation reactions between dihaloaromatic monomers and diethynyl aromatic monomers give poly(arylene ethynylene) type π-conjugated polymers (Scheme 1.7c) [38]. In this reaction, the Cu(I) halides react with the terminal alkyne to produce a Cu(I) acetylide, which serves as an activated species for the coupling reaction. Transmetalation proceeded in the usual manner for the Pd complex (Scheme 1.1). A Cu-co-catalyzed mechanism for the Sonogashira coupling reaction was also proposed since the rate of the reaction was affected by the nature of the substituent and the halide of the copper(I) salt as well as the aryl halide [39].

The Mizoroki–Heck coupling is a Pd-catalyzed cross-coupling reaction between aryl halides and alkenes in the presence of a base to form aryl alkenes [12]. The reaction differs from other Pd-catalyzed cross-coupling reactions because it does

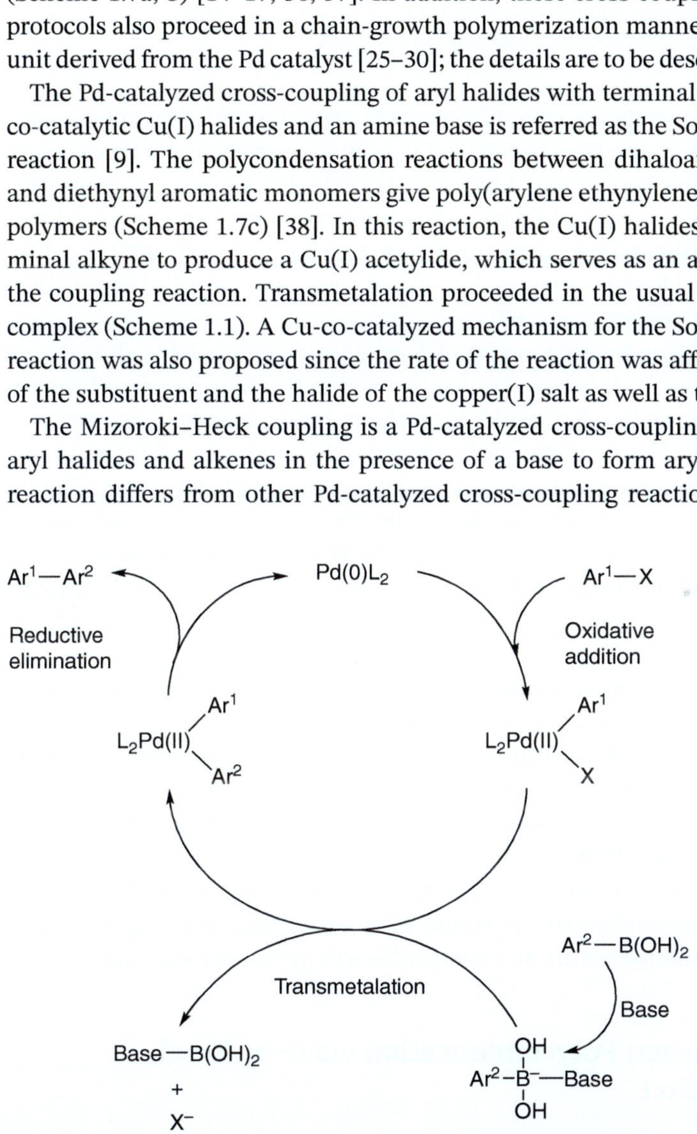

Scheme 1.6 Reaction mechanism of the Suzuki–Miyaura coupling reaction. Ar^1 and Ar^2 denote aromatic units and X denotes halogen or pseudo-halogen.

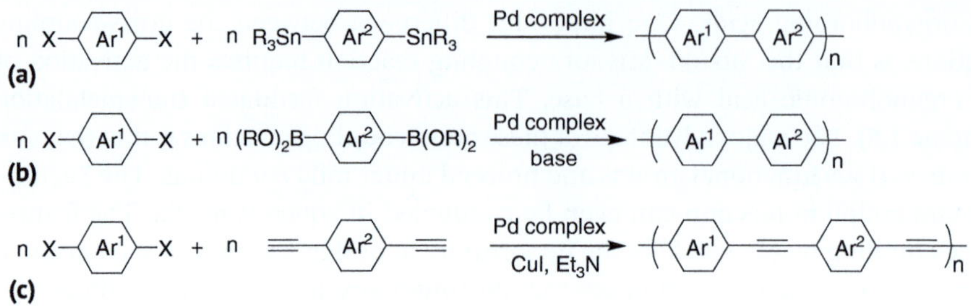

Scheme 1.7 Pd-catalyzed polycondensation via (a) Migita–Kosugi–Stille, (b) Suzuki–Miyaura, and (c) Sonogashira coupling reactions.

not involve a transmetalation step. Alternatively, after the oxidative addition step, the alkene inserts into the Pd—C bond, and the coupling products are produced via a β-hydrogen elimination (Scheme 1.8a). The Mizoroki–Heck reaction is compatible with the synthesis of several poly(*p*-phenylene vinylene) (PPV) derivatives and their analogs, namely poly(arylene vinylene)s (PAVs, Scheme 1.8b) [40]. PPV and PAVs have attracted considerable attention because of their high electrical conductivity and optoelectronic properties [41, 42].

Pd-catalyzed carbon–heteroatom bond formation reactions have also become valuable synthetic tools. In particular, the Pd-catalyzed amination of aryl halides, commonly known as the Buchwald–Hartwig coupling reaction, has attracted attention [43, 44]. The catalytic cycle involves the oxidative addition of aryl halides, followed by the coordination and deprotonation of the amine. The use of bases such as sodium *tert*-butoxide is effective for the deprotonation of the reacting amine. Finally, the reductive elimination of the *N*-aryl product occurs (Scheme 1.9). The aryl amination reaction relies on the design of the catalyst systems, particularly the choice of a ligand, to prevent a side reaction through β-hydrogen elimination.

The Buchwald–Hartwig aryl amination reaction is an effective tool for the synthesis of polyaniline analogs and poly(triarylamine)s, which are difficult to prepare via conventional oxidative polymerization (Scheme 1.10) [45]. It is essential to select a suitable ligand for the desired polymer structure. While 2-(di-*tert*-butylphosphino) biphenyl or 2-dicyclohexylphosphino-2′,4′,6′-triisopropylbiphenyl (X-Phos) is a suitable ligand for the synthesis of polyaniline analogs bearing a secondary amine unit (Scheme 1.10a–c) [46–48], tri-*tert*-butylphosphine (P^tBu_3) is one of the optimal ligands for the synthesis of poly(triarylamine)s (Scheme 1.10d, e) [49–52]. Analogous Ni-catalyzed aryl aminations are also compatible with polycondensation [53].

1.4 Pd-catalyzed Polycondensation via C—H Bond Functionalization

As mentioned above, various π-conjugated polymers have been synthesized by metal-catalyzed cross-coupling reactions. Recently, direct arylation reactions using the C—H bonds of aromatic compounds as the reaction points have attracted

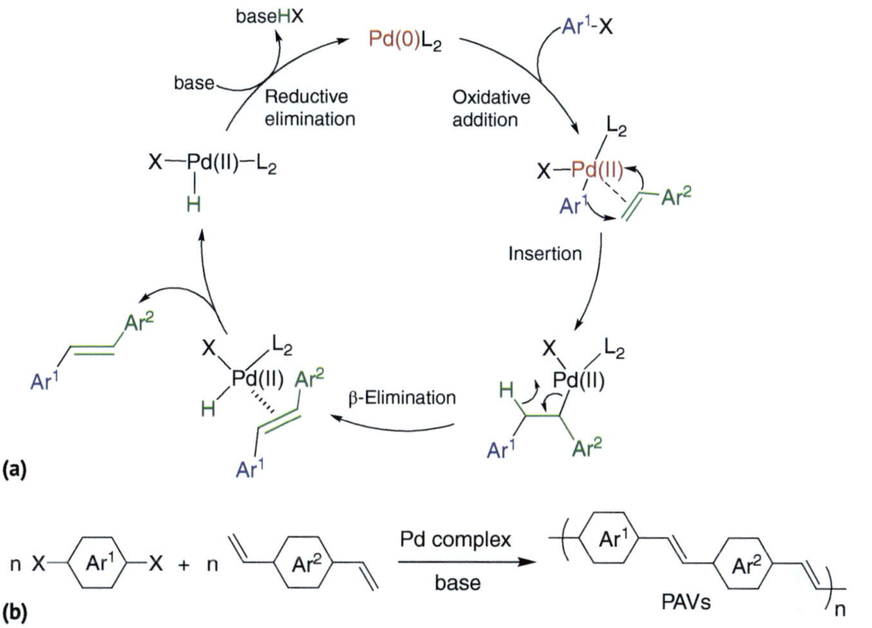

Scheme 1.8 (a) Reaction mechanism of the Mizoroki–Heck coupling reaction. Ar^1 and Ar^2 denote aromatic units and X denotes halogen or pseudo-halogen. (b) Synthesis of PAVs via the Mizoroki–Heck coupling reaction.

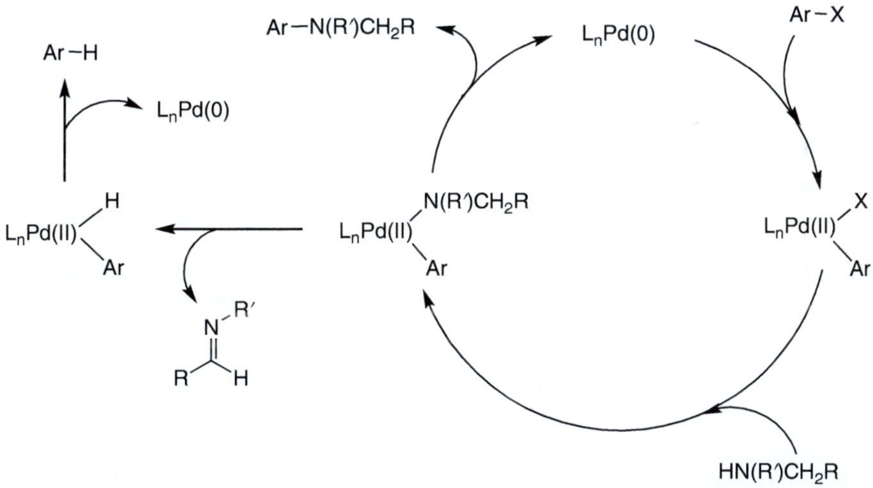

Scheme 1.9 Reaction mechanism of the Buchwald–Hartwig aryl amination reaction. Ar denotes aromatic units, R and R′ denote alkyl or aryl groups, and X represents halogen or pseudo-halogen.

attention as alternatives to conventional cross-coupling reactions (Scheme 1.11) [54–56]. This reaction does not require the introduction of an organometallic moiety into the reactant, thus reducing the number of synthetic steps. Furthermore, while cross-coupling reactions produce metal-containing by-products, the formal

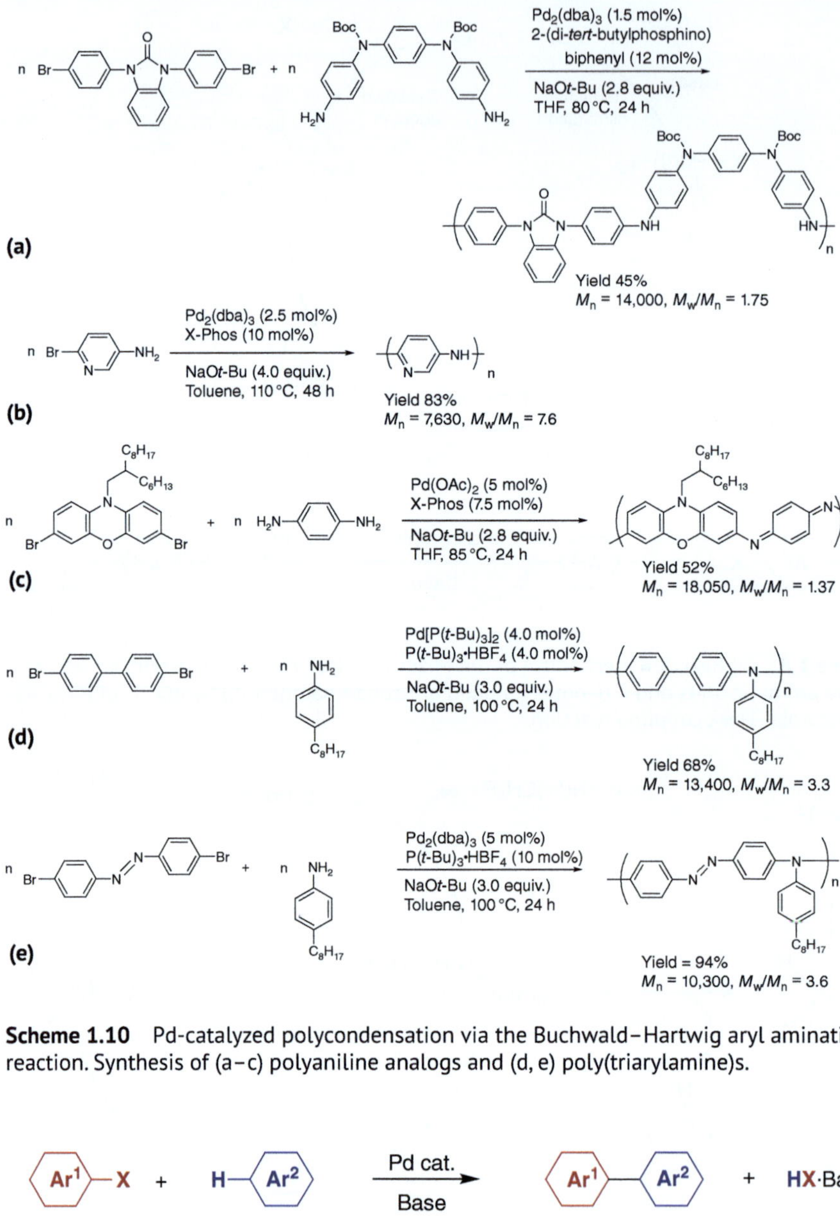

Scheme 1.10 Pd-catalyzed polycondensation via the Buchwald–Hartwig aryl amination reaction. Synthesis of (a–c) polyaniline analogs and (d, e) poly(triarylamine)s.

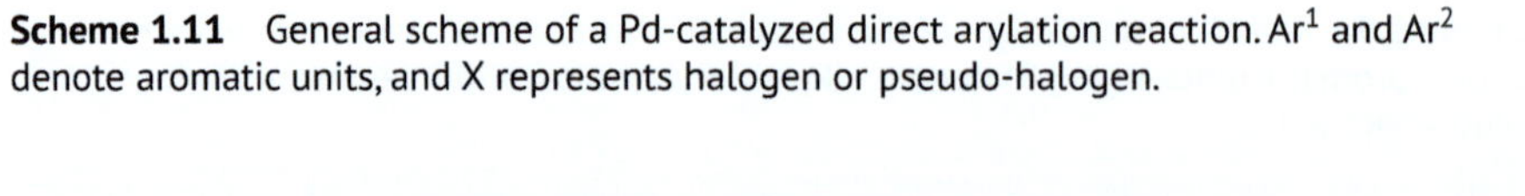

Scheme 1.11 General scheme of a Pd-catalyzed direct arylation reaction. Ar^1 and Ar^2 denote aromatic units, and X represents halogen or pseudo-halogen.

by-product of the direct arylation reaction is hydrogen halides (HX), which reduces the environmental impact. Owing to these advantages, polycondensation reactions involving direct arylation have been actively developed in recent years [57–61].

Lemaire et al. reported the first direct arylation polycondensation of 2-bromo-3-octylthiophene for the synthesis of poly(3-octylthiophene) [62, 63]. Although these studies pioneered a novel synthetic methodology, the regioregularities and molecular weights of the resulting polymers were inferior to those prepared by conventional cross-coupling polycondensation. In 2010, Ozawa et al. reported the successful synthesis of high-molecular-weight poly(3-hexylthiophene) with high regioregularity from 2-bromo-3-hexylthiophene, using their original catalytic system (Scheme 1.12) [64].

Subsequently, the direct arylation polycondensation of 1,2,4,5-tetrafluorobenzene with 2,7-dibromo-9,9-dioctylfluorene afforded the corresponding polymer with a high molecular weight in good yield (Scheme 1.13) [65]. This molecular weight is higher than that prepared by conventional polycondensation using the Suzuki–Miyaura coupling [66]. The polymer exhibited efficacy as a hole-blocking material in organic light-emitting diodes, which can be attributed to the low HOMO energy levels resulting from the presence of multiple F substituents [67].

Scheme 1.14 shows the general reaction mechanism for the direct arylation [68]. As in conventional cross-coupling reactions, direct arylation is initiated by the oxidative addition across the carbon–halogen bond. Subsequently, ligand exchange occurs between the halogen and carboxylate, which is generated by the reaction of a carboxylic acid additive with a base. The carboxylato-Pd complex undergoes C—H bond cleavage with the assistance of the carboxylato ligand and Pd—C bond formation in a concerted manner. This process is referred as the concerted metallation–deprotonation (CMD) pathway [69]. The reductive elimination of the biaryl complex results in the formation of the product and the regeneration of the active Pd catalyst. Carboxylic acid additives are important for direct arylation because they affect the transition state of CMD.

The effects of carboxylic acid additives were investigated in a model reaction using 3,4-ethylenedioxythiophene (EDOT) as the C—H monomer (Scheme 1.15) [70]. Weak carboxylic acids with bulky substituents, such as 1-adamantanecarboxylic acid (1-AdCOOH) and pivalic acid, are effective additives because of the high molecular weight of the resultant polymers and high yield. Weak acidity is advantageous because a weak acid acts as a strong conjugate base in the coordination sphere of the transition state of the CDM pathway. In addition, bulkiness might prevent the aggregation of the Pd catalyst, which facilitates the reaction [71]. Owing to the high efficiency of the catalytic system with 1-AdCOOH, only 1 mol% of $Pd(OAc)_2$ was

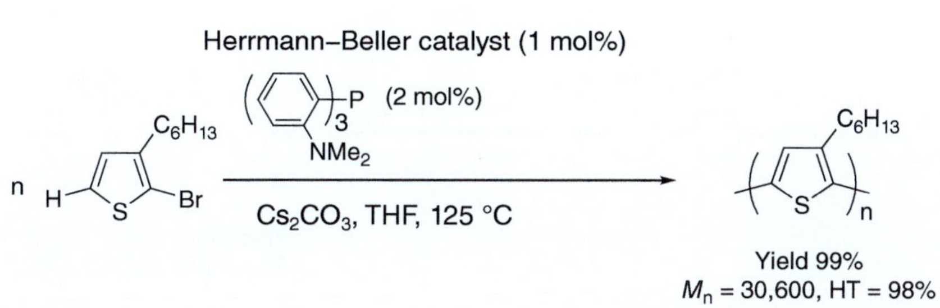

Scheme 1.12 Synthesis of poly(3-hexylthiophene) by direct arylation polycondensation.

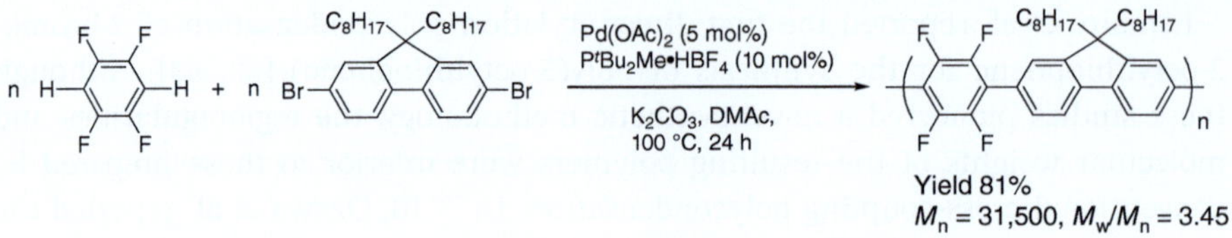

Scheme 1.13 Direct arylation polycondensation of tetrafluorobenzene.

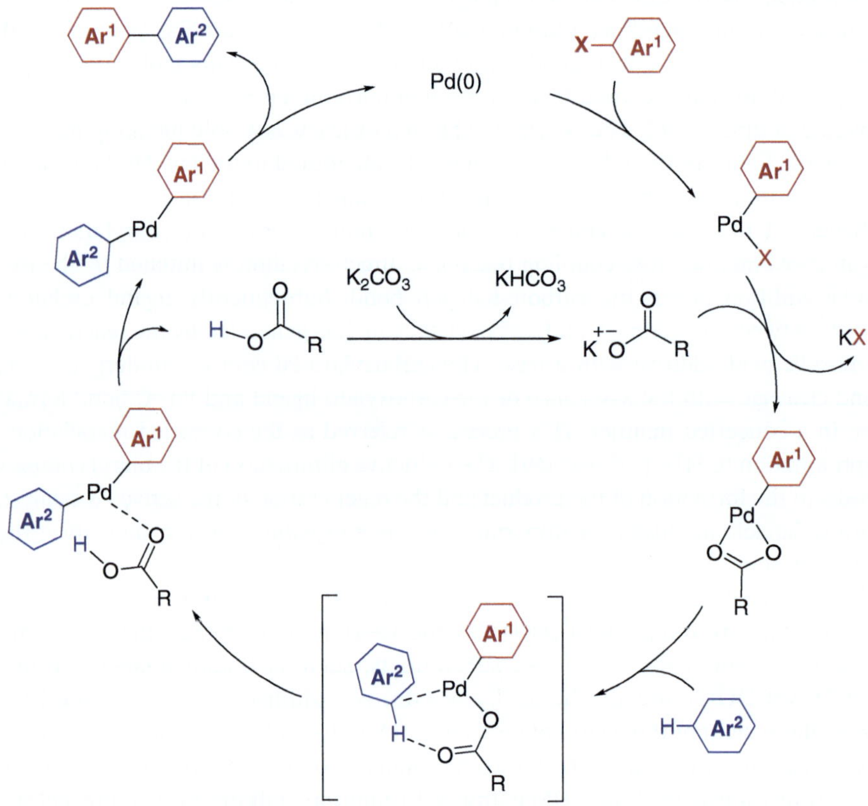

Scheme 1.14 Reaction mechanism of a Pd-catalyzed direct arylation reaction. Ar^1 and Ar^2 denote aromatic units, R denotes an alkyl group, and X represents halogen or pseudo-halogen.

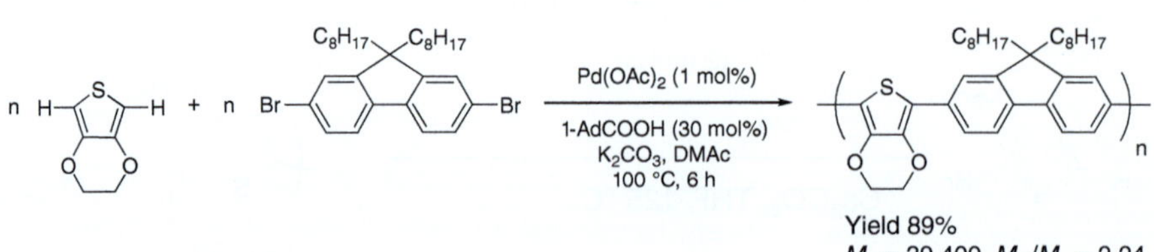

Scheme 1.15 Direct arylation polycondensation of EDOT.

required to afford a high-molecular-weight polymer in good yield (Scheme 1.15) [70]. The advantages of using bulky carboxylic acids, such as neodecanoic acid, have also been reported [72, 73]. Direct arylation polycondensation of thiophene monomers bearing electron-donating substituents is generally conducted in polar solvents such as *N*,*N*-dimethylacetamide (DMAc) [74–76].

Scheme 1.16 shows the direct arylation polycondensation reaction for the synthesis of a donor–acceptor-type conjugated polymer suitable for organic photovoltaic applications [77]. The optimal reaction conditions for the direct arylation of acceptor monomers bearing electron-withdrawing substituents are different from those for donor monomers such as EDOT, as shown in Scheme 1.15 [78]. For acceptor monomers, the combination of a phosphine ligand (PCy_3) and low-polarity solvent (toluene) is effective for obtaining the targeted polymer.

The catalytic system with the phosphine and nitrogen ligands demonstrated high selectivity and yielded polymers without structural defects (Scheme 1.17) [79]. In direct arylation reactions, there are potential concerns regarding side reactions involving C—H bonds at undesired positions and C—H/C—H and C—Br/C—Br homo-coupling reactions. The mixed-ligand approach minimizes these side reactions and prevents the formation of branching, crosslinking, and homo-coupling defects, even in difficult-to-control monomers, such as unsubstituted bithiophene [80]. The obtained polymers could serve as high-performance materials for organic photovoltaics.

Polycondensation using the C—H/C—H homo-coupling reaction is a facile method for synthesizing homopolymers because there is no need for reactive functional groups such as halogens and metal moieties. This method is referred to as oxidative direct arylation polymerization [81]. Scheme 1.18 shows the oxidative direct arylation polymerization of thiazole-based monomers using a Pd catalyst and metal oxidants such as $Cu(OAc)_2$ [82, 83]. Further investigations have led to the development of an environmentally friendly oxidative direct arylation polymerization method that does not utilize a metal oxidant and employs oxygen as the oxidant [84].

Polycondensation reactions using cross-dehydrogenative coupling (C—H/C—H cross-coupling) allow the synthesis of polymers that link two monomers in a short synthetic process [85–87]. However, to achieve this, it is necessary to suppress undesired homo-coupling reactions because both reaction points are C—H bonds [88].

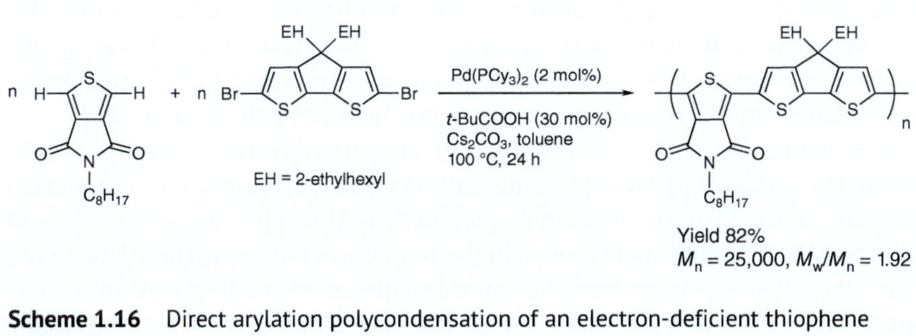

Scheme 1.16 Direct arylation polycondensation of an electron-deficient thiophene derivative.

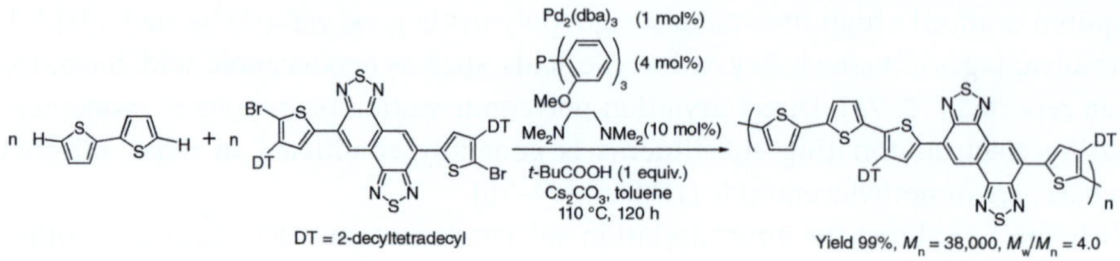

Scheme 1.17 Synthesis of a defect-free conjugated polymer by direct arylation polycondensation.

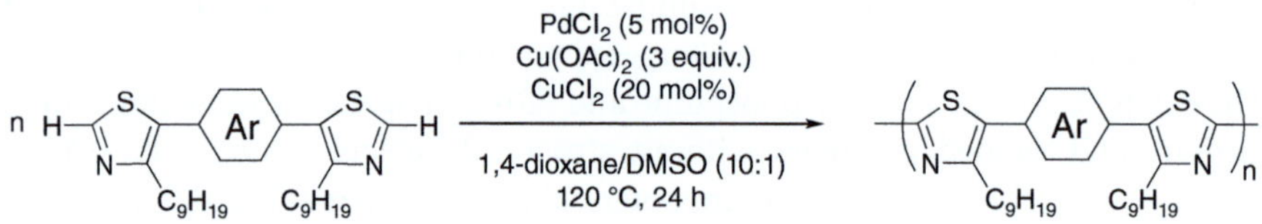

Scheme 1.18 Oxidative direct arylation polycondensation.

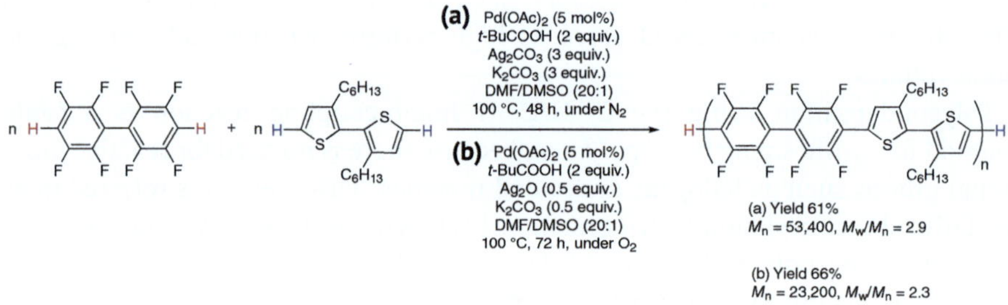

Scheme 1.19 Cross-dehydrogenative-coupling polycondensation.

Scheme 1.19 shows cross-dehydrogenative-coupling polycondensation of 2,2′,3,3′,5,5′,6,6′-octafluorobiphenyl with 3,3′-dihexyl-2,2′-bithiophene [89]. The resultant polymer had a highly alternating structure with only minor homo-coupling defects (approximately 4%). Silver salts are responsible for cleaving the C—H bonds of octafluorobiphenyl, in addition to acting as oxidants. The combined action of the Pd catalyst and the Ag salt prevents homo-coupling [90, 91]. The reaction also proceeded in air, and the use of O_2 as a terminal oxidant resulted in a lower Ag oxidant loading (Scheme 1.18) [89]. Reports on the associated direct sp^2 C—H coupling polycondensation of thiophene derivatives have also been published [92–96].

The dehydrogenative alkenylation of arenes, known as direct alkenylation, is applicable to the synthesis of PAVs (Scheme 1.20) [97–99]. This protocol is a promising alternative to the Mizoroki–Heck-type polycondensation, allowing straightforward access to PAVs without bromo groups in the monomers [40, 100]. The Pd-catalyzed direct alkenylation polycondensation afforded the corresponding PAV in a *trans* configuration (Scheme 1.20) [101]. The polymer contained only 3% homo-coupling defects. The obtained polymer was used as the emitting material in an OLED device.

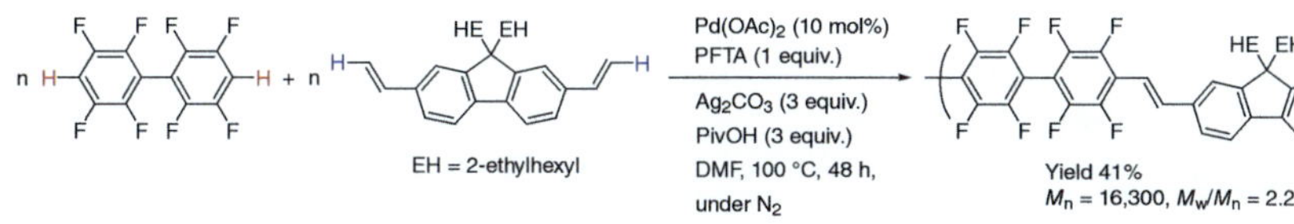

Scheme 1.20 Direct alkenylation polycondensation.

In addition, the direct alkenylation polymerization of diketopyrrolopyrrole monomers [102] and polymerization using directing groups have also been reported [103].

1.5 Cu- and Fe-catalyzed Polycondensation

Because Cu is an earth-abundant metal with lower toxicity than Pd and Ni, there has been increased interest in Cu-catalyzed homo- and cross-coupling reactions. The classical Ullmann coupling reactions involve the synthesis of symmetric biaryl compounds. While the traditional Ullmann coupling reaction has some drawbacks, including the necessity of high temperatures and large amounts of Cu, the use of appropriate ligands can modulate the reactivity and achieve an effective and versatile catalytic system [104–107]. Microwave-assisted Ullmann homo-coupling reaction has been applied to the synthesis of acceptor-type π-conjugated polymers (Scheme 1.21) [108].

The Ullmann-type cross-coupling reactions involve nucleophilic aromatic substitutions of aryl halides with various nucleophiles, such as amines. Cu can participate in cross-coupling reactions in a manner similar to that of Pd. An important difference with respect to Cu is the easy accessibility of the four oxidation states (0 to +3) of the metal; the cross-coupling catalytic cycle with Cu is most likely serviced by the +1/+3 oxidation states (Scheme 1.22a). The Cu-catalyzed aryl amination reaction can also be used to prepare poly(triarylamine)s containing light-harvesting dyes with low bandgaps (Scheme 1.22b) [109, 110].

It has been reported that CuI/PPh_3 or phenanthroline-based catalysts promote the direct arylation of heteroaromatic compounds with aryl halides [111, 112]. Cu-catalyzed direct arylation polycondensation has also been reported by Thompson et al. (Scheme 1.23a) [113]. While the Cu catalyst does not show high reactivity and broad availability compared with those of the Pd catalyst at present, the use of CuBr(phen)(PPh_3) as a precatalyst could overcome these issues (Scheme 1.23b) [114]. Furthermore, Cu-catalyzed direct arylation achieved C—H bond activation at the 2-position in thiazole-based monomers (Scheme 1.23c). This is in contrast to Pd-catalyzed direct arylation, in which the C—H bond at the 5-position of the thiazole moiety is highly reactive [115, 116]. The site selectivity could apply to the unique molecular design of π-conjugated polymers.

The 2-position of the thiazole moiety is also highly reactive in Cu-catalyzed aerobic oxidative C—H/C—H homo-coupling reactions [117]. Cu-catalyzed aerobic oxidative homo-coupling polycondensation affords thiazole- and benzodiimidazole based π-conjugated polymers (Scheme 1.24) [118, 119]. The synthetic method is a

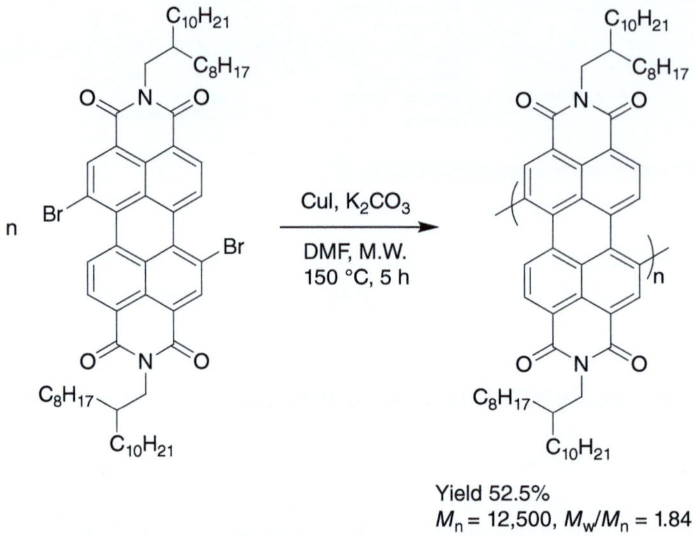

Scheme 1.21 Ullmann homo-coupling polycondensation.

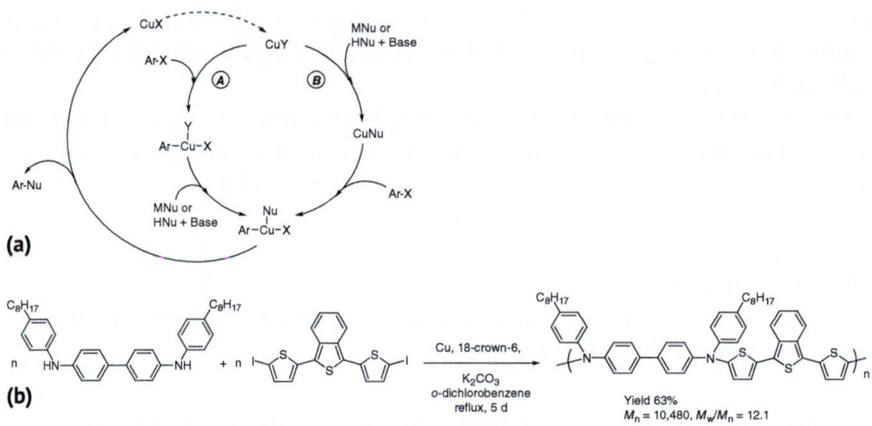

Scheme 1.22 (a) Reaction mechanism of the Ullmann-type cross-coupling reaction. Ar denotes aromatic unit, Nu denotes nucleophilic species, and X and Y represent halogen or pseudo-halogen. (b) Synthesis of poly(triarylamine) by the Ullmann-type cross-coupling polycondensation.

low-cost and environmentally friendly process because of the naturally abundant feedstock of dioxygen and harmless by-products, such as water.

Iron is the best alternative to precious metals as an earth-abundant catalyst. Nakamura et al. demonstrated the Fe-catalyzed regioselective C—H/C—H homo-coupling polycondensation of thiophene derivatives (Scheme 1.25a) [120]. The use of conjugated tridentate phosphine as a ligand suppresses catalyst deactivation, whereas diethyl oxalate, used together with $AlMe_3$, acts as a two-electron acceptor in C—H activation. The polycondensation reaction proceeds under

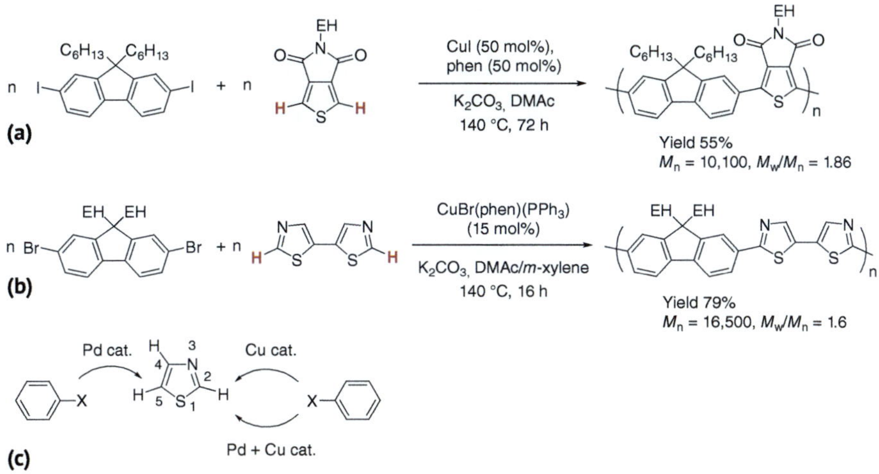

Scheme 1.23 Cu-catalyzed direct arylation polycondensation of (a) thieno[3,4-c]-pyrrole-4,6-dione and (b) 5,5′-bithiazole. (c) Selective direct arylation on thiazole unit.

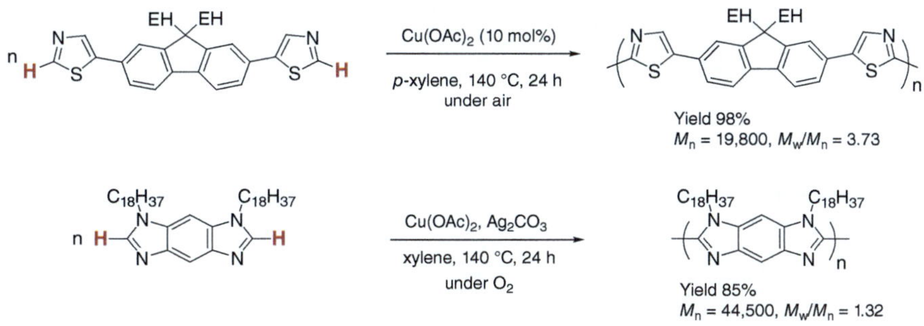

Scheme 1.24 Cu-catalyzed aerobic oxidative homo-coupling polycondensation.

mild conditions and affords a variety of thiophene-based π-conjugated polymers bearing electron-donative π motifs. The cooperative catalytic systems using two earth-abundant metals, Fe and Al, are also compatible with C—H/C—H cross-coupling reaction between thiophenes and enamines and its application to polycondensation of bisenamine and bisthiophene derivatives (Scheme 1.25b) [121].

1.6 Other Transition Metal-catalyzed Polymerization

The acyclic diene metathesis (ADMET) polycondensation of divinyl aromatic monomers is another promising synthetic method for PAVs [122–125]. The catalytic cycle begins with the metathesis of the terminal olefin of a divinyl monomer using a metal alkylidene catalyst. The newly formed metal alkylidene species then undergoes metathesis with another terminal olefin of the monomer to recover the metal methylidene species and produce an internal olefin (Scheme 1.26a). Because ADMET is an equilibrium reaction, the removal of ethylene by-products from the reaction

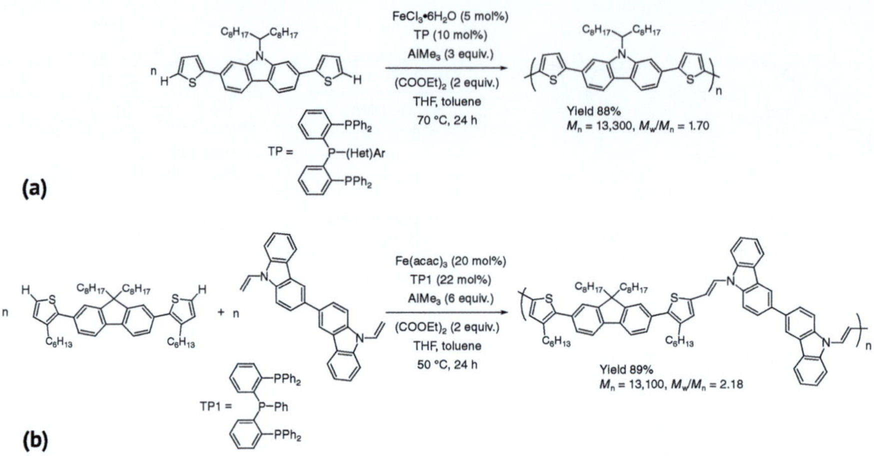

Scheme 1.25 Fe-catalyzed dehydrogenative (a) homo-coupling and (b) cross-coupling polycondensation.

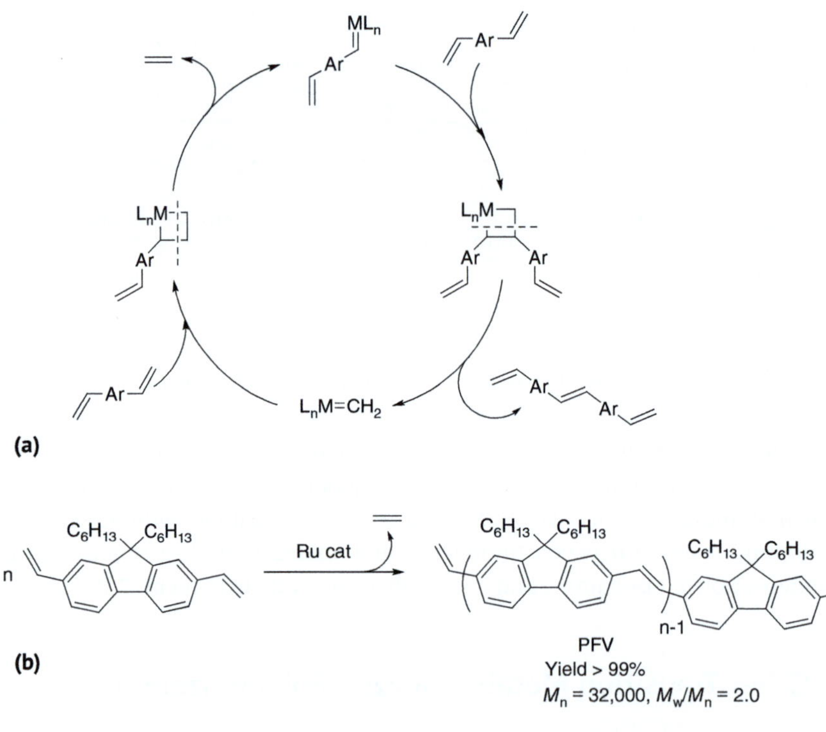

Scheme 1.26 (a) Reaction mechanism of an acyclic diene metathesis (ADMET) polycondensation. Ar denotes aromatic unit. (b) Ru-catalyzed ADMET polycondensation.

medium is a prerequisite for obtaining PAVs with high molecular weights. ADMET polycondensation of 2,7-divinyl-9,9-dialkylfluorenes by Ru-carbene catalysts provides all-*trans* poly(9,9-di-*n*-octylfluorene-2,7-vinylene) (PFV) with a high molecular weight, defect-free, and well-defined chain ends (Scheme 1.26b) [122, 125].

PPV derivatives have also been prepared by the ring-opening metathesis polymerization (ROMP) of strained cyclophanedienes initiated by Ru-carbene catalysts (the second-generation Grubbs catalyst) (Scheme 1.27) [126]. Once the cyclophanediene has been consumed completely, the polymer chain end is still active, and the addition of further cyclophanediene gives block copolymers. This protocol was also applicable to the synthesis of block copolymers of PAVs consisting of donor–acceptor fractions [127].

Polyaddition via the hydroarylation of aromatic diynes with arenes is a promising alternative synthetic methodology for PAVs because the production of by-products from monomers can be eliminated. The synthesis of PAVs via the Rh-catalyzed hydroarylation of internal diynes has been reported [128], whereas the Cp*Co(III)-catalyzed hydroarylation polyaddition of aromatic terminal diynes to arenes bearing directing groups provides the corresponding PAVs under mild reaction conditions (Scheme 1.28) [129–132]. The protocol includes an *ortho*-metalation process, which facilitates site-selective reaction. However, the reaction sometimes produces a 1,1-vinylidene unit from a side reaction during the migratory insertion of the alkyne unit into the Co catalyst, which lowers the regioselectivity of PAVs. The presence of 1,1-vinylidene unit serves as a defect in π-conjugated polymer materials that decreases the semiconducting device performance [129]. This difficulty has been overcome by designing targeted aromatic monomers bearing appropriate directing groups, resulting in site-specific and regioselective polymerization. The synthesized PAVs serve as emitting materials and p-type semiconductors for organic optoelectronic devices [131, 132]. Mn-catalyzed hydroarylation is compatible with the polyaddition protocol, affording pyrrole-based PAVs with quantitative site- and regioselectivity [133].

Transition metal-catalyzed C—H bond activation followed by alkyne insertion is a useful method for the synthesis of fused aromatic rings. Rh-catalyzed C—H bond cleavage of the acetanilide and alkyne insertion forms indoles [134]. The annulation reaction applied to polycondensation of 1,4-bis(acetylamino)benzene and various diynes, affording π-conjugated polymers containing benzodipyrrole moieties with high regioselectivity (Scheme 1.29a) [135]. Oxidative Rh-catalyzed successive C—H activation–annulation reactions of 1-phenylimidazole with alkynes were also applicable to polycondensation, affording main chain-charged conjugated

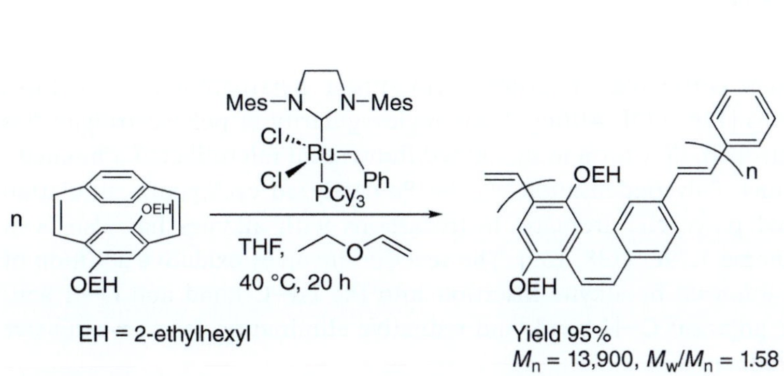

Scheme 1.27 Ru-catalyzed ring-opening metathesis polymerization.

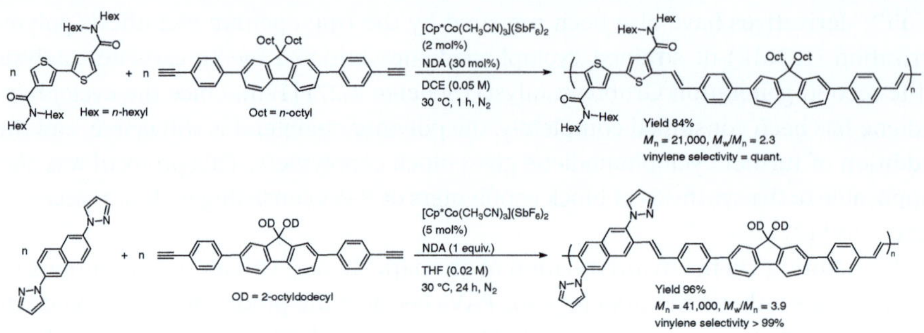

Scheme 1.28 Co-catalyzed hydroarylation polyaddition.

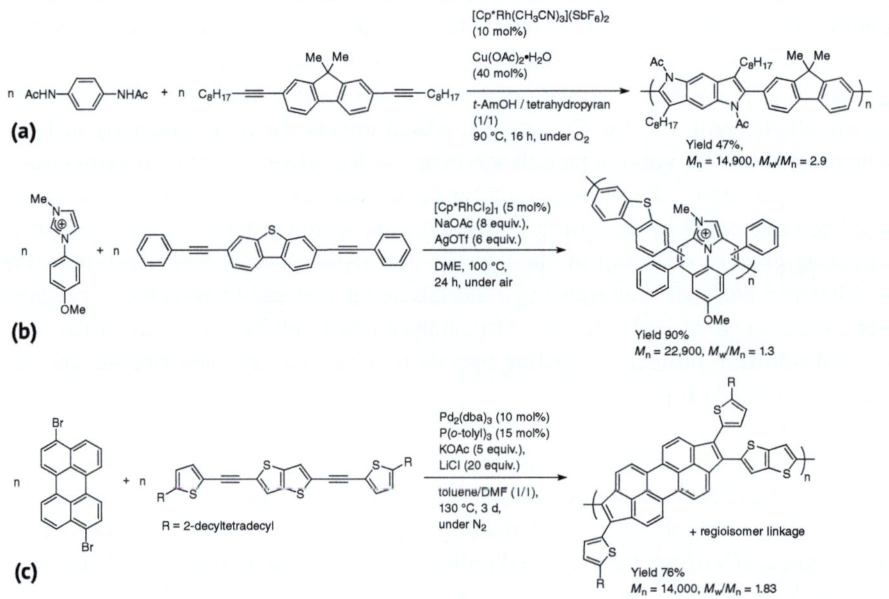

Scheme 1.29 Successive C—H activation–annulation polycondensation of (a) 1,4-bis (acetylamino)benzene, (b) 1-phenyl-imidazolium compound, and (c) 3,9-dibromoperylene with aromatic diynes.

polyelectrolytes consisting of cationic ring-fused heteroaromatic structures (Scheme 1.29b) [136, 137]. Although the regioregularity of polyelectrolytes has not been controlled, they form multicolored fluorescent microfibers for biomedical applications. Polycondensation via the Pd-catalyzed cyclopenta-annulation of brominated polycyclic aromatic hydrocarbons with alkynes has also been reported (Scheme 1.29c) [138, 139]. The reaction involves oxidative addition of C—Br bond, followed by alkyne insertion into the Pd—C bond and C—H activation of the adjacent C—H bond, and reductive elimination forms π-extensive annulated products.

1.7 Summary

Transition metal-catalyzed homo- and cross-coupling reactions are essential tools for the synthesis of varied π-conjugated polymers using diverse monomer materials. To produce high-performance organic optoelectronic materials, the synthesized π-conjugated polymers must have a high molecular weight, defect-free structure, and high purity [16, 140]. Thus, polymerization reactions should proceed efficiently without side reactions because regioirregular structures in the polymer chain cannot be removed from the polymer backbone, even in the purification step. Therefore, appropriate monomer choices and detailed optimization of the reaction conditions are vital. Conversely, the chemical structures of state-of-the-art π-conjugated polymer materials tend to be too complex to meet the requirements for qualifying as materials for organic optoelectronic devices. Complex structures bearing multi-component units require numerous synthesis steps. The synthetic routes of monomers have to be more sophisticated in terms of a reaction step because a successful practical application of π-conjugated polymers will depend on the overall synthetic route [141–145]. Polymerization through direct C—H bond functionalization should contribute to the practical use of π-conjugated polymer materials. In addition, catalysts that enable highly efficient and new coupling reactions with less environmental impact should be developed and polymerization protocols must be updated to incorporate them.

References

1 Miyaura, N. and Suzuki, A. (1995). Palladium-catalyzed cross-coupling reactions of organoboron compounds. *Chemical Reviews* 9 (7): 2457–2483. https://doi.org/10.1021/cr00039a007.

2 Rosen, B.M., Quasdorf, K.W., Wilson, D.A. et al. (2011). Nickel-catalyzed cross-couplings involving carbon—oxygen bonds. *Chemical Reviews* 111: 1346–1416.

3 Martin, R. and Buchwald, S.L. (2008). Palladium-catalyzed Suzuki—Miyaura cross-coupling reactions employing dialkylbiaryl phosphine ligands. *Accounts of Chemical Research* 41 (11): 1461–1473. https://doi.org/10.1021/ar800036s.

4 Suzuki, A. (1999). Recent advances in the cross-coupling reactions of organoboron derivatives with organic electrophiles, 1995–1998. *Journal of Organometallic Chemistry* 576: 147–168.

5 Tamao, K., Sumitani, K., Kiso, Y. et al. (1969). Nickel-phosphine complex-catalyzed Grignard coupling. I. Cross-coupling of alkyl, aryl, and alkenyl Grignard reagents with aryl and alkenyl halides: general scope and limitations. *Bulletin of the Chemical Society of Japan* 49: 1958–1969.

6 Negishi, E. (2007). Transition metal-catalyzed organometallic reactions that have revolutionized organic synthesis. *Bulletin of the Chemical Society of Japan* 80: 233–257.

7 Fugami, K. and Kosugi, M. (2002). Organotin compounds. In: *Topics in Current Chemistry* (ed. N. Miyaura), 87–130. Berlin, Heidelberg: Springer.

8 Hiyama, T. and Shirakawa, E. (2002). Organosilicon compounds. In: *Topics in Current Chemistry* (ed. N. Miyaura), 61–85. Berlin, Heidelberg: Springer.

9 Sonogashira, K. (2002). Development of Pd–Cu catalyzed cross-coupling of terminal acetylenes with sp2-carbon halides. *Journal of Organometallic Chemistry* 653: 46–49.

10 Suzuki, A. (2011). Cross-coupling reactions of organoboranes: an easy way to construct C–C bonds (Nobel Lecture). *Angewandte Chemie International Edition* 50 (30): 6722–6737. https://doi.org/10.1002/anie.201101379.

11 Negishi, E. (2011). Magical power of transition metals: past, present, and future (Nobel Lecture). *Angewandte Chemie International Edition* 50 (30): 6738–6764. https://doi.org/10.1002/anie.201101380.

12 Heck, R. (2006). Cobalt and palladium reagents in organic synthesis: the beginning. *Synlett* 2006 (18): 2855–2860. https://doi.org/10.1055/s-2006-951536.

13 Johansson Seechurn, C.C.C., Kitching, M.O., Colacot, T.J. et al. (2012). Palladium-catalyzed cross-coupling: a historical contextual perspective to the 2010 Nobel Prize. *Angewandte Chemie International Edition* 51, 5062–5085.

14 Yamamoto, T. (2010). Synthesis of π-conjugated polymers by organometallic polycondensation. *Bulletin of the Chemical Society of Japan* 83: 431–455.

15 Xu, S., Kim, E.H., Wei, A. et al. (2014). Pd-and Ni-catalyzed cross-coupling reactions in the synthesis of organic electronic materials. *Science and Technology of Advanced Materials* 15: 044201.

16 Lo, C.K., Wolfe, R.M.W., and Reynolds, J.R. (2021). From monomer to conjugated polymer: a perspective on best practices for synthesis. *Chemistry of Materials* 33 (13): 4842–4852. https://doi.org/10.1021/acs.chemmater.1c01142.

17 Wen, J., Zhu, L., and Li, M. (2020). C—C coupling reactions for the synthesis of two-dimensional conjugated polymers. *ChemPlusChem* 85: 2636–2651. https://doi.org/10.1002/cplu.202000643.

18 Yamamoto, T. (1992). Electrically conducting and thermally stable π-conjugated poly (arylene) s prepared by organometallic processes. *Progress in Polymer Science* 17: 1153–1205.

19 Yamamoto, T. and Yamamoto, A. (1977). A novel type of polycondensation of polyhalogenated organic aromatic compounds producing thermostable polyphenylene type polymers promoted by nickel complexes. *Chemistry Letters* 6: 353–356.

20 Yamamoto, T., Sanechika, K., and Yamamoto, A. (1983). Preparation and characterization of poly (thienylene) s. *Bulletin of the Chemical Society of Japan* 56: 1497–1502.

21 McCullough, R.D. and Lowe, R.D. (1992). Enhanced electrical conductivity in regioselectively synthesized poly (3-alkylthiophenes). *Journal of the Chemical Society, Chemical Communications* 1: 70–72. https://doi.org/10.1039/c39920000070.

22 Chen, T., Wu, X., and Rieke, R.D. (1995). Regiocontrolled synthesis of poly (3-alkylthiophenes) mediated by Rieke zinc: their characterization and solid-state properties. *Journal of the American Chemical Society* 117 (1): 233–244. https://doi.org/10.1021/ja00106a027.

23 Shibuya, Y. and Mori, A. (2020). Dehalogenative or deprotonative? The preparation pathway to the organometallic monomer for transition-metal-catalyzed catalyst-transfer-type polymerization of thiophene derivatives. *Chemistry – A European Journal* 26: 6976–6987.

24 Amna, B., Siddiqi, H.M., Hassan, A. et al. (2020). Recent developments in the synthesis of regioregular thiophene-based conjugated polymers for electronic and optoelectronic applications using nickel and palladium-based catalytic systems. *RSC Advances* 10 (8): 4322–4396. https://doi.org/10.1039/c9ra09712k.

25 Yokozawa, T. and Yokoyama, A. (2009). Chain-growth condensation polymerization for the synthesis of well-defined condensation polymers and π-conjugated polymers. *Chemical Reviews* 109 (11): 5595–5619. https://doi.org/10.1021/cr900041c.

26 Osaka, I. and McCullough, R.D. (2008). Advances in molecular design and synthesis of regioregular polythiophenes. *Accounts of Chemical Research* 41 (9): 1202–1214. https://doi.org/10.1021/ar800130s.

27 Yokoyama, A., Miyakoshi, R., and Yokozawa, T. (2004). Chain-growth polymerization for poly (3-hexylthiophene) with a defined molecular weight and a low polydispersity. *Macromolecules* 37: 1169–1171. https://doi.org/10.1021/ma035396o.

28 Iovu, M.C., Sheina, E.E., Gil, R.R. et al. (2005). Experimental evidence for the quasi-"living" nature of the Grignard metathesis method for the synthesis of regioregular poly (3-alkylthiophenes). *Macromolecules* 38: 8649–8656.

29 Tamba, S., Shono, K., Sugie, A. et al. (2011). C–H functionalization polycondensation of chlorothiophenes in the presence of nickel catalyst with stoichiometric or catalytically generated magnesium amide. *Journal of the American Chemical Society* 133 (25): 9700–9703. https://doi.org/10.1021/ja2033525.

30 Baker, M.A., Tsai, C., and Noonan, K.J.T. (2018). Diversifying cross-coupling strategies, catalysts and monomers for the controlled synthesis of conjugated polymers. *Chemistry – A European Journal* 24: 13078–13088.

31 Ueda, M., Miyaji, Y., Ito, T. et al. (1991). Synthesis of poly (3-phenyl-2, 5-thiophene) by nickel-catalyzed coupling polymerization of 3-phenyl-2, 5-dichlorothiophene. *Macromolecules* 24: 2694–2697.

32 Saito, N., Kanbara, T., Nakamura, Y. et al. (1994). Electrochemical and chemical preparation of linear. pi.-conjugated poly (quinoline-2, 6-diyl) using nickel complexes and electrochemical properties of the polymer. *Macromolecules* 27: 756–761.

33 Yamamoto, T., Kashiwazaki, A., and Kato, K. (1989). Polymers and oligomers with substituted 2, 5-thienylene units. Preparation and electrical conductivity properties. *Die Makromolekulare Chemie* 190: 1649–1654.

34 Yamamoto, T., Hayashida, N., and Maruyama, T. (1997). Ni-catalyzed dehalogenation polycondensation of dihaloaromatic compounds with NaH. *Macromolecular Chemistry and Physics* 198: 341–351.

35 Kitada, K. and Ozaki, S. (1995). Reductive polymerization of halothiophene. *Polymer Journal* 27 (12): 1161–1166. https://doi.org/10.1295/polymj.27.1161.

36 Cheng, Y.-J. and Luh, T.-Y. (2004). Synthesizing optoelectronic heteroaromatic conjugated polymers by cross-coupling reactions. *Journal of Organometallic Chemistry* 689: 4137–4148.

37 Carsten, B., He, F., Son, H.J. et al. (2011). Stille polycondensation for synthesis of functional materials. *Chemical Reviews* 111 (3): 1493–1528. https://doi.org/10.1021/cr100320w.

38 Amna, B., Ateş, A., and Ozturk, T. (2023). Pd/Cu-catalyzed sonogashira cross-coupling polycondensation: a promising approach for synthesizing conjugated polymers with useful applications. *European Polymer Journal* 196: 112275.

39 Wang, X., Song, Y., Qu, J. et al. (2017). Mechanistic insights into the copper-cocatalyzed Sonogashira cross-coupling reaction: key role of an anion. *Organometallics* 36: 1042–1048. https://doi.org/10.1021/acs.organomet.7b00010.

40 Blayney, A.J., Perepichka, I.F., Wudl, F. et al. (2014). Advances and challenges in the synthesis of poly (p-phenylene vinylene)-based polymers. *Israel Journal of Chemistry* 54: 674–688.

41 Grimsdale, A.C., Chan, K.L., Martin, R.E. et al. (2009). Synthesis of light-emitting conjugated polymers for applications in electroluminescent devices. *Chemical Reviews* 109 (3): 897–1091. https://doi.org/10.1021/cr000013v.

42 Banerjee, J. and Dutta, K. (2021). A short overview on the synthesis, properties and major applications of poly (p-phenylene vinylene). *Chemical Papers* 75 (10): 5139–5151. https://doi.org/10.1007/s11696-020-01492-9.

43 Dorel, R., Grugel, C.P., and Haydl, A.M. (2019). The Buchwald–Hartwig amination after 25 years. *Angewandte Chemie International Edition* 58 (48): 17118–17129. https://doi.org/10.1002/anie.201904795.

44 Heravi, M.M., Kheilkordi, Z., Zadsirjan, V. et al. (2018). Buchwald-Hartwig reaction: an overview. *Journal of Organometallic Chemistry* 861: 17–104.

45 Giri, H., Dowell, T.J., Almtiri, M. et al. (2023). Polyaniline derivatives and their applications. In: *Trends and Developments in Modern Applications of Polyaniline* (ed. F. Năstase). Rijeka: IntechOpen.

46 Ward, R.E. and Meyer, T.Y. (2003). *o,p*-Polyaniline: a new form of a classic conducting polymer. *Macromolecules* 36: 4368–4373. https://doi.org/10.1021/ma021343f.

47 Kuwabara, J., Mori, H., Teratani, T. et al. (2009). Regioregulated syntheses of poly (aminopyridine) s by Pd-catalyzed amination reaction. *Macromolecular Rapid Communications* 30 (12): 997–1001. https://doi.org/10.1002/marc.200900033.

48 Almtiri, M., Dowell, T.J., Chu, I. et al. (2021). Phenoxazine-containing polyaniline derivatives with improved electrochemical stability and processability. *ACS Applied Polymer Materials* 3 (6): 2988–2997. https://doi.org/10.1021/acsapm.1c00152.

49 Kodama, S., Kuwabara, J., Jiang, X. et al. (2019). Effect of purification solvent on polymer impurities and device performance. *ACS Applied Polymer Materials* 1 (8): 2083–2088. https://doi.org/10.1021/acsapm.9b00385.

50 Yamada, H., Kukino, M., Wang, Z.A. et al. (2015). Preparation and characterization of green reflective films of polyaniline analogs containing azobenzene units. *Journal of Applied Polymer Science* 132: 41275.

51 Takase, N., Kuwabara, J., Choi, S.J. et al. (2014). Microwave-assisted polycondensation of 4-octylaniline with dibromoarylene. *Journal of Polymer Science Part A: Polymer Chemistry* 53 (4): 536–542. https://doi.org/10.1002/pola.27469.

52 Kukino, M., Kuwabara, J., Matsuishi, K. et al. (2010). Synthesis and metal-like luster of novel polyaniline analogs containing azobenzene unit. *Chemistry Letters* 39: 1248–1250.

53 Kanbara, T., Miyazaki, Y., Hasegawa, K. et al. (2000). Preparation of meta-polyaniline and its related poly (iminoarylene)s by nickel-catalyzed polycondensation of aryl dichlorides with aryl primary diamines. *Journal of Polymer Science Part A: Polymer Chemistry* 38: 4194–4199.

54 Satoh, M. and Miura, T. (2007). Catalytic direct arylation of heteroaromatic compounds. *Chemistry Letters* 36: 200–205.

55 Rossi, R., Bellina, F., Lessi, M. et al. (2014). Cross-coupling of heteroarenes by C–H functionalization: recent progress towards direct arylation and heteroarylation reactions involving heteroarenes containing one heteroatom. *Advanced Synthesis & Catalysis* 356 (1): 17–117. https://doi.org/10.1002/adsc.201300922.

56 Segawa, Y., Maekawa, T., and Itami, K. (2015). Synthesis of extended π-systems through C–H activation. *Angewandte Chemie International Edition* 54: 66–81.

57 Pouliot, J.-R., Grenier, F., Blaskovits, J.T. et al. (2016). Direct (hetero) arylation polymerization: simplicity for conjugated polymer synthesis. *Chemical Reviews* 116 (22): 14225–14274. https://doi.org/10.1021/acs.chemrev.6b00498.

58 Wakioka, M. and Ozawa, F. (2018). Highly efficient catalysts for direct arylation polymerization (DArP). *Asian Journal of Organic Chemistry* 7 (7): 1206–1216. https://doi.org/10.1002/ajoc.201800227.

59 Kuwabara, J. (2018). Direct arylation polycondensation for synthesis of optoelectronic materials. *Polymer Journal* 50 (12): 1099–1106. https://doi.org/10.1038/s41428-018-0101-3.

60 Kuwabara, J. and Kanbara, T. (2019). Facile synthesis of π-conjugated polymers via direct arylation polycondensation. *Bulletin of the Chemical Society of Japan* 92: 152–161.

61 Pankow, R.M. and Thompson, B.C. (2020). Approaches for improving the sustainability of conjugated polymer synthesis using direct arylation polymerization (DArP). *Polymer Chemistry* 11 (3): 630–640. https://doi.org/10.1039/c9py01534e.

62 Sévignon, M., Papillon, J., Schulz, E. et al. (1999). New synthetic method for the polymerization of alkylthiophenes. *Tetrahedron Letters* 40: 5873–5876.

63 Hassan, J., Schulz, E., Gozzi, C. et al. (2003). Palladium-catalyzed synthesis of oligo (alkylthiophenes). *Journal of Molecular Catalysis A: Chemical* 195: 125–131.

64 Wang, Q., Takita, R., Kikuzaki, Y. et al. (2010). Palladium-catalyzed dehydrohalogenative polycondensation of 2-bromo-3-hexylthiophene: an efficient approach to head-to-tail poly(3-hexylthiophene). *Journal of the American Chemical Society* 132 (33): 11420–11421. https://doi.org/10.1021/ja105767z.

65 Lu, W., Kuwabara, J., and Kanbara, T. (2011). Polycondensation of dibromofluorene analogues with tetrafluorobenzene via direct arylation. *Macromolecules* 44: 1252–1255. https://doi.org/10.1021/ma1028517.

66 Giovanella, U., Botta, C., Galeotti, F. et al. (2014). Perfluorinated polymer with unexpectedly efficient deep blue electroluminescence for full-colour OLED displays and light therapy applications. *Journal of Materials Chemistry C* 2 (34): 5322–5329. https://doi.org/10.1039/c3tc31139b.

67 Lu, W., Kuwabara, J., Iijima, T. et al. (2012). Synthesis of π-conjugated polymers containing fluorinated arylene units via direct arylation: efficient synthetic method of materials for OLEDs. *Macromolecules* 45: 4128–4133.

68 Liégault, B., Lapointe, D., Caron, L. et al. (2009). Establishment of broadly applicable reaction conditions for the palladium-catalyzed direct arylation of heteroatom-containing aromatic compounds. *The Journal of Organic Chemistry* 74 (5): 1826–1834. https://doi.org/10.1021/jo8026565.

69 Lapointe, D. and Fagnou, K. (2010). Overview of the mechanistic work on the concerted metallation–deprotonation pathway. *Chemistry Letters* 39: 1118–1126.

70 Yamazaki, K., Kuwabara, J., and Kanbara, T. (2013). Detailed optimization of polycondensation reaction via direct C–H arylation of ethylenedioxythiophene. *Macromolecular Rapid Communications* 34: 69–73.

71 Wakioka, M., Nakamura, Y., Montgomery, M. et al. (2015). Remarkable ligand effect of $P(2\text{-}MeOC_6H_4)_3$ on palladium-catalyzed direct arylation. *Organometallics* 34: 198–205.

72 Rudenko, A.E. and Thompson, B.C. (2015). Influence of the carboxylic acid additive structure on the properties of poly (3-hexylthiophene) prepared via direct arylation polymerization (DArP). *Macromolecules* 48: 569–575. https://doi.org/10.1021/ma502131k.

73 Dudnik, A.S., Aldrich, T.J., Eastham, N.D. et al. (2016). Tin-free direct C–H arylation polymerization for high photovoltaic efficiency conjugated copolymers. *Journal of the American Chemical Society* 138 (48): 15699–15709. https://doi.org/10.1021/jacs.6b10023.

74 Choi, S.J., Kuwabara, J., and Kanbara, T. (2013). Microwave-assisted polycondensation via direct arylation of 3, 4-ethylenedioxythiophene with 9, 9-dioctyl-2, 7-dibromofluorene. *ACS Sustainable Chemistry & Engineering* 1: 878–882.

75 Kuwabara, J., Yasuda, T., Choi, S.J. et al. (2014). Direct arylation polycondensation: a promising method for the synthesis of highly pure, high-molecular-weight conjugated polymers needed for improving the performance of organic photovoltaics. *Advanced Functional Materials* 24: 3226–3233.

76 Hayashi, S. and Koizumi, T. (2015). Chloride-promoted Pd-catalyzed direct C–H arylation for highly efficient phosphine-free synthesis of π-conjugated polymers. *Polymer Chemistry* 6 (28): 5036–5039. https://doi.org/10.1039/c5py00871a.

77 Kuwabara, J., Fujie, Y., Maruyama, K. et al. (2016). Suppression of homocoupling side reactions in direct arylation polycondensation for producing high performance OPV materials. *Macromolecules* 49: 9388–9395.

78 Kuwabara, J., Yamazaki, K., Yamagata, T. et al. (2015). The effect of a solvent on direct arylation polycondensation of substituted thiophenes. *Polymer Chemistry* 6 (6): 891–895. https://doi.org/10.1039/c4py01387e.

79 Iizuka, E., Wakioka, M., and Ozawa, F. (2015). Mixed-ligand approach to palladium-catalyzed direct arylation polymerization: synthesis of donor–acceptor polymers with dithienosilole (DTS) and thienopyrroledione (TPD) units. *Macromolecules* 48: 2989–2993. https://doi.org/10.1021/acs.macromol.5b00526.

80 Wakioka, M., Morita, H., Ichihara, N. et al. (2020). Mixed-ligand approach to palladium-catalyzed direct arylation polymerization: synthesis of donor–acceptor polymers containing unsubstituted bithiophene units. *Macromolecules* 53: 158–164.

81 Gobalasingham, N.S., Noh, S., and Thompson, B.C. (2016). Palladium-catalyzed oxidative direct arylation polymerization (Oxi-DArP) of an ester-functionalized thiophene. *Polymer Chemistry* 7 (8): 1623–1631. https://doi.org/10.1039/c5py01973g.

82 Guo, Q., Jiang, R., Wu, D. et al. (2016). Rapid access to 2,2′-bithiazole-based copolymers via sequential palladium-catalyzed C–H/C–X and C–H/C–H coupling reactions. *Macromolecular Rapid Communications* 37 (9): 794–798. https://doi.org/10.1002/marc.201600028.

83 Shi, Y., Zhang, X., Du, T. et al. (2023). A high-performance n-type thermoelectric polymer from C—H/C—H oxidative direct arylation polycondensation. *Angewandte Chemie International Edition* 62 (20): e202219262. https://doi.org/10.1002/anie.202219262.

84 Guo, Q., Wu, D., and You, J. (2016). Oxidative direct arylation polymerization using oxygen as the sole oxidant: facile, green access to bithiazole-based polymers. *ChemSusChem* 9: 2765–2768.

85 Chakraborty, B. and Luscombe, C.K. (2023). Cross-dehydrogenative coupling polymerization via C—H activation for the synthesis of conjugated polymers. *Angewandte Chemie International Edition* 62 (21): e202301247. https://doi.org/10.1002/anie.202301247.

86 Yang, Y., Wu, Y., Bin, Z. et al. (2024). Discovery of organic optoelectronic materials powered by oxidative Ar–H/Ar–H coupling. *Journal of the American Chemical Society* 146 (2): 1224–1243. https://doi.org/10.1021/jacs.3c12234.

87 Albano, G. (2024). Palladium-catalyzed cross-dehydrogenative coupling of (hetero) arenes. *Organic Chemistry Frontiers* 11 (5): 1495–1622. https://doi.org/10.1039/d3qo01925j.

88 Gobalasingham, N.S., Pankow, R.M., and Thompson, B.C. (2017). Synthesis of random poly (hexyl thiophene-3-carboxylate) copolymers via oxidative direct arylation polymerization (oxi-DArP). *Polymer Chemistry* 8 (12): 1963–1971. https://doi.org/10.1039/c7py00181a.

89 Aoki, H., Saito, H., Shimoyama, Y. et al. (2018). Synthesis of conjugated polymers containing octafluorobiphenylene unit via Pd-catalyzed cross-dehydrogenative-coupling reaction. *ACS Macro Letters* 7: 90–94.

90 Shimoyama, Y., Kuwabara, J., and Kanbara, T. (2020). Mechanistic study of Pd/Ag dual-catalyzed cross-dehydrogenative coupling of perfluoroarenes with thiophenes. *ACS Catalysis* 10 (5): 3390–3397. https://doi.org/10.1021/acscatal.9b05326.

91 Xing, L., Liu, J.-R., Hong, X. et al. (2022). An exception to the carothers equation caused by the accelerated chain extension in a Pd/Ag cocatalyzed cross dehydrogenative coupling polymerization. *Journal of the American Chemical Society* 144 (5): 2311–2322. https://doi.org/10.1021/jacs.1c12599.

92 Zhang, Q., Chang, M., Lu, Y. et al. (2018). A direct C–H coupling method for preparing π-conjugated functional polymers with high regioregularity. *Macromolecules* 51: 379–388.

93 Kang, L.J., Xing, L., and Luscombe, C.K. (2019). Exploration and development of gold- and silver-catalyzed cross dehydrogenative coupling toward donor–acceptor π-conjugated polymer synthesis. *Polymer Chemistry* 10 (4): 486–493. https://doi.org/10.1039/c8py01162a.

94 Tanaka, C., Kuwabara, J., Yasuda, T. et al. (2019). Synthesis of conjugated polymer consisting of three-component aromatic units via two-step cross-dehydrogenative-coupling reactions. *Synthetic Metals* 254: 180–183.

95 Li, J., Han, D., Zhang, Q. et al. (2021). Synthesis and properties of fluorinated benzotriazole-based donor-acceptor-type conjugated polymers via Pd-catalyzed direct C–H/C– H coupling polymerization. *Journal of Polymer Science* 59: 240–250.

96 Onda, N., Sato, R., Kuwabara, J. et al. (2023). One-pot two-step cross-dehydrogenative-coupling polycondensation for synthesis of tetrafluorobenzene-based conjugated polymer. *Synthetic Metals* 293: 117279.

97 Moritani, I. and Fujiwara, Y. (1967). Aromatic substitution of styrene-palladium chloride complex. *Tetrahedron Letters* 8 (12): 1119–1122.

98 Yeung, C.S. and Dong, V.M. (2011). Catalytic dehydrogenative cross-coupling: forming carbon—carbon bonds by oxidizing two carbon—hydrogen bonds. *Chemical Reviews* 111 (3): 1215–1292. https://doi.org/10.1021/cr100280d.

99 Chen, S., Ranjan, P., Voskressensky, L.G. et al. (2020). Recent developments in transition-metal catalyzed direct C–H alkenylation, alkylation, and alkynylation of azoles. *Molecules* 25: 4970.

100 Cho, H.N., Kim, J.K., Kim, D.Y. et al. (1999). Statistical copolymers for blue-light-emitting diodes. *Macromolecules* 32: 1476–1481. https://doi.org/10.1021/ma981340w.

101 Saito, H., Kuwabara, J., Yasuda, T. et al. (2018). Synthesis of polyfluoro arylene-based poly (arylenevinylene) s via Pd-catalyzed dehydrogenative direct alkenylation. *Macromolecular Rapid Communications* 39 (18): 1800414. https://doi.org/10.1002/marc.201800414.

102 Lee, J., Park, H.J., Joo, J.M. et al. (2019). Synthesis and characterization of DPP-based conjugated polymers via dehydrogenative direct alkenylation polycondensation. *Macromolecular Research* 27 (2): 115–118. https://doi.org/10.1007/s13233-019-7070-z.

103 Saito, H., Kuwabara, J., Yasuda, T. et al. (2016). Synthesis of pyrrole-based poly (arylenevinylene) s via Rh-catalyzed dehydrogenative direct alkenylation. *Polymer Chemistry* 7 (16): 2775–2779. https://doi.org/10.1039/c6py00016a.

104 Rao, H. and Fu, H. (2011). Copper-catalyzed coupling reactions. *Synlett* 2011 (6): 745–769. https://doi.org/10.1055/s-0030-1259919

105 Ma, D. and Cai, Q. (2008). Copper/amino acid catalyzed cross-couplings of aryl and vinyl halides with nucleophiles. *Accounts of Chemical Research* 41 (11): 1450–1460. https://doi.org/10.1021/ar8000298.

106 Beletskaya, I.P. and Cheprakov, A.V. (2004). Copper in cross-coupling reactions: the post-Ullmann chemistry. *Coordination Chemistry Reviews* 248: 2337–2364.

107 Das, P., Sharma, D., Kumar, M. et al. (2010). Copper promoted CN and CO type cross-coupling reactions. *Current Organic Chemistry* 14: 754–783.

108 Li, Z., Chen, Y., Ye, P. et al. (2019). Microwave-assisted classic Ullmann C–C coupling polymerization for acceptor-acceptor homopolymers. *Polymers* 11: 1741.

109 Kisselev, R. and Thelakkat, M. (2004). Synthesis and characterization of poly (triarylamine) s containing isothianaphthene moieties. *Macromolecules* 37: 8951–8958. https://doi.org/10.1021/ma0495947.

110 Thelakkat, M., Hagen, J., Haarer, D. et al. (1999). Poly (triarylamine) s-synthesis and application in electroluminescent devices and photovoltaics. *Synthetic Metals* 102: 1125–1128.

111 Jia, N.-N., Tian, X.-C., Qu, X.-X. et al. (2017). Copper-catalyzed direct 2-arylation of benzoxazoles and benzoimidazoles with aryl bromides and cytotoxicity of products. *Scientific Reports* 7 (1): 43758. https://doi.org/10.1038/srep43758.

112 Do, H.-Q. and Daugulis, O. (2008). Copper-catalyzed arylation and alkenylation of polyfluoroarene C—H bonds. *Journal of the American Chemical Society* 130 (4): 1128–1129. https://doi.org/10.1021/ja077862l.

113 Pankow, R.M., Ye, L., and Thompson, B.C. (2018). Copper catalyzed synthesis of conjugated copolymers using direct arylation polymerization. *Polymer Chemistry* 9 (30): 4120–4124. https://doi.org/10.1039/c8py00913a.

114 Ye, L., Schmitt, A., Pankow, R.M. et al. (2020). An efficient precatalyst approach for the synthesis of thiazole-containing conjugated polymers via Cu-catalyzed direct arylation polymerization (Cu-DArP). *ACS Macro Letters* 9 (10): 1446–1451. https://doi.org/10.1021/acsmacrolett.0c00626.

115 Kuramochi, M., Kuwabara, J., Lu, W. et al. (2014). Direct arylation polycondensation of bithiazole derivatives with various acceptors. *Macromolecules* 47: 7378–7385. https://doi.org/10.1021/ma5014397.

116 Lu, W., Kuwabara, J., and Kanbara, T. (2012). Synthesis of 4,4′-dinonyl-2,2′-bithiazole-based copolymers via Pd-catalyzed direct C–H arylation. *Polymer Chemistry* 3 (12): 3217–3219. https://doi.org/10.1039/c2py20539d.

117 Li, Y., Jin, J., Qian, W. et al. (2010). An efficient and convenient Cu (OAc)2/air mediated oxidative coupling of azoles via C–H activation. *Organic & Biomolecular Chemistry* 8 (2): 326–330. https://doi.org/10.1039/b919396k.

118 Faradhiyani, A., Zhang, Q., Maruyama, K. et al. (2018). Synthesis of bithiazole-based semiconducting polymers *via* Cu-catalysed aerobic oxidative coupling. *Materials Chemistry Frontiers* 2 (7): 1306–1309. https://doi.org/10.1039/c7qm00584a.

119 Huang, Q., Qin, X., Li, B. et al. (2014). Cu-catalysed oxidative C–H/C–H coupling polymerisation of benzodiimidazoles: an efficient approach to regioregular polybenzodiimidazoles for blue-emitting materials. *Chemical Communications* 50 (89): 13739–13741. https://doi.org/10.1039/c4cc06291d.

120 Doba, T., Ilies, L., Sato, W. et al. (2021). Iron-catalysed regioselective thienyl C–H/C–H coupling. *Nature Catalysis* 4 (7): 631–638. https://doi.org/10.1038/s41929-021-00653-7.

121 Doba, T., Shang, R., and Nakamura, E. (2022). Iron-catalyzed C-H activation for heterocoupling and copolymerization of thiophenes with enamines. *Journal of the American Chemical Society* 144 (47): 21692–21701. https://doi.org/10.1021/jacs.2c09470.

122 Haque, T. and Nomura, K. (2015). Acyclic diene metathesis (ADMET) polymerization for precise synthesis of defect-free conjugated polymers with well-defined chain ends. *Catalysts* 5: 500–517.

123 Chen, Y., Abdellatif, M.M., and Nomura, K. (2018). Olefin metathesis polymerization: some recent developments in the precise polymerizations for synthesis of advanced materials (by ROMP, ADMET). *Tetrahedron* 74: 619–643.

124 Caire da Silva, L., Rojas, G., Schulz, M.D. et al. (2017). Acyclic diene metathesis polymerization: history, methods and applications. *Progress in Polymer Science* 69: 79–107.

125 Miyashita, T., Kunisawa, M., Sueki, S. et al. (2017). Synthesis of poly (arylene vinylene) s with different end groups by combining acyclic diene metathesis polymerization with Wittig-type couplings. *Angewandte Chemie International Edition* 56 (19): 5288–5293. https://doi.org/10.1002/anie.201703019.

126 Yu, C.-Y., Horie, M., Spring, A.M. et al. (2010). Homopolymers and block copolymers of p-phenylenevinylene-2, 5-diethylhexyloxy-p-phenylenevinylene and m-phenylenevinylene-2, 5-diethylhexyloxy-p-phenylenevinylene by ring-opening metathesis polymerization. *Macromolecules* 43: 222–232.

127 Chang, S.-W. and Horie, M. (2015). A donor–acceptor conjugated block copolymer of poly (arylenevinylene) s by ring-opening metathesis polymerization. *Chemical Communications* 51 (44): 9113–9116. https://doi.org/10.1039/c5cc00498e.

128 Selmani, S., Vanderzwet, L., Kukor, A. et al. (2018). Synthesis of poly (heteroarylenevinylene) derivatives via rhodium-catalyzed hydroarylation of alkynes. *Synlett* 29: 2552–2556.

129 Iwamori, R., Sato, R., Kuwabara, J. et al. (2021). Synthesis of pyrrole-based poly (arylenevinylene) s via co-catalyzed hydroarylation of alkynes. *Macromolecular Rapid Communications* 42 (16): 2100283. https://doi.org/10.1002/marc.202100283.

130 Iwamori, R., Sato, R., Kuwabara, J. et al. (2022). Nonstoichiometric hydroarylation polyaddition for synthesis of pyrrole-based poly (arylenevinylene) s. *Polymer Chemistry* 13 (3): 379–382. https://doi.org/10.1039/d1py01425k.

131 Iwamori, R., Kuwabara, J., Yasuda, T. et al. (2023). Regioselective synthesis of thiophene-based poly (arylenevinylene) s via co-catalyzed hydroarylation polyaddition. *Macromolecules* 56: 5407–5414.

132 Iwamori, R., Kuwabara, J., Yasuda, T. et al. (2024). Molecular design of naphthalene- and carbazole-based monomers for regiospecific synthesis of poly (arylenevinylene) s via co-catalyzed hydroarylation polyaddition. *Macromolecular Rapid Communications* 45 (16): 2400168. https://doi.org/10.1002/marc.202400168.

133 Tsukahara, K., Iwamori, R., Kuwabara, J. et al. (2024). Regioselective synthesis of pyrrole-based poly (arylenevinylene) s via Mn-catalyzed hydroarylation polyaddition. *Macromolecular Rapid Communications* 45 (20): 2400456. https://doi.org/10.1002/marc.202400456.

134 Stuart, D.R., Alsabeh, P., Kuhn, M. et al. (2010). Rhodium(III)-catalyzed arene and alkene C–H bond functionalization leading to indoles and pyrroles. *Journal of the American Chemical Society* 132 (51): 18326–18339. https://doi.org/10.1021/ja1082624.

135 Tokoro, Y., Sato, H., and Fukuzawa, S.I. (2015). Synthesis of π-conjugated polymers containing benzodipyrrole moieties in the main chain through cleavage of C–H bonds in 1,4-bis (acetylamino) benzene. *ACS Macro Letters* 4 (7): 689–692. https://doi.org/10.1021/acsmacrolett.5b00343.

136 Wang, K., Yan, S., Han, T. et al. (2022). Cascade C–H-activated polyannulations toward ring-fused heteroaromatic polymers for intracellular pH mapping and cancer cell killing. *Journal of the American Chemical Society* 144 (26): 11788–11801. https://doi.org/10.1021/jacs.2c04032.

137 Wang, K., Liu, J., Liu, P. et al. (2023). Multifunctional fluorescent main-chain charged polyelectrolytes synthesized by cascade C–H activation/annulation polymerizations. *Journal of the American Chemical Society* 145 (7): 4208–4220. https://doi.org/10.1021/jacs.2c12654.

138 Uddin, A. and Plunkett, K.N. (2020). Donor-acceptor copolymers from cyclopentannulation polymerizations with dicyclopenta [cd, jk] pyrene and dicyclopenta [cd, lm] perylene acceptors. *Journal of Polymer Science* 58: 3165–3169.

139 Bheemireddy, S.R., Hautzinger, M.P., Li, T. et al. (2017). Conjugated ladder polymers by a cyclopentannulation polymerization. *Journal of the American Chemical Society* 139 (16): 5801–5807. https://doi.org/10.1021/jacs.6b12916.

140 Kuwabara, J., Yasuda, T., Takase, N. et al. (2016). Effects of the terminal structure, purity, and molecular weight of an amorphous conjugated polymer on its photovoltaic characteristics. *ACS Applied Materials & Interfaces* 8: 1752–1758.

141 Osedach, T.P., Andrew, T.L., and Bulović, V. (2013). Effect of synthetic accessibility on the commercial viability of organic photovoltaics. *Energy & Environmental Science* 6 (3): 711–718. https://doi.org/10.1039/c3ee24138f.

142 Po, R., Bernardi, A., Calabrese, A. et al. (2014). From lab to fab: how must the polymer solar cell materials design change? – an industrial perspective. *Energy & Environmental Science* 7 (3): 925–943. https://doi.org/10.1039/c3ee43460e.

143 Po, R., Bianchi, G., Carbonera, C. et al. (2015). "All that glisters is not gold": an analysis of the synthetic complexity of efficient polymer donors for polymer solar cells. *Macromolecules* 48: 453–461.

144 Kuwabara, J. and Kanbara, T. (2021). Step-economical synthesis of conjugated polymer materials composed of three components: donor, acceptor, and π units. *Macromolecular Rapid Communications* 42: 2000493.

145 Kuwabara, J. and Kanbara, T. (2024). Synthesis of organic optoelectronic materials using direct C–H functionalization. *ChemPlusChem* 89: e202300400.

2

Living Conjugated Polymers

Kevin J. T. Noonan

Department of Chemistry, Carnegie Mellon University, Pittsburgh, PA, USA

2.1 Introduction

The evolution of π-conjugated polymer synthesis has been shaped by two central demands: precise control over molecular architecture to tune material properties and the ability to produce polymers with high reproducibility. This chapter outlines methods for preparing quasi-living, regioregular conjugated polymers with well-defined chain lengths and termini. For a living polymerization (which proceeds with no termination [1]), a chain-growth mechanism must be operative.

The key step in the chain-growth cross-coupling polymerization of an aromatic monomer bearing two different functional groups, as shown in Figure 2.1, is the hypothesized coordination of the catalyst to the growing polymer after each monomer enchainment. By maintaining the coordination between the transition metal catalyst and the growing polymer chain, propagation proceeds where one catalyst controls the growth of one chain. This step is the key event to the quasi-living reaction, which is named catalyst-transfer polymerization (CTP) [2–6]. CTP is beneficial in conjugated polymer synthesis because of the repeatability for the process, with precise control over degree of polymerization (DP), regioregularity, and defects [7–17]. Polymer architecture can also be controlled (e.g. block and star copolymers are possible). Given the 7 criteria of a living polymerization [1], the CTP reaction is sometimes described as "quasi-living," since the expected pseudo first-order kinetic profile for monomer consumption is often not followed after 50–60% conversion, as shown by McCullough in their early work [2].

Evidence to support the CTP mechanism for P3AT synthesis typically includes linear molecular weight (M_n) versus conversion plots, a set of well-defined polymer end groups as evidenced by ^{1}H NMR spectroscopy and MALDI-TOF mass

Conjugated Polymer Synthesis and Materials Applications, First Edition. Edited by Yoshiki Chujo.
© 2026 Wiley-VCH GmbH. All rights reserved, including rights for text and data mining and training of artificial intelligence technologies or similar technologies. Published 2026 by Wiley-VCH GmbH.

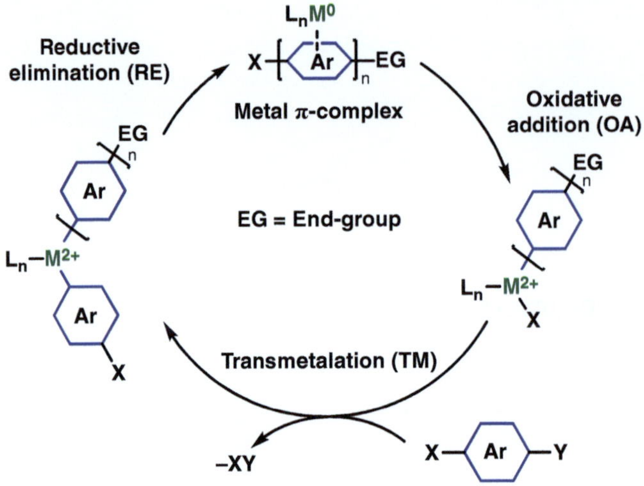

Figure 2.1 The chain-growth mechanism to prepare conjugated polymers generally involves an additional step in cross-coupling. It is hypothesized that selective activation of the halide (X group) on the polymer chain end occurs via coordination of the catalyst to the growing polymer after the reductive elimination step (M = metal). The transmetalating group is generically defined as Y above.

spectrometry of the sample, narrow molecular weight distributions (MWDs) as determined using size exclusion gel permeation chromatography (GPC), and sequential monomer addition of reactive monomers to afford block copolymers [7–17]. CTP was first reported by the McCullough and Yokozawa groups in 2004 [2–6]. They synthesized well-defined poly(3-alkylthiophenes) (P3ATs) using Kumada–Corriu cross-coupling and postulated on aspects of the mechanism in that work [2–6]. Since that time, interest in CTP for conjugated polymer synthesis has grown rapidly. Expansion beyond Kumada cross-coupling (Y = MgX) to Negishi (Y = ZnX) as well as Suzuki–Miyaura (Y = $B(OR)_2$) and beyond. Reviews have appeared on this topic [7–17], and the emphasis in this chapter will be on recent developments [18].

2.2 Metal π-Binding in CTP

McCullough suggested a non-diffusive associated pair forms between the metal catalyst and polymer to explain the chain-growth mechanism [2]. Yokozawa's group suggested the halide group at the polymer chain end was either more reactive than the monomer or that the catalyst binds to the growing chain for selective end group activation [5]. In 2008, Kiriy noted the non-diffuse associated pair and catalyst-binding proposals were both consistent with reports of Ni(0) species that can ring-walk on different π-systems [19], resulting in selective activation of aryl halides [20]. Kiriy indicated that the CTP mechanism logically hinged on this ring-walking mechanism [19], and a large body of evidence has been gathered, which supports this intramolecular transfer via chain-walking along the π-system and selective activation of the halogen chain end. Kiriy's work using NMR spectroscopy to examine end groups and defects was key to establishing that chain-growth can be bidirectional if the polymer has halide end groups on each terminus [21, 22]. Kiriy and Noonan have also demonstrated that CTP is possible when polymerizing dimer, trimer, and tetramer monomers where the chain-walking distance will be longer [19, 23], and McNeil has illustrated that both Ni and Pd catalysts with specific ligands could chain walk >35 rings for poly(3-dodecylthiophene) [24].

Kennepohl's report focusing on the first structural characterization of a η^2-Ni(0)–bithiophene complex provided further evidence for the proposed CTP mechanism [25]. In the solid-state molecular structure, the metal center was bound to the *exo* double bond of the bithiophene ring (Figure 2.2) [25]. In THF at low temperatures, the authors noted the *exo*-bound Ni species is in equilibrium with a C—S insertion product (temperature dependent) and a minor *endo*-bound isomer (Figure 2.2 – bottom) [25]. This key finding supports the mechanism that the catalyst can coordinate with the growing polyarene and "walk" to different positions before activation of the reactive chain end. Computational studies have also provided insight into the metal binding to the polymer chain [25–30].

Small molecule cross-coupling studies of di- or polybrominated arenes have been used as indirect evidence to support the CTP mechanism [2]. For certain catalysts and substrates, it has been observed that stoichiometry of the cross-coupling partner does not significantly impact product formation, and exhaustive substitution is observed. The observed selectivity in these small-molecule couplings suggests a persistent metal–arene interaction, consistent with the proposed intramolecular transfer mechanism central to CTP. The earliest report on this

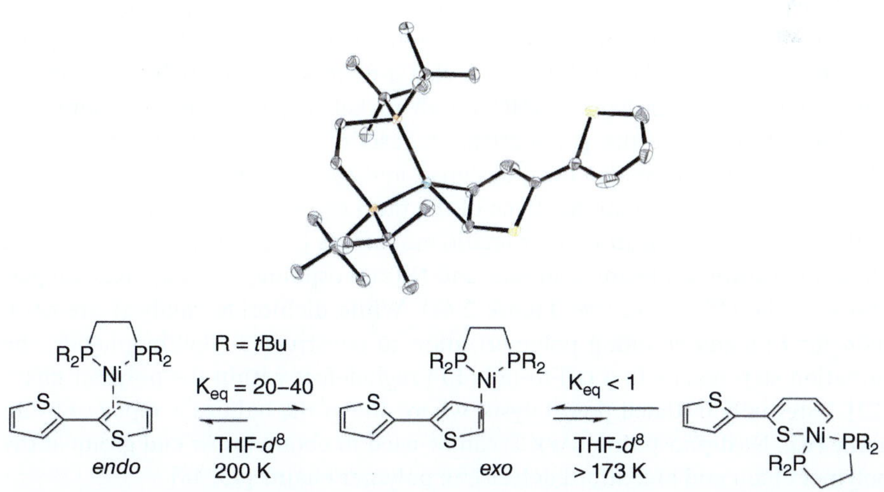

Figure 2.2 Solid-state molecular structure of η^2-Ni(0)–bithiophene complex reported in [25], along with the illustration of the different species present in equilibrium at low temperatures in THF.

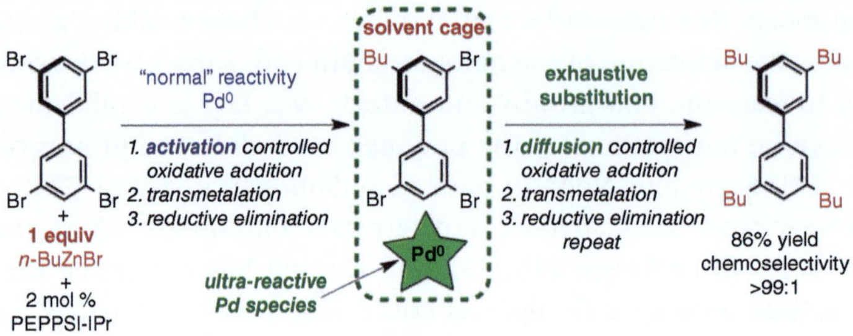

Figure 2.3 Cross-coupling of a polybrominated biphenyl with *n*-butylzinc bromide leading to exhaustive substitution. *Source*: [35]/American Chemical Society.

phenomenon (though not fully understood at the time) appeared from Kumada in 1976 when 1 equiv. of *n*BuMgCl was mixed with an excess of dichlorobenzenes in the presence of Ni(dppp)Cl_2 and Ni(dppe)Cl_2. The authors did not obtain the expected arene with a chloro and butyl group but noted conversion was primarily to dibutyl benzene [31]. These model studies can be used to identify catalysts for successful chain-growth polymerization of aromatics [32–37]. For example, Larrosa and Goldup reported on this type of experiment with Pd-PEPPSI-IPr catalysts for Negishi cross-coupling in 2011 (Figure 2.3) [35]. They demonstrated with a number of different di- and polybrominated arenes that exhaustive substitution is observed as the primary product, which is thought to proceed via the ring-walking mechanism (Figure 2.3).

2.3 Catalysts for CTP

The CTP process hinges on the interaction between the metal center and its ancillary ligands which, together, govern the strength of binding to the growing polymer chain (Figure 2.4) [8, 12]. Electron-rich σ-donors such as trialkylphosphines and N-heterocyclic carbenes (NHCs) with nickel or palladium are commonly employed to promote the polymerization reaction, and a number of different bidentate ligands have also been designed and synthesized to examine in CTP [38–44]. Figure 2.4b illustrates some of the most common ancillary ligands used in the living polymerization of aromatic monomers [7–17]. Precatalysts that are air and moisture stable are desirable, and Ni(diphosphine)Cl_2 complexes are pervasive in the CTP literature (Figure 2.4a). While dichloride catalysts are common for Kumada coupling polymerization to construct 3-alkylthiophenes, the initiation step results in a tail-to-tail (TT) regiodefect within the polymer chain [22]. Externally initiated precatalysts, where one of the halides is replaced by an arene (e.g. Ni(diphosphine)(Ar)Cl) can be used to obtain an Ar end group at the polymer chain end and afford defect-free polymer chains [45, 46].

Examples of catalysts used for CTP

Ni(dppe)Cl_2, Ni(dppp)Cl_2, Ni(diphosphine)(Ar)Cl, IPrNi(PPh_3)Cl_2
RuPhosPd(Ar)Br, (tBu)$_3$PPd(Ar)Br, IPrPd(py-3Cl)Cl_2
(a)

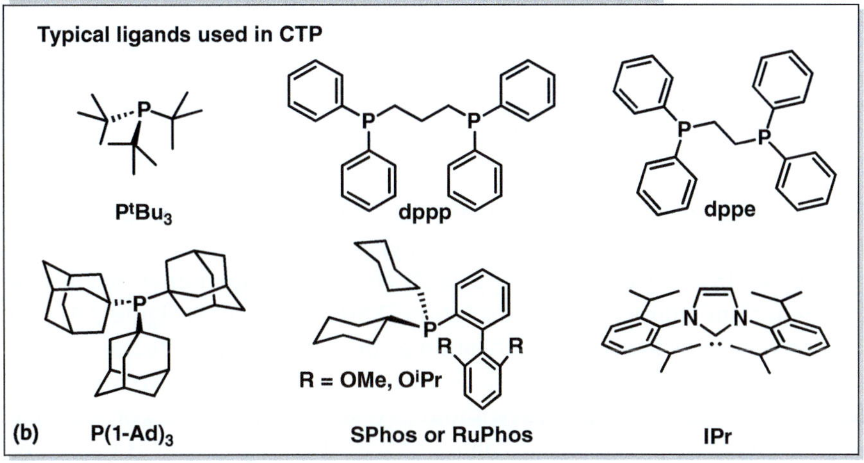

Figure 2.4 (a) Precatalysts commonly used in CTP. (b) Some of the common ancillary ligands employed in CTP.

2.4 Kumada–Corriu Catalyst-transfer Polymerization

Kumada–Corriu catalyst-transfer polymerization (KCTP) is the method of choice to construct poly(3-alkylthiophenes) as well as many other conjugated polymers [2–6, 22, 47]. Magnesium halogen exchange of 2,5-dihalo-3-alkylthiophenes, or deprotonation of 2-bromo-3-alkylthiophenes can be used to create the active monomer for polymerization (Figure 2.5). Activation of 2,5-dibromo-3-alkylthiophenes with alkyl Grignards [48, 49] and activated zinc [50, 51] leads to a mixture of two regioisomers, with functionalization of the 5-position being strongly favored when the R group is an alkyl chain (Figure 2.5). Regioregular poly(3-alkylthiophenes) or *rr*-P3ATs with minimal defects (>95 – 98% head-to-tail couplings) are formed when polymerizing this isomeric mixture using Ni(dppp)Cl_2, as the reaction of the minor regioisomer is sluggish [49]. It is important to note that the presence of LiCl in the reaction mixture can increase the rate of reaction for the minor regioisomer which leads to additional defects and decreases the regioregularity of the final polymer [52]. The rate of the reaction also increases with LiCl present in the polymerization mixture [52, 53]. Though the preparation of 2-bromo-3-alkyl-5-iodothiophene requires an additional synthetic step as compared to 2,5-dibromo-3-alkylthiophene, the main benefit is that activation can be carried out selectively at lower temperatures, affording only one active regioisomer for polymerization [4, 5].

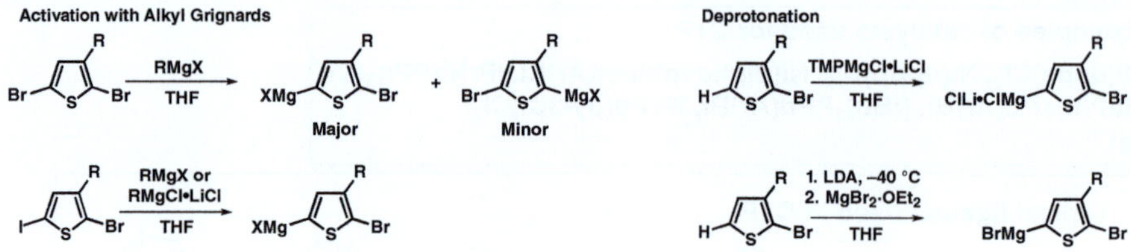

Figure 2.5 Left – activation of dihalo-3-alkylthiophenes using alkyl Grignard reagents. Right – activation of 2-bromo-3-alkylthiophenes using metal amide bases.

The 2-halo-3-substituted thiophenes can be deprotonated using non-nucleophilic bases such as lithium diisopropylamide (followed by exchange with $MgBr_2 \cdot OEt_2$) [54, 55] or using the 2,2,6,6-tetramethyl piperidinyl magnesium chloride–lithium chloride complex solution (TMPMgCl · LiCl) (Figure 2.5) [56]. This approach leads to selective activation at the 5-position, producing a single regioisomer for polymerization [57]. A specific benefit of the TMPMgCl · LiCl base is that it can be employed without cryogenic conditions. For example, 2-bromo-3-hexylthiophene can be combined with TMPMgCl · LiCl at room temperature for 1 h to prepare the activated monomer nearly quantitatively, and high molecular weight *rr*-P3HT (>98%) can be prepared by addition of different nickel catalysts to the active monomer solution [58–60].

A selection of the different types of conjugated polymers prepared using KCTP is shown in Figure 2.6. McCullough reported on poly(3-alkoxythiophenes) or P3AOTs in 2005 and noted that oligoether side chains were critical to achieving a high molecular weight polymer (perhaps due to increased solubility, $M_n = 18\,\text{kg/mol}$, $Đ = 1.6$) [61]. Koeckelberghs and coworkers used a deprotonative method to afford regioregular P3AOTs with a pendant chiral side chain [62]. The regioregularity of the polymer was critical to the strong chiroptical response observed in the solid state, as evidenced by circular dichroism (CD) studies and comparison with the regioirregular analog [62]. Koeckelberghs synthesized regioregular poly(3-octyloxythiophene)s the following year (Figure 2.6) [63], and has reported that P3AOT synthesis using Ni(dppp) does not proceed by a chain-growth mechanism because M_n does not change as expected with varying catalyst concentration [64].

An alkyl thiophene with an oxoethylene side chain was polymerized with Ni(dppe)Cl_2 at 0 °C, resulting in 86% conversion to polymer ($M_n = 8.6\,\text{kg/mol}$, $Đ = 1.15$) (Figure 2.6) [65]. Yokozawa and coworkers discovered the Ni(dppe)Cl_2 catalyst was key to the observed living behavior as other Ni(diphosphines) did not result in efficient chain-growth. Seferos has recently found that a water-scavenging Grignard reagent in the reaction mixture can facilitate the synthesis of oligoether-tethered, water soluble polythiophenes [66]. A polythiophene with a hexyl acetate group has also been polymerized using KCTP, and base-induced hydrolysis afforded a regioregular polythiophene-3-acetic acid [67].

The Swager and Higashihara groups have both reported on poly((3-alkylthio)thiophene) (P3ATT) synthesis using KCTP (Figure 2.6) [68, 69]. Both groups found that branched alkyl side chains were key to forming highly soluble polymers [68, 69].

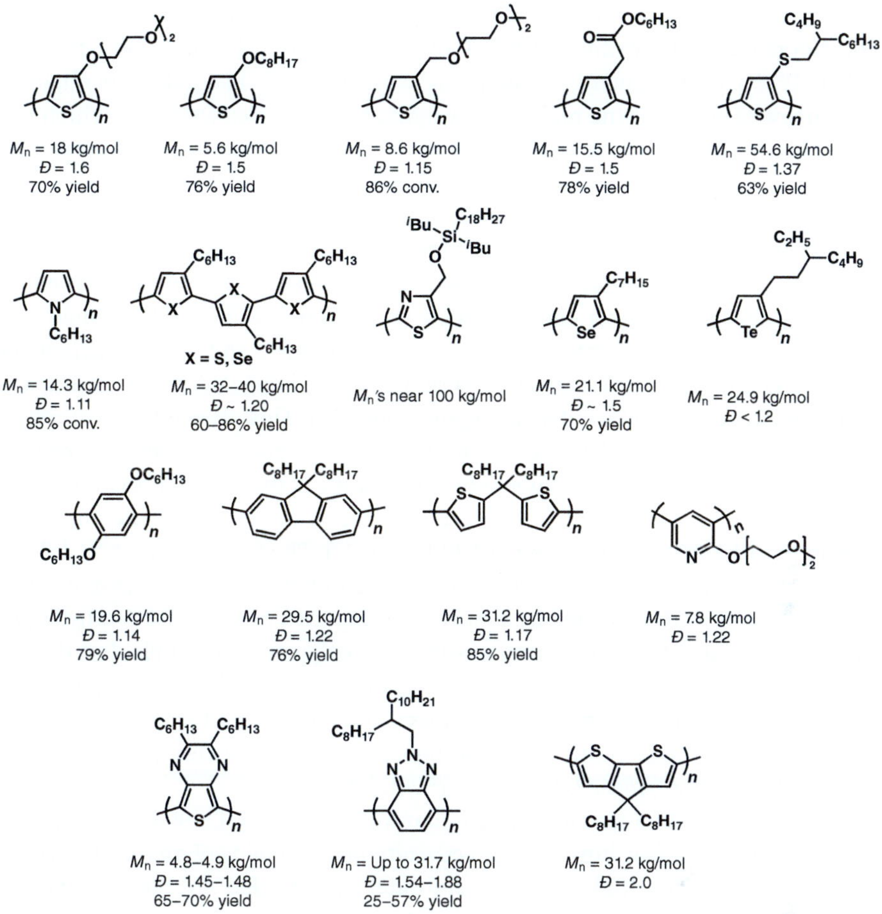

Figure 2.6 A selection of the different polymer types that have been prepared using the KCTP method.

Activation is best accomplished by deprotonation of 2-bromo-3-alkylthiothiophene monomers with TMPMgCl·LiCl to afford regioregular P3ATTs, as irregular materials are formed upon activation of 2,5-dihalo-3-alkylthiothiophenes (where the alkyl group is hexyl or dodecyl) [69]. Higashihara noted that the activation of 2,5-dibromo-3-alkylthiothiophenes with isopropylmagnesium chloride lithium chloride leads to a near 1 : 1 ratio of regioisomers with the thioether side chain, which is much less selective as compared to 2,5-dihalo-3-alkylthiophenes. The authors hypothesized this contributes to the regiodefects in the polymer chain [69]. Using the deprotonative method, Higashihara synthesized a 2-ethylhexyl functionalized P3ATT with an $M_n = 26.5$ kg/mol and $Đ = 1.4$ using Ni(dppe)Cl_2 [69]. Swager used the same catalyst and a 2-butyloctyl side chain, producing a P3ATT an $M_n = 54.6$ kg/mol and $Đ = 1.37$ [68].

Well-defined poly(*N*-hexylpyrrole) was prepared from 2,5-dibromo-1-hexylpyrrole (activated with isopropylmagnesium chloride) and Ni(dppe)Cl_2 at 0 °C (Figure 2.6) [70]. The authors found that an additional equivalent of dppe in the reaction mixture suppressed formation of low molecular weight oligomers [70]. Noonan has also examined KCTP of 2-bromo-3-hexylfuran (after activation with TMPMgCl · LiCl), using Ni(dppp)Cl_2 as the catalyst. The polymers obtained had narrow MWDs, but the molecular weight of the samples were low, which was thought to be due to strong aggregation of the poly(3-hexylfuran) and premature precipitation from the reaction solution during polymerization [71].

Poly(3-alkylselenophenes) (P3ASe) have attracted attention due to the lower bandgap as compared to P3ATs and they can be synthesized by activation of 2,5-dihalo-3-alkylselenophene monomers with isopropylmagnesium chloride, followed by polymerization with nickel catalysts such as Ni(dppp)Cl_2 [72–75]. Noonan also reported on the polymerization of mixed 3-alkylthiophene and 3-alkylselenophene dimers and trimers to examine how periodicity in conjugated polymers impacts electronic properties and solid-state organization [23]. This offered straightforward access to well-defined, high molecular weight polymers with precise repeating sequences of thiophene and selenophene repeat units in the polymer backbone [23]. Seferos has also examined the KCTP of the heavier 3-alkyltellurophene analogs to prepare poly(3-alkyltellurophenes) or P3ATe [76, 77]. The Seferos team noted that a specific branched side chain is critical to obtaining well-defined polymers and ensuring the chain-growth mechanism is operative (Figure 2.6) [77]. The 3-ethylheptyl side chain on the tellurophene ring ensured the polymer was soluble, but did not impede the polymerization reaction(M_n~25 kg/mol, Đ~1.2) [77]. Several reports have appeared on 4-substituted thiazole monomers, and the choice of side chains are key to ensuring polymer solubility such that high molecular polymers can be obtained and that the reaction proceeds by chain-growth [78–80].

Moving beyond the 5-membered heteroaromatic rings, KCTP of 4-bromo-2,5-bis(hexyloxy)phenylmagnesium chloride was examined to obtain poly(*para*-phenylene) or PPP (Figure 2.6) [81]. Lithium chloride was needed in the polymerization mixture, which the authors supposed reduced monomer aggregation, and Ni(dppe) was the most effective Ni(diphosphine) catalyst tested [81]. The preparation of poly(9,9-dialkylfluorenes) or PFs using KCTP has been examined extensively [82–85], and Geng noted that Ni(acac)$_2$/dppp was a unique catalyst system that enabled formation of high molecular weight PFs with narrow MWDs (e.g. M_n ~ 62.2 kg/mol, Đ ~ 1.23) [84]. The KCTP of bithienylmethane and pyridine monomers has also been examined [86, 87]. Polymerization of thieno[3,4-*b*]pyrazine, cyclopentadithiophene and benzotriazole monomers using KCTP has been reported as well (Figure 2.6) [42, 88–91], but each monomer brings some challenges with respect to achieving a well-behaved chain-growth process. The chain-growth, metal-free, photopolymerization of a magnesiated benzotriazole has been communicated [92, 93]. Recently, chain-growth polymerization of benzotriazoles and thienopyrazines using a dinuclear nickel catalyst has been described [94]. Attempts to prepare poly(*N*-alkyl-2,7-carbazole)s and poly(*N*-alkyl-3,6-carbazole)s using KCTP have appeared [95, 96], but low molecular weights were noted for the 2,7 derivative and catalyst dissociation was considered a significant issue for the 3,6-variant as determined by end group analysis of the polymer samples.

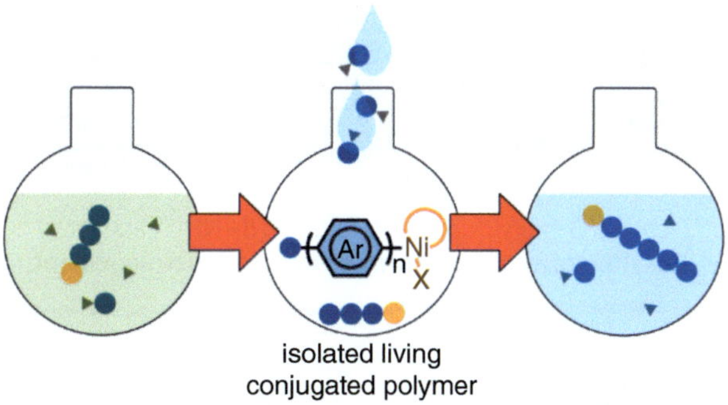

Figure 2.7 Preparation of metal-capped living polymer chains. *Source:* [97]/American Chemical Society.

Living polymer chains with a metal-capped chain end have been reported recently (Figure 2.7) [97]. This includes, P3AT, P3ASe, P3ATe, and PPP chains. A significant benefit of this work is the metal at the chain-end can be used to promote chain extension or alternatively, it can be directly reacted with a Grignard reagent for end-capping. As an example, Seferos prepared a living poly(3-hexylthiophene) chain by reaction of the activated monomer with Ni(dppe)(*o*-tolyl)Br followed by solvent removal from the reaction mixture and repeated washing and sonication with cold methanol to isolate the metal-functionalized polymer. The functional P3HT-Ni(dppe)Br chain end can be converted into a specific end groups via treatment with a Grignard reagent, or it can be combined with other monomers for chain extension to build block copolymers [97].

2.5 Catalyst-transfer Polymerization with Organolithium Reagents

Cross-coupling of organolithium reagents has also been employed to afford well-defined conjugated polymers via a chain-growth mechanism. The commercially available NHC-Ni catalyst, [1,3-bis(2,6-diisopropylphenyl)imidazole-2-ylidene]triphenylphosphine nickel(II) dichloride ($NiCl_2(PPh_3)IPr$) has been used in the synthesis of P3ATs (Figure 2.8) [98]. Mori demonstrated that 2-chloro-3-hexylthiophene could be selectively deprotonated with *n*-butyllithium at −78 °C without any competing lithium-halogen exchange, and the active monomer was polymerized with $NiCl_2(PPh_3)IPr$ in cyclopentyl methyl ether (CPME) [98]. The Ni—NHC catalyst is necessary to ensure the polymerization proceeds effectively [98], and a linear M_n versus conversion plot was observed, providing evidence for the chain-growth mechanism [98]. A "halogen dance" polymerization was also described by Mori, where the poly(3-hexyl-4-bromothiophene) that was synthesized had a narrow MWD ($M_n \sim 7.9$ kg/mol, $Đ \sim 1.28$). The low dispersity could suggest a possible chain-growth mechanism, though this was not explicitly stated (Figure 2.8b) [98, 99].

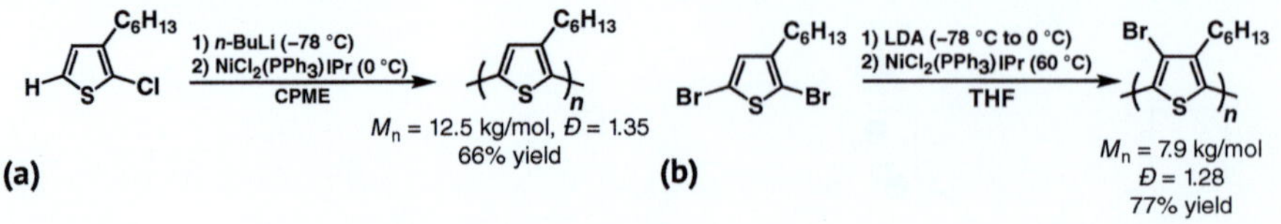

Figure 2.8 (a) Polymerization of 2-chloro-3-hexylthiophene using a Ni–NHC catalyst. (b) Preparation of poly(3-bromo-4-hexylthiophene) via deprotonation and rearrangement.

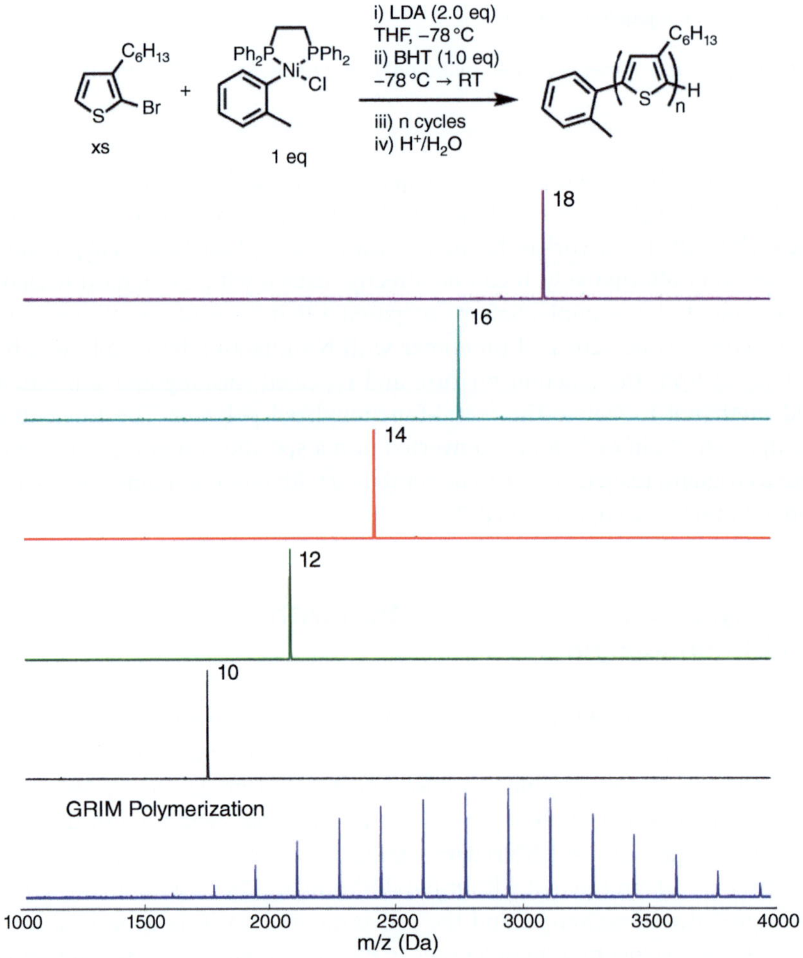

Figure 2.9 Synthesis of monodisperse oligo-3-alkylthiophenes. *Source*: [100]/American Chemical Society.

Efforts to build monodisperse oligo- and poly-3-alkylthiophenes has been reported [100–102]. The Seferos team developed a low-temperature approach where they lithiate 2-bromo-3-hexylthiophene using excess LDA in the presence of (dppe)Ni(*o*-tolyl)Br. A transmetalation occurs, and the resultant (dppe)Ni(*o*-tolyl)

(2-thienyl-Br) is stable at low temperature. Upon warming, reductive elimination occurs with formation of the new C—C bond, followed by walking to the thiophene C-Br chain end, and the process can be repeated [100]. Butylated hydroxytoluene (BHT) is used to quench any excess lithiated monomers after each cycle, without affecting the reaction process since ensuring the perfect stoichiometry is very difficult [100]. The low temperature (−78 °C) required to separate the process into discrete steps tends to put an upper limit on the length of the oligomer, and the authors demonstrated that octadecamers can be obtained (Figure 2.9) [100]. The method has been utilized recently to create sequence-defined oligomers with different 3-substituted thiophene monomers [103].

2.6 Negishi Catalyst-transfer Polymerization

Cross-coupling of organozinc reagents is a popular approach to C—C bond formation [104] and is beneficial in polymerization, as the scope of potential functional groups present on the aromatic monomers can increase as compared to cross-coupling of organomagnesium and organolithium reagents. While Riecke reported on Negishi cross-coupling for the synthesis of P3ATs and polyarylenes in the early 1990s [50, 51, 105, 106], Higashihara recently reported on the quasi-living polymerization of a zincated 3-alkylthiophene monomer [107–109]. The activated monomer was obtained by combining dilithium tetra(*tert*-butyl) zincate (tBu_4ZnLi_2) and 2-bromo-3-hexyl-5-iodothiophene (Figure 2.10) [107]. In that work, $Ni(dppe)Cl_2$ was used as the polymerization precatalyst, and well-defined *rr*-P3HTs were formed at 60 °C (typical reaction time 2 h). Higashihara also demonstrated that a pendant alcohol group could be present on the monomer without impacting the polymerization reaction (Figure 2.9) [107]. Higashihara has also reported on the preparation of PPPs via activation of 1-bromo-2,5-bis(hexyloxy)-4-iodobenzene with tBu_4ZnLi_2 [110].

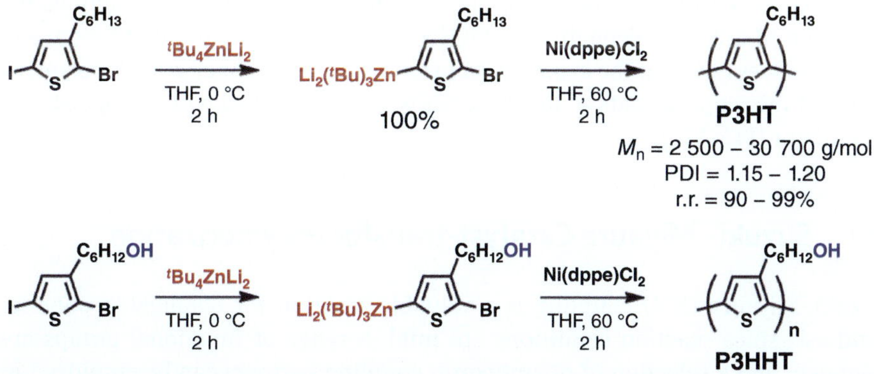

Figure 2.10 Dilithium tetra(*tert*-butyl) zincate activation of 2-bromo-3-hexyl-5-iodothiophene, followed by polymerization with $Ni(dppe)Cl_2$. *Source:* [107]/American Chemical Society.

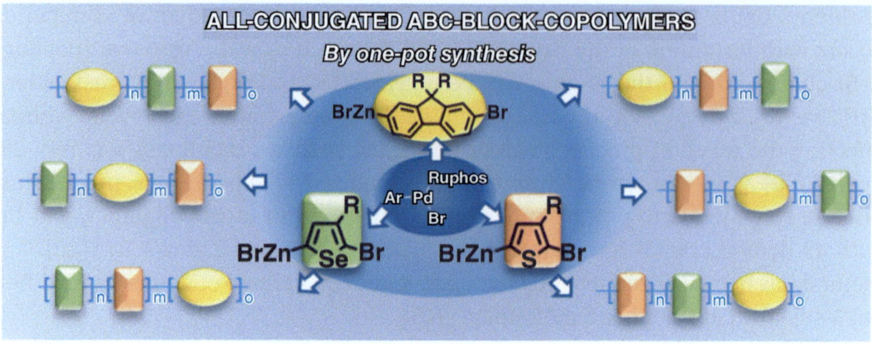

Figure 2.11 NCTP to prepare triblock copolymers where order of addition does not matter. *Source*: [111]/American Chemical Society.

A RuPhosPd complex has been examined for potential chain-growth polymerization of aromatic monomers [111, 112]. Koeckelberghs reported that the reaction deviates from the typical catalyst-transfer mechanism, and that chain-growth is attributed to a more reactive aryl halide at the polymer chain end and that the aryl halide of the zincated monomer is deactivated. The three zinc monomers (3-alkylthiophene, 3-alkylselenophene, and 9,9-dialkylfluorene), as shown in Figure 2.11, were formed from alkyl Grignard reagents, followed by treatment with $ZnBr_2$. The polymerization was accomplished with RuPhosPd(Ar)Br as the catalyst, and triblock copolymers could be synthesized starting from any monomer (Figure 2.11) [111]. Kiriy has also shown that $(t\text{Bu})_3$PPd(Ph)Br can be used as the catalyst for an activated fluorene monomer [113].

A dithienosilole monomer was polymerized to form poly(4,4-bis(2-ethylhexyl) dithienosilole) (PDTS) with controllable molecular weights, though chain-transfer is clearly an issue, as the MWDs became broader as the mass of the polymer increased [114]. A P3HT-*block*-PDTS copolymer was synthesized, where the authors transferred from a Kumada-type organomagnesium monomer when synthesizing the P3HT block to an organozinc for the PDTS block [114]. Kiriy has examined the activation of a naphthalene diimide bithiophene monomer using Rieke zinc, and though it does not form the expected Zn insertion product, chain-growth polymerization of the proposed radical anion species was accomplished with a (dppe)Ni(Ar) Br initiator [115, 116].

2.7 Suzuki–Miyaura Catalyst-transfer Polymerization

Suzuki–Miyaura (SM) coupling is a valuable reaction for conjugated polymer synthesis since reaction conditions are mild, a range of functional groups are tolerated, and a selection of organoboron coupling partners can be employed in the reaction. Several book chapters [117, 118] and reviews [119, 120] offer more details on the reaction itself and mechanism. The first report of a Suzuki–Miyaura catalyst-transfer polymerization (SMCTP) reaction appeared in 2007 from the

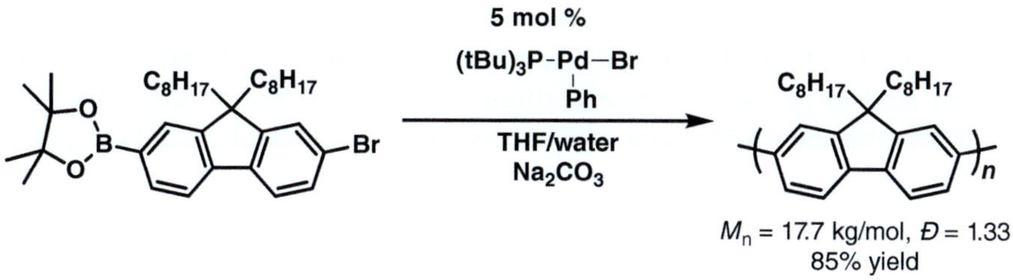

Figure 2.12 SMCTP of a 9,9-dialkylfluorene. *Source*: Adapted from [121].

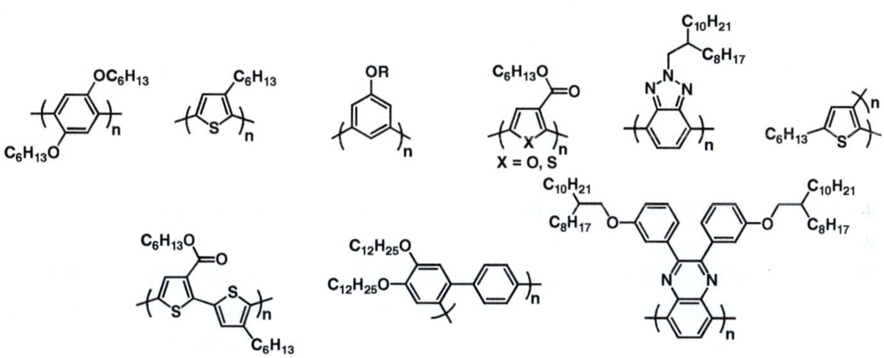

Figure 2.13 Polymers synthesized via SMCTP.

Yokozawa group, when they examined polymerization of a 9,9-dialkylfluorene monomer using the $(t\text{Bu})_3$PPd(Ph)Br catalyst (Figure 2.12) [121]. The polymerization was determined to proceed by a chain-growth mechanism due to the fairly narrow MWDs observed and the linear dependence of M_n with conversion [121]. The preparation of alkyl substituted PFs using SMCTP has been described in a range of reports with various organoboron monomers and catalyst pairings [122–132].

Some of the examples of different homopolymers prepared using SMCTP are highlighted in Figure 2.13. After the work on PFs in 2007, Yokozawa reported on SMCTP of 1,4-dialkoxy-phenylene and 3-alkylthiophene monomers in 2010 and 2011 (Figure 2.13) [133, 134]. The reaction conditions were key to chain-growth, where the polymerization was conducted at 0 °C with CsF in the presence of 18-crown-6 [133, 134]. A number of other catalysts have been examined for P3AT synthesis using SMCTP over the last 10 years [39, 135–140], and it should be noted that Ni(dppp)Cl_2, which is common in KCTP, has also been successfully used in SMCTP [138, 140].

The utility of the SMCTP reaction is clear from the different polymers that have been synthesized when compared to the polymers from KCTP as shown in Figure 2.6. A *m*-phenylene monomer with a bromo and pinacol ester group was polymerized in 2012, where high molecular weight open-chain polymers were obtained with a Buchwald-type (SPhos)Pd catalyst [141]. Regioregular poly(3-alkylester)furans and thiophenes with controllable molecular weights have been obtained using SMCTP with $NiCl_2(PPh_3)IPr$ as the catalyst (Figure 2.13) [140, 142]. The SMCTP reaction has been used to synthesize *ortho*-substituted polymers including the poly(*ortho, para*-alternating-phenylenes) and helical poly(5-alkyl-2,3-thiophene)s [143, 144]. Molecular weight was dependent on catalyst loading in both instances [143, 144]. Benzotriazoles have also been prepared using SMCTP of two different types of organoboron monomers with (SPhos)Pd and (RuPhos)Pd [145, 146]. As has been noted in multiple reports [70, 138], free ligand can be beneficial in chain-growth polymerization to prevent chain-transfer. Both Choi and Noonan noted that free ligand is key to preventing chain-transfer in the SMCTP of benzotriazole monomers [145, 146]. Choi has also reported on SMCTP to afford poly(5,8-quinoxalines) (Figure 2.13) [146], with preparation of statistical and block copolymers. The SMCTP reaction was limited when targeting poly(3,6-phenanthrenes), poly(phenylene-vinylenes), and a benzothiadiazole-containing copolymer [147–150].

Donor–acceptor (D-A) copolymers have also been described recently using SMCTP with thiophene and selenophene donors paired with either quinoxaline, diketopyrrolopyrrole, and benzotriazole as the acceptor building blocks [151]. High molecular weight polymers could be prepared (9.2 – 40.0 kg/mol) with narrow MWDs ($Đ = 1.05$–1.29) using a (RuPhos)Pd catalyst [151]. This report highlights the broad potential of CTP to make more complicated polymer constructs, particularly using SMCTP. The authors found that appropriate pairing of donor and acceptor building blocks resulted in band gap modulation from 1.29 to 1.77 eV for the series [151].

2.8 Stille Catalyst-transfer Polymerization (Stille CTP)

The cross-coupling of organostannanes and halide groups is arguably still the most employed reaction for construction of conjugated polymers [152], but Stille CTP is not very common. In 2013, an iodo and tin-functionalized aryl ethynylene monomer (shown in Figure 2.14a) was polymerized using $(t\text{Bu})_3PPd(Ph)Br$ to prepare well-defined poly(phenylene ethynylenes) or PPEs [153]. The authors noted the importance of additional PPh_3 in the reaction mixture for the CTP mechanism to be operative. Bielawski recently expanded on this work, examining chain-growth copolymerization of *para*- and *meta*-substituted aryl ethynylene monomers using Stille CTP [154].

Only a single report on Stille CTP for P3AT synthesis has appeared (Figure 2.14b) [155]. In that work, 2-bromo-3-hexyl-5-trimethylstannylthiophene was polymerized with the commercially available Pd-PEPPSI-IPr catalyst with CsF as an additive. A linear M_n versus conversion plot suggested the chain-growth mechanism was operative, and chain extension could be accomplished by adding a second monomer feed to the reaction mixture after consumption of the first feed. MWDs did increase with decreasing catalyst loading, and very high molecular weights could be obtained using this method ($M_n = 73\,\text{kg/mol}, Đ = 1.53$).

Normally, the combination of AA and BB monomers leads to a step-growth polymerization reaction when functional groups have equal reactivity. Stoichiometry

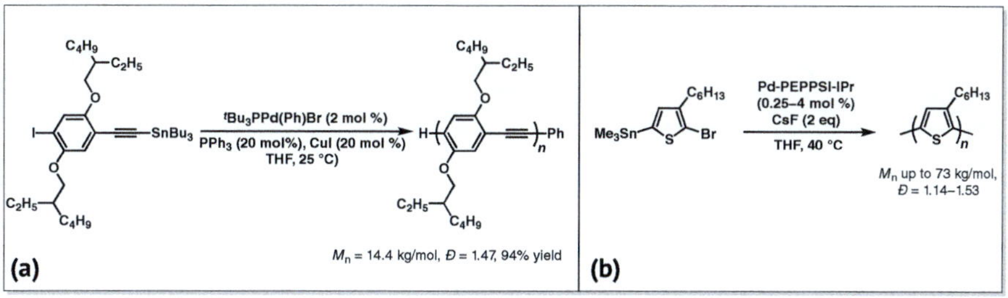

Figure 2.14 Examples of Stille CTP.

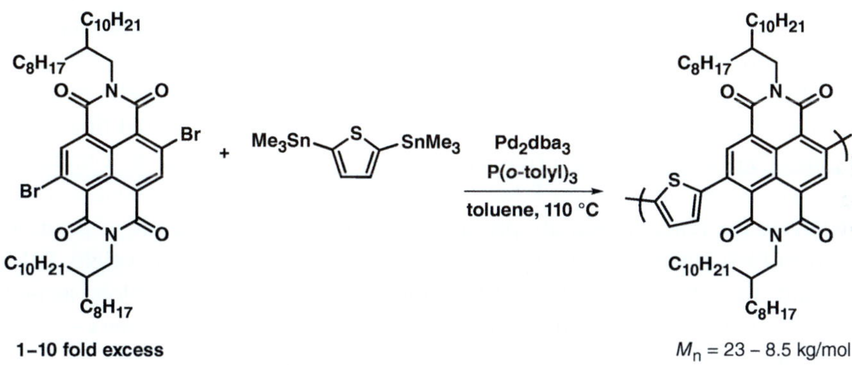

Figure 2.15 Non-stoichiometric coupling of a naphthalene diimide and 2,5-di(tributylstannyl)thiophene. *Source*: Adapted from [157].

is critical to achieve high molecular weight polymers in this instance [156]. Deviations are possible if functional group activation changes in the reaction (e.g. if the chain end of the growing polymer is more reactive than a monomer unit). Higashihara reported that a 10-fold excess of dibromo naphthalene diimide monomer with 2,5-trimethylstannylthiophene still led to polymers (Figure 2.15), which is attributed to intramolecular transfer of the metal after the first coupling step [157]. Model studies with 2-trimethylstannylthiophene were used to provide supporting evidence of the intramolecular transfer step. This phenomenon has also been observed with *N*-octyl-3.6-dibromophthalimide [158] and ester-functionalized dibromobenzenes [159]. Non-stoichiometric polymerizations have also been reported with Suzuki–Miyaura cross-coupling [160–163].

2.9 Buchwald–Hartwig Amination and Direct Arylation Polymerization (C—H Activation)

Poly(arylamines) were synthesized over 30 years ago [164–167], but the first report of chain-growth polymerization to prepare a triarylamine polymer appeared in 2019 (Figure 2.16a) [168]. The aniline monomers are shown in Figure 2.16a, and $(tBu)_3PPd(Ph)I$ was used to initiate the polymerizations. The authors tried three different

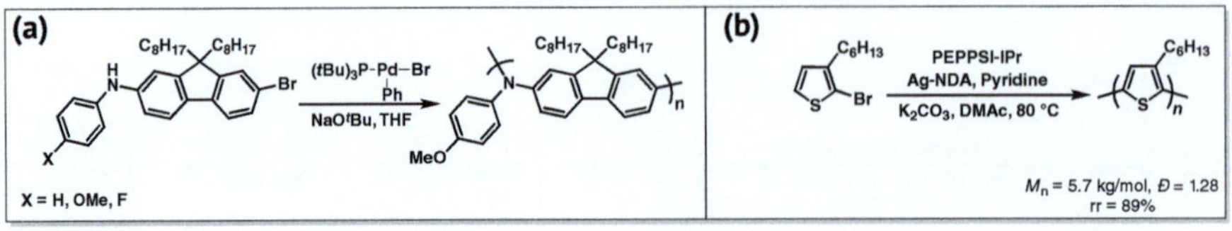

Figure 2.16 (a) Preparation of poly(arylamine) using Buchwald–Hartwig amination. *Source*: Adapted from [168]. (b) C—H activation polymerization for P3HT synthesis. *Source*: [173]/American Chemical Society/CC BY 4.0.

monomer variants (X = H, OMe and F). Most experiments were carried out with X = H, and room temperature reactions were to be critical for chain-growth. All three variants of the monomer could be polymerized with 5 mol % of the catalyst to ~10 kg/mol with narrow MWDs (1.1–1.3).

The direct activation of C—H bond for conjugated polymer synthesis is highly desirable, as it minimizes the number of synthetic steps to the desired final product [169–172]. Luscombe reported on the direct polymerization of 2-bromo-3-hexylthiophene using a dual Ag/Pd system (Figure 2.16b) [173]. Silver neodecanoate (AgNDA) in the reaction mixture presumably activates the C—H bond (5 position) and Pd-PEPPSI-IPr brings about cross-coupling. Lower molecular weight P3HT with relatively narrow MWD could be obtained using this method, so long as pyridine is present in the reaction mixture. Monomodal GPC traces were observed when AgNDA and pyridine were both present in the reaction mixture.

2.10 Chain-growth Polymerization of Aromatics Without Metal Catalysts

A cationic chain-growth polymerization of 2-chloroalkylenedioxythiophenes has been reported using $SnCl_4$ as a Lewis acid (Figure 2.17) [174]. Linear M_n versus conversion plots and moderately narrow MWDs were noted for the polymerization of chloro dibutyl propylene dioxythiophene (Cl-ProDOT) [174]. The monomer scope for this cationic polymerization was examined, with several different 3,4-dialkoxy and a 3-alkoxy functionalized thiophene monomers being investigated [174].

Trifluoromethanesulfonic acid has been examined for polymerization of 3-alkoxy-, 3-alkylthio-, and 3-alkylthiophenes [175]. The 2-bromo-3-substituted-thiophenes are more reactive in the cationic polymerization, but improved regioregularity is noted for the final polymer if 2-chloro-3-substituted-thiophenes are used as the monomers [175]. Recently, 2-fluoro-3-alkoxythiophenes were prepared and polymerized in a similar manner, with 10% trifluoromethanesulfonic acid at 90 °C in 1,1,2,2-tetrachloroethane as the solvent [176]. Branched side chains improved the polymerization behavior. For example, the combination of 2-fluoro-3-(2-ethylhexyl) oxythiophene with triflic acid resulted in a polymer with M_n = 13.5 kg/mol, and the highest regioregularity (96% HT) [176].

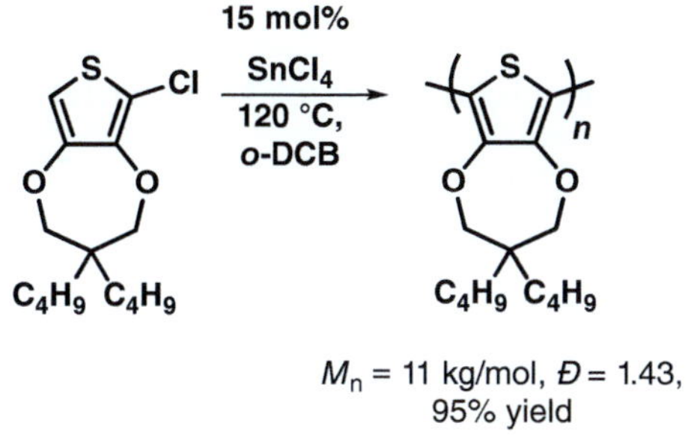

Figure 2.17 Polymerization of chlorodibutylpropylenedioxythiophene (Cl-ProDOT) using $SnCl_4$. *Source*: Adapted from [174].

2.11 Conclusion

The living polymerization of aromatic monomers offers exceptional control and opportunities to control backbone structure, chain length, and ultimately properties and performance of conjugated polymers in devices. Recent reports on living donor–acceptor copolymers using SMCTP [151], monodisperse oligothiophene synthesis using KCTP [103], and non-stoichiometric polymerization with AA/BB monomers [157] illustrate the potential of harnessing this ring-walking mechanism to prepare well defined polymers. This is not to mention that block and star conjugated copolymers can be prepared using CTP, and though block copolymers were not discussed extensively above, in many of the reports these are common targets with these chemistries. Further advances in CTP will come from new monomers and new catalysts, with inspiration from conjugated polymer design and small molecule cross-coupling. Living or chain-growth C—H activation is an exciting future goal.

Acknowledgment

The authors are grateful to NSF for support of their work in synthetic polymerization techniques and conjugated polymers and macrocycles (CHE-1455136, CHE-2109065 and 2404367).

Abbreviations

Polymers

rr – regioregular
irr – regioirregular
P3AT – poly(3-alkylthiophene)
P3HT – poly(3-hexylthiophene)
PPP – poly(*para*-phenylene)
PF – polyfluorene
PAEs – poly(arylene ethynylenes)

Metal Catalysts

Pd-PEPPSI-IPr – [1,3-bis(2,6-diisopropylphenyl)imidazol-2-ylidene](3-chloropyridyl) dichloropalladium(II)
$Ni(PPh_3)Cl_2IPr$ – [1,3-bis(2,6-diisopropylphenyl)imidazol-2-ylidene]triphenylphosphine Nickel(II) Dichloride

Ligands

PPh_3 – triphenylphosphine
$P(tBu)_3$ – tri-*tert*-butylphosphine
SPhos – 2-dicyclohexylphosphino-2′,6′-dimethoxybiphenyl
RuPhos – 2-dicyclohexylphosphino-2′,6′-diisopropoxybiphenyl
dppp – 1,3-bis(diphenylphosphino)propane
dppe – 1,2-bis(diphenylphosphino)ethane
dppf – 1,1′-bis(diphenylphosphino)ferrocene
IPr – 1,3-bis(2,6-diisopropylphenyl)imidazol-2-ylidene
Dtbpe – 1,2-bis(di-tert-butylphosphino)ethane

Miscellaneous Reagents and Terms

NHC – N-heterocyclic carbene
CTP – catalyst-transfer polymerization
nBuLi – *n*-butyllithium
LDA – lithium diisopropylamide
DP_n – number-average degree of polymerization
M_n – number-average molecular weight
M_w – weight-average molecular weight
MWD – molecular weight distribution

References

1 Quirk, R.P. and Lee, B. (1992). Experimental criteria for living polymerizations. *Polymer International* 27: 359–367.
2 Sheina, E.E., Liu, J., and Iovu, M.C. et al. (2004). Chain growth mechanism for regioregular nickel-initiated cross-coupling polymerizations. *Macromolecules* 37: 3526–3528.
3 Iovu, M.C., Sheina, E.E., Gil, R.R. et al. (2005). Experimental evidence for the quasi-"living" nature of the Grignard metathesis method for the synthesis of regioregular poly(3-alkylthiophenes). *Macromolecules* 38: 8649–8656.
4 Miyakoshi, R., Yokoyama, A., and Yokozawa, T. (2004). Synthesis of poly(3-hexylthiophene) with a narrower polydispersity. *Macromolecular Rapid Communications* 25: 1663–1666.

5 Yokoyama, A., Miyakoshi, R., and Yokozawa, T. (2004). Chain-growth polymerization for poly(3-hexylthiophene) with a defined molecular weight and a low polydispersity. *Macromolecules* 37: 1169–1171.

6 Miyakoshi, R., Yokoyama, A., and Yokozawa, T. (2005). Catalyst-transfer polycondensation. Mechanism of Ni-catalyzed chain-growth polymerization leading to well-defined poly(3-hexylthiophene). *Journal of the American Chemical Society* 127: 17542–17547.

7 Cheng, S., Zhao, R., and Seferos, D.S. (2021). Precision synthesis of conjugated polymers using the Kumada methodology. *Accounts of Chemical Research* 54: 4203–4214.

8 Leone, A.K., Mueller, E.A., and McNeil, A.J. (2018). The history of palladium-catalyzed cross-couplings should inspire the future of catalyst-transfer polymerization. *Journal of the American Chemical Society* 140: 15126–15139.

9 Baker, M.A., Tsai, C.-H., and Noonan, K.J.T. (2018). Diversifying cross-coupling strategies, catalysts and monomers for the controlled synthesis of conjugated polymers. *Chemistry – A European Journal* 24: 13078–13088.

10 Verheyen, L., Leysen, P., Van den Eede, M.-P. et al. (2017). Advances in the controlled polymerization of conjugated polymers. *Polymer* 108: 521–546.

11 Yokozawa, T. and Ohta, Y. (2016). Transformation of step-growth polymerization into living chain-growth polymerization. *Chemical Reviews* 116: 1950–1968.

12 Leone, A.K. and McNeil, A.J. (2016). Matchmaking in catalyst-transfer polycondensation: optimizing catalysts based on mechanistic insight. *Accounts of Chemical Research* 49: 2822–2831.

13 Grisorio, R. and Suranna, G.P. (2015). Intramolecular catalyst transfer polymerisation of conjugated monomers: from lessons learned to future challenges. *Polymer Chemistry* 6: 7781–7795.

14 Bryan, Z.J. and McNeil, A.J. (2013). Conjugated polymer synthesis via catalyst-transfer polycondensation (CTP): mechanism, scope, and applications. *Macromolecules* 46: 8395–8405.

15 Okamoto, K. and Luscombe, C.K. (2011). Controlled polymerizations for the synthesis of semiconducting conjugated polymers. *Polymer Chemistry* 2: 2424–2434.

16 Kiriy, A., Senkovskyy, V., and Sommer, M. (2011). Kumada catalyst-transfer polycondensation: mechanism, opportunities, and challenges. *Macromolecular Rapid Communications* 32: 1503–1517.

17 Yokozawa, T. and Yokoyama, A. (2009). Chain-growth condensation polymerization for the synthesis of well-defined condensation polymers and π-conjugated polymers. *Chemical Reviews* 109: 5595–5619.

18 Noonan, K.J.T. (2025). Controlled/living polymerization toward conjugated polymers. In: *Springer Handbook of Functional Polymers, Springer Handbooks* (ed. Y. Chujo). Springer.

19 Beryozkina, T., Senkovskyy, V., Kaul, E. et al. (2008). Kumada catalyst-transfer polycondensation of thiophene-based oligomers: robustness of a chain-growth mechanism. *Macromolecules* 41: 7817–7823.

20 Zenkina, O.V., Karton, A., Freeman, D. et al. (2008). Directing aryl–I versus aryl–Br bond activation by nickel via a ring walking process. *Inorganic Chemistry* 47: 5114–5121.

21 Tkachov, R., Senkovskyy, V., Komber, H. et al. (2010). Random catalyst walking along polymerized poly(3-hexylthiophene) chains in Kumada catalyst-transfer polycondensation. *Journal of the American Chemical Society* 132: 7803–7810.

22 Kohn, P., Huettner, S., Komber, H. et al. (2012). On the role of single regiodefects and polydispersity in regioregular poly(3-hexylthiophene): defect distribution, synthesis of defect-free chains, and a simple model for the determination of crystallinity. *Journal of the American Chemical Society* 134: 4790–4805.

23 Tsai, C.-H., Fortney, A., Qiu, Y. et al. (2016). Conjugated polymers with repeated sequences of group 16 heterocycles synthesized through catalyst-transfer polycondensation. *Journal of the American Chemical Society* 138: 6798–6804.

24 Leone, A.K., Goldberg, P.K., and McNeil, A.J. (2018). Ring-walking in catalyst-transfer polymerization. *Journal of the American Chemical Society* 140: 7846–7850.

25 He, W., Patrick, B.O., and Kennepohl, P. (2018). Identifying the missing link in catalyst transfer polymerization. *Nature Communications* 9: 3866.

26 Sontag, S.K., Bilbrey, J.A., Huddleston, N.E. et al. (2014). π-Complexation in nickel-catalyzed cross-coupling reactions. *Journal of Organic Chemistry* 79: 1836–1841.

27 Bilbrey, J.A., Bootsma, A.N., Bartlett, M.A. et al. (2017). Ring-walking of zerovalent nickel on aryl halides. *Journal of Chemical Theory and Computation* 13: 1706–1711.

28 Mikami, K., Nojima, M., Masumoto, Y. et al. (2017). Catalyst-dependent intrinsic ring-walking behavior on π-face of conjugated polymers. *Polymer Chemistry* 8: 1708–1713.

29 Leone, A.K., Souther, K.D., Vitek, A.K. et al. (2017). Mechanistic insight into thiophene catalyst-transfer polymerization mediated by nickel diimine catalysts. *Macromolecules* 50: 9121–9127.

30 Pollit, A.A., Ye, S., and Seferos, D.S. (2020). Elucidating the role of catalyst steric and electronic effects in controlling the synthesis of π-conjugated polymers. *Macromolecules* 53: 138–148.

31 Tamao, K., Sumitani, K., Kiso, Y. et al. (1976). Nickel-phosphine complex-catalyzed Grignard coupling. I. Cross-coupling of alkyl, aryl, and alkenyl Grignard reagents with aryl and alkenyl halides: general scope and limitations. *Bulletin of the Chemical Society of Japan* 49: 1958–1969.

32 Sugita, H., Kamigawara, T., Miyazaki, S. et al. (2023). Intramolecular palladium catalyst transfer on benzoheterodiazoles as acceptor monomers and discovery of catalyst transfer inhibitors. *Chemistry – A European Journal* 29: e202301242.

33 Tokita, Y., Katoh, M., Ohta, Y. et al. (2016). Mechanistic investigation of catalyst-transfer Suzuki-Miyaura condensation polymerization of thiophene-pyridine biaryl monomers with the aid of model reactions. *Chemistry – A European Journal* 22: 17436–17444.

34 Tanaka, S., Tatsuta, G., Sugie, A. et al. (2013). Studies on the preference of multiple coupling in the introduction of thiophene ring into poly-halogenated aromatic compounds with nickel NHC catalyst. *Tetrahedron Letters* 54: 1976–1979.

35 Larrosa, I., Somoza, C., Banquy, A. et al. (2011). Two flavors of PEPPSI-IPr: activation and diffusion control in a single NHC-ligated Pd catalyst? *Organic Letters* 13: 146–149.

36 Dong, C.-G. and Hu, Q.-S. (2005). Preferential oxidative addition in palladium(0)-catalyzed Suzuki cross-coupling reactions of dihaloarenes with arylboronic acids. *Journal of the American Chemical Society* 127: 10006–10007.

37 Weber, S.K., Galbrecht, F., and Scherf, U. (2006). Preferential oxidative addition in Suzuki cross-coupling reactions across one fluorene unit. *Organic Letters* 8: 4039–4041.

38 Schiefer, D., Wen, T., Wang, Y. et al. (2014). Nickel catalyst with a hybrid P, N ligand for Kumada catalyst transfer polycondensation of sterically hindered thiophenes. *ACS Macro Letters* 3: 617–621.

39 Baker, M.A., Ayuso-Carrillo, J., Koos, M.R.M. et al. (2020). A robust nickel catalyst with an unsymmetrical propyl-bridged diphosphine ligand for catalyst-transfer polymerization. *Polymer Journal* 52: 83–92.

40 Lanni, E.L., Locke, J.R., Gleave, C.M. et al. (2011). Ligand-based steric effects in Ni-catalyzed chain-growth polymerizations using bis(dialkylphosphino)ethanes. *Macromolecules* 44: 5136–5145.

41 Senkovskyy, V., Tkachov, R., Beryozkina, T. et al. (2009). "Hairy" poly(3-hexylthiophene) particles prepared via surface-initiated Kumada catalyst-transfer polycondensation. *Journal of the American Chemical Society* 131: 16445–16453.

42 Bridges, C.R., McCormick, T.M., Gibson, G.L. et al. (2013). Designing and refining Ni(II)diimine catalysts toward the controlled synthesis of electron-deficient conjugated polymers. *Journal of the American Chemical Society* 135: 13212–13219.

43 Bridges, C.R., Yan, H., Pollit, A.A. et al. (2014). Controlled synthesis of fully π-conjugated donor–acceptor block copolymers using a Ni(II) diimine catalyst. *ACS Macro Letters* 3: 671–674.

44 Lee, S.R., Bryan, Z.J., Wagner, A.M. et al. (2012). Effect of ligand electronic properties on precatalyst initiation and propagation in Ni-catalyzed cross-coupling polymerizations. *Chemical Science* 3: 1562–1566.

45 Boyd, S.D., Jen, A.K.-Y., and Luscombe, C.K. (2009). Steric stabilization effects in nickel-catalyzed regioregular poly(3-hexylthiophene) synthesis. *Macromolecules* 42: 9387–9389.

46 Bronstein, H.A. and Luscombe, C.K. (2009). Externally initiated regioregular P3HT with controlled molecular weight and narrow polydispersity. *Journal of the American Chemical Society* 131: 12894–12895.

47 Tkachov, R., Senkovskyy, V., Komber, H. et al. (2011). Influence of alkyl substitution pattern on reactivity of thiophene-based monomers in Kumada catalyst-transfer polycondensation. *Macromolecules* 44: 2006–2015.

48 Loewe, R.S., Khersonsky, S.M., and McCullough, R.D. (1999). A simple method to prepare head-to-tail coupled, regioregular poly(3-alkylthiophenes) using Grignard metathesis. *Advanced Materials* 11: 250–253.

49 Loewe, R.S., Ewbank, P.C., Liu, J. et al. (2001). Regioregular, head-to-tail coupled poly(3-alkylthiophenes) made easy by the GRIM method: investigation of the reaction and the origin of regioselectivity. *Macromolecules* 34: 4324–4333.

50 Chen, T.A. and Rieke, R.D. (1992). The first regioregular head-to-tail poly (3-hexylthiophene-2,5-diyl) and a regiorandom isopolymer: nickel versus palladium catalysis of 2(5)-bromo-5(2)-(bromozincio)-3-hexylthiophene polymerization. *Journal of the American Chemical Society* 114: 10087–10088.

51 Chen, T.-A., Wu, X., and Rieke, R.D. (1995). Regiocontrolled synthesis of poly(3-alkylthiophenes) mediated by Rieke zinc: their characterization and solid-state properties. *Journal of the American Chemical Society* 117: 233–244.

52 Wu, S., Huang, L., Tian, H. et al. (2011). LiCl-promoted chain growth Kumada catalyst-transfer polycondensation of the "reversed" thiophene monomer. *Macromolecules* 44: 7558–7567.

53 Lamps, J.P. and Catala, J.M. (2011). Kinetic study, by UV–vis spectroscopy, on the strong effect of LiCl on the controlled polymerization of 2-bromo-3-hexyl-5-iodothiophene and 2-iodo-3-hexyl-5-bromothiophene: determination of the propagation rate constants, application to the synthesis of high molecular weight polydodecylthiophene. *Macromolecules* 44: 7962–7968.

54 McCullough, R.D. and Lowe, R.D. (1992). Enhanced electrical-conductivity in regioselectively synthesized poly (3-alkylthiophenes). *Journal of the Chemical Society, Chemical Communications* 70–72.

55 McCullough, R.D., Lowe, R.D., Jayaraman, M. et al. (1993). Design, synthesis, and control of conducting polymer architectures: structurally homogeneous poly (3-alkylthiophenes). *Journal of Organic Chemistry* 58: 904–912.

56 Krasovskiy, A., Krasovskaya, V., and Knochel, P. (2006). Mixed Mg/Li amides of the type $R_2NMgCl{\cdot}LiCl$ as highly efficient bases for the regioselective generation of functionalized aryl and heteroaryl magnesium compounds. *Angewandte Chemie International Edition* 45: 2958–2961.

57 Shibuya, Y. and Mori, A. (2020). Dehalogenative or deprotonative? The preparation pathway to the organometallic monomer for transition-metal-catalyzed catalyst-transfer-type polymerization of thiophene derivatives. *Chemistry – A European Journal* 26: 6976–6987.

58 Tamba, S., Tanaka, S., Okubo, Y. et al. (2011). Nickel-catalyzed dehydrobrominative polycondensation for the practical preparation of regioregular poly (3-substituted thiophene) s. *Chemistry Letters* 40: 398–399.

59 Tamba, S., Shono, K., Sugie, A. et al. (2011). C–H functionalization polycondensation of chlorothiophenes in the presence of nickel catalyst with stoichiometric or catalytically generated magnesium amide. *Journal of the American Chemical Society* 133: 9700–9703.

60 Tamba, S., Mitsuda, S., Tanaka, F. et al. (2012). Studies on the generation of metalating species equivalent to the Knochel–Hauser Base in the dehydrobrominative polymerization of thiophene derivatives. *Organometallics* 31: 2263–2267.

61 Sheina, E.E., Khersonsky, S.M., Jones, E.G. et al. (2005). Highly conductive, regioregular alkoxy-functionalized polythiophenes: a new class of stable, low band gap materials. *Chemistry of Materials* 17: 3317–3319.

62 Koeckelberghs, G., Vangheluwe, M., Samyn, C. et al. (2005). Regioregular poly (3-alkoxythiophene)s: toward soluble, chiral conjugated polymers with a stable oxidized state. *Macromolecules* 38: 5554–5559.

63 Koeckelberghs, G., Vangheluwe, M., Van Doorsselaere, K. et al. (2006). Regioregularity in poly (3-alkoxythiophene)s: effects on the faraday rotation and polymerization mechanism. *Macromolecular Rapid Communications* 27: 1920–1925.

64 Van den Bergh, K., De Winter, J., Gerbaux, P. et al. (2011). Ni-catalyzed polymerization of poly (3-alkoxythiophene)s. *Macromolecular Chemistry and Physics* 212: 328–335.

65 Adachi, I., Miyakoshi, R., Yokoyama, A. et al. (2006). Synthesis of well-defined polythiophene with oxyethylene side chain: effect of phosphine ligands on catalyst-transfer polycondensation. *Macromolecules* 39: 7793–7795.

66 Cheng, S., Ye, S., Apte, C.N. et al. (2021). Improving the Kumada catalyst transfer polymerization with water-scavenging Grignard reagents. *ACS Macro Letters* 10: 697–701.

67 Vallat, P., Lamps, J.-P., Schosseler, F. et al. (2007). Quasi-controlled polymerization through a nickel catalyst process of a functionalized thiophene monomer: kinetic studies and application to the synthesis of regioregular poly (thiophene-3-acetic acid). *Macromolecules* 40: 2600–2602.

68 Wang, P., Jeon, I., Lin, Z. et al. (2018). Insights into magneto-optics of helical conjugated polymers. *Journal of the American Chemical Society* 140: 6501–6508.

69 Lin, P.-S., Shoji, Y., Afraj, S.N. et al. (2021). Controlled synthesis of poly [(3-alkylthio)thiophene] s and their application to organic field-effect transistors. *ACS Applied Materials & Interfaces* 13: 31898–31909.

70 Yokoyama, A., Kato, A., Miyakoshi, R. et al. (2008). Precision synthesis of poly (*N*-hexylpyrrole) and its diblock copolymer with poly (*p*-phenylene) via catalyst-transfer polycondensation. *Macromolecules* 41: 7271–7273.

71 Qiu, Y., Fortney, A., Tsai, C.-H. et al. (2016). Synthesis of polyfuran and thiophene-furan alternating copolymers using catalyst-transfer polycondensation. *ACS Macro Letters* 5: 332–336.

72 Hollinger, J., Jahnke, A.A., Coombs, N. et al. (2010). Controlling phase separation and optical properties in conjugated polymers through selenophene–thiophene copolymerization. *Journal of the American Chemical Society* 132: 8546–8547.

73 Gao, D., Hollinger, J., Jahnke, A.A. et al. (2014). Influence of selenophene-thiophene phase separation on solar cell performance. *Journal of Materials Chemistry A* 2: 6058–6063.

74 Palermo, E.F. and McNeil, A.J. (2012). Impact of copolymer sequence on solid-state properties for random, gradient and block copolymers containing thiophene and selenophene. *Macromolecules* 45: 5948–5955.

75 Heeney, M., Zhang, W., Crouch, D.J. et al. (2007). Regioregular poly (3-hexyl) selenophene: a low band gap organic hole transporting polymer. *Chemical Communications* 5061–5063.

76 Jahnke, A.A., Djukic, B., McCormick, T.M. et al. (2013). Poly (3-alkyltellurophene) s are solution-processable polyheterocycles. *Journal of the American Chemical Society* 135: 951–954.

77 Ye, S., Steube, M., Carrera, E.I. et al. (2016). What limits the molecular weight and controlled synthesis of poly (3-alkyltellurophene)s? *Macromolecules* 49: 1704–1711.

78 Pammer, F. and Passlack, U. (2014). Head-to-tail regioregular polythiazole prepared via Kumada-coupling polycondensation. *ACS Macro Letters* 3: 170–174.

79 Pammer, F., Jäger, J., Rudolf, B. et al. (2014). Soluble head-to-tail regioregular polythiazoles: preparation, properties, and evidence for chain-growth behavior in the synthesis via Kumada-coupling polycondensation. *Macromolecules* 47: 5904–5912.

80 Jäger, J., Tchamba Yimga, N., Urdanpilleta, M. et al. (2016). Toward n-type analogues to poly (3-alkylthiophene)s: influence of side-chain variation on bulk-morphology and electron transport characteristics of head-to-tail regioregular poly (4-alkylthiazole) s. *Journal of Materials Chemistry C* 4: 2587–2597.

81 Miyakoshi, R., Shimono, K., Yokoyama, A. et al. (2006). Catalyst-transfer polycondensation for the synthesis of poly (*p*-phenylene) with controlled molecular weight and low polydispersity. *Journal of the American Chemical Society* 128: 16012–16013.

82 Javier, A.E., Varshney, S.R., and McCullough, R.D. (2010). Chain-growth synthesis of polyfluorenes with low polydispersities, block copolymers of fluorene, and end-capped polyfluorenes: toward new optoelectronic materials. *Macromolecules* 43: 3233–3237.

83 Huang, L., Wu, S., Qu, Y. et al. (2008). Grignard metathesis chain-growth polymerization for polyfluorenes. *Macromolecules* 41: 8944–8947.

84 Sui, A., Shi, X., Wu, S. et al. (2012). Controlled synthesis of polyfluorenes via Kumada Catalyst transfer polycondensation with $Ni(acac)_2$/dppp as the catalyst. *Macromolecules* 45: 5436–5443.

85 Sui, A., Shi, X., Wang, Y. et al. (2015). Kumada catalyst transfer polycondensation for controlled synthesis of polyfluorenes using 1,3-bis(diarylphosphino) propanes as ligands. *Polymer Chemistry* 6: 4819–4827.

86 Wu, S., Sun, Y., Huang, L. et al. (2010). Grignard metathesis chain-growth polymerization for poly (bithienylmethylene)s: Ni catalyst can transfer across the nonconjugated monomer. *Macromolecules* 43: 4438–4440.

87 Nanashima, Y., Yokoyama, A., and Yokozawa, T. (2012). Synthesis of well-defined poly (2-alkoxypyridine-3,5-diyl) via Ni-catalyst-transfer condensation polymerization. *Macromolecules* 45: 2609–2613.

88 Wen, L., Duck, B.C., Dastoor, P.C. et al. (2008). Poly (2,3-dihexylthieno[3,4-*b*]pyrazine) via GRIM polymerization: simple preparation of a solution processable, low-band-gap conjugated polymer. *Macromolecules* 41: 4576–4578.

89 Willot, P., Moerman, D., Leclère, P. et al. (2014). One-pot synthesis and characterization of all-conjugated poly (3-alkylthiophene)-*block*-poly (dialkylthieno[3,4-*b*]pyrazine). *Macromolecules* 47: 6671–6678.

90 Willot, P., Govaerts, S., and Koeckelberghs, G. (2013). The controlled polymerization of poly (cyclopentadithiophene) s and their all-conjugated block copolymers. *Macromolecules* 46: 8888–8895.

91 Pollit, A.A., Obhi, N.K., Lough, A.J. et al. (2017). Evaluation of an external initiating Ni(II) diimine catalyst for electron-deficient π-conjugated polymers. *Polymer Chemistry* 8: 4108–4113.

92 Woods, E.F., Berl, A.J., and Kalow, J.A. (2020). Photocontrolled synthesis of n-type conjugated polymers. *Angewandte Chemie International Edition* 59: 6062–6067.

93 Woods, E.F., Berl, A.J., Kantt, L.P. et al. (2021). Light directs monomer coordination in catalyst-free Grignard photopolymerization. *Journal of the American Chemical Society* 143: 18755–18765.

94 King, A.J. and Zhukhovitskiy, A.V. (2022). A chain-growth mechanism for conjugated polymer synthesis facilitated by dinuclear complexes with redox-active ligands. *Angewandte Chemie International Edition* 61 (29): e202206044.

95 Schroot, R., Schubert, U.S., and Jäger, M. (2016). Poly (*N*-alkyl-3,6-carbazole) s via Kumada catalyst transfer polymerization: impact of metal–halogen exchange. *Macromolecules* 49: 8801–8811.

96 Stefan, M.C., Javier, A.E., Osaka, I. et al. (2009). Grignard metathesis method (GRIM): toward a universal method for the synthesis of conjugated polymers. *Macromolecules* 42: 30–32.

97 Ye, S., Cheng, S., Pollit, A.A. et al. (2020). Isolation of living conjugated polymer chains. *Journal of the American Chemical Society* 142: 11244–11251.

98 Fuji, K., Tamba, S., Shono, K. et al. (2013). Murahashi coupling polymerization: nickel(II)–N-heterocyclic carbene complex-catalyzed polycondensation of organolithium species of (hetero)arenes. *Journal of the American Chemical Society* 135: 12208–12211.

99 Shono, K., Sumino, Y., Tanaka, S. et al. (2014). Polythiophene synthesis via halogen dance. *Organic Chemistry Frontiers* 1: 678–682.

100 McKeown, G.R., Ye, S., Cheng, S. et al. (2019). Homogenous synthesis of monodisperse high oligomers of 3-hexylthiophene by temperature cycling. *Journal of the American Chemical Society* 141: 17053–17056.

101 Koch, F.P.V., Smith, P., and Heeney, M. (2013). "Fibonacci's route" to regioregular oligo (3-hexylthiophene)s. *Journal of the American Chemical Society* 135: 13695–13698.

102 Kim, Y., Park, H., Abdilla, A. et al. (2020). Chain-length-dependent self-assembly behaviors of discrete conjugated oligo (3-hexylthiophene). *Chemistry of Materials* 32: 3597–3607.

103 Xu, H., Ye, S., Zhao, R. et al. (2022). Homogeneous synthesis of monodisperse sequence-defined conjugated oligomers by temperature cycling. *Angewandte Chemie International Edition* 61: e202210340.

104 Haas, D., Hammann, J.M., Greiner, R. et al. (2016). Recent developments in Negishi cross-coupling reactions. *ACS Catalysis* 6: 1540–1552.

105 Chen, T.A., O'Brien, R.A., and Rieke, R.D. (1993). Use of highly reactive zinc leads to a new, facile synthesis for polyarylenes. *Macromolecules* 26: 3462–3463.

106 Chen, T.A. and Rieke, R.D. (1993). Polyalkylthiophenes with the smallest bandgap and the highest intrinsic conductivity. *Synthetic Metals* 60: 175–177.

107 Higashihara, T., Goto, E., and Ueda, M. (2012). Purification-free and protection-free synthesis of regioregular poly (3-hexylthiophene) and poly (3-(6-hydroxyhexyl)thiophene) using a zincate complex of $^{t}Bu_4ZnLi_2$. *ACS Macro Letters* 1: 167–170.

108 Goto, E., Nakamura, S., Kawauchi, S. et al. (2014). Precision synthesis of regioregular poly (3-hexylthiophene) with low dispersity using a zincate complex catalyzed by nickel with the ligand of 1,2-bis(dicyclohexylphosphino) ethane. *Journal of Polymer Science Part A: Polymer Chemistry* 52: 2287–2296.

109 Higashihara, T. and Goto, E. (2014). Controlled synthesis of low-polydisperse regioregular poly (3-hexylthiophene) and related materials by zincate-complex metathesis polymerization. *Polymer Journal* 46: 381–390.

110 Ochiai, Y., Goto, E., and Higashihara, T. (2017). Controlled synthesis of poly (*p*-phenylene) using a zincate complex, $^tBu_4ZnLi_2$. *Macromolecular Rapid Communications* 38: 1700155.

111 Verswyvel, M., Steverlynck, J., Mohamed, S.H. et al. (2014). All-conjugated ABC-*block*-copolymer formation with a varying sequence via an unassociated catalyst. *Macromolecules* 47: 4668–4675.

112 Verswyvel, M., Verstappen, P., De Cremer, L. et al. (2011). Development of a universal chain-growth polymerization protocol of conjugated polymers: toward a variety of all-conjugated block-copolymers. *Journal of Polymer Science Part A: Polymer Chemistry* 49: 5339–5349.

113 Tkachov, R., Senkovskyy, V., Beryozkina, T. et al. (2014). Palladium-catalyzed chain-growth polycondensation of AB-type monomers: high catalyst turnover and polymerization rates. *Angewandte Chemie International Edition* 53: 2402–2407.

114 Erdmann, T., Back, J., Tkachov, R. et al. (2014). Dithienosilole-based all-conjugated block copolymers synthesized by a combination of quasi-living Kumada and Negishi catalyst-transfer polycondensations. *Polymer Chemistry* 5: 5383–5390.

115 Senkovskyy, V., Tkachov, R., Komber, H. et al. (2011). Chain-growth polymerization of unusual anion-radical monomers based on naphthalene diimide: a new route to well-defined n-type conjugated copolymers. *Journal of the American Chemical Society* 133: 19966–19970.

116 Liu, W., Tkachov, R., Komber, H. et al. (2014). Chain-growth polycondensation of perylene diimide-based copolymers: a new route to regio-regular perylene diimide-based acceptors for all-polymer solar cells and n-type transistors. *Polymer Chemistry* 5: 3404–3411.

117 Krishna, A., Lunchev, A.V., and Grimsdale, A.C. (2017). Suzuki polycondensation. In: *Synthetic Methods for Conjugated Polymers and Carbon Materials* (ed. M. Leclerc and J.-F. Morin), 59–95. Wiley-VCH, Wiley Online Books.

118 Sakamoto, J., Rehahn, M., and Schlüter, D. (2010). Suzuki polycondensation: a powerful tool for polyarylene synthesis. In: *Design and Synthesis of Conjugated Polymers* (ed. M. Leclerc and J.-F. Morin), 45–90. Wiley-VCH, Wiley Online Books.

119 Sakamoto, J., Rehahn, M., Wegner, G. et al. (2009). Suzuki polycondensation: polyarylenes à la carte. *Macromolecular Rapid Communications* 30: 653–687.

120 Bautista, M.V., Varni, A.J., Ayuso-Carrillo, J. et al. (2021). Pairing Suzuki–Miyaura cross-coupling and catalyst transfer polymerization. *Polymer Chemistry* 12: 1404–1414.

121 Yokoyama, A., Suzuki, H., Kubota, Y. et al. (2007). Chain-growth polymerization for the synthesis of polyfluorene via Suzuki-Miyaura coupling reaction from an externally added initiator unit. *Journal of the American Chemical Society* 129: 7236–7237.

122 Kobayashi, S., Fujiwara, K., Jiang, D.H. et al. (2020). Suzuki-Miyaura catalyst-transfer polycondensation of triolborate-type fluorene monomer: toward rapid access to polyfluorene-containing block and graft copolymers from various macroinitiators. *Polymer Chemistry* 11: 6832–6839.

123 Tokita, Y., Sugawara, K., Awayama, R. et al. (2019). Synthesis of molecular-weight-controlled polyfluorene with boronate at one end. *Journal of Polymer Science Part A: Polymer Chemistry* 57: 2498–2504.

124 Huber, S. and Mecking, S. (2019). Straightforward synthesis of conjugated block copolymers by controlled Suzuki-Miyaura cross-coupling polymerization combined with ATRP. *Macromolecules* 52: 5917–5924.

125 Zhang, Z., Hu, P., Li, X. et al. (2015). Investigation of Suzuki-Miyaura catalyst-transfer polycondensation of AB-type fluorene monomer using coordination-saturated aryl Pd(II) halide complexes as initiators. *Journal of Polymer Science Part A: Polymer Chemistry* 53: 1457–1463.

126 Grisorio, R., Mastrorilli, P., and Suranna, G.P. (2014). A $Pd(AcO)_2/t\text{-}Bu_3P/K_3PO_4$ catalytic system for the control of Suzuki cross-coupling polymerisation. *Polymer Chemistry* 5: 4304–4310.

127 Grisorio, R. and Suranna, G.P. (2017). Impact of precatalyst activation on Suzuki–Miyaura catalyst-transfer polymerizations: new mechanistic scenarios for pre-transmetalation events. *ACS Macro Letters* 6: 1251–1256.

128 Zhang, H.-H., Xing, C.-H., and Hu, Q.-S. (2012). Controlled $Pd(0)/t\text{-}Bu_3P$-catalyzed Suzuki cross-coupling polymerization of AB-type monomers with $PhPd(t\text{-}Bu_3P)I$ or $Pd_2(dba)_3/t\text{-}Bu_3P/ArI$ as the initiator. *Journal of the American Chemical Society* 134: 13156–13159.

129 Zhang, H.-H., Xing, C.-H., Hu, Q.-S. et al. (2015). Controlled $Pd(0)/t\text{-}Bu_3P$-catalyzed Suzuki cross-coupling polymerization of AB-type monomers with $ArPd(t\text{-}Bu_3P)X$ or $Pd_2(dba)_3/t$-Bu3P/ArX as the initiator. *Macromolecules* 48: 967–978.

130 Zhang, H.H., Peng, W., Dong, J. et al. (2016). $t\text{-}Bu_3P$-Coordinated 2-phenylaniline-based palladacycle complex/ArBr as robust initiators for controlled Pd(0)/*t*-Bu3P-catalyzed Suzuki cross-coupling polymerization of AB-type monomers. *ACS Macro Letters* 5: 656–660.

131 Dong, J., Guo, H., and Hu, Q.-S. (2017). Controlled $Pd(0)/Ad_3P$-catalyzed Suzuki cross-coupling polymerization of AB-type monomers with Ad_3P-coordinated acetanilide-based palladacycle complex as initiator. *ACS Macro Letters* 6: 1301–1304.

132 Zhang, K., Tkachov, R., Ditte, K. et al. (2020). AB- versus AA+BB-Suzuki polycondensation: a palladium/tris(*tert*-butyl)phosphine catalyst can outperform conventional catalysts. *Macromolecular Rapid Communications* 41: 1900521.

133 Yokozawa, T., Kohno, H., Ohta, Y. et al. (2010). Catalyst-transfer Suzuki–Miyaura coupling polymerization for precision synthesis of poly (*p*-phenylene). *Macromolecules* 43: 7095–7100.

134 Yokozawa, T., Suzuki, R., Nojima, M. et al. (2011). Precision synthesis of poly (3-hexylthiophene) from catalyst-transfer Suzuki–Miyaura coupling polymerization. *Macromolecular Rapid Communications* 32: 801–806.

135 Seo, K.-B., Lee, I.-H., Lee, J. et al. (2018). A rational design of highly controlled Suzuki-Miyaura catalyst-transfer polycondensation for precision synthesis of polythiophenes and their block copolymers: marriage of palladacycle precatalysts with MIDA-boronates. *Journal of the American Chemical Society* 140: 4335–4343.

136 Lee, J., Park, H., Hwang, S.-H. et al. (2020). RuPhos Pd precatalyst and MIDA boronate as an effective combination for the precision synthesis of poly (3-hexylthiophene): systematic investigation of the effects of boronates, halides, and ligands. *Macromolecules* 53: 3306–3314.

137 Kosaka, K., Uchida, T., Mikami, K. et al. (2018). AmPhos Pd-catalyzed Suzuki–Miyaura catalyst-transfer condensation polymerization: narrower dispersity by mixing the catalyst and base prior to polymerization. *Macromolecules* 51: 364–369.

138 Baker, M.A., Zahn, S.F., Varni, A.J. et al. (2018). Elucidating the role of diphosphine ligand in nickel-mediated Suzuki-Miyaura polycondensation. *Macromolecules* 51: 5911–5917.

139 Liversedge, I.A., Higgins, S.J., Giles, M. et al. (2006). Suzuki route to regioregular polyalkylthiophenes using Ir-catalysed borylation to make the monomer, and Pd complexes of bulky phosphanes as coupling catalysts for polymerisation. *Tetrahedron Letters* 47: 5143–5146.

140 Qiu, Y., Worch, J.C., Fortney, A. et al. (2016). Nickel-catalyzed Suzuki polycondensation for controlled synthesis of ester-functionalized conjugated polymers. *Macromolecules* 49: 4757–4762.

141 Hohl, B., Bertschi, L., Zhang, X. et al. (2012). Suzuki polycondensation toward high molecular weight poly (*m*-phenylene) s: mechanistic insights and end-functionalization. *Macromolecules* 45: 5418–5426.

142 Varni, A.J., Fortney, A., Baker, M.A. et al. (2019). Photostable helical polyfurans. *Journal of the American Chemical Society* 141: 8858–8867.

143 Zhang, H.-H., Zhu, Y.-X., Wang, W. et al. (2018). Controlled synthesis of *ortho, para*-alternating linked polyarenes via catalyst-transfer Suzuki coupling polymerization. *Polymer Chemistry* 9: 3342–3346.

144 Zhang, H.-H., Ma, C., Bonnesen, P.V. et al. (2016). Helical poly (5-alkyl-2,3-thiophene)s: controlled synthesis and structure characterization. *Macromolecules* 49: 4691–4698.

145 Bautista, M.V., Varni, A.J., Ayuso-Carrillo, J. et al. (2020). Chain-growth polymerization of benzotriazole using Suzuki-Miyaura cross-coupling and dialkylbiarylphosphine palladium catalysts. *ACS Macro Letters* 9: 1357–1362.

146 Lee, J., Kim, H., Park, H. et al. (2021). Universal Suzuki-Miyaura catalyst-transfer polymerization for precision synthesis of strong donor/acceptor-based conjugated polymers and their sequence engineering. *Journal of the American Chemical Society* 143: 11180–11190.

147 Verswyvel, M., Hoebers, C., De Winter, J. et al. (2013). Study of the controlled chain-growth polymerization of poly (3,6-phenanthrene). *Journal of Polymer Science Part A: Polymer Chemistry* 51: 5067–5074.

148 Nojima, M., Ohta, Y., and Yokozawa, T. (2014). Investigation of catalyst-transfer condensation polymerization for synthesis of poly (*p*-phenylenevinylene). *Journal of Polymer Science Part A: Polymer Chemistry* 52: 2643–2653.

149 Nojima, M., Kamigawara, T., Ohta, Y. et al. (2019). Catalyst-transfer Suzuki-Miyaura condensation polymerization of stilbene monomer: different polymerization behavior depending on halide and aryl group of ArPd(tBu$_3$P)X initiator. *Journal of Polymer Science Part A: Polymer Chemistry* 57: 297–304.

150 Komber, H., Senkovskyy, V., Tkachov, R. et al. (2011). Ring walking versus trapping of nickel(0) during Kumada catalyst transfer polycondensation using externally initiated electron-accepting thiophene–benzothiadiazole–thiophene precursors. *Macromolecules* 44: 9164–9172.

151 Kim, H., Lee, J., Kim, T. et al. (2022). Precision synthesis of various low-bandgap donor-acceptor alternating conjugated polymers via living Suzuki-Miyaura catalyst-transfer polymerization. *Angewandte Chemie International Edition* 61: e202205828.

152 Carsten, B., He, F., Son, H.J. et al. (2011). Stille polycondensation for synthesis of functional materials. *Chemical Reviews* 111: 1493–1528.

153 Kang, S., Ono, R.J., and Bielawski, C.W. (2013). Controlled catalyst transfer polycondensation and surface-initiated polymerization of a *p*-phenyleneethynylene-based monomer. *Journal of the American Chemical Society* 135: 4984–4987.

154 Kang, S., Todd, A.D., Paul, A. et al. (2018). Controlled syntheses of poly (phenylene ethynylene) s with regiochemically-tuned optical band gaps and ordered morphologies. *Macromolecules* 51: 5972–5978.

155 Qiu, Y., Mohin, J., Tsai, C.-H. et al. (2015). Stille catalyst-transfer polycondensation using Pd-PEPPSI-IPr for high-molecular-weight regioregular poly (3-hexylthiophene). *Macromolecular Rapid Communications* 36: 840–844.

156 Odian, G. (2004). *Principles of Polymerization*. Wiley-Interscience.

157 Goto, E., Ando, S., Ueda, M. et al. (2015). Nonstoichiometric Stille coupling polycondensation for synthesizing naphthalene-diimide-based π-conjugated polymers. *ACS Macro Letters* 4: 1004–1007.

158 Terayama, K., Goto, E., and Higashihara, T. (2018). Nonstoichiometric Stille coupling polycondensation via an intramolecular Pd(0) catalyst transfer using excess phthalimide monomer. *Macromolecular Chemistry and Physics* 219: 1800175.

159 Terayama, K., Liu, C.-W., and Higashihara, T. (2022). Catalyst-transfer system in stoichiometry-independent AA+BB-type Migita–Kosugi–Stille coupling polycondensation using ester-functionalized dibromo monomer. *Polymer Journal* 54: 143–150.

160 Sugita, H., Nojima, M., Ohta, Y. et al. (2017). Unusual cyclic polymerization through Suzuki–Miyaura coupling of polyphenylene bearing diboronate at both ends with excess dibromophenylene. *Chemical Communications* 53: 396–399.

161 Sugita, H., Nojima, M., Ohta, Y. et al. (2019). Unstoichiometric Suzuki–Miyaura cyclic polymerization of extensively conjugated monomers. *Polymer Chemistry* 10: 1182–1185.

162 Sugita, H., Ohta, Y., and Yokozawa, T. (2020). Synthesis of cyclic aromatic polymer containing thiophene or pyridine by means of unstoichiometric Suzuki–Miyaura cyclic polycondensation: effect of the position of bromine of heteroarylenes on cyclic polycondensation. *Journal of Polymer Science* 58: 1236–1240.

163 Nojima, M., Kosaka, K., Kato, M. et al. (2016). Alternating intramolecular and intermolecular catalyst-transfer Suzuki–Miyaura condensation polymerization: synthesis of boronate-terminated π-conjugated polymers using excess dibromo monomers. *Macromolecular Rapid Communications* 37: 79–85.

164 Kanbara, T., Honma, A., and Hasegawa, K. (1996). Preparation of novel poly (aryleneamine) s by palladium complex catalyzed polycondensation of dibromobenzenes with diamines. *Chemistry Letters* 25: 1135–1136.

165 Kanbara, T., Izumi, K., Nakadani, Y. et al. (1997). Preparation of poly (imino-1, 3-phenylene) and its related polymer by palladium-catalyzed polycondensation. *Chemistry Letters* 26: 1185–1186.

166 Kanbara, T., Izumi, K., Narise, T. et al. (1998). Preparation of poly (arylenediamine)s by palladium-catalyzed polycondensation of aryl dibromides with secondary diamines. *Polymer Journal* 30: 66–68.

167 Kanbara, T., Oshima, M., Imayasu, T. et al. (1998). Preparation of new polymers containing an azobenzene group in the side chain by palladium-catalyzed polymer reaction and polycondensation and characterization of the polymers. *Macromolecules* 31: 8725–8730.

168 Grisorio, R. and Suranna, G.P. (2019). Catalyst-transfer polymerization of arylamines by the Buchwald-Hartwig cross-coupling. *Polymer Chemistry* 10: 1947–1955.

169 Gobalasingham, N.S. and Thompson, B.C. (2018). Direct arylation polymerization: a guide to optimal conditions for effective conjugated polymers. *Progress in Polymer Science* 83: 135–201.

170 Pankow, R.M. and Thompson, B.C. (2020). Approaches for improving the sustainability of conjugated polymer synthesis using direct arylation polymerization (DArP). *Polymer Chemistry* 11: 630–640.

171 Pouliot, J.-R., Grenier, F., Blaskovits, J.T. et al. (2016). Direct (hetero) arylation polymerization: simplicity for conjugated polymer synthesis. *Chemical Reviews* 116: 14225–14274.

172 Bura, T., Blaskovits, J.T., and Leclerc, M. (2016). Direct (hetero) arylation polymerization: trends and perspectives. *Journal of the American Chemical Society* 138: 10056–10071.

173 Lee, J.A. and Luscombe, C.K. (2018). Dual-catalytic Ag-Pd system for direct arylation polymerization to synthesize poly (3-hexylthiophene). *ACS Macro Letters* 7: 767–771.

174 Bonillo, B. and Swager, T.M. (2012). Chain-growth polymerization of 2-chlorothiophenes promoted by Lewis acids. *Journal of the American Chemical Society* 134: 18916–18919.

175 Balasubramanian, A., Ku, T.-C., Shih, H.-P. et al. (2014). Chain-growth cationic polymerization of 2-halogenated thiophenes promoted by Bronsted acids. *Polymer Chemistry* 5: 5928–5941.

176 Chiang, J.-W., Lu, P.-Y., and Han, C.-C. (2022). Facile and effective syntheses of 2-fluoro-3-alkoxythiophenes and their uses in making highly regioregular poly (3-alkoxythiophene)s via transition-metal-free cationic chain-growth polymerization. *Macromolecules* 55: 7469–7480.

3

Optically Active Through-space Conjugated Polymers Based on Planar Chiral [2.2]Paracyclophane

Yasuhiro Morisaki

Department of Applied Chemistry for Environment, School of Biological and Environmental Sciences, Kwansei Gakuin University, Sanda, Hyogo 669-1330, Japan

3.1 Introduction

Conjugated polymers and oligomers have attracted considerable attention in polymer chemistry and material chemistry, because of their potential applications in optoelectronic devices, including electroluminescence devices [1], field effect transistors (FETs) [2], and photovoltaic cells [3]. Unlike inorganic optoelectronic materials, organic conjugated polymers and oligomers exhibit the distinct advantage of versatile processability. For instance, a thin film of these materials can be easily fabricated using spin-coating, spray-coating, or casting techniques. Additionally, they are lightweight and flexible and allow for easy tuning of optical properties by modifying the π-conjugated units. In these materials, charges and energy migrate within the layers and are transferred across the conjugated backbone via a hopping mechanism. While a wide range of π-conjugated compounds [4] has been synthesized to date, only a few studies have focused on the development of through-space conjugated polymers and oligomers composed of layered π-electron systems. One approach to constructing through-space conjugated systems is to employ the π-stacked compound as a building block. From this perspective, the [2.2]paracyclophane skeleton, characterized by two benzene rings stacked in a face-to-face configuration in proximity [5], is a logical choice for synthesizing through-space conjugated polymers and oligomers. Moreover, substituted [2.2]paracyclophanes are planar chiral compounds without chiral centers, with their chirality depending on the substitution positions [6]. This is because the two benzene rings in [2.2]paracyclophane are rigidly connected by ethylene chains at the para positions, and their rotatory motions are suppressed. Therefore, through-space conjugated polymers consisting of the planar chiral [2.2]paracyclophane units in the main chain are optically active; these polymers demonstrate remarkable chiroptical properties, such as circular dichroism (CD) [7] with high molar ellipticity and

Conjugated Polymer Synthesis and Materials Applications, First Edition. Edited by Yoshiki Chujo.
© 2026 Wiley-VCH GmbH. All rights reserved, including rights for text and data mining and training of artificial intelligence technologies or similar technologies. Published 2026 by Wiley-VCH GmbH.

circularly polarized luminescence (CPL) [8] with a large anisotropy factor. This chapter explores the concept of through-space conjugation in [2.2]paracyclophane-based conjugated polymers and highlights the synthesis and properties of optically active through-space conjugated polymers and oligomers composed of the planar chiral [2.2]paracyclophane skeletons.

3.2 Through-space Conjugation of [2.2]Paracyclophane-based Conjugated Polymers

Pseudo-*para*-disubstituted [2.2]paracyclophanes can serve as monomers for producing through-space conjugated polymers. Palladium-catalyzed cross-coupling reactions, such as Sonogashira–Hagihara [9], Mizoroki–Heck [10], and Suzuki–Miyaura cross-coupling [11], have proven to be effective protocols for synthesizing through-space conjugated polymers, including poly(*p*-arylene-ethynylene)s, poly(*p*-arylenevinylene)s, and poly(*p*-arylene)s, respectively. Figure 3.1 illustrates the synthesis of π-stacked polymers **1** via Pd-catalyzed Sonogashira–Hagihara cross-coupling between pseudo-*para*-diethynyl[2.2]paracyclophane and a *p*-diiodobenzene derivative [12]. The resulting polymer exhibited good solubility in common organic solvents due to the presence of dodecyloxy chains, enabling the formation of thin film through spin-coating and casting techniques.

By incorporating [2.2]paracyclophane units into the conjugated polymer backbones, such as in polymer **1**, π-electron systems stack along the terminal benzene rings, as illustrated in Figure 3.1. Most conjugated polymers reported to date consist of monomeric units connected by chemical bond(s). In contrast, π-stacked polymers, which are composed of stacked π-electron systems, have been rarely synthesized.

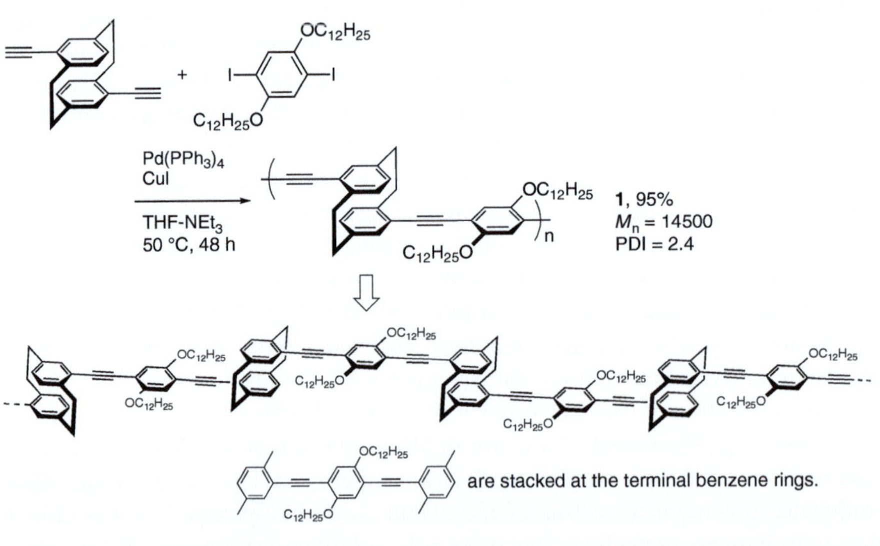

Figure 3.1 Synthetic route to pseudo-*para*-disubstituted [2.2]paracyclophane-based through-space conjugated polymer.

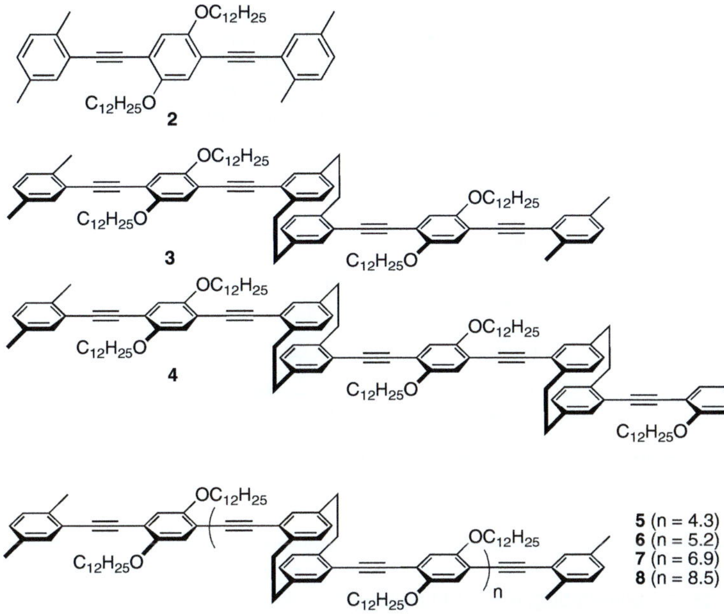

Figure 3.2 Pseudo-*para*-disubstituted [2.2]paracyclophane-based through-space conjugated oligomers and polymers.

Therefore, the optical properties of **1** were thoroughly investigated and compared with those of well-defined oligomers as well as polymers.

Model compound **2** and the π-stacked dimer **3** and trimer **4** (Figure 3.2) were prepared stepwise using the Sonogashira–Hagihara cross-coupling [12b]. Additionally, the π-stacked polymer was fractioned based on molecular weight into four fractions (polymers 5–8) using gel permeation chromatography (GPC) columns (Figure 3.2). Figure 3.3 presents their normalized UV-vis absorption spectra of these compounds in a dilute $CHCl_3$ solution. All molecules **2–8** displayed $\pi - \pi^*$ transition bands characteristics of *p*-arylene-ethynylenes, with their peak wavelengths gradually red-shifting as the number of stacked π-electron systems increased. This red-shift is a typical feature of through-bond π-conjugation systems. Based on this observation, through-space conjugation is suggested for the [2.2]paracyclophane-based π-stacked oligomers and polymers.

As well-known, the absorption maximum $\lambda_{max}\left(E_g \text{cm}^{-1}\right)$ of typical conjugated polymers can be estimated from the relationship between the number of repeating units and $\lambda_{max}\left(E_g \text{cm}^{-1}\right)$ for well-defined π-conjugated oligomers [13]. This principle was applied to the [2.2]paracyclophane-based through-space conjugation system. Figure 3.4 illustrates the relation between the peak top energies $\left(E_g \text{cm}^{-1}\right)$ and the inverse of the number of repeating units (n). As shown in Figure 3.4, a linear correlation with a coefficient of determination (R^2) of 0.99 is observed. For a sufficiently long polymer chain (infinite π-stacked units, n = ∞), the peak top energy is expected to reach a limiting value of E_g = 25,800 cm^{-1}. This result is consistent

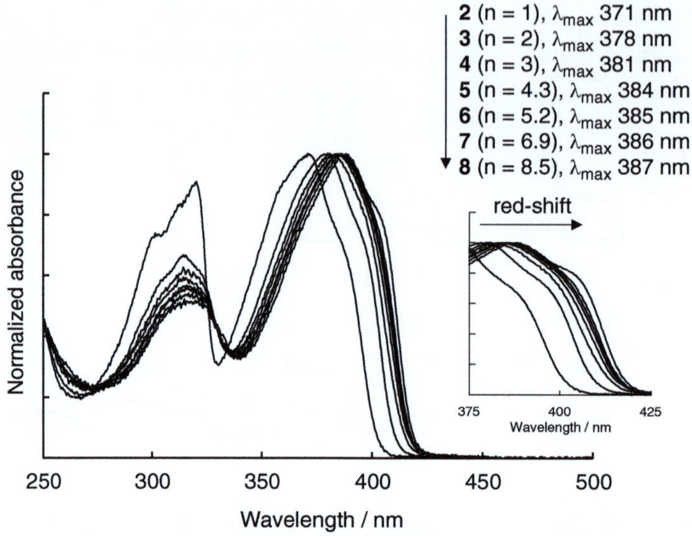

Figure 3.3 UV-vis absorption spectra of pseudo-*para*-disubstituted [2.2]paracyclophane-based through-space conjugated oligomers and polymers.

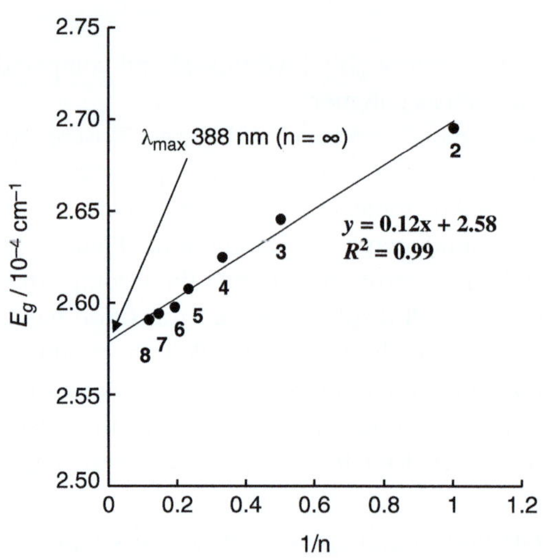

Figure 3.4 Correlation between the λ_{max} $\left(E_g \text{cm}^{-1}\right)$ and the inverse of repeating unit (1/n) of pseudo-*para*-disubstituted [2.2]paracyclophane-based through-space conjugated oligomers and polymers.

with the experimental value of the peak maximum (387 nm) observed for polymer **9** with a number-average molecular weight (M_n) of 14,500, corresponding to approximately 21 stacked π-electron systems. The effective conjugation repeating unit number (n) was calculated to be 11.4 using the empirical power–law relationship ($E = 1/\lambda_{max} \approx n^{-x}$). The electronic interactions among stacked π-electron systems of the [2.2]paracyclophane-based through-space conjugated polymer are effective through 11 π-electron systems in the ground state.

3.3 Optically Active Through-space Conjugated Polymers Based on Planar Chiral [2.2]Paracyclophane

[2.2]Paracyclophane comprises two face-to-face benzene rings in close proximity with completely suppressed rotational movement. Therefore, [2.2]paracyclophane compounds with substituent(s) are planar chiral compounds without chiral centers, with their chirality depending on the substitution position(s). Since the first synthesis of chiral [2.2]paracyclophane in 1955 [6a], various optically active [2.2]paracyclophane compounds have been synthesized primarily for use as chiral auxiliaries and chiral ligands in synthetic organic and organometallic chemistry [6b-l]. Despite their unique structural and chiroptical properties, planar chiral [2.2]paracyclophane compounds have largely been overlooked in the fields of polymer chemistry and materials chemistry.

For pseudo-*para*-disubstituted [2.2]paracyclophane, when the two substituents are identical, the molecule is achiral (not chiral); therefore, polymers and oligomers **3–9** described in Chapter 2 are not chiral. In contrast, pseudo-*ortho*-disubstituted [2.2]paracyclophane compounds are chiral. Consequently, when a pseudo-*ortho*-disubstituted [2.2]paracyclophane compound is synthesized, a racemic mixture is obtained without optical resolution. Through-space conjugated polymer *rac*-**10** (Figure 3.5), consisting of racemic pseudo-*ortho*-disubstituted [2.2]paracyclophane skeletons, was synthesized [14]. In this polymer, π-electron systems were stacked in a zigzag manner. The optical properties of polymer *rac*-**10** with pseudo-*ortho*-disubstituted [2.2]paracyclophane units were nearly identical to those of polymer **1** based on pseudo-*para*-disubstituted [2.2]paracyclophane units.

In 2012, optically active through-space conjugated polymers (R_p)-**10** and (S_p)-**10** were synthesized using enantiopure disubstituted [2.2]paracyclophane as the monomer [15]. As depicted in Figure 3.6, Sonogashira–Hagihara cross-coupling of (R_p)-pseudo-*ortho*-diethynyl[2.2]paracyclophane with a diiodobenzene derivative yielded the corresponding through-space conjugated polymer (R_p)-**10**, along with a triangle-shaped cyclic molecule (R_p)-**11**, which were separated by GPC. Figure 3.6b shows the CPL spectra of both enantiomers of polymer **10** in dilute

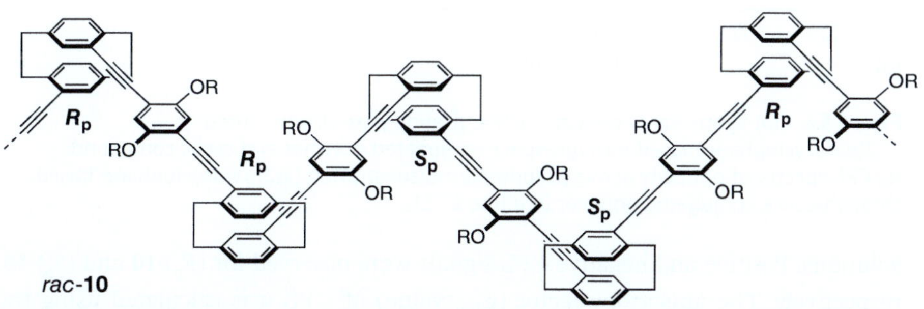

rac-**10**

Figure 3.5 Structure of pseudo-*ortho*-disubstituted [2.2]paracyclophane-based through-space conjugated polymer.

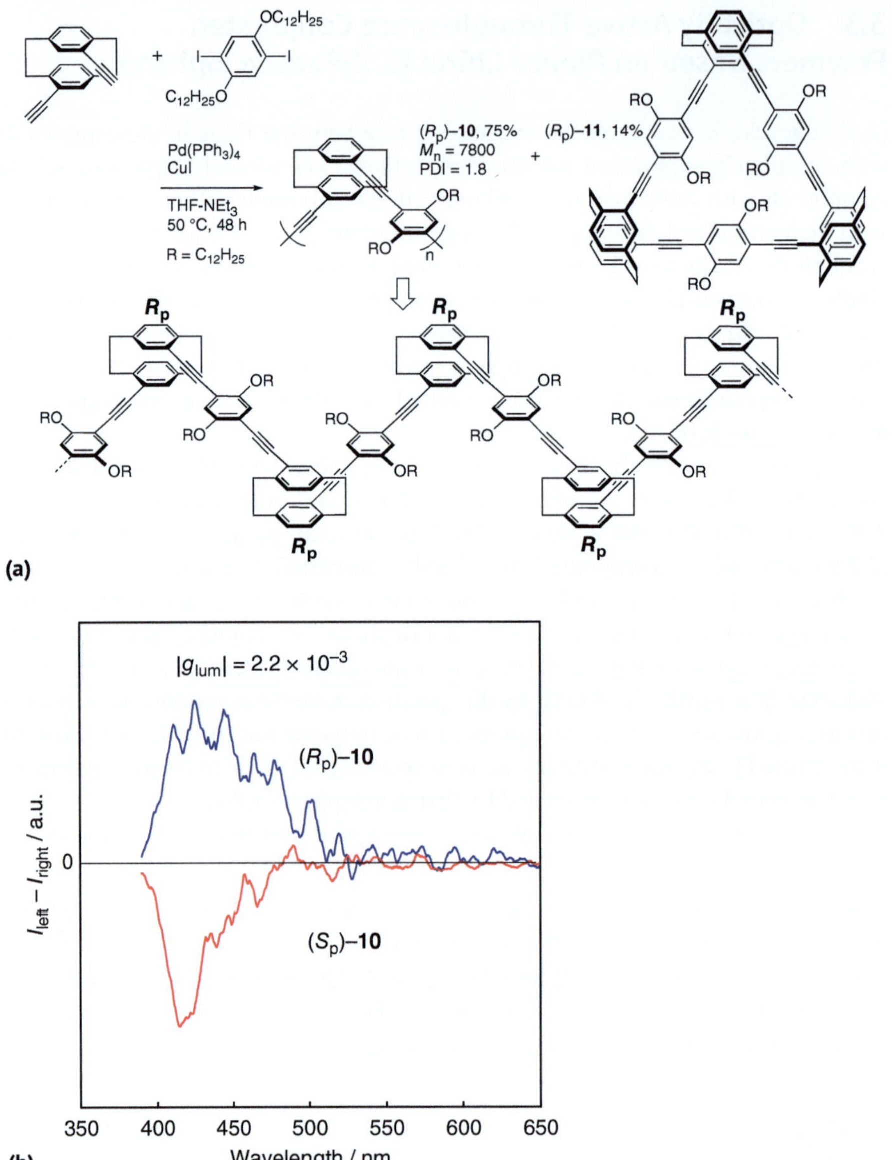

Figure 3.6 (a) Synthesis of optically active pseudo-*para*-disubstituted [2.2]paracyclophane-based through-space conjugated polymer and cyclic compound. (b) CPL spectra of optically active pseudo-*para*-disubstituted [2.2]paracyclophane-based through-space conjugated polymer in dilute $CHCl_3$.

solutions. Positive and negative CPL signals were observed for (R_p)-**10** and (S_p)-**10**, respectively. The anisotropy factor (g_{lum} value) of CPL was calculated using the equation $g_{lum} = 2(I_{left} - I_{right})/(I_{left} + I_{right})$, where I_{left} and I_{right} are intensities of left- and right-handed CPL, respectively [8]. The $|g_{lum}|$ value of polymer **10** was estimated to be 2.2×10^{-3}. Notably, a g_{lum} value of the order of 10^{-3} is considered high

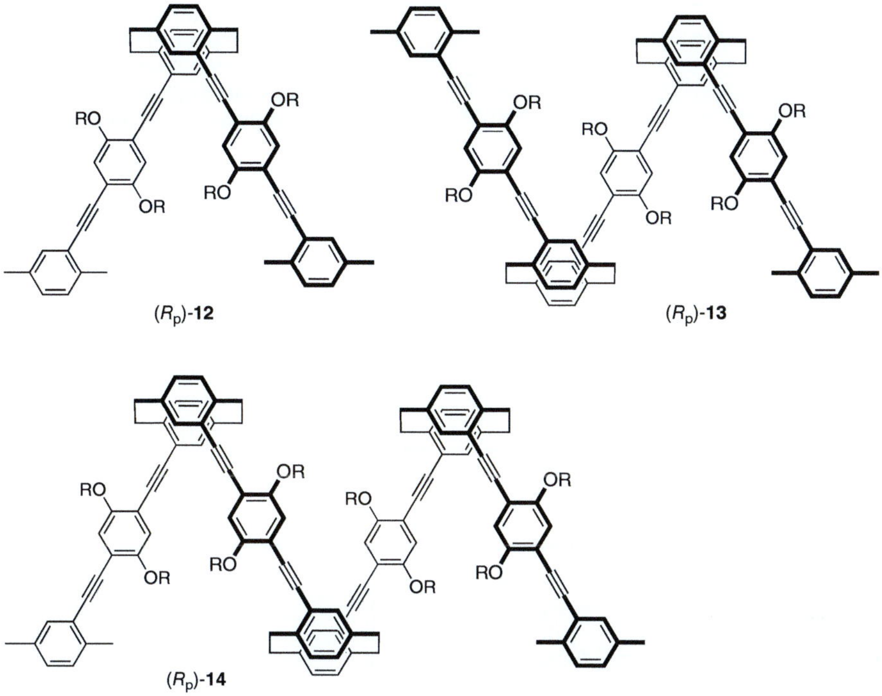

Figure 3.7 Structures of pseudo-*ortho*-disubstituted [2.2]paracyclophane-based through-space conjugated oligomers.

for the organic molecules in their dispersion states. This marks the first reported instance of a planar chiral molecule emitting CPL.

The CPL behavior of the planar chiral [2.2]paracyclophane-based through-space conjugated polymer was studied using its corresponding oligomers, dimer **12**, trimer **13**, and tetramer **14** (Figure 3.7), as well as the triangle-shaped cyclic trimer **11** (Figure 3.6a) [16].

CPL spectra for both enantiomers of **12–14** have been recorded in dilute solutions, revealing intense CPL signals, as shown in Figure 3.8. The CPL spectra of the enantiomers are perfect mirror images of each other. The $|g_{lum}|$ values are also displayed in Figure 3.8, with calculated values of 1.8×10^{-3}, 2.2×10^{-3}, and 2.6×10^{-3} for **12**, **13**, and **14**, respectively. As the number of the stacked π-electron systems increased, the $|g_{lum}|$ values showed a corresponding gradual increase. Interestingly, since the $|g_{lum}|$ value is a normalized parameter, it would typically be expected to remain consistent among the oligomers, irrespective of the number of the stacked π-electron systems. This anomaly suggests additional factors at play in the system. Figure 3.9 illustrates the CD spectra with g_{abs} charts. The anisotropy factor (g_{abs} value) of absorbance is obtained by the equation $g_{abs} = 2(\varepsilon_{left} - \varepsilon_{right})/(\varepsilon_{left} + \varepsilon_{right})$, where ε_{left} and ε_{right} are molar extinction coefficients of left- and right-handed circularly polarized light, respectively. As shown in Figure 3.9b, the $|g_{abs}|$ values of **12–14** were nearly identical. These results indicate some chirality is derived from

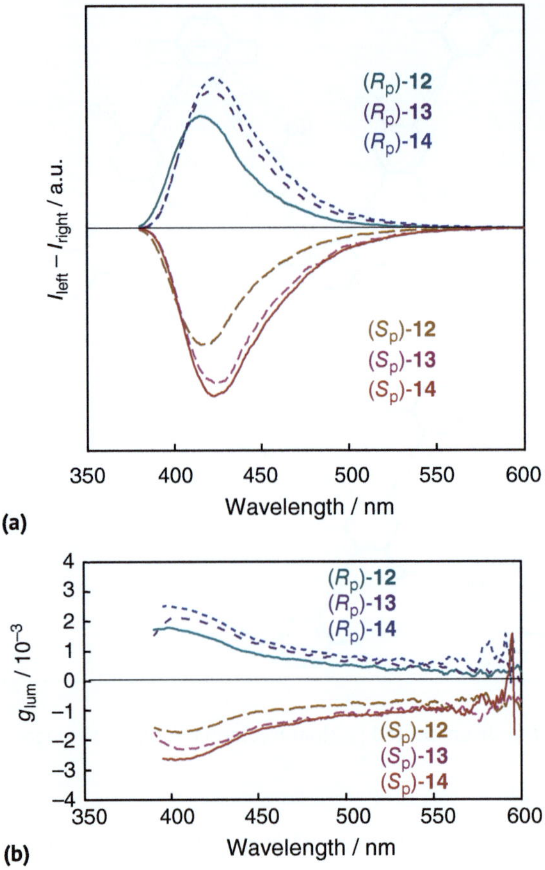

Figure 3.8 (a) CPL spectra of optically active pseudo-*ortho*-disubstituted [2.2]paracyclophane-based through-space conjugated oligomers in dilute $CHCl_3$ and (b) their g_{lum} charts.

the higher-ordered structure besides the planar chirality of the [2.2]paracyclophane skeleton.

CPL spectra of trimer **13** and cyclic trimer **11** exhibited identical CPL profiles despite their structural differences (Figure 3.10). The cyclic trimer **11** adopts a chiral second-ordered structure, specifically a chiral triangle-shaped configuration, in addition to the planar chirality of its [2.2]paracyclophane moieties. The observation that linear trimer **13** and cyclic trimer **11** displayed identical CPL profiles suggests that linear trimer **13** forms a chiral second-ordered structure, such as a one-handed helical structure, in addition to its planar chirality in the excited stated, as shown in Figure 3.11. In contrast, the anisotropy factors of absorbance (g_{abs} value) for linear trimer **13** and cyclic trimer **11** were estimated to be 3.2×10^{-3} and 3.7×10^{-3}, respectively, with the $|g_{abs}|$ of **13** being lower than that of **11** (Figure 3.12). In the ground state, trimer **13** forms various conformations due to the free rotation of its carbon–carbon bonds, resulting in a random conformation. Conversely, cyclic trimer **11** maintains a fixed chiral triangle structure. Chiral higher-ordered structures, such as the triangle-shaped structure in **11** and the helical structure hypothesized for **13**, contribute to the enhancement of $|g_{abs}|$ and $|g_{lum}|$ values in the ground and excited states, respectively.

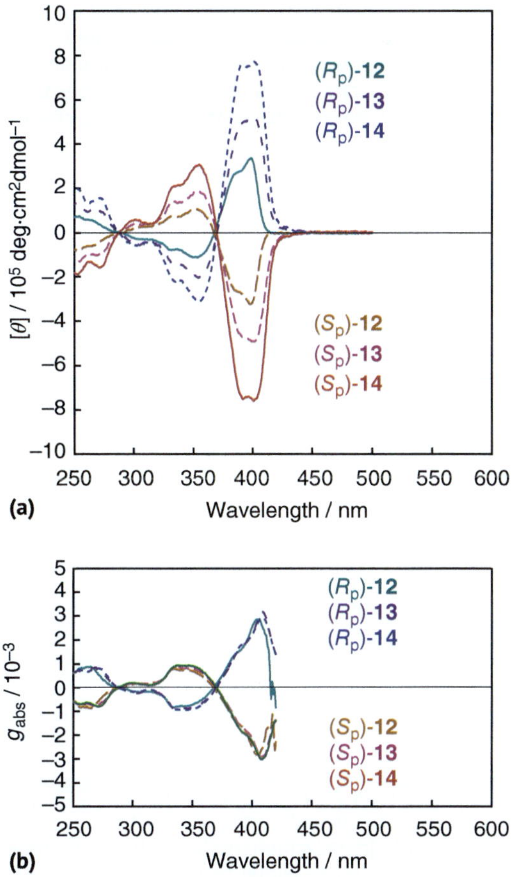

Figure 3.9 (a) CD spectra of optically active pseudo-*ortho*-disubstituted [2.2]paracyclophane-based through-space conjugated oligomers in dilute $CHCl_3$ and (b) their g_{abs} charts.

The use of enantiopure pseudo-*meta*-disubstituted [2.2]paracyclophane as a monomer leads to the synthesis of the corresponding optically active through-space conjugated polymer **15**, in which π-electron systems are stacked at the angle of 120° (Figure 3.13) [17]. There are no intrinsic differences in the optical properties of through-space conjugated polymers composed of pseudo-*para*-, pseudo-*ortho*-, and pseudo-*meta*-disubstituted [2.2]paracyclophanes because the stacked π-electron systems are identical.

The optical and chiroptical properties of optically active through-space conjugated dimers **12** (Figure 3.7) and 16 (Figure 3.14) based on pseudo-*ortho*- and pseudo-*meta*-disubstituted [2.2]paracyclophanes, respectively, were studied [17b]. No intrinsic differences were observed between pseudo-*ortho*- and pseudo-*meta*-disubstituted [2.2]paracyclophanes in their optical properties. Figure 3.14a shows the CPL spectra of dimer **16**. Although the CPL spectra of **12** and **16** were nearly identical, the $|g_{lum}|$ values differed; the $|g_{lum}|$ for **12** was 1.8×10^{-3}, while that for **16** was 1.3×10^{-3}. According to thorough investigations on chiroptical properties of various π-stacked dimers based on planar chiral [2.2]paracyclophanes, $|g_{lum}|$ values of dimers containing pseudo-*ortho*-disubstituted [2.2]paracyclophane are

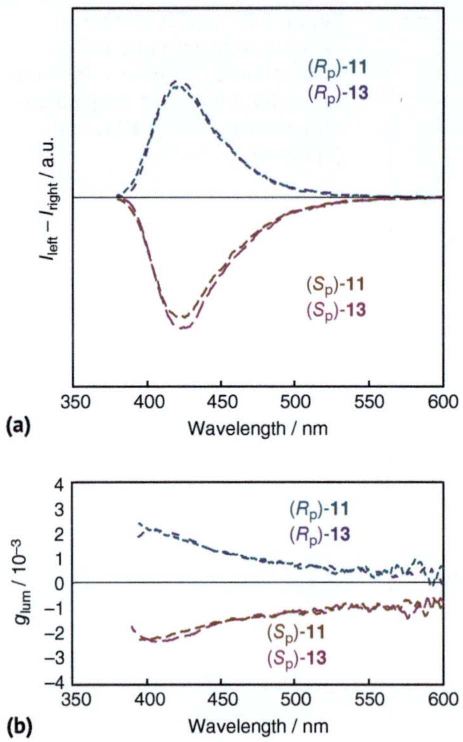

Figure 3.10 (a) CPL spectra of optically active pseudo-*ortho*-disubstituted [2.2]paracyclophane-based through-space conjugated trimer and cyclic trimer in dilute $CHCl_3$ and (b) their g_{lum} charts.

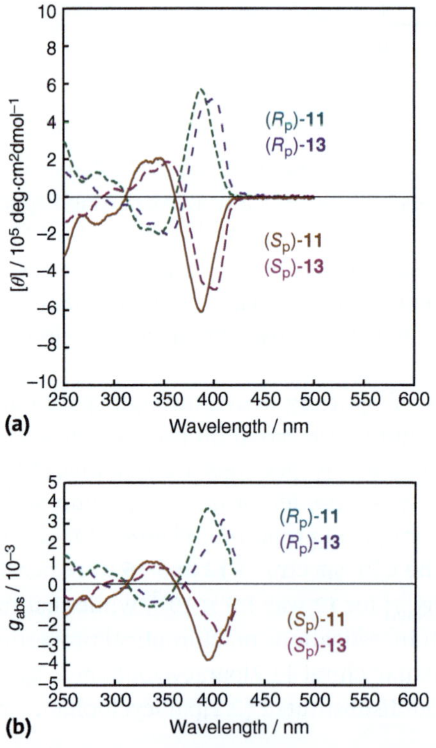

Figure 3.11 (a) CD spectra of optically active pseudo-*ortho*-disubstituted [2.2]paracyclophane-based through-space conjugated trimer and cyclic trimer in dilute $CHCl_3$ and (b) their g_{abs} charts.

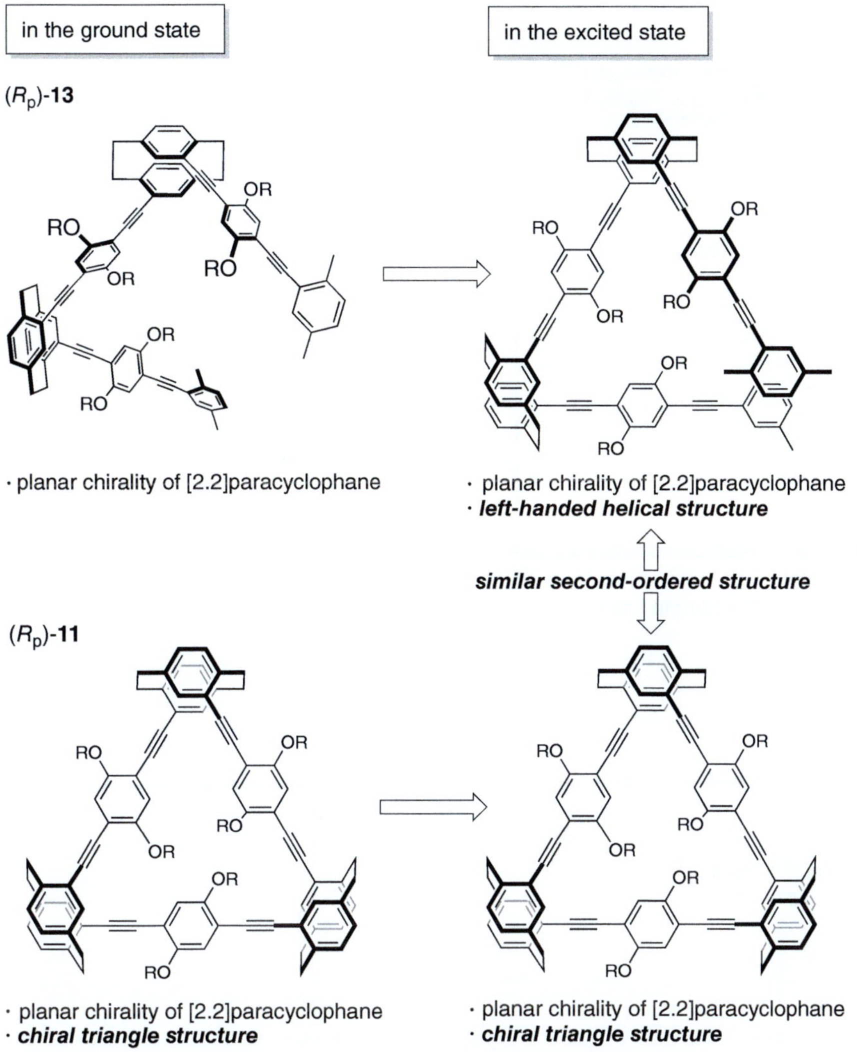

Figure 3.12 Conformation of optically active pseudo-*ortho*-disubstituted [2.2]paracyclophane-based through-space conjugated trimer and cyclic trimer in the ground and excited states.

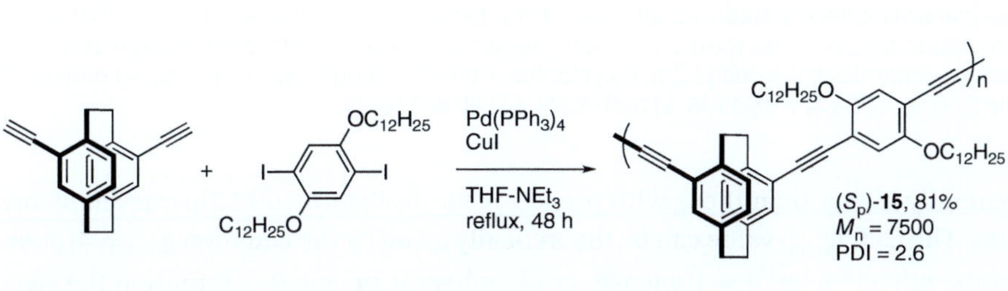

Figure 3.13 Synthetic route to optically active pseudo-*meta*-disubstituted [2.2]paracyclophane-based through-space conjugated polymer.

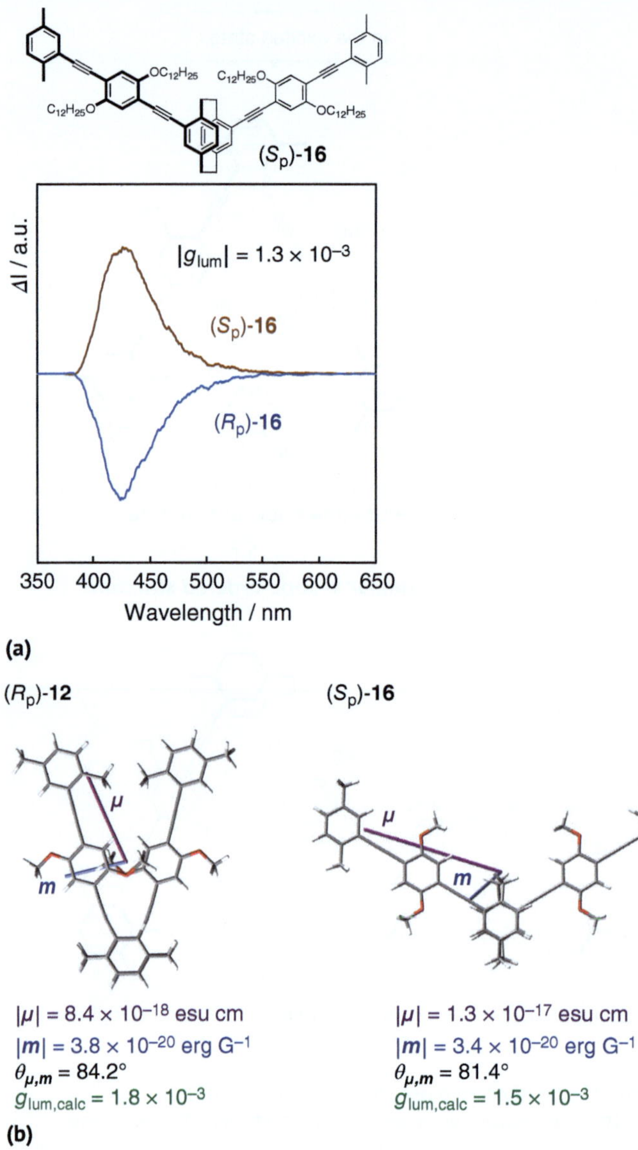

Figure 3.14 (a) CPL spectra of optically active pseudo-*meta*-disubstituted [2.2]paracyclophane-based through-space conjugated dimers in dilute $CHCl_3$ and (b) the simulation results of the transition dipole moments of pseudo-*meta*-disubstituted and pseudo-*ortho*-disubstituted [2.2]paracyclophane-based through-space conjugated dimers in the S_1 state (TD-CAM-B3LYP/6-31G(d)//CAM-B3LYP/6-31G(d)).

generally higher than those with pseudo-*meta*-disubstituted [2.2]paracyclophane [18]. The CPL $|g_{lum}|$ value can be theoretically given by the equation $g_{lum} = 4|\boldsymbol{\mu}||\boldsymbol{m}| \cos\theta\boldsymbol{\mu}, \boldsymbol{m}/(|\boldsymbol{\mu}|^2 + |\boldsymbol{m}|^2) = 4|\boldsymbol{m}|\cos\theta_{\mu,m}/|\boldsymbol{\mu}|)$, where $\boldsymbol{\mu}$, $\boldsymbol{m}$, and $\theta_{\mu,m}$ represent the electronic transition dipole moment, the magnetic transition dipole moment, and the angle between $\boldsymbol{\mu}$ and $\boldsymbol{m}$. Simulation results support the experimental observation, indicating that the dimer with pseudo-*ortho*-disubstituted [2.2]paracyclophane exhibits a higher $|g_{lum}|$ value than those with pseudo-*meta*-disubstituted [2.2]paracyclophane (Figure 3.14b).

3.4 Optically Active Through-space Conjugated Cyclic Oligomers Based on Planar Chiral [2.2]Paracyclophane

In addition to cyclic trimer **11**, enantiopure [2.2]paracyclophane-containing cyclic oligomers have been prepared, as shown in Figure 3.15. Cyclic dimers **16** and **17**, which are polyphenylene-based cyclic molecules, were CPL-emissive, with $|g_{lum}|$ values estimated to be 1.6×10^{-3} [19]. Thiophene- and selenophene-containing cyclic dimers **19** and **20** were prepared [20]. These compounds exhibited distinct Cotton effects with $|g_{abs}|$ values of 1.5×10^{-3} and 4×10^{-3}, respectively, although their CPL behaviors were not reported. The Sonogashira–Hagihara cross-coupling

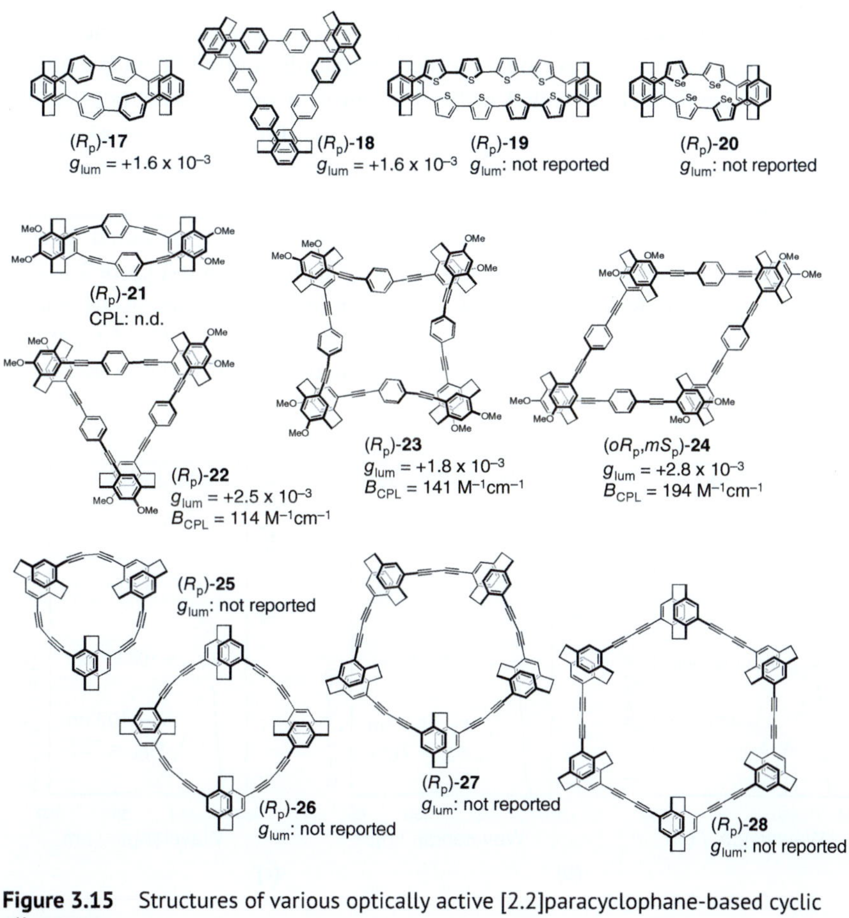

Figure 3.15 Structures of various optically active [2.2]paracyclophane-based cyclic oligomers.

of enantiopure pseudo-*ortho*-diethynyl[2.2]paracyclophane with diiodobenzene in diluted solutions yielded cyclic dimer **21**, trimer **22**, and tetramer **23** [21]. Additionally, stepwise Sonogashira–Hagihara cross-coupling using enantiopure pseudo-*ortho*- and pseudo-*meta*-disubstituted [2.2]paracyclophanes produced a #-shaped cyclic tetramer **23** [22]. Although cyclic dimer **20** did not emit CPL, cyclic trimer and tetramers **22–24** were CPL-emissive compounds with $|g_{lum}|$ values calculated to be 2.5×10^{-3}, 1.8×10^{-3}, and 2.8×10^{-3}, respectively. Recently, the B_{CPL} (brightness of CPL) value has been introduced, defined by the equation: $B_{CPL} = 2 \cdot \varepsilon \cdot \Phi_{PL} \cdot |g_{lum}|$, where ε and Φ_{PL} are molar extinction coefficient and PL quantum efficiency, respectively [23]. The B_{CPL} values of **22–24** were calculated to be 114, 141, and 194 $M^{-1}cm^{-1}$, respectively. These values, all exceeding 100 $M^{-1}cm^{-1}$, represent sufficiently high B_{CPL} values for the organic CPL emitters. Optically active cyclic trimer **24**, tetramer **25**, pentamer **26**, and hexamer **27** were prepared through oxidative coupling reactions using pseudo-*meta*-diethynyl[2.2]paracyclophane as a chiral building block [24]. Although their CPL profiles remain unknown, their CD behaviors were investigated, and their $|g_{abs}|$ values were high, estimated to be of the order of 10^{-2}.

Figure 3.16 shows the PL and CPL spectra of cyclic trimer **22** [21], cyclic tetramer **23** [21], and #-shaped cyclic tetramer **24** [22]. No significant differences were observed within the PL spectra, and all three compounds emitted PL with high Φ_{PL}. Intense CPL signals were detected in the PL emission ranges, with $|g_{lum}|$ values estimated to be 2.5×10^{-3}, 1.8×10^{-3}, and 2.8×10^{-3} for **22**, **23**, and **24**, respectively; therefore, their B_{CPL} values were calculated to be 114, 141, and 194 $M^{-1}cm^{-1}$, respectively.

Molecular orbitals in the S_1 states of cyclic tetramers **23–25**, obtained through time-dependent density functional theory (TD-DFT), are shown in Figure 3.17. The lowest unoccupied molecular orbitals (LUMOs) are localized on one of the stacked π-electron systems, indicating that the emitting species are identical in **23–25**, regardless of the number of the stacked π-electron systems or their orientation. This observation

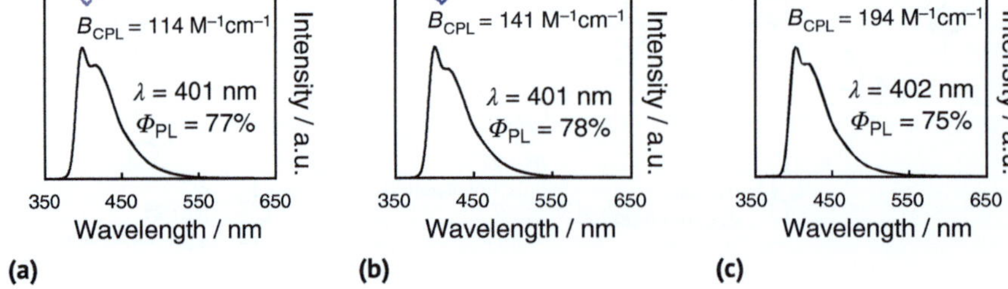

Figure 3.16 PL and CPL spectra of optically active (a) cyclic trimer, (b) cyclic tetramer, and (c) #-shaped cyclic tetramer in dilute $CHCl_3$.

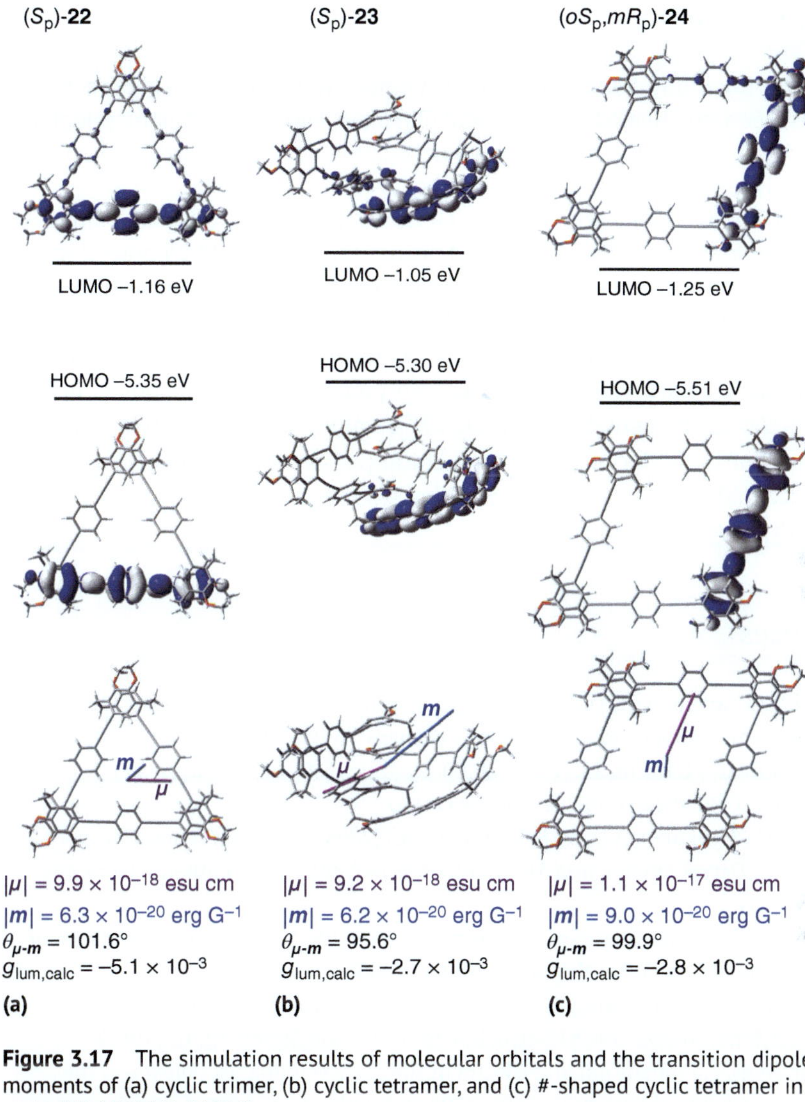

Figure 3.17 The simulation results of molecular orbitals and the transition dipole moments of (a) cyclic trimer, (b) cyclic tetramer, and (c) #-shaped cyclic tetramer in the S_1 state (TD-MN15/6-31G(d)).

explains why their PL spectra are identical. Such a PL behavior was suggested by Bazan et al.; namely, emission occurs from one of the two π-electron systems, when the π-extended conjugation systems are partially stacked by the [2.2]paracyclophane [25]. Thus, even if three or four π-electron systems are stacked, only one unit emits when it is a π-extended conjugation system. Figure 3.17 also presents the simulated dipole moments ($\boldsymbol{\mu}$) and ($\boldsymbol{m}$) of **23–25** in the S_1 states. The $\boldsymbol{\mu}$ vectors are aligned along the long axis of one of the π-electron systems, while the $\boldsymbol{m}$ vectors are oriented as normal vectors to the corresponding π-electron systems. The calculated $|g_{lum}|$ values closely match the experimental observed $|g_{lum}|$ values, confirming the accuracy of the simulations.

3.5 Conclusion

In summary, this study described the brief synthetic routes and optical properties of through-space π-conjugated polymers and oligomers based on [2.2]paracyclophane. When π-electron systems were partially stacked, their absorption behavior changed progressively with the number of the stacked π-electron systems, similar to a typical through-bond π-conjugation system. Optically active through-space conjugated polymers were successfully synthesized using enantiopure [2.2]paracyclophane compounds as chiral monomers. These optically active polymers consisting of planar chiral pseudo-*ortho*- or pseudo-*meta*-linked [2.2]paracyclophane exhibited CPL with good $|g_{lum}|$ values on the order of 10^{-3}. Optically active cyclic oligomers with planar chiral [2.2]paracyclophanes were also prepared. These oligomers demonstrated excellent CPL emission, with high B_{CPL} values exceeding 100 $M^{-1}cm^{-1}$ and $|g_{lum}|$ values of the order of 10^{-3}. Planar chiral [2.2]paracyclophane thus presents an ideal scaffold for producing CPL-emitting polymers, oligomers, and small molecules with intense CPL properties.

References

1 Müellen, K. and Scherf, U. (eds.) (2006). *Organic Light Emitting Devices: Synthesis, Properties and Application*. Weinheim, Germany: Wiley-VCH.

2 Groza, J.R. and Locklin, J.J. (eds.) (2007). *Organic Field-Effect Transistors*. New York: CRC Press Taylor & Francis Group.

3 Brabec, C., Dyakonov, V., and Scherf, U. (eds.) (2008). *Organic Photovoltaics: Materials, Device Physics, and Manufacturing Technologies*. Weinheim, Germany: Wiley-VCH.

4 Skotheim, T.A., Elsenbaumer, R.L., and Reynolds, J.R. (eds.) (2006). *Handbook of Conducting Polymers*, 3e. New York: Marcel Dekker.

5 a) Brown, C.J. and Farthing, A.C. (1949). Preparation and structure of Di-*p*-xylylene. *Nature* 164: 915–916, b) Cram, D.J. and Steinberg, H. (1951). Macro rings. I. Preparation and spectra of the paracyclophanes. *Journal of the American Chemical Society* 73: 5691–5704, c) Vögtle, F. (ed.) (1993). *Cyclophane Chemistry: Synthesis, Structures and Reactions*. Chichester: John Wiley & Sons, d) Gleiter, R. and Hopf, H. (eds.) (2004). *Modern Cyclophane Chemistry*. Weinheim: Wiley-VCH, e) Hopf, H. (2008). [2.2]Paracyclophanes in polymer chemistry and materials science. *Angewandte Chemie International Edition* 47: 9808–9812.

6 a) Cram, D.J. and Laainger, N.L. (1955). Macro rings. XII. Stereochemical consequences of steric compression in the smallest paracyclophane. *Journal of the American Chemical Society* 77: 6289–6294, b) Rozenberg, V., Sergeeva, E., and Hopf, H. (2004). Cyclophanes as templates in stereoselective synthesis. In: *Modern Cyclophane Chemistry* (ed. R. Gleiter and H. Hopf), 435–462. Weinheim, Germany: Wiley-VCH, c) Rowlands, G.J. (2008). The synthesis of enantiomerically pure [2.2]paracyclophane derivatives. *Organic and Biomolecular Chemistry* 6: 1527–1534, d) Gibson, S.E. and Knight, J.D. (2003). [2.2]Paracyclophane derivatives

in asymmetric catalysis. *Organic and Biomolecular Chemistry* 1: 1256–1269, e) Aly, A.A. and Brown, A.B. (2009). Asymmetric and fused heterocycles based on [2.2]paracyclophane. *Tetrahedron* 65: 8055–8089, f) Paradies, J. (2011). [2.2]Paracyclophane derivatives: synthesis and application in catalysis. *Synthesis* 3749–3766, g) Delcourt, M.-L., Felder, S., Turcaud, S. et al. (2019). Highly enantioselective asymmetric transfer hydrogenation: a practical and scalable method to efficiently access planar chiral [2.2]paracyclophanes. *Journal of Organic Chemistry* 84: 5369–5382, h) Vorontsova, N.V., Rozenberg, V.I., Sergeeva, E.V. et al. (2008). Symmetrically tetrasubstituted [2.2]paracyclophanes: their systematization and regioselective synthesis of several types of bis-bifunctional derivatives by double electrophilic substitution. *Chemistry: A European Journal* 14: 4600–4617, i) David, O.R.P. (2012). Syntheses and applications of disubstituted [2.2]paracyclophanes. *Tetrahedron* 68: 8977–8993, j) Hassan, Z., Spluling, E., Knoll, D.M. et al. (2018). Planar chiral [2.2]paracyclophanes: from synthetic curiosity to applications in asymmetric synthesis and materials. *Chemical Society Reviews* 47: 6947–6963, k) Hassan, Z., Spuling, E., Knoll, D.M. et al. (2020). Regioselective functionalization of [2.2]paracyclophanes: recent synthetic progress and perspectives. *Angewandte Chemie International Edition* 59: 2156–2170, l) Felder, S., Wu, S., Brom, J. et al. (2021). Enantiopure planar chiral [2.2]paracyclophanes: synthesis and applications in asymmetric organocatalysis. *Chirality* 33: 506–527.

7 a) Berova, N., Nakanishi, K., and Woody, R.W. (eds.) (2000). *Circular Dichroism*, 2e. Toronto: Wiley-VCH, b) Riehl, J.P. and Muller, F. (2012). *Comprehensive Chiroptical Spectroscopy*. New York: Wiley and Sons.

8 a) Richardson, F.S. and Riehl, J.P. (1977). Circularly polarized luminescence spectroscopy. *Chemical Reviews* 77: 773–792, b) Riehl, J.P. and Richardson, F.S. (1986). Circularly polarized luminescence spectroscopy. *Chemical Reviews* 86: 1–16.

9 a) Tohda, Y., Sonogashira, K., and Hagihara, N. (1975). A convenient synthesis of acetylenes: catalytic substitutions of acetylenic hydrogen with bromoalkenes, iodoarenes and bromopyridines. *Tetrahedron Letters* 16: 4467–4470, b) Sonogashira, K. (2002). Historical background of organopalladium chemistry. In: *Handbook of Organopalladium Chemistry for Organic Synthesis* (ed. E. Negishi), 493–529. New York: Wiley-Interscience.

10 a) Mizoroki, T., Mori, K., and Ozaki, A. (1971). Arylation of olefin with aryl iodide catalyzed by palladium. *Bulletin of the Chemical Society of Japan* 44: 581, b) Heck, R.F. and Nolley Jr., J.P. (1972). Palladium-catalyzed vinylic hydrogen substitution reactions with aryl, benzyl, and styryl halides. *Journal of Organic Chemistry* 37: 2320–2322.

11 a) Miyaura, N., Yamada, K., and Suzuki, A. (1979). A new stereospecific cross-coupling by the palladium-catalyzed reaction of 1-alkenylboranes with 1-alkenyl or 1-alkynyl halides. *Tetrahedron Letters* 20: 3437–3440, b) Miyaura, M. and Suzuki, A. (1995). Palladium-catalyzed cross-coupling reactions of organoboron compounds. *Chemical Reviews* 95: 2457–2483.

12 a) Morisaki, Y. and Chujo, Y. (2002). Synthesis of novel π-conjugated polymers having [2.2]paracyclophane skeleton in the main chain. Extension of π-conjugated length via the through-space. *Macromolecules* 35: 587–589,

b) Morisaki, Y., Ueno, S., Saeki, A. et al. (2012). π-Electron-system-layered polymer: through-space conjugation and properties as a single molecular wire. *Chemistry: A European Journal* 18: 4216–4224.

13 Müllen, K. and Wegner, G. (eds.) (1998). *Electronic Materials: The Oligomer Approach*. Weinheim: Wiley-VCH.

14 Morisaki, Y., Wada, N., Arita, M. et al. (2009). Synthesis of through-space conjugated polymers containing the pseudo-ortho-linked [2.2]paracyclophane moiety. *Polymer Bulletin* 62: 305–314.

15 Morisaki, Y., Hifumi, R., Lin, L. et al. (2012). Through-space conjugated polymers consisting of planar chiral pseudo-ortho-linked [2.2]paracyclophane. *Polymer Chemistry* 3: 2727–2730.

16 Morisaki, Y., Inoshita, K., and Chujo, Y. (2014). Planar-chiral through-space conjugated oligomers: synthesis and characterization of chiroptical properties. *Chemistry: A European Journal* 20: 8386–8390.

17 a) Maeda, H., Kameda, M., Hatakeyama, T. et al. (2018). π-Stacked polymer consisting of a pseudo-*meta*-[2.2]paracyclophane skeleton. *Polymers* 10: 1140, b) Maeda, H., Inoue, R., Saeki, A. et al. (2023). Synthesis of optically active through-space conjugated polymers consisting of planar chiral pseudo-*meta*-disubstituted [2.2]paracyclophane. *Polymer Journal* 55: 537–545.

18 a) Miki, N., Maeda, H., Inoue, R. et al. (2021). Syntheses and chiroptical properties of optically active V-shaped molecules based on planar chiral [2.2]paracyclophane. *ChemistrySelect* 6: 12970–12974, b) Asakawa, R., Tabata, D., Miki, N. et al. (2021). Syntheses of optically active V-shaped molecules: relationship between their chiroptical properties and the orientation of the stacked π-electron system. *European Journal of Organic Chemistry* 2021: 5725–5731, c) Miki, N., Inoue, R., and Morisaki, Y. (2021). Synthesis of optically active V-shaped molecules: studies on the orientation of the stacked π-electron systems and their chiroptical properties. *Bulletin of the Chemical Society of Japan* 94: 451–453, d) Sawada, R., Gon, M., Chujo, Y. et al. (2022). Synthesis of optically active π-stacked molecules: effect of π-stacking position on the chiroptical properties. *Bulletin of the Chemical Society of Japan* 95: 1353–1359, e) Matsumura, K., Inoue, R., and Morisaki, Y. (2023). Synthesis of two optically active V-shaped molecules: investigating the correlation between the stacking angle and chiroptical properties. *Tetrahedron* 138: 133406.

19 Hasegawa, M., Ishida, Y., Sasaki, H. et al. (2021). Helical oligophenylene linked with [2.2]paracyclophane: stereogenic π-conjugated dye for highly emissive chiroptical properties. *Chemistry: A European Journal* 27: 16225–16231.

20 a) Hasegawa, M., Kobayakawa, K., Matsuzawa, H. et al. (2017). Macrocyclic oligothiophene with stereogenic [2.2]paracyclophane scaffolds: chiroptical properties from π-transannular interactions. *Chemistry: A European Journal* 23: 3267–3271, b) Hasegawa, M., Kobayakawa, K., Nojima, Y. et al. (2019). Synthesis and chiroptical properties of stereogenic cyclic dimers based on 2,2′-biselenophene and [2.2]paracyclophane. *Organic & Biomolecular Chemistry* 17: 8822–8826.

21 Tanaka, K., Inoue, R., and Morisaki, Y. (2022). Optically active cyclic oligomers based on planar chiral [2.2]paracyclophane. *Chemistry: An Asian Journal* 17: e202101267.

22 Morisaki, A., Inoue, R., and Morisaki, Y. (2023). Synthesis of two novel optically active #-shaped cyclic tetramers based on planar chiral [2.2]paracyclophanes. *Chemistry: A European Journal* 29: e202203533.

23 Arrico, L., di Bari, L., and Zinna, F. (2021). Quantifying the overall efficiency of circularly polarized emitters. *Chemistry: A European Journal* 27: 2920–2934.

24 Sidler, E., Zwick, P., Kress, C. et al. (2022). Intense molar circular dichroism in fully conjugated all-carbon macrocyclic 1,3-butadiyne linked pseudo-meta [2.2]paracyclophanes. *Chemistry: A European Journal* 28: e202201764.

25 a) Wang, S., Bazan, G.C., Tretiak, S. et al. (2000). Oligophenylenevinylene phane dimers: probing the effect of contact site on the optical properties of bichromophoric pairs. *Journal of the American Chemical Society* 122: 1289–1297, b) Bartholomew, G.P. and Bazan, G.C. (2001). Bichromophoric paracyclophanes: models for interchromophore delocalization. *Accounts of Chemical Research* 34: 30–39, c) Bartholomew, G.P. and Bazan, G.C. (2002). Strategies for the synthesis of 'through-space' chromophore dimers based on [2.2]paracyclophane. *Synthesis* 2002: 1245–1255, d) Hong, J.W., Woo, H.Y., and Bazan, G.C. (2005). Solvatochromism of distyrylbenzene pairs bound together by [2.2]paracyclophane: evidence for a polarizable "through-space" delocalized state. *Journal of the American Chemical Society* 127: 7435–7443, e) Bazan, G.C. (2007). Novel organic materials through control of multichromophore interactions. *Journal of Organic Chemistry* 72: 8615–8635.

4

Development of Solid-state Luminescent Conjugated Polymers Containing Boron Element-blocks and Their Applications

Kazuo Tanaka[1,2], *Masayuki Gon*[1,2], *and Shunichiro Ito*[1,2]

[1]*Department of Polymer Chemistry, Graduate School of Engineering, Kyoto University, Katsura, Kyoto, Japan*
[2]*Department of Technology and Ecology, Graduate School of Global Environmental Studies, Kyoto University, Katsura, Kyoto, Japan*

4.1 Introduction

We have proposed the new concept for material design based on "element-blocks," which are minimum functional units containing heteroatoms, and we developed functional polymers by the combination with element-blocks [1–5]. Especially, we discovered unique optical properties from π-conjugated polymers involving electronically functional element-blocks. π-Conjugated polymers have attracted tremendous attention as a key material for constructing optical and/or electronic materials because of their unique properties. Originating from extended π-conjugated system through polymer main chains, light absorption and luminescent properties are obtained. Additionally, by the combination with a comonomer unit, various optical properties, such as magnitude of absorption and emission intensity, can be tuned in the resulting copolymers. Moreover, by employing reactive or environment-sensitive units, their electronic properties can be dynamically modulated. For the fabrication of sensing materials, much effort has been directed not only to explore stimuli-responsive materials but also to control their optoelectronic properties.

Compared to color sensors, there are many advantages to utilizing changes in luminescence properties. Because of the high visibility of fluorescent dyes, it is possible to obtain signals from nano-sized microscopic areas using these dyes as a fluorescent probe. In addition, since it is easy to obtain spectra, further information can be obtained. By monitoring changes in the luminescent properties, various information can be obtained from optical measurements, such as luminescence intensity, color, and lifetime, and detailed measurements are also possible. Moreover, in the case of polymers, it is relatively easy to observe multiple emission

Conjugated Polymer Synthesis and Materials Applications, First Edition. Edited by Yoshiki Chujo.
© 2026 Wiley-VCH GmbH. All rights reserved, including rights for text and data mining and training of artificial intelligence technologies or similar technologies. Published 2026 by Wiley-VCH GmbH.

bands in the different wavelength regions from the single molecule. Based on these multiple emission properties, quantitative detections are feasible. Therefore, stimuli-responsive luminescent materials are regarded as a scaffold for realizing future sensing technologies.

However, similarly to luminescent organic dyes, emission properties of π-conjugated polymers can be usually observed in the diluted solution, meanwhile weak luminescence and unexpected color change are often obtained in film due to aggregation-caused quenching (ACQ), which is commonly induced by nonspecific intermolecular interactions in the condensed state. Especially, due to extended π-conjugated system through polymer main chains, ACQ severely occurs in polymers, followed by emission quenching in film. Thus, it is necessary to disturb intermolecular interactions in the condensed state.

One of the conventional strategies for suppressing ACQ is the modification of π-conjugated system by steric functional groups. For instance, by tethering steric cluster compounds, named as carboranes, to the aromatic conjugated systems, intense emission was obtained even in the solid states [6]. Consequently, various types of AIE or almost quantitative emission efficiencies in solid have been obtained. As another instance, it is effective to blend luminophores with conventional polymers or silicate [7]. By introducing multiple dyes with different luminescent colors into the silica matrix, it was possible to obtain the respective luminescent colors in the bulk material. As a result, the organic–inorganic hybrid materials that emit white light were able to be obtained.

However, another problem is rising when additional components are employed for improving solid-state luminescent properties. As one of further applications of luminescent organic materials, the development of film-type sensors has been focused. For this purpose, it is necessary to show both stimuli responsiveness and solid-state luminescence. The suppression of intermolecular interactions followed by the isolation of luminescent molecules from the surroundings is advantageous for preserving luminescence in solid, while since the influence of the environment is not transmitted to the electronic state, the responsiveness to environmental changes tends to be spoiled. Therefore, it becomes difficult to express the stimuli-responsiveness that is essential for sensor applications with these materials. In other words, there is a trade-off relationship between solid-state luminescence properties and stimuli-responsiveness.

To solve the ACQ problem, we have focused on the aggregation-induced emission (AIE) property. AIE-active molecules can show emission annihilation in solution, whereas enhanced emission can be observed in the solid state. These unique behaviors mean that they have environmental sensitivity as well as solid-state luminescence. Moreover, their solid-state emission properties, such as luminescent color, can be varied corresponding to environmental alterations in solid, such as morphology changes between crystal and amorphous states. Therefore, we have studied discovery of AIE-active molecules and regulation of their solid-state luminescent properties.

Boron complexes have rigid and planar structures, which are favorable for expanding π-conjugated system. Therefore, various types of luminescent materials have been obtained based on boron complexes. In addition, by using luminescent boron complexes as a monomer, conjugated polymers having superior luminescent properties have been developed. Extended π-conjugated systems through polymer main chains have narrower energy gaps between frontier molecular orbitals (FMOs) than those of small molecules, followed by deep red and/or near infrared (NIR) emission. Moreover, by introducing environment-sensitive substituents and comonomers into the conjugated polymers, stimuli-responsivity can be loaded. These materials have been applied as a key unit for preparing highly sensitive sensing materials.

In this chapter, we summarize optical properties including AIE and crystallization-induced emission (CIE) properties of boron element-blocks. It will be demonstrated that some of flexible boron element-blocks containing π-conjugated ligands can show stimuli-responsive luminochromic behaviors as well as AIE and CIE properties [8]. By utilizing these element-blocks as a monomer, the resulting conjugated polymers having environment sensitivity can be obtained. Next, we will explain design for such flexible element-blocks. From quantum calculations, the optimized conformations in ground and excited states were obtained. From the degree of the difference between them, we estimated the degree of structural relaxation after photoexcitation. Based on this protocol, we explored several boron element-blocks and found that they can act as an AIE material. We named these complexes as excitation-driven complexes, and in this chapter, the chemistry of azomethine (Am) and azobenzene (Az) complexes is illustrated. We also describe the conversion of AIE-active dyes to NIR-luminescent AIE molecules by the aza-substitution of the skeletal carbon to nitrogen at the specific spot. These design strategies will be illustrated here.

4.2 Boron Ketoiminates

4.2.1 Transformation of the Conventional Luminescent Complex to AIE-active One

β-Diketones, also known as acetylacetone (2,4-pentanedione), are one of the most versatile ligand frameworks for the construction of main group and transition metal complexes (Figure 4.1). So far, various luminescent boron complexes have been synthesized by extending the π-conjugated system based on boron diketonates and modifying the substituents [9]. However, similar to conventional organic luminescent dyes, many of these boron complexes exhibit ACQ due to their high planarity. Despite these backgrounds, we found that the boron complex with the diketone ligand can be converted to an AIE-active molecule through element substitution at the coordinating atom with boron.

Comparing to the B—O bond, the bond energy of B—N is smaller. Therefore, we presumed that flexibility might be changed by replacing one of oxygen to nitrogen in boron diketonate. Based on this idea, a series of boron ketoiminates were prepared, and influence of the replacement on their luminescence properties was examined (Figure 4.1) [10]. In summary, it was found that boron ketoiminates have AIE properties. As one of the representative examples, boron diketonate (**BDK**) showed intense emission in solution (quantum yield $\Phi_{PL,solution}$: 0.91), while weak emission was observed in solid ($\Phi_{PL,solid}$: 0.36), similarly to conventional organic luminescent dyes. In contrast, boron ketoiminates, **BKI-H**, **BKI-Me**, and **BKI-iPr**, exhibited very weak luminescence in the diluted solution ($\Phi_{PL,solution}$: ~0.01),

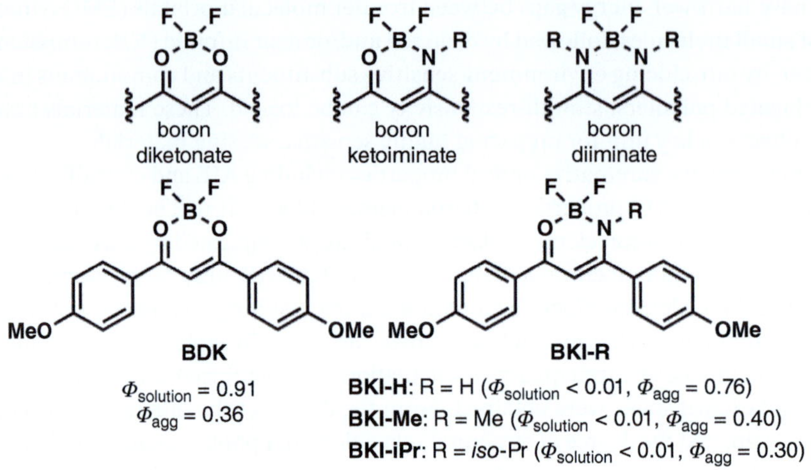

Figure 4.1 Chemical structures of boron complexes and their optical properties.

meanwhile significant intense luminescence was observed in aggregate ($\Phi_{PL,solid}$: 0.76 for **BKI-H**, 0.40 for **BKI-Me**, and 0.30 for **BKI-iPr**), clearly indicating AIE behaviors.

Since the solidification-induced emission enhancement of boron ketoiminates, it is assumable that molecular motions in the excited state could play a key role in such environment-dependent emission properties. Therefore, initially, the photoluminescence (PL) spectra at the liquid nitrogen temperature (77 K), where molecular motions would be frozen, were monitored. As a result, the significant increase in emission intensity was observed compared to measurements at room temperature. This fact clearly indicates that molecular motions in the excited state should be responsible for emission annihilation in solution. Furthermore, the PL spectra under viscous conditions were measured, and more intense emission was obtained, significantly supporting that deactivation channel should be opened in the excited state by molecular motions. As proposed above, B—N bonds (~445.6 kJ mol^{-1} or less) are known to be weaker than B—O bonds (~536 kJ mol^{-1}). Therefore, it is likely that more vigorous molecular motions should be acceptable in both ground and excited states of boron ketoiminates. Thus, emission annihilation through the nonradiative decay processes tends to frequently occur in ketoiminates in solution. In the solid state, these molecular motions would be suppressed by structural restriction. Furthermore, according to quantum calculations, it was proposed that the distributions of molecular orbitals (MOs), especially highest occupied molecular orbital (HOMO), are localized around the oxygen side in boron ketominates, comparing to that of boron diketonate. This fact supports that intermolecular interactions could be disturbed in the condensed state of boron ketoiminate. From the above data, it is summarized that the AIE behavior is achieved by a combination of solution quenching, suppression of molecular motion in the solid, and inhibition of intermolecular interactions in boron ketoiminates.

4.2.2 Modulation of the Direction of Color Changes in Mechanochromic Luminescence Behaviors by the Substituent Effect

From the structural feature of boron ketoiminates, we aimed to regulate luminescent properties in the solid state. Since AIE-active boron ketoiminates have relatively high planarity, it is presumable that their photophysical properties such as luminescent color might be critically varied by molecular morphology in the solid state. In addition, by the extension of the π-electron plane, environmental sensitivity could be enhanced in the condensed state. To evaluate the validity of this idea, we attempted to develop mechanochromic luminescent materials based on boron ketoiminate derivatives [11]. When mechanical stimuli such as compression, grinding, and crushing can change the emission color of luminescent molecules, this behavior is called mechanochromic or piezochromic luminescence and is expected to be applicable for pressure sensors and optical recording/memory devices. In order to realize mechanochromic luminescence, we designed the boron ketoiminate derivative having expanded π-conjugated systems.

The triad molecule **BKM-H** consisting of two boron ketoiminates connected with the bithiophene linker was synthesized (Figure 4.2). From the optical measurements in the crystalline state, the typical mechanochromic luminescent behavior from red to orange was observed. Subsequently, by heating, luminescent color returned to red. Interestingly, the BKM derivative having iodine groups at both ends of the molecular skeleton showed the opposite behavior: the yellow luminescence from the pristine crystal was turned to orange by grinding, and initial yellow color was observed after heating. To clarify these behaviors, we synthesized a series of BKM derivatives with various types of end groups and examined their optical properties. All derivatives showed similar absorption and emission spectra in solution, while **BKM-F**, which has relatively smaller substituents, showed similar behaviors with **BKM-H** in crystal. On the other hand, the compounds with relatively large substituents, **BKM-Cl**, **BKM-Br**, and **BKM-I**, showed bathochromic shifts of emission by mechanical treatments. The emission color of the pristine samples could be restored when the ground samples were heated. From the comparison with the data from powder X-ray diffraction and differential scanning calorimetry, it was revealed that the grinding process increased the amorphous domain content of the samples. These observations lead to the following scenario on the opposing mechanochromic luminescent behaviors depending on the size of end substituents (Figure 4.2). For the triad structures with smaller substituents, the molecules interact with face-to-face motifs to achieve a tightly packed structure in the crystalline state. As a result, the emission bands appear in the relatively long wavelength region in the initial crystalline state. When such tight packing is broken, random molecular distribution occurs, resulting in a hypsochromic shift. Meanwhile, steric hindrance of larger substituents can disrupt such face-to-face stacking structures in crystals. In the amorphous state after the mechanical treatments, π–π interactions are more frequently promoted, leading to a bathochromic shift. This is one example that provides the environment-sensitive solid-state emission of AIE-active complexes.

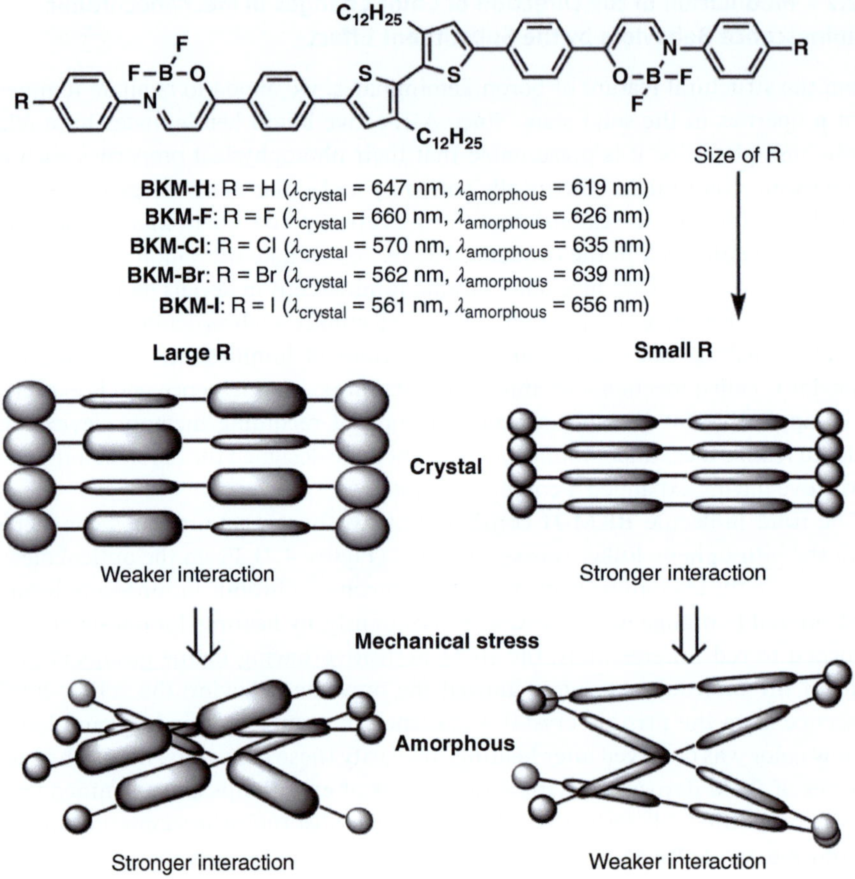

Figure 4.2 Mechanochemical luminescent properties of boron ketoiminate derivatives and their plausible models.

4.2.3 Requirement of Heat-insensitive Mechanochromic Luminescent Materials

As one of the practical applications, mechanochromic compounds are utilized as a paint-type sensor for detecting pressure at the surface of the model in wind-tunnel tests. If we use the emissive paints which can show luminescent chromism triggered by pressure, it is possible to monitor the magnitude of mechanical forces at the tiny spots on various shapes of models, as mentioned above. However, conventional mechanochromic luminescent compounds generally exhibit recovery of color change by heating after mechanochromic behaviors. This means that the significant mechanochromic luminescent responses, which are generated by the detection target, could be crucially perturbed or somehow spoiled by fraction heating. Therefore, heat-insensitive mechanochromic luminescent materials are strongly needed to overcome the disturbance from friction heating. In general, mechanochromic luminescent molecules form a crystalline state because the crystal packing is a thermodynamically stable distribution. The application of mechanical force reduces the fraction of thermodynamically stable ordered structures by increasing the surface area and introducing defects and disorder, eventually producing an amorphous-like random distribution in most places. Compared with the initial crystalline state, these relatively disordered states are energetically unfavorable at the molecular level, and therefore annealing or solvent vapor-assisted rearrangement can restore the ordered distribution and subsequently obtain molecular environment similar to the initial state. Thus, to obtain heat-insensitive mechanochromic luminescent properties, the stabilization of the disordered states produced by mechanical forces is essential.

4.2.4 Development of Heat-insensitive Mechanochromic Luminescent Materials

To fulfill the purpose, chemistry of organic–inorganic hybrid materials was applied [12]. By mixing organic materials such as polymers with inorganic components at the molecular level, it is possible to obtain materials that combine the advantages of the organic with the durability of the inorganic. These mixed materials are called organic–inorganic hybrids and are widely known to be effective design strategies, particularly for improving thermal stability and durability of polymers. Conventional hybrid materials are prepared through a sol–gel reaction. Under the general reaction conditions, polar solvents are commonly used, and hydrophobic organic materials cannot be often applied due to low solubility. In addition, hybrids are produced by heating for a long time, and decomposition often occurs by the acids or bases used as catalysts. To solve these problems, we devised the use of polyhedral oligomeric silsesquioxane (POSS, Figure 4.3) as a building block for constructing hybrids according to the preprogrammed design without sol–gel reactions.

POSS has the rigid silica cube at the core and organic substituents at each vertex. Because of structural features, POSS is regarded as a molecular hybrid. There have been numerous published papers showing that simply adding POSS to polymer materials can improve their heat resistance. Furthermore, by using POSS, we accomplished to hybridize substances that were difficult to hybridize using conventional methods and have succeeded in loading heat resistance on functional materials. For example, by linking the ion pair that is capable of forming an ionic liquid with POSS through covalent bonds, it was possible to increase the thermal decomposition temperature of the ionic pair. In addition, by introducing POSS into the block copolymer that can spontaneously form a lamellar structure when made into a film, it was also possible to improve the thermal stability of the higher-order

Figure 4.3 Chemical structure of POSS.

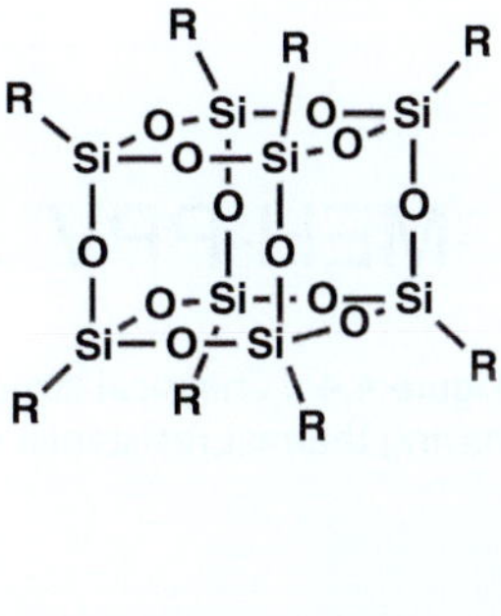

structure, followed by structural color. Here, we present a representative study on thermally stable luminescent hybrids based on the modified POSS.

To obtain thermally stable white light-emitting films, we prepared hybrid materials by mixing a conjugated polymer poly[2-methoxy-5-(2′-ethylhexyloxy)-1,4-phenylene vinylene] (MEH-PPV) and the POSS derivative modified with tetraphenylethene (TPE) (TPE-POSS), a well-known AIE-active molecule (Figure 4.4) [13]. TPE-POSS was loaded as a filler on MEH-PPV, and thin films were obtained with variable concentrations of TPE-POSS through a spin-coating method with the mixture solutions. Owing to the good dispersion of the rigid silica cubes in the matrix, homogeneous films were obtained. The TPE-POSS-containing films showed high thermal stability against molecular rearrangement and thermal decomposition upon annealing, similarly to conventional polymer hybrids. In addition, it was found that TPE-POSS was able to enhance the luminescence efficiency, presumably by suppressing chain aggregation. By adjusting the introduction ratio of TPE-POSS, dual-emission characteristics were achieved, with white emission consisting of cyan and orange emissions from TPE-POSS and MEH-PPV, respectively. It should be noted that these color balances were able to be maintained even in the high-temperature region (425 K). Finally, the white-emitting material with heat resistance was obtained. It is assumed that the rigid core of POSS might suppress thermal motions even in the excited state. As a consequence, luminescence properties of TPE as well as MEH-PPV were able to be preserved even at high temperatures.

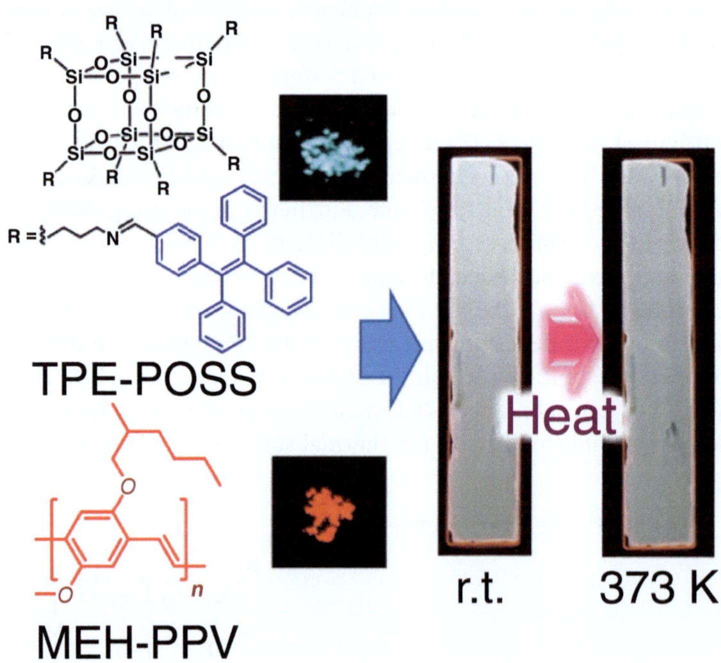

Figure 4.4 Chemical structures of TPE-POSS and MEH-PPV and paintable hybrid films having thermal resistance in white-light luminescent properties. *Source*: Adapted from [13].

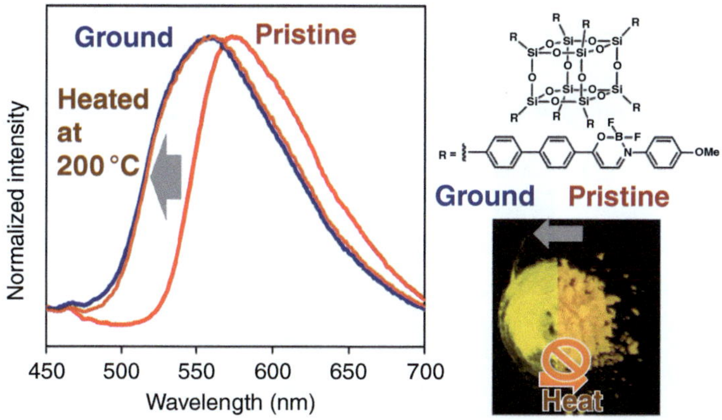

Figure 4.5 Chemical structure and heat-insensitive mechanochromic luminescent material consisting of POSS and boron ketoiminates. *Source*: Reproduced from [14] / with permission of John Wiley & Sons.

By employing POSS, the heat-insensitive mechanochromic luminescent material was designed (Figure 4.5) [14]. Eight boron ketoiminate complexes were connected with the POSS core through the Pd-catalyzed coupling reaction. From optical measurements, it was initially found that the modified POSS showed orange emission, and by adding mechanical forces to the pristine sample, luminescent color was clearly changed to yellow. It is likely that the regular structure at the initial state should be collapsed by the mechanical treatment, and molecular environments would be altered, followed by the luminochromic behavior. Significantly, the yellow-emissive amorphous state can be preserved during heating. Although the ground sample was heated to 200 °C, significant changes in the wavelengths of the PL band were hardly observed. By radially distributing luminescent chromophores on the rigid cube, both optical and thermal properties can be modulated to realize heat-resistant mechanochromic luminescent properties.

4.2.5 Development of AIE-active Conjugated Polymers

AIE-active conjugated polymers were obtained by using boron ketoiminate as a repeating unit (Figure 4.6) [15]. With the conventional Pd-catalyzed cross-coupling reactions, we synthesized π-conjugated polymers containing the ketoimine structure and comonomers, where boron can be captured, to examine influence of boron complexation on the electronic structures of a main-chain π-conjugated system. We prepared the two polymers with or without boron complexation and evaluated their optical properties. From the PL measurements, it was demonstrated that **pBL** exhibits AIE ($\Phi_{PL,solution}$: 0.04, $\Phi_{PL,solid}$: 0.06), whereas **pL** shows typical ACQ ($\Phi_{PL,solution}$: 0.09, $\Phi_{PL,solid}$: 0.02). These results represent that AIE properties of boron ketoiminates can be preserved and expressed even in the main-chain conjugation. Additionally, extension of π-conjugated systems was also confirmed. From **pL**, the PL band was observed with the peak at 519 nm in solution and at 562 nm in film, while the peak wavelengths in the PL spectra of **pBL** were detected at 595 nm in solution and at 646 nm in film. These data mean that

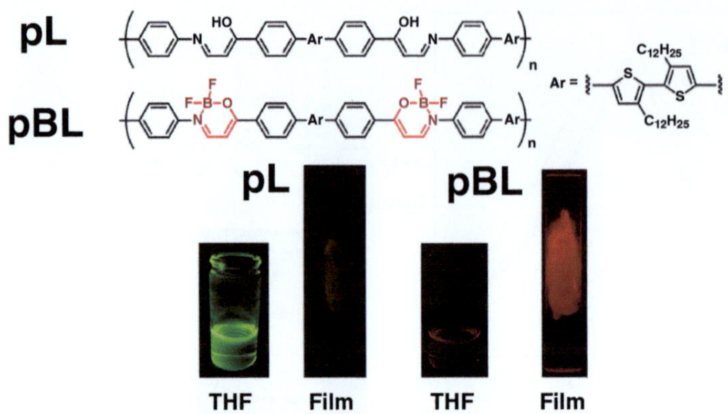

	$\lambda_{PL(THF)}$	$\Phi_{PL(THF)}$	$\lambda_{PL(film)}$	$\Phi_{PL(film)}$
pL	519	0.09	562	0.02 ↓
pBL	595	0.04	646	0.06 ↑

Figure 4.6 Chemical structure and AIE behavior of **pBL**. *Source*: Adapted from [15].

the boron ketoiminate-based polymer has more extended and robust main-chain conjugation than the ketoimine ligand polymer. It is because the electronic structure in the ketoimine units is fixed in the enolimine form, which should be capable to expand main-chain conjugations involving the boron complexation.

In this study, we also synthesized the polymers in which the bithiophene moiety was replaced with fluorene, and the changes in optical properties were monitored (Figure 4.7). As a result, AIE properties ($\Phi_{PL,solution}$: 0.10, $\Phi_{PL,solid}$: 0.13) and extension of the conjugated system by boron complexation were observed, as described above. The PL bands were observed in the shorter wavelength region around 560 nm both in solution and film than those of the bithiophene polymer. This can be explained by the fact that fluorene has a weaker electron-donating property than bithiophene, and therefore has a lower ability to shift the PL band of the monomer unit to longer wavelengths. From the above data, it is shown that the AIE properties of boron ketoiminate complexes can be expressed even in extended π-conjugated systems, and moreover, that the solid-state emission properties can be diverse. We also prepared a series of copolymers by changing the connection points with comonomer units (Figure 4.7) [16]. Interestingly, AIE properties were not observed, and various luminescent colors were obtained from the films. This fact suggests that molecular motions at the phenyl ring connected with the imine carbon should be responsible for emission annihilation in solution.

4.2.6 Enhancement of Solid-state Luminescent Properties

The development of luminochromic materials by utilizing the solid-state luminescence of boron ketoiminates has been illustrated in Section 4.2.5. On the other hand, in these stimuli-responsive materials, critical decreases in luminescence efficiency

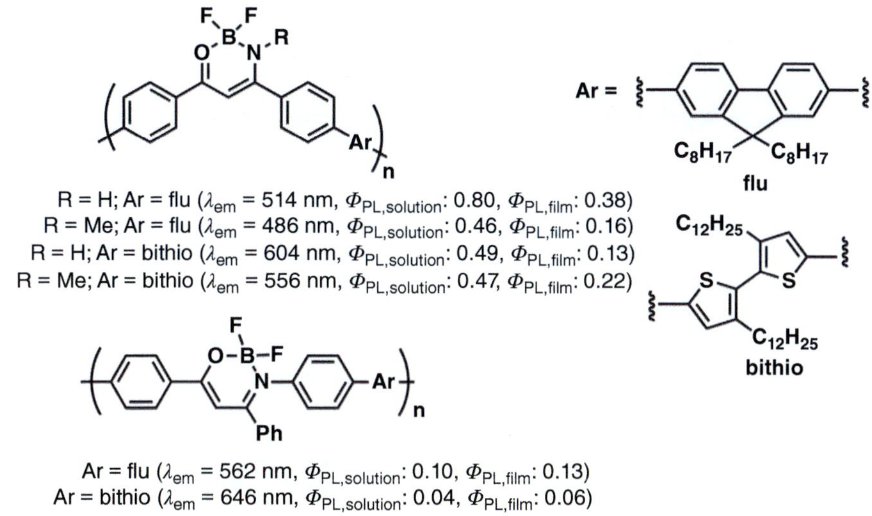

Figure 4.7 Summary of optical properties of boron ketoiminate polymers.

have been accompanied. For example, in the case of heat-insensitive POSS, the $\Phi_{PL,solid}$ value is about 0.01. In addition, the $\Phi_{PL,solid}$ value of the conjugated polymer with the bithiophene units was below 0.1, meaning that a decrease in luminescence ability should be induced. This is probably because the ketoimine complexes, as seen in solution, are prone to exhibit potential molecular motion, and when environmental changes occur, molecular motions are released, and the quenching channel is opened. Therefore, we tried establishing molecular design to achieve both more efficient solid-state luminescence and stimuli-responsiveness.

As mentioned above, it was shown that ACQ-presenting boron diketonate can be transformed to the AIE-active dye by replacing one of the oxygen atoms to nitrogen [10]. It was proposed that excitation deactivation should be caused by molecular motions involving the nitrogen atom. Therefore, emission from the resulting boron ketoiminate complexes was completely annihilated in the solution state. In contrast, intense emission was obtained in solid. It is proposed that localization of MOs might be responsible for avoiding intermolecular interactions in solid, followed by ACQ. From this speculated mechanism of AIE behaviors of boron ketoiminates, we presumed that emission annihilation could be significantly suppressed without losses of solid-state luminescent properties by introducing the fused structure around the nitrogen atom. Furthermore, owing to expanded π-conjugated system around the nitrogen, environmental sensitivity might be enhanced. Based on this idea, we synthesized a series of fused boron ketoiminates (FBKIs) and examined optical properties [17].

Luminescent properties of **1a–1d** were compared in solution and crystal and before and after grinding (Figure 4.8). Slight differences were observed in the Φ_{PL} values between solution and crystalline states in each complex as we expected. In particular, **1d** exhibited highly efficient solid-state emission ($\Phi = 0.81$) without showing ACQ, indicating that FBKIs are no longer AIE-active molecules. It is likely

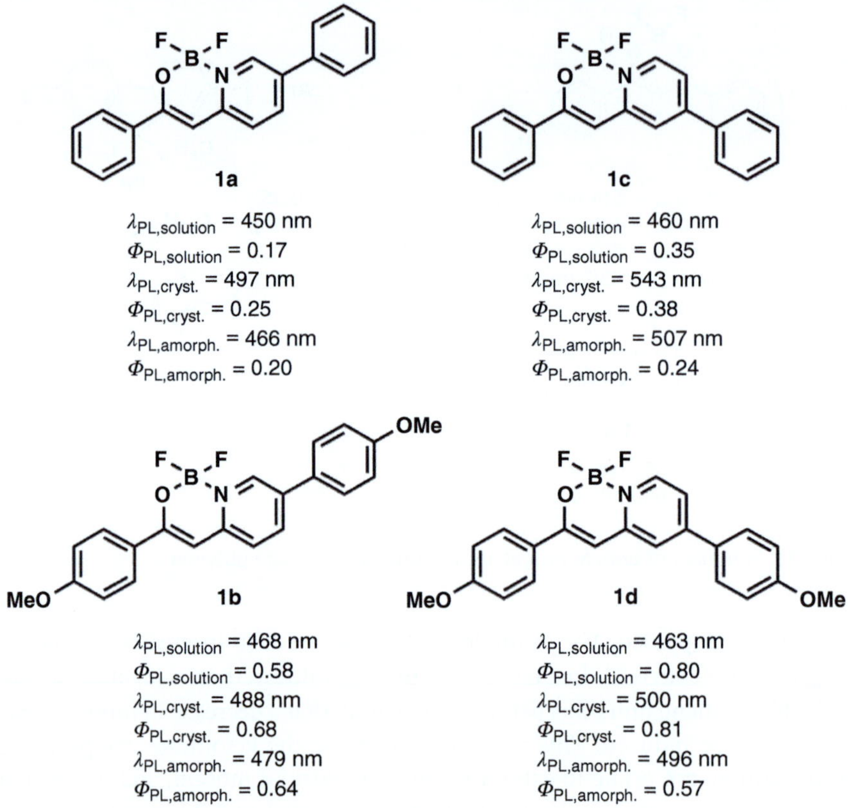

Figure 4.8 Chemical structures and luminescent properties of FBKIs.

that introducing the fused structure should suppress deactivation of excited energy caused by molecular motions. Thus, emission annihilation is suppressed even in solution. These data also indicate that validity of the above discussion on the π-conjugation in the complexes should be confirmed.

Next, we evaluated mechanochromic luminescent behaviors. By applying external forces, luminescent color changes were observed from FBKIs as we also expected. Since blue-shifts were observed in mechanochromic luminescent behaviors, it is suggested that by applying mechanical forces to the crystalline samples, collapses and disordering of regular structures would be induced, leading to random distribution in which the stabilization effect crucially decreased. Consequently, the degree of intermolecular interactions could be lowered in the disordered state. Finally, blue-shifts of emission bands were detected. It should be noted that almost same degrees of the Φ_{PL} values were observed before and after the mechanical treatments. From these data, it can be said that luminescent abilities of FBKIs in crystal can be preserved both in solution and amorphous-like random states. Therefore, stimuli-responsive luminochromic properties, such as mechanochromic behaviors, have been obtained without critical losses of emission efficiency.

4.2.7 Development of the Thermochromic and Mechanochromic Luminescent Polymers

Next, we synthesized FBKI-containing copolymers and found stimuli-responsive luminescent properties from some of the synthesized conjugated polymers (Figure 4.9) [18]. In the $CHCl_3$ solution of **PfBT**, changes in absorption and emission colors with temperature were observed. In both absorption and emission spectra, the disappearance of the long-wavelength peak ($\lambda_{abs} = 547$ nm, $\lambda_{em} = 578$ nm) and the increase of the short-wavelength peak ($\lambda_{abs} = 448$ nm, $\lambda_{em} = 543$ nm) were observed with increasing temperature, indicating that the chromic and luminochromic behaviors should be caused by these changes in spectra. In addition, a pseudoisobestic point (around $\lambda_{abs} = 460$ nm) was confirmed in the absorption spectrum, representing the existence of two systems with different absorption. It was also revealed that the absorption and emission spectra of the **PfBT** solution returned to the spectra before heating by leaving it for a sufficient time after heating. This result indicates that the thermochromic properties of the **PfBT** solution are reversible, suggesting that the mechanism is not due to an irreversible reaction such as the decomposition of the boron complex site. This speculation can be supported by the fact that there were no changes in the ^{1}H and ^{11}B NMR spectra before and after heating.

To clarify the mechanism of these thermochromic properties, dynamic light scattering (DLS) measurements were performed. In the $CHCl_3$ solution after heating, unimodal distributions were observed around $d_H = 40$ nm, while in the $CHCl_3$ solution before heating, size distributions of aggregation were observed around 15 nm and 115 nm. Considering the size of the synthesized **PfBT**, it is suggested that in addition to single **PfBT** chains, aggregates should be formed in the $CHCl_3$ solution before heating. From these data, it was concluded that the newly appeared aggregates cause self-absorption of luminescence, resulting in the thermochromic luminescence behaviors.

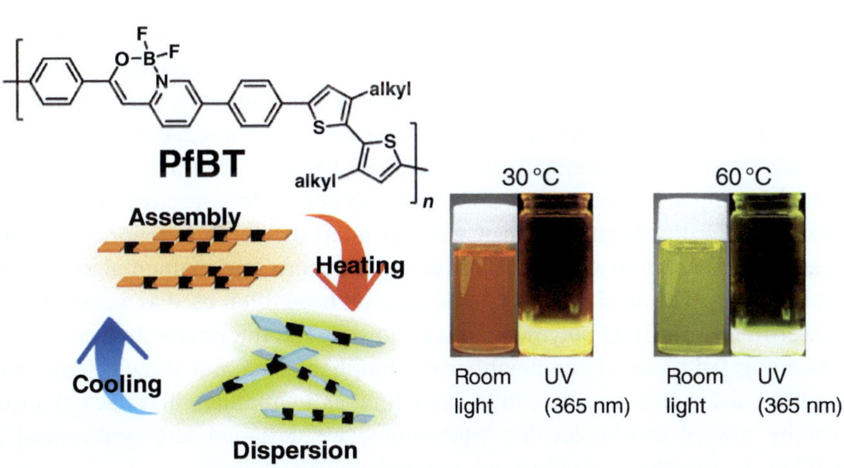

Figure 4.9 Thermochromic luminescent properties of **PfBT**. *Source*: [18]/with permission of Royal Society of Chemistry.

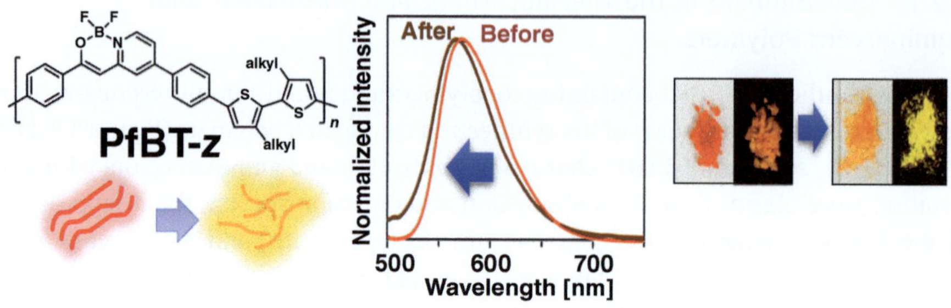

Figure 4.10 Mechanochromic luminescent property of **PfBT-z.** *Source*: [18]/with permission of Royal Society of Chemistry.

By changing the connection point with the comonomer, mechanochromic luminescent properties were obtained (Figure 4.10) [18]. We found that the polymer **PfBT-z** can exhibit mechanochromic luminescence in solid. The emission band of the polymer solid was observed in the orange region (λ_{PL} = ~580 nm, $\Phi_{PL,solid} = 0.14$) in the pristine state, while the blue-shift to the yellow region (λ_{PL} = ~560 nm, $\Phi_{PL,solid} = 0.21$) was induced by grinding. From the mechanistic studies, it is proposed that the luminochromism should be attributed to some losses of ordered structures according to the data from powder X-ray diffraction and differential scanning calorimetry. It is because the FBKI units were able to respond to environmental alterations and reflect to their luminescence properties even within the polymer main chain. Therefore, luminochromic behaviors can be obtained. However, the luminescence chromism was not observed from the film sample. Therefore, for the practical usages, it is essential to realize optical changes in the film state.

4.2.8 Enhancement of Environmental Sensitivity

It was demonstrated that FBKIs have environmental sensitivity as well as solid-state luminescent properties. We next attempted to enhance their environmental sensitivity by chemical modification. The thiophene ring has the characteristic that the π-conjugated plane can be efficiently extended by substituting a thienyl group on an aromatic molecule. The twist angle between the planes is smaller in the thienyl group-substituted benzene ring than in the phenyl group-substituted benzene ring, and the π-conjugated plane can be efficiently extended. Moreover, compared to the phenyl ring, the thiophene ring can work as a strong electron donor. Therefore, increase in the electron density followed by far more environmental sensitivity could be expected. Indeed, the thiophene ring has been often used as a comonomer or spacer in π-conjugated polymers, as we showed. The thiophene ring can extend the π-conjugated plane of the neighboring thiophene ring and show a significant red-shift in absorption and PL bands. Therefore, as a candidate that has enhanced sensitivity toward intermolecular interactions, we designed and synthesized **2**, which has the thiophene ring instead of the phenyl group in **1a**.

The thienyl-substituted compound **2** was prepared through the Pd-catalyzed coupling reactions, and a single crystal was obtained (Figure 4.11) [19]. According to the

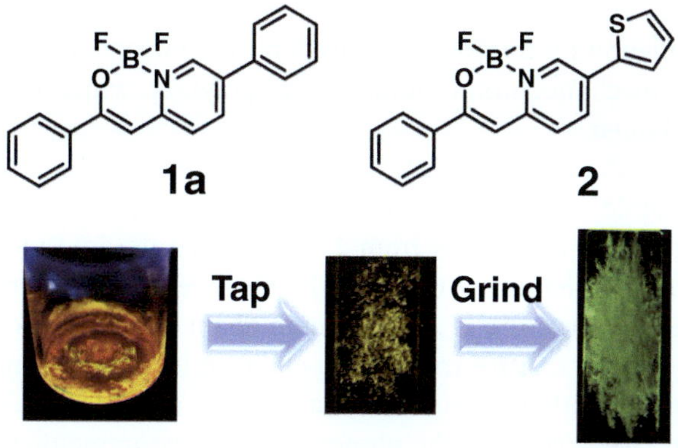

Figure 4.11 Chemical structure and multi-step mechanochromic luminescence of **2**. *Source*: [19]/with permission of Royal Society of Chemistry.

structure comparison between **2** and **1a** with the single crystal X-ray analysis, it was found that **2** possesses the higher planar conformation in the crystal packing. Since the thiophene ring has relatively smaller steric hindrance than the phenyl ring, the planar structure is favorable for extending π-conjugated systems and intermolecular interactions in solid. From the PL measurements in solution, emission bands were obtained in the blue region from both complexes (**2**: $\lambda_{PL,solution} = 466$ nm, **1a**: $\lambda_{PL,solution} = 450$ nm), indicating that the electronic structure of sole molecules should be almost same between **2** and **1a**.

In the solid-state luminescent properties, significant differences were observed. By using methanol as the recrystallization solvent, the crystalline polymorph exhibiting orange luminescence was obtained from **2**. Surprisingly, the luminescence color of this crystal drastically changed to green when it was joust touched with a weak force. Furthermore, the ground powder presenting yellow luminescence was obtained when the greenish luminescent sample was crushed. These results indicate that **2** exhibits multi-step mechanochromic luminescence properties. Detailed analysis of the PL spectra revealed that the new PL bands were generated in the shorter wavelength regions at each mechanical treatment. It is proposed that the ordered structure in the crystal collapses stepwise with each mechanical stimulus, and finally random distribution should be generated. The possible scenario is that **2**, because of its sensitive environmental responsiveness, can respond to these changes, become a different luminescent species, and subsequently exhibit multi-step changes.

4.2.9 Mechanochromic Luminescent Sensor for Marking a Tiny Spot

One of the advantages of fluorescent sensors is the applicability to imaging technologies that visualize changes and stimuli on a microscale. On the other hand, conventional mechanochromic luminescent materials generally require a strong force to induce luminochromic behaviors, and therefore require rubbing or grinding. This inevitably results in a decrease in spatial resolution. If highly sensitive materials can be obtained, we can detect weak stimuli applied to a small point, and this response is expected to be a key phenomenon for improving resolution in the image. In this part, the development of the mechanochromic luminescent polymer, which can meet an above demand, is explained.

In the previous work, it was shown that **PfBT** exhibits not mechanochromic luminescence but thermochromic one. Only **PfBT-z** showed mechanochromic luminescence in solid. According to the results from **2**, to enhance the intermolecular interactions among polymer chains, we removed the phenylene linker from the repeating unit and realized the structure of **2** in the main-chain conjugation of **PBF** (Figure 4.12a) [20]. The powder of **PBF** irreversibly changed its luminescent properties by grinding the powder with an agate mortar and pestle (Figure 4.12b). To understand the mechanism, we performed the photophysical measurements in the dilute solution and neat spin-coated film. The absorption and emission bands of the neat film were detected in the longer wavelength regions than those in the solution state, indicating the existence of low-energy exciton states derived from the intermolecular interactions in the aggregated states. The mechanical force applied to the powder led to the blue shift of the emission bands with a decrease in the Φ_{PL} value and the disappearance of the longest fluorescent lifetime component. These changes suggest that the intermolecular interactions giving rise to low-energy exciton states should be suppressed by grinding. In contrast, **PBPh** showed the red shift of absorption and emission bands by aggregation but no significant changes before and after grinding the powder. The results implied that the bulky phenyl groups on the boron atom might fill the free volume between the polymer main chains and prevent the changes in the intermolecular interactions. Finally, we applied **PBF** to a mechanical force-sensing film (Figure 4.12c). Piercing the neat drop-casted film with a syringe needle made a hole around which the emission color changed from

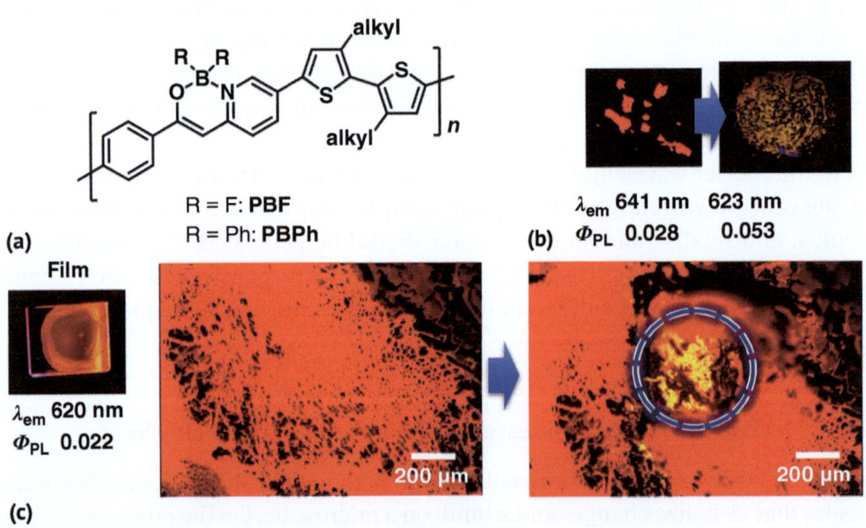

Figure 4.12 Sensitive mechanochromic luminescent behaviors of FBKI-containing polymer. (a) Chemical structures of polymers and (b) mechanochromic luminescence of the powder and (c) film of **PBF**. *Source*: Adapted from [20].

red to orange. Importantly, the color change was maintained even after pulling out the needle, indicating that the **PBF** film can memorize the stress history as the change in the emission color at the tiny spot.

4.3 Rational Design for AIE-active Molecules Based on Excitation-driven Complexes

4.3.1 Idea of Excitation-driven Complexes and Logical Design from Scratch

From the above researches, it can be said that AIE-active boron element-blocks have wide versatility and can be a potential platform for building luminescent functional materials with stimuli-responsiveness and environmental sensitivity. To obtain advanced materials, it is essential to establish the rational design of AIE-active molecules from scratch through theoretical calculations. We presumed that emission quenching in solution might occur in boron complexes that can exhibit large structural relaxation after photo-excitation (Figure 4.13a) [21]. In the aggregation state, these structural transformations would be disturbed because of structural restriction. Based on this scenario, the AIE behavior can be observed. To realize this scenario, we explored boron complexes that could show larger degree of structural relaxation in the excited state with theoretical calculations.

Quantum calculations using density functional theory (DFT) and time-dependent DFT (TD-DFT) provide optimized structures in the ground and excited states, respectively. When the structural difference between these states is large, we assumed that structural relaxation might proceed in the excited state. Based on this idea, we explored the boron complexes and found **BPI** that could show larger degree of bending motion in the excited state (Figure 4.13a) [22]. As a control molecule in which structural relaxation is suppressed, we designed the fused molecule **FBPI**. According to optical measurements, it was found that in the solution state, only **FBPI** exhibited intense emission, whereas **BPI** hardly showed significant luminescence (Figure 4.13b, c). On the other hand, intense emission was observed from the solid sample of **BPI**, while **FBPI** hardly exhibited emission in solid due to ACQ. These data clearly indicate that **BPI** is an AIE- and CIE-active molecule. Importantly, in the optical spectra, it was found that **BPI** showed a larger width of Stokes shift than **FBPI**. This fact strongly supports the speculation that a larger degree of structural relaxation could undergo in the excited state of **BPI**. From the crystal packing of **BPI**, it is suggested that the phenyl substituent should inhibit intermolecular interactions. Therefore, ACQ was able to be suppressed in solid. In the case of conventional luminescent materials, it is essential to form rigid skeletons for suppressing molecular motions followed by intense emission. Correspondingly, **BPI** provided very weak emission in solution. However, significant AIE and CIE were obtained in solid. From these data, flexible boron element-blocks containing conjugated ligands, which can exhibit drastic structural relaxation in the excited state, are a promising platform for constructing AIE-active materials. Thus, we defined these element-blocks as an excitation-driven complex, and we attempted to develop further AIE materials based on these complexes.

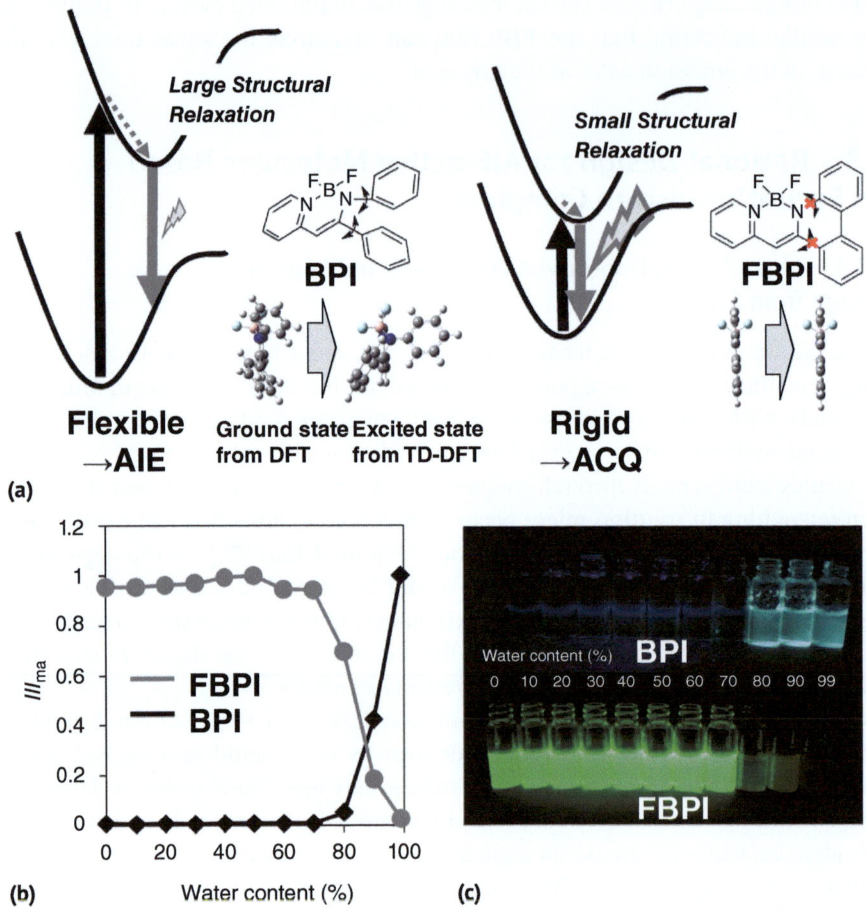

Figure 4.13 (a) Illustrations of structural relaxation of **BPI** and **FBPI** after photoexcitation, (b) monitoring of emission intensity ratios by increasing water contents, and (c) their luminescent behaviors under UV irradiation (365 nm). *Source*: [22]/with permission of Royal Society of Chemistry.

4.3.2 Design for Planar AIE-active Molecules

Many of the AIE molecules reported so far have multiple external benzene rings to efficiently induce solution quenching. Owing to the distorted distributions toward π-plane, intermolecular interactions are disturbed, and subsequently solid-state emission can appear. However, these distortions tend to limit to the extension of π-conjugation. When these units are incorporated into polymer main chains for constructing conjugated polymers, electronic communications should be disturbed due to distortions. Therefore, we designed planar AIE-active molecules that can expand π-conjugated systems. As illustrated above, by comparing the energetically stable conformations obtained from DFT and TD-DFT calculations, we searched the molecular skeleton that has a planar conformation in the ground state but a distorted structure in the excited state. After surveying, it was proposed that the azomethine skeleton could be a suitable candidate (Figure 4.14) [23]. The boron-fused

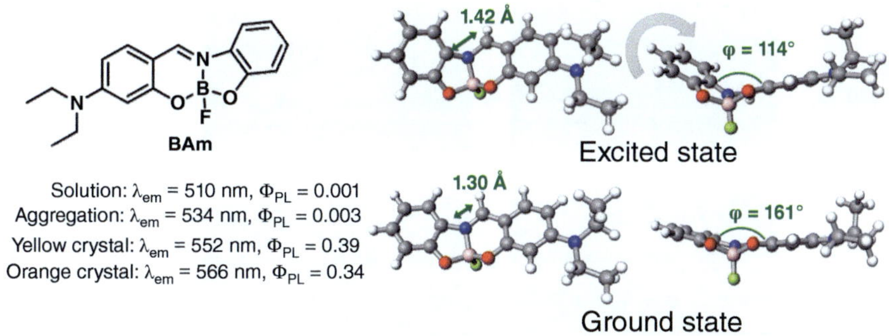

Figure 4.14 Optical properties and calculated conformations of **BAm**. *Source*: [23]/with permission of John Wiley & Sons.

azomethine (**BAm**) was suggested to have high planarity before photoexcitation, and the results of computer modeling were confirmed by X-ray single crystal analysis. Interestingly, a dramatic structural bending was proposed during the structural relaxation in the excited state. From these results, AIE properties were expected.

Boron azomethine complex **BAm** was synthesized, and yellow and orange (green and yellow emission, respectively) crystal polymorphs were generated during the recrystallization process. By picking up them, these polymorphs were able to be isolated. The optical measurements revealed that **BAm** showed not only AIE but also CIE characters, similarly to **KI**. Large Stokes shifts were detected from both complexes in solution, suggesting that larger degree of structural relaxation should proceed in the excited state. From the crystal packing, it was suggested that the fluorine atoms on the boron atoms could interfere with intermolecular interactions and subsequently suppress ACQ. In particular, the yellow-luminescent crystal showed mechanochromic and thermochromic luminescence properties. From these results, it is suggested that the concept for logical design of AIE-active boron complexes based on excitation-driven complexes can be proved.

Although these are ground-state properties, the crystals of **BAm** showed unique physical and luminochromic behaviors upon temperature changes (Figure 4.15). From the differential scanning calorimetry, it was found that the crystal–crystal transitions between the green and the yellow-luminescent states can proceed reversibly. It should be mentioned that these crystals showed thermosalient effect during these transitions by heating as well as cooling. Moreover, thermochromic luminescence was also observed corresponding to the transition. The flexible framework of Az might play a critical role in such physical movements during the structural transition in the crystalline state.

4.3.3 Development of Solid-state Luminescent Polymers

To obtain solid-state emission in the longer wavelength regions owing to the planarity of Am, we synthesized conjugated copolymers with fluorene and bithiophene (BT) as a comonomer unit and constructed a donor–acceptor (D–A) conjugated

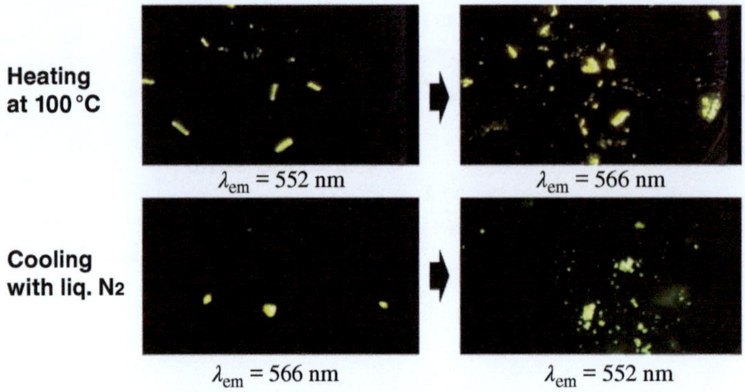

Figure 4.15 Thermosalient effects of **BAm** with thermochromic luminescence. *Source*: Reproduced with permission from [23]/John Wiley & Sons.

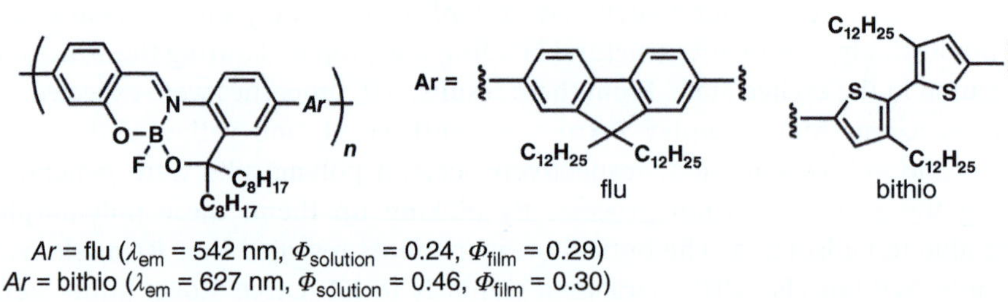

Figure 4.16 Chemical structures and optical properties of boron azomethine polymers.

system involving the boron azomethine structure as the electron acceptor in the main chain (Figure 4.16) [24]. Electrochemical measurements revealed that the electron-accepting ability was enhanced by boron coordination. Because of the strong D–A interaction, the conjugated polymers showed a significant red shift of the optical bands in both UV–vis absorption and PL spectra, comparing to the monomer unit. Especially, these polymers showed intense emission in both solution and film states. It is because structural relaxation in the excited state should be suppressed by the connection with the comonomer units at both ends of the boron azomethine structure. Therefore, emission annihilation in solution should be disturbed, followed by intense constant emission. Yellow emission was observed from the fluorene-containing polymers, while red emission was obtained from the more electron-donating BT. In addition, the ligand polymers without boron showed weak emission. The highly efficient emission properties of the azomethine-containing polymers should be caused not only by the rigid backbone but also by the strong D–A conjugated system derived from the boron-fused structure. In this study, we demonstrated that the combination of Am with a boron atom is a promising strategy to obtain accepting monomer units, leading to the creation of luminescent conjugated polymer materials.

4.3.4 Stimuli-responsive Materials Based on Regioregular Conjugated Polymer

We have also synthesized regioregular and random conjugated polymers based on the boron azomethine complex and demonstrated stimuli-responsive luminochromism (Figure 4.17) [25]. Although luminescence chromism by vapor annealing was able to be detected with the regioregular conjugated polymer with boron diiminate, thermal annealing hardly influenced its optical properties. As mentioned above, we discovered that conjugated polymers containing boron azomethine exhibit highly efficient solid-state luminescence by avoiding ACQ. Therefore, we investigated the solid-state luminescence properties and influence by regioregularity.

Regioregular and random boron azomethine-based conjugated polymers were synthesized, and the effect of regioregularity on the solid-state luminescence properties of the conjugated polymers was evaluated from a series of measurements. Comparison studies with regioregular and random conjugated polymers revealed that a highly ordered structure exists only in the regioregular polymers according to the data from optical spectra and SEM observations. Significantly, following thermal annealing of the regioregular polymer film improved the Φ_{PL} value. According to the spectral changes in the absorption spectra before and after heat treatment, it can be inferred that the collision of the highly ordered structure plays an important role in improving the emission efficiency. These results indicate that controlling the regioregularity can affect not only the photophysical properties of the film state but also the stimuli-responsive ability. Additionally, the regioregularity of conjugated polymers has been studied for applications in organic solar cells and organic field-effect transistors. It has been shown that regulation of orientation of the alternating asymmetric repeat units is a promising approach to

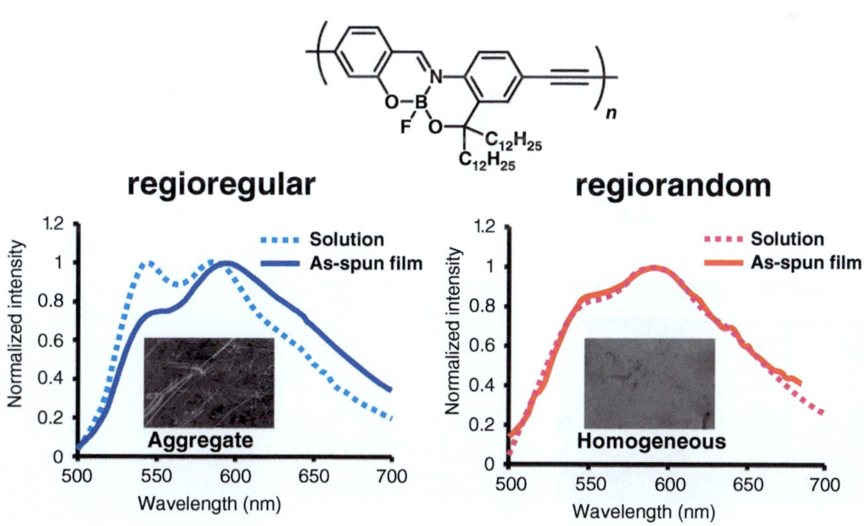

Figure 4.17 Aggregation-induced spectrum change observed only from the regioregular polymer after film formation. *Source*: [25]/with permission of John Wiley & Sons.

enhance device performance compared with their random counterparts because the well-defined structure of regioregular polymers improves the π-stacking orientation and enhances the crystallinity of thin films. Our findings could be also helpful for improving device efficiency.

4.3.5 Injection of Stimuli-responsiveness into Conjugated Polymer by Element Replacement

The study on the introduction of stimuli-responsiveness to the boron azomethine structure is explained. To realize a stimuli-responsive π-conjugated polymer capable of showing luminochromic behaviors in a thin film, we designed the boron complex with an NNO-tridentate ligand obtained by replacing one of the B—O bonds with B—N based on boron azomethine (Figure 4.18) [26]. The polymer film showed a red shift of the emission band (orange → deep red) under acid vapor, and this response was reversible by fuming basic vapor. Furthermore, the bilayer film of the polymer coated with poly(*n*-butyl methacrylate) (PBMA) showed vapor chromic emission with a gradual blue shift (orange → yellow → green) depending on the exposure time upon chloroform annealing. It was suggested that the change in emission characteristics occurred through intermolecular interactions with the newly introduced nitrogen atom after diffusion in PBMA matrices, resulting in stimuli-responsiveness. As an application, we also demonstrated that these changes can be used to evaluate the solvent vapor permeability of membrane materials. As a result, it was shown that the gas barrier ability of parafilm and cellophane tape was higher than that of latex gloves and polytetrafluoroethylene tape. We can show that nitrogen substitution in π-conjugated polymers should be a useful strategy to incorporate stimuli-responsiveness into the electronic properties.

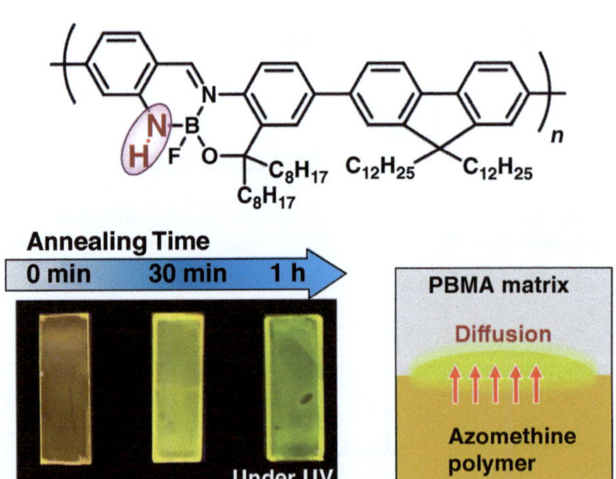

Figure 4.18 Luminochromic behaviors of the boron azomethine polymer with the NNO-ligand by the diffusion through PBMA matrices. *Source*: Adapted from [26].

4.4 AIE-active Materials Possessing NIR Luminescent Properties

4.4.1 Transformation to NIR-luminescent Materials by Aza-substitution at "the Isolated LUMO"

In the last part, the selective modulation of energy level of LUMO for realizing narrow energy gap between FMOs, which can generate NIR luminescence, is explained. Because of high permeability to various media including quarts and water, NIR light (wavelength 700 nm to 2 μm) is conventionally used in information communications. Moreover, NIR-luminescent probes have gathered attention as a new generation bioimaging tool for visualizing biological events at the deep spot inside vital organs. Therefore, the development of NIR-luminescent materials is one of the relevant topics in the optical research field. We have recently established to selectively lower the energy level of LUMO in conjugated molecules, including small molecules and polymers. Herein, the concept and representative examples are described.

To generate NIR emission, narrow energy gaps between FMOs are necessary, and there are several strategies for realizing expected electronic structures. By expanding π-conjugated systems, HOMO and LUMO are elevated and lowered, respectively, and the narrower energy gaps can be obtained. As another strategy, electron-withdrawing and electron-donating groups are introduced into π-conjugated systems, the similar effect can be induced. Phosphorescence, by employing the heavy atom effect, appears in the longer wavelength region than fluorescence. Furthermore, electronic interactions with boron or silicone can lower the LUMO energy level, followed by NIR emissions.

So far, most of the NIR-luminescent dyes have larger sizes of π-conjugated systems than luminescent materials in the visible wavelength region, although there are several design strategies as mentioned above. Extended π-conjugated systems can readily form nonspecific intermolecular interactions in the solid state, and solid-state luminescence is spoiled due to ACQ. Thus, it is strongly desirable to develop a new molecular design strategy that can realize a narrow energy gap structure without expanding the π-conjugated system. We have recently succeeded in meeting this demand and achieving a narrow energy gap structure by selectively lowering the LUMO energy level. The details will be examined.

The design concept for the π-conjugated system having a narrow energy gap between FMOs is illustrated in Figure 4.19 [27]. By replacing a skeletal carbon in π-conjugated system with nitrogen (aza-substitution), the energy level is lowered (Figure 4.19a). In the case of conventional organic molecules, both energy levels of HOMO and LUMO are simultaneously lowered. Meanwhile, we have investigated electronic structures of pentaazaphenalene (5AP, Figure 4.19b) and found that only the LUMO energy level can be critically lowered by the modification with the electron-accepting group (Figure 4.19c) [28]. As a consequence, we received a series of NIR-absorbing molecules involving the 5AP core [28]. From quantum calculation results, it was found that only the LUMO is distributed (indicated by the arrows) on some of skeletal carbon in 5AP (Figure 4.19b), and the electron-accepting groups contribute to selectively lowering LUMO energy level. We named these positions

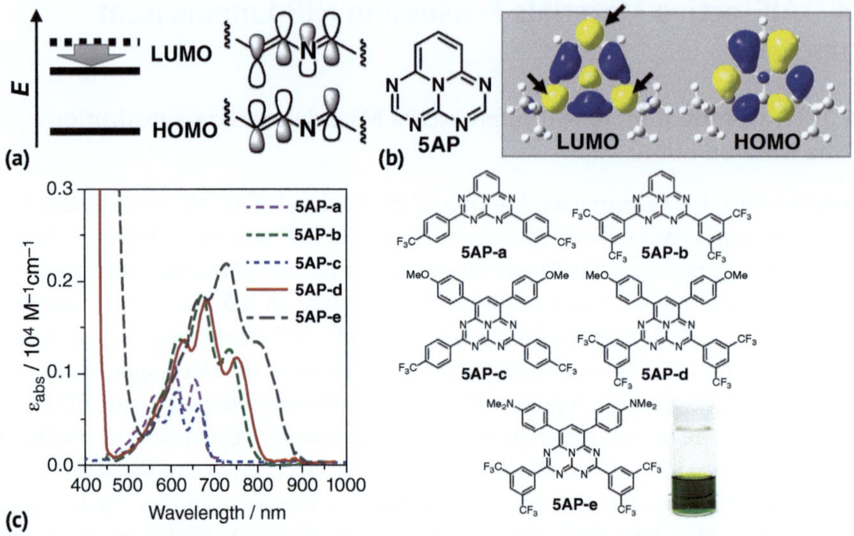

Figure 4.19 (a) Selective lowering of LUMO energy level by the aza-substitution. (b) Electron distributions of FMOs of 5AP (black arrow: isolated LUMO). (c) Optical properties of 5AP derivatives having NIR absorption. *Source*: Reproduced with permission from [28]/John Wiley & Sons.

as "isolated LUMO" and investigated whether it is possible to selectively control only the LUMO energy level by functionalizing the isolated LUMO or by the aza-substitution with other molecules for proving generality of the effect of the aza-substitution at the isolated LUMO on the energy levels. There is an advantage in the aza-substitution method. Since the type of element is converted in π-conjugated system, it is not necessary to expand the size of π-conjugated system. This is favorable for avoiding ACQ. Moreover, material properties, such as solubility and film-formability, can be also improved. To obtain NIR-luminescent AIE materials, we applied the aza-substitution method for previous luminescent materials.

4.4.2 NIR–AIE Molecules and Polymers and Their Applications

Based on solid-state luminescent properties of the boron azomethine complexes, we aimed to develop NIR-luminescent materials through the aza-substitution method as described above. According to the electron distribution of FMOs, it was found that the significant lobe exists at the carbon atom in the imine backbone (Figure 4.20a). Therefore, we designed the aza-substituted complex, azobenzene (Az), in which the skeletal carbon atom is replaced with nitrogen [29]. A series of boron azobenzene (BAz) derivatives were synthesized, and their optical properties were surveyed. Initially, as we expected, both absorption and emission bands were detected from the BAz complexes in the longer wavelength regions, comparing to those of the corresponding azomethine complexes. It should be noted that AIE and CIE were also realized in BAz derivatives, as detected in the azomethine complexes. For example, it was shown that the complex consisting of the BAz dimer structure has the

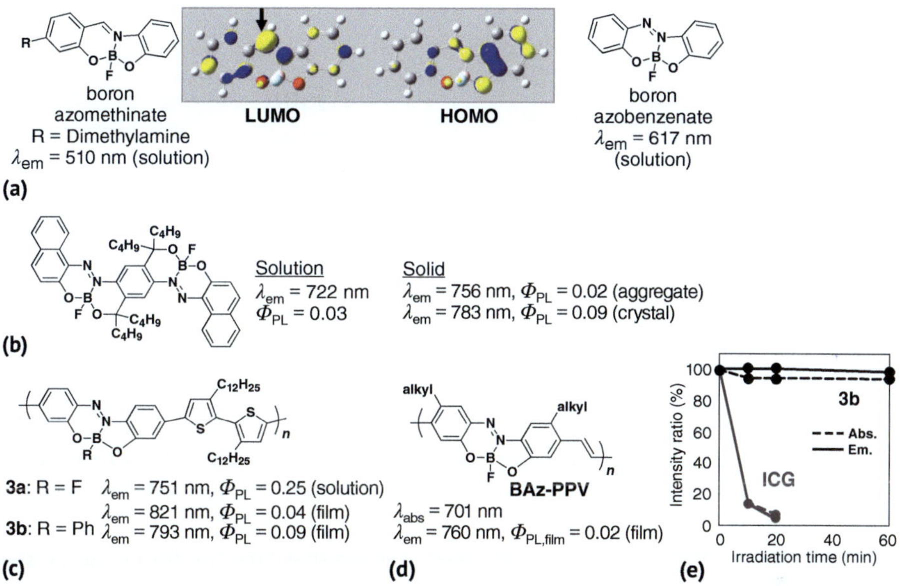

Figure 4.20 (a) Electron distributions of FMOs of boron azomethine (black arrow: isolated LUMO). (b) NIR-luminescent azobenzene complex and its environment-sensitive optical properties. (c) Chemical structures of copolymers involving boron azobenzene complexes and their optical properties. (d) NIR-absorbing and luminescent polymer (**BAz-PPV**). (e) Changes in absorption and emission properties of polymer **3b**-protein complex and ICG under UV irradiation (254 nm) in phosphate buffer saline (pH = 7.5). *Source*: [34]/with permission of Oxford University Press.

CIE property ($\Phi_{PL,crystal} = 0.09$) in the NIR region of 783 nm (Figure 4.20b) [30]. By expanding π-conjugated systems, emission bands were obtained in the longer wavelength region. A series of copolymers were synthesized using BAz derivatives and their optical bands were observed in the NIR region (Figure 4.20c) [31, 32]. The NIR luminescent film ($\lambda_{em,film} = 793$ nm) was also obtained from the copolymer including BAz, and the Φ_{PL} value was determined as 0.09. This result indicates that efficient luminescent films can be obtained.

Poly(p-phenylene vinylene) (PPV) is one of the well-known π-conjugated polymers, and PPV derivatives have been traditionally used in fabricating organic devices. However, very few reports have been reported on NIR-absorbing and emitting PPV derivatives. The polymer containing the PPV-like structure (**BAz-PPV**) was constructed using the conjugated moiety of Baz (Figure 4.20d) [33]. From optical measurements, it was shown that **BAz-PPV** has significant NIR absorption and emission ($\lambda_{abs} = 702$ nm, $\lambda_{PL} = 760$ nm, $\Phi_{PL,film} = 2.0\%$) and high carrier mobility (10^{-2}–10^{-3} cm^2 V^{-1}s^{-1}) measured by the time-of-flight technique.

In addition, the BAz polymer was adsorbed onto proteins and irradiated with UV light (Figure 4.20e) [34]. From the emission spectra, stability was examined and compared with the results from the protein complex with ICG, which is a conventional NIR luminescent dye. The ICG complex was 90% photobleached in 10 minutes, meanwhile the polymer complex showed high stability with no change in luminescence intensity

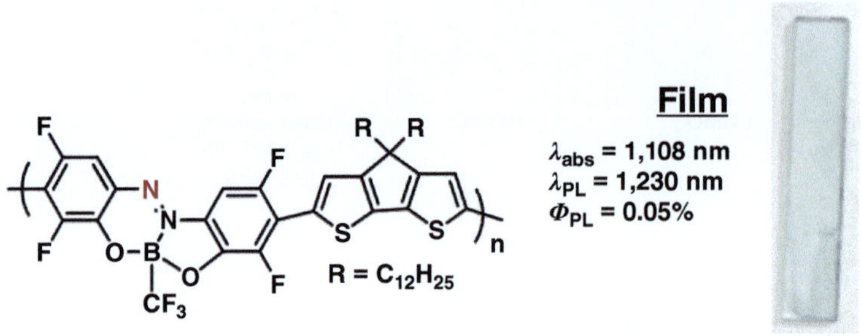

Figure 4.21 NIR-II absorbing and emitting film composed of the copolymer involving the BAz derivative.

even after 1 h of irradiation. Because the HOMO position remained unchanged and only the LUMO was lowered by the aza-substitution, it is suggested that oxidative decomposition followed by photobleaching should be suppressed.

Furthermore, more recently, we succeeded in synthesizing the film material that absorbs and emits in the NIR-II region above 1,000 nm (Figure 4.21) [35]. In particular, we were able to improve the efficiency by more than 10 times, with a molar absorption coefficient of approximately 86,000 $cm^{-1}M^{-1}$, compared to conventional NIR polymers. It is because robust conjugation through polymer main chains should be realized even after the aza-substitution.

4.5 Conclusion

So far, most of the organic luminescent materials have suffered from ACQ, and there have been difficulties in the development of luminescent organic devices with high efficiency. Furthermore, based on the conventional design strategies, it has been extremely difficult to realize highly sensitive luminescent sensors. In this chapter, we explained the new strategy for simultaneously realizing solid-state luminescence and stimuli-responsiveness based on flexible element-blocks, which can show relatively larger degree of structural relaxation in the excited state. By using these complexes, various environmentally sensitive polymers were obtained. These materials are able to present luminochromic properties in response to tiny environmental changes and weak external stimuli and act as a sensor. Therefore, the design strategies are applicable not only for tuning detection properties but also for constructing advanced sensing materials. Moreover, we introduced the series of excitation-driven complexes and illustrated several examples on the design with quantum calculations and development of AIE-active molecules. By the combination with the selective lowering of LUMO energy level by the aza-substitution at the isolated LUMO, NIR-AIE materials were obtained. This means that these materials can be created from scratch with conventional quantum calculations. We expect these findings could be a versatile tool for obtaining advanced sensing materials, which can meet demands in practical usages.

References

1 Chujo, Y. and Tanaka, K. (2015). New polymeric materials based on element-blocks. *Bulletin of the Chemical Society of Japan* 88: 633–643. https://doi.org/10.1246/bcsj.20150081.

2 Gon, M., Tanaka, K., and Chujo, Y. (2018). Recent progress in the development of advanced element-block materials. *Polymer Journal* 50: 109–126. https://doi.org/10.1038/pj.2017.56.

3 Tanaka, K. and Chujo, Y. (2020). Modulation of the solid-state luminescent properties of conjugated polymers by changing the connecting points of flexible boron element-blocks. *Polymer Journal* 52: 555–566. https://doi.org/10.1038/s41428-020-0316-y.

4 Tanaka, K. and Chujo, Y. (2023). Frustrated element-blocks: a new platform for constructing unique stimuli-responsive luminescent materials. *Polymer Journal* 55: 353–364. https://doi.org/10.1038/s41428-022-00709-w.

5 Ito, S., Gon, M., Tanaka, K. et al. (2021). Recent developments in stimuli-responsive luminescent polymers composed of boron compounds. *Polymer Chemistry* 12: 6372–6380. https://doi.org/10.1039/D1PY01170G.

6 Ochi, J., Tanaka, K., and Chujo, Y. (2020). Recent progress in the development of solid-state luminescent o-carboranes with stimuli responsivity. *Angewandte Chemie International Edition* 59: 9841–9855. https://doi.org/10.1002/anie.201916666.

7 Kajiwara, Y., Nagai, A., Tanaka, K. et al. (2013). Efficient simultaneous emission from RGB-emitting organoboron dyes incorporated into organic-inorganic hybrids and preparation of white light-emitting materials. *Journal of Materials Chemistry C* 1: 4437–4444. https://doi.org/10.1039/C3TC30276H.

8 Gon, M., Tanaka, K., and Chujo, Y. (2019). Concept of excitation-driven boron complexes and their applications for functional luminescent materials. *Bulletin of the Chemical Society of Japan* 92: 7–18. https://doi.org/10.1246/bcsj.20180245.

9 Tanaka, K. and Chujo, Y. (2015). Recent progress of optical functional nanomaterials based on organoboron complexes with β-diketonate, ketoiminate and diiminate. *NPG Asia Materials* 7: e223. https://doi.org/10.1038/am.2015.118.

10 Yoshii, R., Nagai, A., Tanaka, K. et al. (2013). Highly emissive boron ketoiminate derivatives as a new class of aggregation-induced emission fluorophores. *Chemistry: A European Journal* 19: 4506–4512. https://doi.org/10.1002/chem.201203703.

11 Yoshii, R., Suenaga, K., Tanaka, K. et al. (2015). Mechanofluorochromic materials based on aggregation-induced emission-active boron ketoiminates: regulation of the direction of the emission color changes. *Chemistry: A European Journal* 21: 7231–7237. https://doi.org/10.1002/chem.201500033.

12 Gon, M., Tanaka, K., and Chujo, Y. (2017). Creative synthesis of organic–inorganic molecular hybrid materials. *Bulletin of the Chemical Society of Japan* 90: 463–474. https://doi.org/10.1246/bcsj.20170005.

13 Gon, M., Saotome, M., Tanaka, K. et al. (2021). Paintable hybrids with thermally stable dual emission composed of tetraphenylethene-integrated POSS and MEH-PPV for heat-resistant white-light luminophores. *ACS Applied Materials & Interfaces* 13: 12483–12490. https://doi.org/10.1021/acsami.0c22298.

14 Suenaga, K., Tanaka, K., and Chujo, Y. (2017). Heat-resistant mechanoluminescent chromism of the hybrid molecule based on boron ketoiminate modified octa-substituted polyhedral oligomeric silsesquioxane. *Chemistry: A European Journal* 23: 1409–1414. https://doi.org/10.1002/chem.201604662.

15 Yoshii, R., Tanaka, K., and Chujo, Y. (2014). Conjugated polymers based on tautomeric units: regulation of main-chain conjugation and expression of aggregation induced emission property via boron-complexation. *Macromolecules* 47: 2268–2278. https://doi.org/10.1021/ma500082e.

16 Yoshii, R., Nagai, A., Tanaka, K. et al. (2014). Boron ketoiminate-based polymers: fine-tuning of the emission color and expression of strong emission both in the solution and film state. *Macromolecular Rapid Communications* 35: 1315–1319. https://doi.org/10.1002/marc.201400198.

17 Suenaga, K., Tanaka, K., and Chujo, Y. (2017). Design and luminescence chromism of fused boron complexes having constant emission efficiencies in solution and in the amorphous and crystalline states. *European Journal of Organic Chemistry* 5191–5196. https://doi.org/10.1002/ejoc.201700704.

18 Suenaga, K., Uemura, K., Tanaka, K. et al. (2020). Stimuli-responsive luminochromic polymers consisting of multi-state emissive fused boron ketoiminate. *Polymer Chemistry* 11: 1127–1133. https://doi.org/10.1039/c9py01733j.

19 Saotome, S., Suenaga, K., Tanaka, K. et al. (2020). Design for multi-step mechanochromic luminescence property by enhancement of environmental sensitivity in a solid-state emissive boron complex. *Materials Chemistry Frontiers* 4: 1781–1788. https://doi.org/10.1039/c9qm00719a.

20 Aoyama, Y., Ito, S., and Tanaka, K. (2024). Mechanochromic luminescence of π-conjugated polymer based on pyridylenolate boron complexes: molecular design and application to shear-force memory. *Macromolecules* 57 (14): 6559–6567. https://doi.org/10.1021/acs.macromol.4c01149.

21 Tanaka, K., Nishino, K., Ito, S. et al. (2017). Development of the solid-state emissive o-carborane and theoretical investigation for mechanism of aggregation-induced emission behaviors of organoboron "element-blocks". *Faraday Discussions* 196: 31–42. https://doi.org/10.1039/C6FD00155F.

22 Yamaguchi, M., Ito, S., Hirose, A. et al. (2017). Control of aggregation-induced emission versus fluorescence aggregation-caused quenching by bond existence at a single site in boron pyridinoiminate complexes. *Materials Chemistry Frontiers* 1: 1573–1579. https://doi.org/10.1039/C7QM00076F.

23 Ohtani, S., Gon, M., Tanaka, K. et al. (2017). A flexible, fused, azomethine–boron complex: thermochromic luminescence and thermosalient behavior in structural transitions between crystalline polymorphs. *Chemistry: A European Journal* 23: 11827–11833. https://doi.org/10.1002/chem.201702309.

24 Ohtani, S., Gon, M., Tanaka, K. et al. (2019). Construction of the luminescent donor-acceptor conjugated systems based on boron-fused azomethine acceptor. *Macromolecules* 52: 3387–3393. https://doi.org/10.1021/acs.macromol.9b00259.

25 Ohtani, S., Gon, M., Tanaka, K. et al. (2024). Synthesis of regioregular and random boron-fused azomethine conjugated polymers for film morphology control. *Chemistry: An Asian Journal* 19: e202301136. https://doi.org/10.1002/asia.202301136.

26 Gon, M., Kanjo, M., Ohtani, S. et al. (2023). Selective modulation of energy levels of frontier orbitals in solid-state luminescent boron-fused azomethine polymers with orthogonal orientation to the main-chains. *Polymer Chemistry* 14: 2893–2901. https://doi.org/10.1039/D3PY00335C.

27 Tanaka, K. and Chujo, Y. (2021). New idea for narrowing an energy gap by selective perturbation for one of the frontier molecular orbitals. *Chemistry Letters* 50: 269–279. https://doi.org/0.1246/cl.200756.

28 Watanabe, H., Ochi, J., Tanaka, K. et al. (2020). Tuning the NIR absorption properties of 1,3,4,6,9b-pentaazaphenalene derivatives through the spatially separated frontier molecular orbitals. *European Journal of Organic Chemistry* 2020 (7): 777–783. https://doi.org/10.1002/ejoc.201901569.

29 Gon, M., Tanaka, K., and Chujo, Y. (2018). A highly-efficient near-infrared-emissive copolymer with N=N double-bond π-conjugated system based on a fused azobenzene-boron complex. *Angewandte Chemie International Edition* 57: 6546–6551. https://doi.org/10.1002/anie.201803013.

30 Nakamura, M., Gon, M., Natsuda, S. et al. (2022) Development of NIR emissive fully-fused bisboron complexes with π-conjugated systems including multiple azo groups. *Dalton Transactions* 51: 74–84. https://doi.org/10.1039/D1DT03652A.

31 Nakamura, M., Gon, M., Tanaka, K. et al. (2023). Solid-state near-infrared emission of π-conjugated polymers consisting of boron complexes with vertically projected steric substituents. *Macromolecules* 56: 2709–2718. https://doi.org/10.1021/acs.macromol.2c02578.

32 Nakamura, M., Yamauchi, M., Gon, M. et al. (2023). UV-to-NIR wavelength conversion of π-conjugated polymers based on pyrene-substituted boron-fused azobenzene complexes. *Macromolecules* 56: 7571–7578. https://doi.org/10.1021/acs.macromol.3c01124.

33 Wakabayashi, J., Gon, M., Tanaka, K. et al. (2020). A near-infrared absorptive and emissive poly(p-phenylene vinylene) derivative containing azobenzene-boron complexes. *Macromolecules* 53: 4524–4532. https://doi.org/10.1021/acs.macromol.0c00745.

34 Yoo, D., Nakamura, M., Kanjo, M. et al. (2023) Facile preparation of near infrared-luminescent protein complexes with conjugated polymers consisting of boron azobenzene units. *Bulletin of the Chemical Society of Japan* 96: 659–662. https://doi.org/10.1246/bcsj.20230083.

35 Nakamura, M., Kanetani, I., Gon, M. et al. (2024). NIR-II absorption/fluorescence of D–A π-conjugated polymers composed of strong electron acceptors based on boron-fused azobenzene complexes. *Angewandte Chemie International Edition* 63: e202404178. https://doi.org/10.1002/anie.202404178.

5

Arsenic-containing Conjugated Polymers

Hiroaki Imoto and Kensuke Naka

Faculty of Molecular Chemistry and Engineering, Kyoto Institute of Technology, Goshokaido-cho, Kyoto, Japan

5.1 Introduction

Arsenic, discovered by Albertus Magnus in the thirteenth century, derives its name from the yellow mineral "arsenikon" [1]. Inorganic arsenic compounds, such as arsenic oxides and sulfoxides, are well-known for their high toxicity. Arsenic trioxide (As_2O_3), for instance, is highly toxic and easily introduced into food or drink, earning it the nickname "poudre de succession" (inheritance powder) in France. In contrast, the history of organoarsenic compounds is relatively brief. Nonetheless, they have played significant roles in both science and human history. The first organoarsenic compound, "cacodyl" ($Me_2AsAsMe_2$), was synthesized by Cadet de Gassicourt in 1760. Its chemical structure was identified in the nineteenth century, marking it as one of the earliest organometallic compounds. In 1907, Salvarsan was developed as the first effective treatment for syphilis, though its cyclic structure was only clarified in 2005 [2, 3]. Despite early advances, organoarsenic compounds were misused during World War I as chemical weapons. Agents like adamsite (diphenylaminechlorarsine), lewisite (2-chlorovinyldichloroarsine), and chlorodiphenylarsine caused severe blistering upon skin exposure. Improper disposal of these arsenic-based weapons continues to cause serious soil and marine pollution today.

Beyond this negative historical context, the lack of effective synthetic methods has hindered the development of organoarsenic compounds for materials science, including conjugated polymers. Traditional methods rely on toxic and volatile arsenic chlorides and hydrides. For example, arsole derivatives have been synthesized from dichlorophenylarsine [4] or phenylarsine [5], posing significant risks to researchers from toxic gas exposure. Consequently, experimental studies on functional organoarsenic materials have been limited. Instead, advances in computational chemistry have facilitated the simulation of structural and electronic properties of organoarsenic compounds, especially conjugated molecules. As a result, arsenic-containing conjugated polymers have remained largely unexplored in the broader field of heteroatom-fused conjugated materials.

Conjugated Polymer Synthesis and Materials Applications, First Edition. Edited by Yoshiki Chujo.
© 2026 Wiley-VCH GmbH. All rights reserved, including rights for text and data mining and training of artificial intelligence technologies or similar technologies. Published 2026 by Wiley-VCH GmbH.

5.2 Arsenic in Organic Compounds

The term "pnictogen (Pn)" is often used to refer to group 15 elements (excluding nitrogen), akin to how group 16 elements are called "chalcogens." Moving down the group from phosphorus to bismuth, electronegativity decreases, and metallic character increases. Neutral pnictogen compounds in their +III oxidation state typically adopt a pyramidal structure, with a lone pair occupying the fourth position. Their bond angles are smaller than the ideal tetrahedral angle of 109.5°, and the Lewis acidity of +III pnictogens arises from available empty d orbitals. For heavier pnictogens, bond angles approach 90° [6].

In arsenic–carbon (As—C) bonds, the arsenic atom's lone pair exhibits higher s-character compared to phosphorus. Consequently, the coordination direction of arsenic is less defined, allowing organoarsenic ligands to form diverse complexes. However, the lone pair's s-character and its large molecular orbital reduce the σ-donating and π-accepting abilities of organoarsenic ligands. The weaker σ-donation is attributed to the increased energy gap between the valence s and p orbitals, which enhances the p-character of the As—C bond. The reduced electronegativity and diffuse orbitals of arsenic account for its weaker π-acceptance. Despite these limitations, arsenic ligands have proven valuable in transition metal catalysis. For instance, Farina demonstrated that triphenylarsine significantly enhances the Stille coupling reaction compared to the traditionally used triphenylphosphine, owing to its weaker σ-donation. Similarly, Dong reported that thiophene difunctionalization via C—H activation is uniquely facilitated by triphenylarsine, as phosphine ligands fail to show activity.

Trivalent phosphorus compounds are highly reactive toward oxygen, with phosphines readily oxidizing to phosphine oxides under ambient conditions. In contrast, organoarsenic compounds are much more resistant to oxidation, making arsenic an attractive candidate for designing trivalent-pnictogen-based organic materials. Additionally, arsenic atoms exhibit high stereochemical stability due to their significant inversion barriers, similar to phosphorus. For example, inversion barriers for NH_3 and AsH_3 are 6.0 and 39.2 kcal/mol, respectively. Unlike NH_3, which undergoes rapid umbrella inversion even at room temperature, the heavier hydrides maintain rigid pyramidal structures. Although acid-catalyzed inversion [7, 8] or cationic metal complexation [9–11] can lead to racemization at arsenic centers, free tertiary arsines are generally stereochemically stable under non-acidic, ambient conditions. This stability has enabled the use of As-stereogenic compounds in optically active transition metal complexes.

5.3 As—C Bond Formation

5.3.1 Conventional Methodologies

Traditional methods for forming As—C bonds primarily involve electrophilic or nucleophilic substitution, yielding a wide variety of arsenic-containing compounds. Trichloroarsine is a commonly used precursor in this context (Scheme 5.1a). Dichlorophenylarsine has been employed in the synthesis of functional

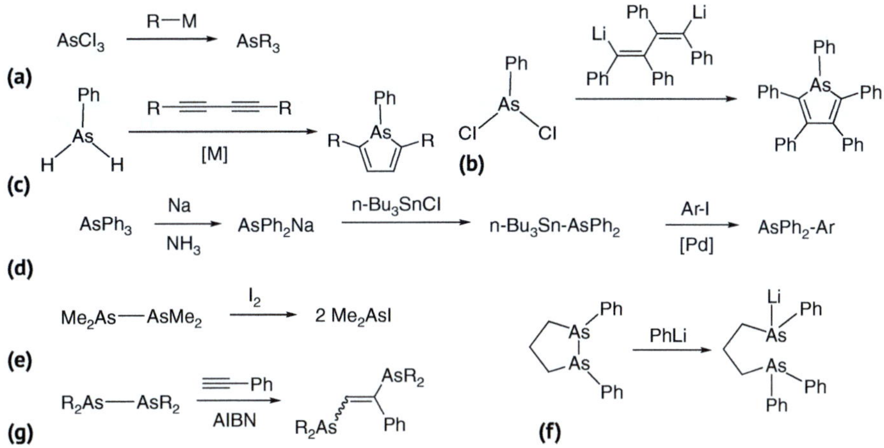

Scheme 5.1 Conventional C—As bond formation reactions.

organoarsenic compounds, such as arsole derivatives (Scheme 5.1b) [4]. Arsole skeletons can also be constructed via the lithiation of phenylarsine followed by addition to 1,4-diphenylbutadiyne (Scheme 5.1c) [5]. Despite their utility, arsenic halides and hydrides are hazardous due to their volatility and toxicity, discouraging extensive experimental studies in organoarsenic chemistry. A Pd-catalyzed cross-coupling reaction between aryl iodides and stannylarsanes has been reported (Scheme 5.1d) [12]. In this reaction, triphenylarsine is reacted with sodium in liquid ammonia to generate a Ph_2As^- anion, which undergoes nucleophilic substitution with tributyltin chloride to produce n-$Bu_3SnAsPh_2$. Subsequent cross-coupling of n-$Bu_3SnAsPh_2$ with aryl iodides in the presence of Pd catalyst yields various triarylarsines. As—As bond cleavage serves as an effective strategy for generating reactive arsenic species. For example, Me_2As—$AsMe_2$ can be converted to Me_2AsI by reacting with iodine (Scheme 5.1e) [13]. Kauffmann et al. demonstrated the preparation of an arsenic nucleophile via As—As bond cleavage of a diarsine compound using phenyllithium (Scheme 5.1f) [14]. Bismetallation, a process involving the cleavage of an element–element bond and the simultaneous addition of two heteroatoms to an organic compound, is a particularly attractive methodology. For instance, radical reactions of tetraphenyldiphosphine with phenylacetylene, initiated by azobisisobutyronitrile (AIBN) or irradiation, yield the corresponding 1,2-disubstituted ethylenes (Scheme 5.1g) [15].

5.3.2 In Situ Generation of Organoarsenic Electrophile from Arsenic Homocycles

In situ iodination of arsenic homocycles represents a practical and efficient method for As—C bond formation. This approach involves the quantitative generation of diiodomethylarsine and diiodophenylarsine from *cyclo*-$(AsMe)_5$ and *cyclo*-$(AsPh)_6$, respectively, with iodine (Scheme 5.2a) [16]. The key advantage of this method lies in the avoidance of isolating and purifying the volatile diiodoarsine intermediates, as no salts or by-products are formed. The resulting solution can be directly utilized in subsequent substitution reactions with anionic reagents such as organometallic

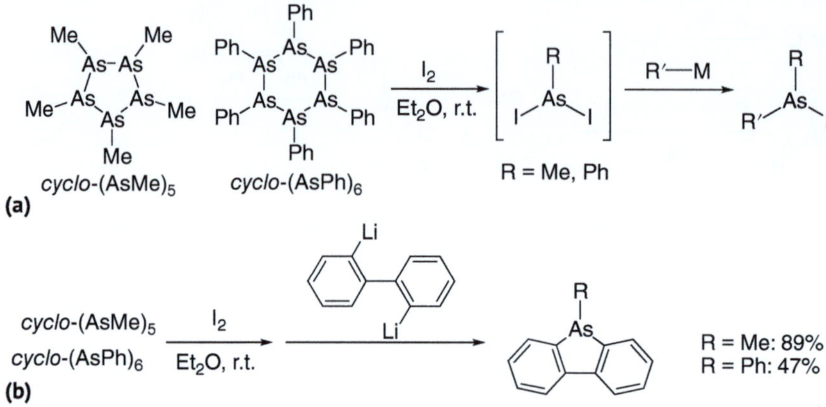

Scheme 5.2 Generation and utilization of diiodoarsines from *cyclo*-$(AsMe)_5$ and *cyclo*-$(AsPh)_6$.

reagents. Importantly, this synthetic route eliminates the need for volatile intermediates derived from starting materials like arsonic acids, enhancing safety and practicality. For example, the reaction of diiodomethylarsine and diiodophenylarsine with 2,2′-dilithiobiphenyl efficiently produces 9-arsafluorene derivatives, such as 9-methyl-9-arsafluorene (89% yield) and 9-phenyl-9-arsafluorene (47% yield) (Scheme 5.2b). This protocol has also been extended to polymer synthesis, leading to the preparation of arsole-containing polymers (see Sections 5.4 and 5.6 for details).

5.3.3 In Situ Generation of Organoarsenic Nucleophile from Arsenic Homocycles

While in situ iodination of arsenic homocycles effectively generates diiodoarsines for synthesizing symmetrical $R^1R^2_2As$ compounds, it is less suitable for constructing unsymmetrical ($R^1R^2R^3As$) or bidentate ($R^1R^2As—R^3—AsR^1R^2$) structures, which are commonly used as ligands. An alternative approach involves generating organoarsenic nucleophiles via As—As bond cleavage of organoarsenic homocycles with organolithium or Grignard reagents (Scheme 5.3) [17]. For example, dropwise addition of a THF solution of phenyllithium into a dispersion of *cyclo*-$(AsPh)_6$ in THF produces a homogeneous wine-red solution. Subsequent reaction with benzyl chloride yields the desired product. Various organolithium and Grignard reagents can be employed for this methodology.

5.3.4 Preparation and Utilization of Tribromoarsine

Trichloroarsine ($AsCl_3$) is conventionally used to synthesize organoarsenic compounds through nucleophilic substitution. However, its high volatility (vapor pressure 10 mmHg at 23.5 °C) [18] poses safety concerns. Tribromoarsine ($AsBr_3$) and triiodoarsine (AsI_3) are promising alternatives due to their significantly lower volatility (vapor pressures: 1 mmHg for $AsBr_3$ at 41.8 °C and 2.9 mmHg for AsI_3 at 164.0 °C) [18, 19]. A preparation method was developed to synthesize $AsBr_3$ and AsI_3 from As_2O_3 using hydrobromic acid (HBr) and hydroiodic acid (HI),

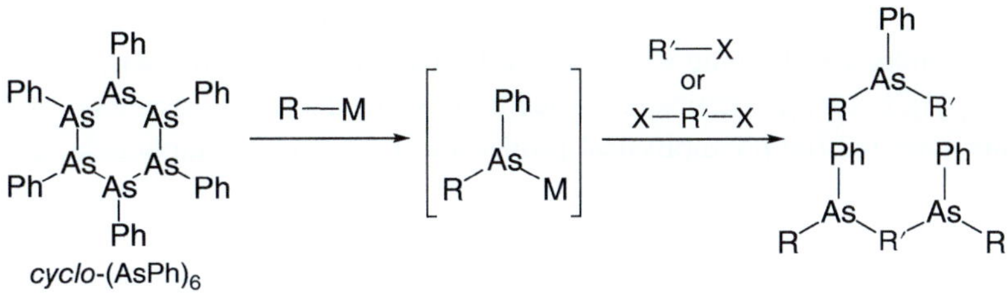

Scheme 5.3 In situ generation of nucleophilic organoarsine and subsequent nucleophilic substitution.

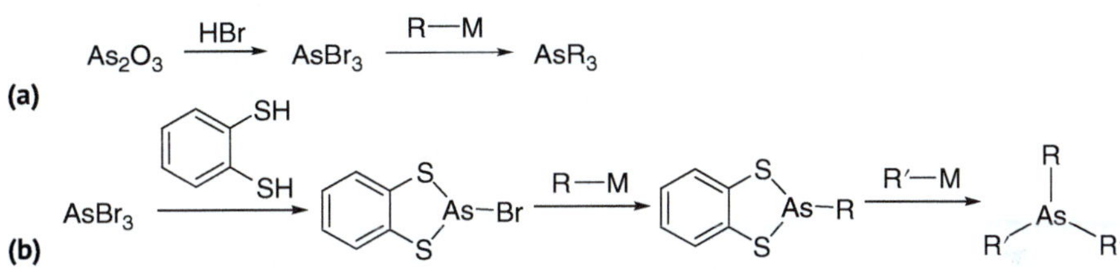

Scheme 5.4 Synthesis and utilization of tribromoarsine ($AsBr_3$).

respectively [20]. $AsBr_3$ can be extracted with *n*-hexane, yielding a high isolated yield of 92%. In contrast, the low solubility of AsI_3 necessitates filtration to collect the precipitate, resulting in a lower yield of 33%. Unlike AsI_3, $AsBr_3$ dissolves readily in various organic solvents (e.g. hexane, Et_2O, THF, toluene), making it a non-volatile precursor for organoarsenic compounds. $AsBr_3$ is particularly suitable for synthesizing A_3-type (C_3-symmetric) compounds, introducing three identical substituents (Scheme 5.4a).

It is challenging to introduce two different substituents to produce AB_2-type (C_s-symmetric) compounds. For phosphines, stepwise nucleophilic substitution of trichlorophosphine is often successful for AB_2-type synthesis. In contrast, the relatively long As—C bond reduces steric hindrance, leading to over-reaction and formation of A_3 products. A novel strategy overcomes this limitation by leveraging the high reactivity of arsenic halides with thiols to form As—S bonds (Scheme 5.4b) [21]. For example, $AsBr_3$ reacts with benzenedithiol to produce a dithiaarsole bromide intermediate. This bromide can undergo stepwise substitution: the first nucleophile reacts with the bromide moiety, followed by substitution of the chelating benzenedithiol group with a second nucleophile. This approach achieves complete selectivity, enabling efficient synthesis of AB_2-type arsines, including arsa-Buchwald ligands [22].

5.4 Arsenic-containing π-conjugated Systems

5.4.1 Arsole Derivatives

Five-membered heterocyclic compounds known as "heteroles" have been extensively studied to incorporate heteroatoms into π-conjugated systems. The arsenic derivative of heteroles is termed "arsole." Historically, arsoles were synthesized using volatile and toxic arsenic precursors, limiting detailed studies of their properties [4, 5]. For example, arsafluorene was first synthesized from dichlorophenylarsine in the 1920s [23], but its properties remained largely unexplored. Reciszewski and Yasuike later reported the photophysical properties of 1,2,3,4,5-pentaphenylarsole [24] and benzoarsoles [25], respectively, though they also relied on volatile dichlorophenylarsine as a precursor.

Safer and more practical methods have since been developed. Using diiodoarsines generated *via* iodine addition to cyclooligoarsines, 9-phenylarsafluorene and 9-methylarsafluorene can be synthesized under safer conditions [16]. Additionally, 1,2,5-triphenylarsole and its derivatives have been prepared from titanacyclopentadienes, which are generated by reacting phenylacetylene derivatives with low-valent titanium(II) complexes (Scheme 5.5) [26, 27]. Oxidative stability studies showed that 1,2,5-triphenylarsole is more resistant to oxidation than its phosphole analogue. After exposure to dry air for three hours, only 2% of 1,2,5-triphenylarsole was oxidized, compared to 12% for 1,2,5-triphenylphosphole. Photophysical studies revealed emission maxima of 458 and 465 nm for the arsole and phosphole, respectively.

Arsoles with extended π-conjugation, such as 4-phenyldithieno[3,2-*b*:2′,3′-*d*]arsoles, were reported by Baumgartner and collaborators in 2016 (Scheme 5.6) [28]. Dibromination of dithienoarsoles using *N*-bromosuccinimide (NBS), followed by Pd-catalyzed coupling reactions, yielded π-extended dithienoarsoles, including polymers (see Section 5.6 for detail). These compounds exhibit intense emission properties, with their colors tunable via oxidation or coordination at the arsenic atom.

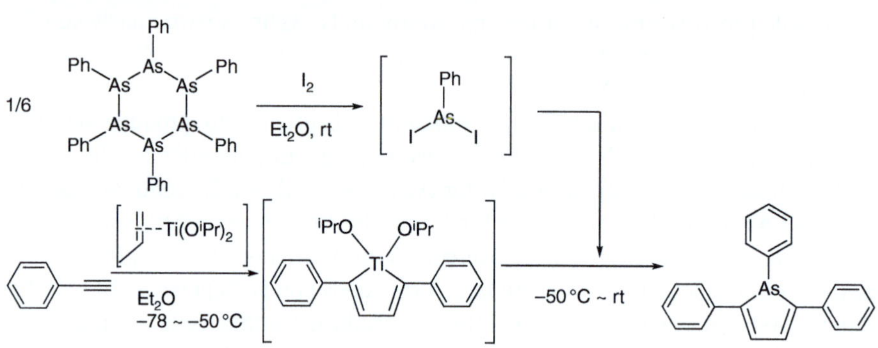

Scheme 5.5 Synthesis of 1,2,5-triphenylarsole.

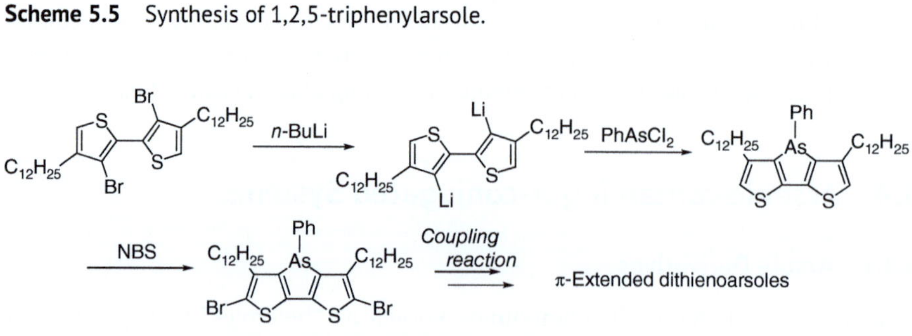

Scheme 5.6 Synthesis and modification of dithienoarsole.

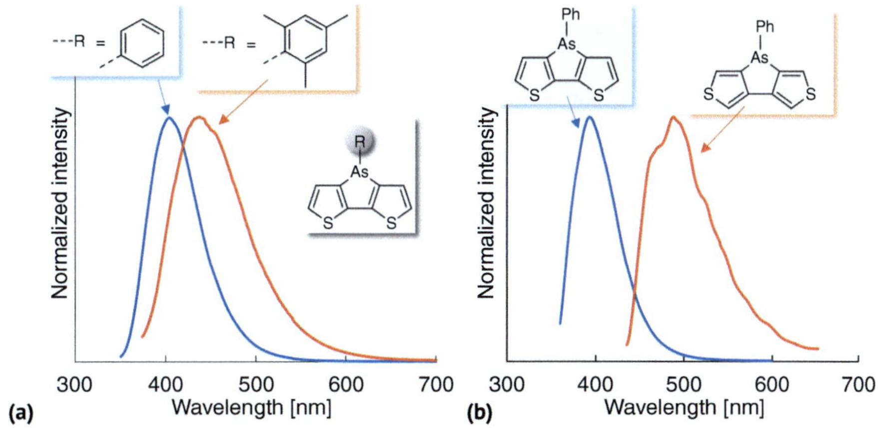

Figure 5.1 Photoluminescence spectra (solutions) of dithienoarsoles for examining (a) substituent of arsenic atom (298 K) and (b) fusing manner (77 K).

Computational studies revealed that arsenic primarily contributes to the lowest unoccupied molecular orbital (LUMO) via σ*–π* interaction, similar to phospholes. Oxidation or coordination at the arsenic center stabilizes the LUMO, resulting in a bathochromic shift in absorption and emission spectra.

The substituents of dithienoarsole significantly affect the photophysical properties [29–32]. In particular, the introduction of sterically hindered substituents changes the photophysical properties of dithienoarsoles (Figure 5.1a) [31]. Bulky substituents distort the conjugated system, leading to a bathochromic shift in emission spectra, as the distortion relaxes upon photoexcitation due to elongation of As—C bonds. Furthermore, dithieno[3,4-*b*:3′,4′-*d*]arsoles, isomers of dithieno[3,2-*b*:2′,3′-*d*]arsoles, exhibit phosphorescence at 77 K, whereas the latter only shows fluorescence (Figure 5.1b) [32].

Solid-state luminescence of arsoles has also been explored. 1,2,3,4,5-Pentaarylarsoles demonstrate aggregation-induced emission (AIE) in the solid state (Figure 5.2a) [33]. While 1,2,3,4,5-pentaphenylarsole shows low quantum yield in solution ($\Phi = 0.04$) due to exciton deactivation by phenyl group motion, solid-state restriction of these motions enhances quantum yield ($\Phi = 0.61$) to values comparable to pentaphenylsilole. Cycloalka[*c*]arsoles have been synthesized via the reaction of $PhAsI_2$ with zirconacyclopentadienes [34, 35]. The stability of these arsoles is influenced by the fused cycloalkane structure; for example, cyclopentane-fused arsoles are stable under ambient conditions, whereas cyclohexane-fused arsoles decompose in solution (Figure 5.2b). Decomposition occurs via reaction with singlet oxygen $(^1O_2)$ generated under light, releasing phenylarsonic acid and forming diketones. This decomposition pathway allows for the chemical recycling of arsenic components to synthesize the original arsoles (Figure 5.2c) [36].

Dipyridinoarsoles exhibit unique chemoselective reactivity [37]. The arsenic atom is preferentially oxidized or coordinated, while the nitrogen atoms undergo quaternization to form viologen structures. The resulting dipyridinoarsoles exhibit reversible electrochromism, in contrast to their less stable phosphorus analogues

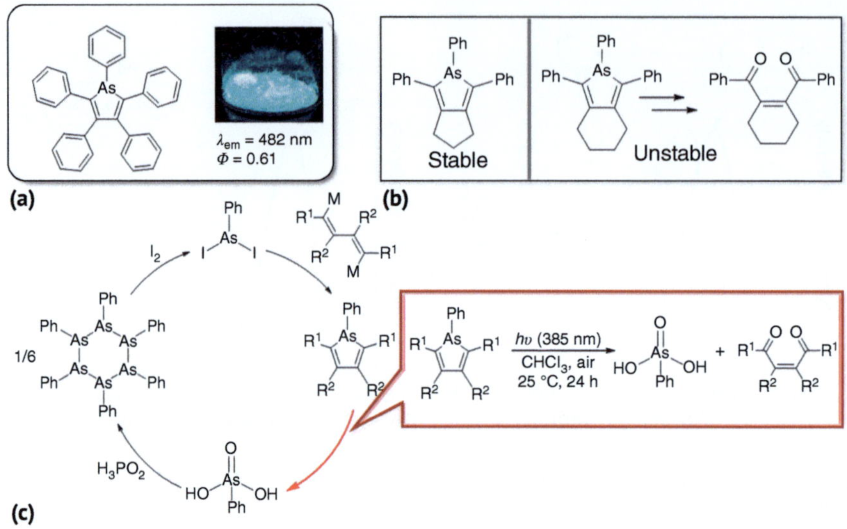

Figure 5.2 (a) Chemical structure and solid-state emission of 1,2,3,4,5-pentaphenylarsole. (b) Stability of cycloalka[*c*]arsoles. (c) Chemical recycling of arsenic component in arsole.

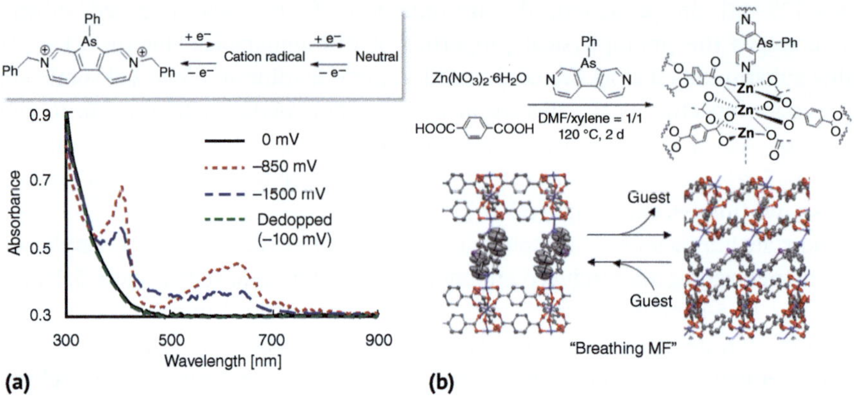

Figure 5.3 (a) Electrochromic behavior of arsenic-bridged viologen. (b) Breathing MOF constructed by dipyridinoarsole.

(Figure 5.3a). Additionally, the nitrogen atoms' hard Lewis basicity facilitates selective coordination to Zn^{2+} ions, enabling the construction of metal-organic frameworks (MOFs) without involving the arsenic atom [38]. A phenyl-substituted dipyridinoarsole forms a "breathing" MOF, where pores open and close in response to guest molecule interactions (Figure 5.3b). In contrast, methyl-substituted dipyridinoarsoles and phenyl-substituted dipyridinophospholes fail to form breathing MOFs, likely due to insufficient steric hindrance required to stabilize the closed-pore structure.

5.4.2 Other Arsenic-containing π-conjugated Systems

Heteropins, seven-membered heterocyclic compounds, include an arsenic derivative known as "arsepin." Unlike arsole, which has a 6π-electron system, arsepin features an 8π-electron system [39]. In the ground state, arsepin adopts a bent structure in accordance with Hückel's aromaticity rule. However, upon photoexcitation, the molecule planarizes due to Baird aromaticity [40], a phenomenon extensively studied in cyclooctatetraene derivatives. Dibenzoarsepin and its gold(I) chloride complex have been synthesized. Dibenzoarsepin exhibits dual fluorescence (371 and 557 nm), with emissions originating from the Franck–Condon structure and the planarized structure, respectively (Figure 5.4). In contrast, the gold(I) chloride complex shows only short-wavelength emission (375 nm). This difference arises because the arsenic lone pair in the complex is involved in coordination, preventing structural planarization and associated long-wavelength emission. The energy barrier between the two excited-state structures for dibenzoarsepin was determined to be 3.2 kJ/mol based on temperature-dependent photoluminescence spectra. Following this discovery, various Baird-aromatic heteropins, such as oxepins and azepins, have been reported.

Arsenic-bridged diphenyl sulfone exhibits unique excited-state dynamics [41]. Like triarylphosphines, triarylarsines planarize upon photoexcitation due to a reduction in electron density around the arsenic atom. This electron density decrease enhances the Lewis acidity of the arsenic atom, enabling photoinduced Lewis pair formation when a suitable Lewis base is incorporated into the molecular framework. In arsenic-bridged diphenyl sulfone, the arsenic (Lewis acid) and sulfone (Lewis base) moieties are positioned to interact upon excitation. The As···O distance is 3.18 Å in the ground state, shortening to 2.27 Å in the singlet excited state (Figure 5.5). This interaction significantly bends the central six-membered ring upon excitation. An energy barrier between the Franck–Condon and planarized

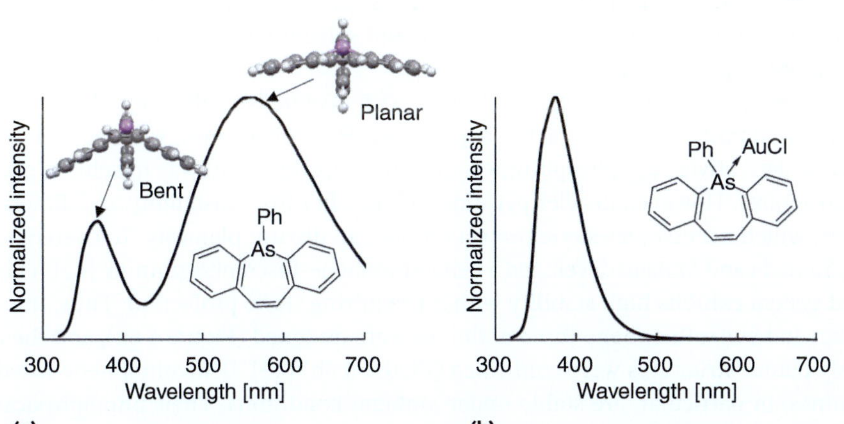

Figure 5.4 Photoluminescence spectra (solutions at 298 K) of (a) dibenzoarsepin and (b) its AuCl complex. The structures optimized in the S_0 (bent) and S_1 (planar) states are shown.

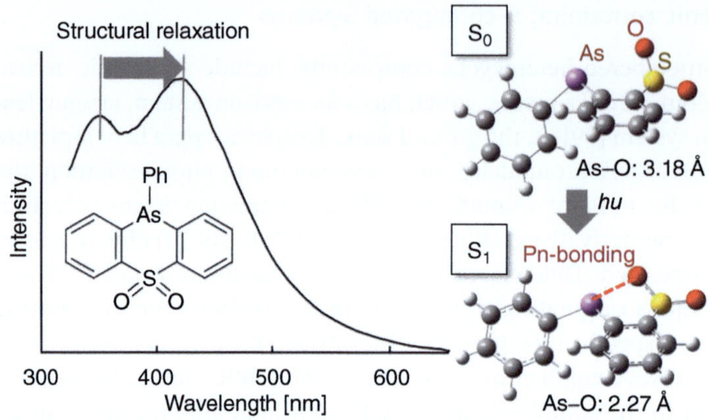

Figure 5.5 Photoluminescence spectrum of arsenic-bridged diphenyl sulfone and its structures optimized in the S_0 and S_1 states. As···O distances are shown.

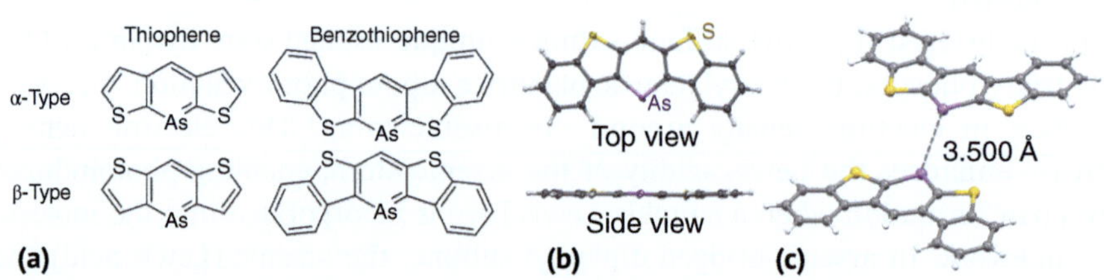

Figure 5.6 (a) Chemical structures of thiophene- and benzothiophene-fused arsinines. (b) Results of X-ray crystallography of β-type benzothiophene-fused arsinine and (c) α-type benzothiophene-fused arsinine dimer.

structures results in dual fluorescence. The emission spectrum is highly sensitive to environmental factors, such as temperature and solvent viscosity, offering tunable photophysical properties.

Designing planar π-conjugated molecules with arsenic is challenging due to the trigonal pyramidal structure of tertiary arsines. Heterobenzenes, such as heavier pyridine derivatives (e.g. phosphinine and arsinine), offer a pathway to achieve planar structures. However, heavier pyridines often suffer from instability and dimerization, which necessitates steric protection but can disrupt planarity. To overcome this, Yamada and Matsuo developed a benzothiophene-fused phosphinine [42]. This fused system exhibits high stability without requiring steric protection. Thus, thiophene- and benzothiophene-fused arsinines were designed (Figure 5.6a), and their highly planar structures were confirmed (Figure 5.6b) [43]. Benzothiophene-fused arsinines, in particular, are stable under ambient conditions. Their photophysical properties, including emission behavior, vary depending on the fused aromatic ring and the mode of fusion. In the solid state, α-type benzothiophene-fused arsinine displays intermolecular As···As interactions (Figure 5.6c), which are absent in their phosphorus analogues. This distinction arises from the larger atomic radius of arsenic compared to phosphorus. These intermolecular interactions provide unique opportunities for designing advanced materials with arsenic-containing conjugated systems.

5.5 Poly(vinylene-arsine)s

5.5.1 Synthesis of Poly(vinylene-arsine)s

Among various types of heteroatom-containing π-conjugated polymers, the simplest one is heteroatom-containing polyvinylenes. Poly(vinylene-arsine)s can be obtained by ring-collapsed radical alternating copolymerization (RCRAC) of the cyclic organoarsenic compounds, such as *cyclo*-$(MeAs)_5$ or *cyclo*-$(PhAs)_6$, with 5 and 6 equivalents of phenylacetylene (**1a**) against the As atom, respectively, using 2,2′-azobisisobutyronitrile (AIBN) as a radical initiator or irradiation with xenon lamp at room temperature (Scheme 5.7) (Table 5.1, runs 1, 2, and 3) [44]. These polymers are well-soluble in common organic solvents such as THF, chloroform, and benzene. The polymerization with 3 equivalents of **1a** to *cyclo*-$(PhAs)_6$ yielded a polymer with the same molecular weight and structure as when that with 6 equivalents of **1a** was used. Even the polymerization with one or 12 equivalents of **1a** to *cyclo*-$(PhAs)_6$ also gave polymers of almost the same molecular weight and structure. The alternating copolymer structures, poly(vinylene-arsine)s, form independently of the feed ratios of the two monomers, *cyclo*-$(PhAs)_6$ and **1a**.

Dichalcogenides, group 16 elements such as disulfide, diselenide, and diteruride, undergo chalcogen–chalcogen bond homolysis upon light irradiation or heating to generate the corresponding chalcogen radicals. It is well-known that in the presence of alkynes, biscalcogenation proceeds by a chain mechanism involving the formation of vinyl radicals by alkyne addition of chalcogen radicals and their capture reaction by dichalcogenides [45, 46]. Similarly, it is known that the radical reaction

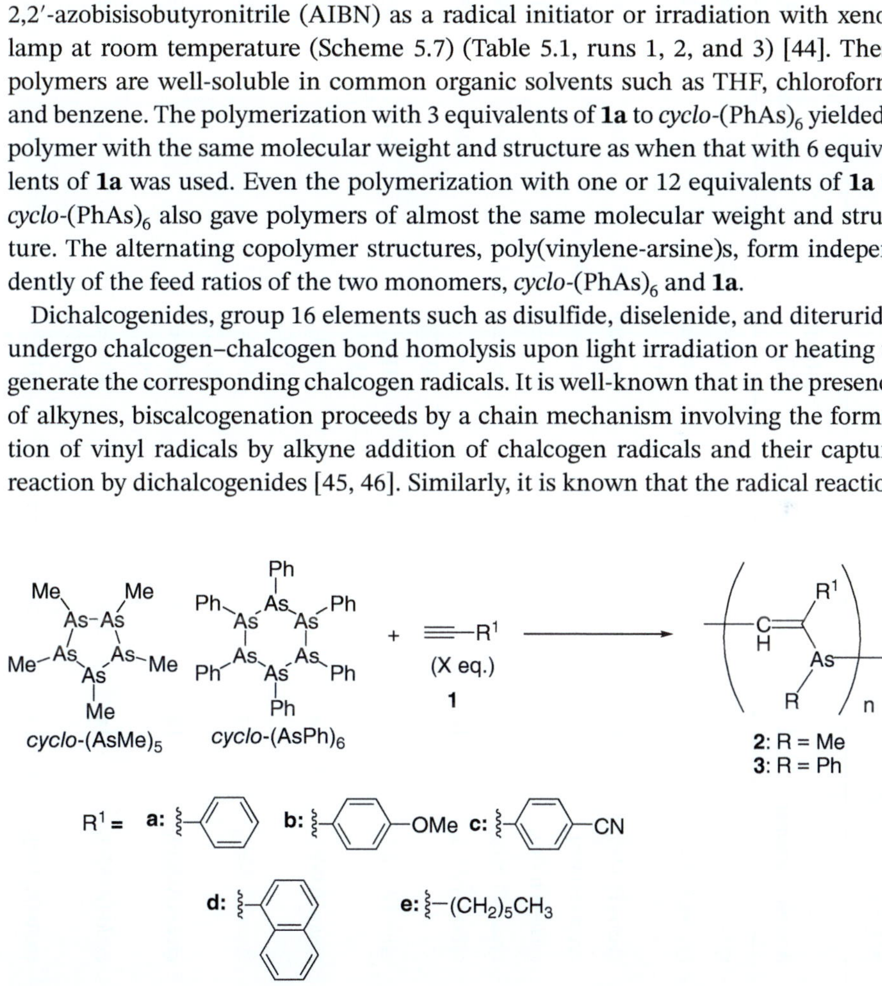

Scheme 5.7 Ring-collapsed radical alternating copolymerization (RCRAC) of cyclic organoarsenic compounds with mono-substituted acetylenic compounds (**1**).

Table 5.1 Results of polymerization and acetylenic compounds.

Run	Arsenic monomer	1	x[a]	Initiator[d]	Solvent (°C)	Polymer	M_w[b]	M_n[b]	M_w/M_n[b]	Yield (%)[c]
1	*cyclo*-$(AsMe)_5$	**1a**	5.0	AIBN	Benzene (reflux)	**2a**	48700	11500	4.3	46
2[e]	*cyclo*-$(AsMe)_5$	**1a**	5.0	hυ	Benzene (r.t.)	**2a**	11100	3400	303	48
3	*cyclo*-$(PhMe)_6$	**1a**	6.0	AIBN	Benzene (reflux)	**3a**	5600	3900	1.4	35
4	*cyclo*-$(PhMe)_6$	**1b**	6.0	AIBN	Benzene (reflux)	**3b**	15100	8400	1.8	29
5	*cyclo*-$(PhMe)_6$	**1c**	6.0	AIBN	Benzene (reflux)	**3c**	5100	3700	1.4	22
6	*cyclo*-$(PhMe)_6$	**1d**	6.0	AIBN	Benzene (reflux)	**3d**	9900	7600	1.3	33
7	*cyclo*-$(AsMe)_5$	**1e**	6.0	AIBN	None (60 °C)	**2e**	6500	7800	1.2	45
8	*cyclo*-$(AsMe)_5$	**1a**	5.0	no	$CHCl_3$ (25 °C)	**2a**	23300	11100	2.1	64
9	*cyclo*-$(AsMe)_5$	**1a**	5.0	no	$CHCl_3$ (0 °C)	**2a**	20800	9900	2.1	56
10	*cyclo*-$(AsMe)_5$	**1a**	5.0	no	$CHCl_3$ (50 °C)	**2a**	16700	8800	1.9	52
11	*cyclo*-$(AsMe)_5$	**1a**	5.0	no	Toluene (25 °C)	**2a**	13000	31200	2.4	60
12	*cyclo*-$(AsMe)_5$	**1a**	5.0	no	Benzene (25 °C)	**2a**	11300	24900	2.2	59
13	*cyclo*-$(AsMe)_5$	**1a**	5.0	no	THF (25 °C)	**2a**	15000	42000	2.8	58

14	*cyclo*-$(AsMe)_5$	**1a**	1.7	no	$CHCl_3$ (25 °C)	**2a**	13500	5200	2.6	17
15	*cyclo*-$(AsMe)_5$	**1a**	3.4	no	$CHCl_3$ (25 °C)	**2a**	14000	5200	2.7	42
16	*cyclo*-$(AsMe)_5$	**1a**	7.5	no	$CHCl_3$ (25 °C)	**2a**	30100	13100	2.3	52
17	*cyclo*-$(MeP)_5$	**1a**	5.0	AIBN	Benzene (reflux)	**4a**	2500	3200	1.3	46
18	*cyclo*-$(PhSb)_6$	**1f**	6.0	AIBN	THF (reflux)	**5f**	4300	6300	1.5	26

a. Molar ratio of **1** to the cyclic organoarsenic compounds.
b. GPC ($CHCl_3$). Polystyrene standards.
c. Isolated yields after reprecipitation into hexane based on the molar amount of 1.
d. 3 mol% of AIBN was employed.
e. Xenon lamp.

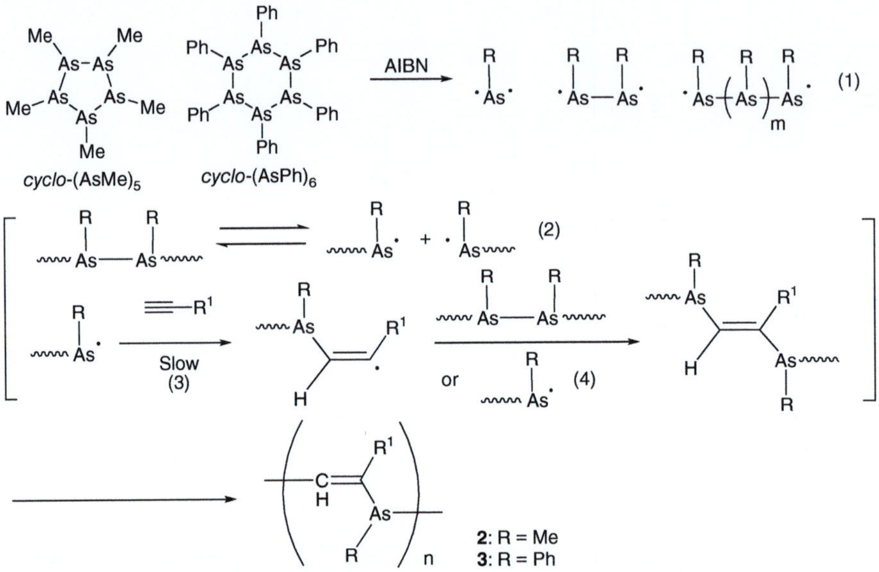

Scheme 5.8 Proposed reaction mechanism of ring-collapsed radical alternating copolymerization (RCRAC) of the cyclic organoarsenic compounds with acetylenic compounds (**1**).

of group 15 elements, tetraphenyldiphosphine and tetraphenyldialysine, with phenylacetylene yields the corresponding 1,2-substituted ethylene [15, 46–48]. Based on the mechanism of these radical bismetalations, the RCRAC mechanism has been proposed as follows (Scheme 5.8). AIBN or UV irradiation cleaved arsenic–arsenic bonds of the cyclooligoarsine to produce arsenic radicals. The homolysis of the other arsenic–arsenic bonds may proceed spontaneously due to instability caused by disruption of the stable five- or six-membered ring structure. In competition with this reaction, arsenic radicals add to acetylenic compounds to form vinyl radicals, which react with arsenic–arsenic bonds or with arsenic radicals to form new carbon–arsenic bonds. Repeated formation of arsine radicals by cleavage of the remaining arsenic–arsenic bonds and addition to the acetylene compound yields poly(vinylene-arsenic).

The aromatic stabilization of the vinyl radical is essential for the copolymerization. The copolymerization of *cyclo*-$(PhAs)_6$ with cyano-substituted phenylacetylene **1b** (Table 5.1, run 4) or naphthyl-substituted one **1d** (Table 5.1, run 6) was faster than that with non-substituted one **1a,** because the aromatic stabilization of the vinyl radical by 4-cyanophenyl or naphthyl group is stronger than that by phenyl group [49]. The more effectively aromatic stabilization acts on the vinyl radical, the more smoothly the copolymerization progresses. The copolymerization of *cyclo*-$(PhAs)_6$ with methoxy-substituted one **1c** (Table 5.1, run 5) was slightly slower than that with **1a**, suggesting that electrostatic effect of the substituent might affect the reaction between arsenic radical and carbon–carbon triple bond. The arsenic radical may prefer to react with electron-accepting monomers rather than electron-donating monomers due to the lone pair on the arsenic atom. Both conjugative and electrostatic effects of 4-cyanophenyl group made the fastest consumption of **1b** during the copolymerization with *cyclo*-$(PhAs)_6$ among the acetylenic monomers employed here. The polymerization proceeded to reach nearly quantitative conversion. The RCRAC of *cyclo*-$(MeAs)_5$ with aliphatic acetylenes in the presence of AIBN or an incandescent lamp irradiation has been achieved to produce the poly(vinylene-arsine)s with no aromatic substitution (Table 5.1, run 7), while *cyclo*-$(PhAs)_6$ did not copolymerize with aliphatic acetylenes [50]. The progress of the copolymerization of *cyclo*-$(MeAs)_5$ with aliphatic acetylenes was much slower than that with phenylacetylenes due to the lack of the conjugative substitution to the ethynyl group.

The arsenic ring compound with *cyclo*-$(MeAs)_5$ reacted with **1a** in chloroform at 25 °C in the absence of any radical initiators or catalysts to produce the corresponding poly(vinylene-arsine) in moderate yield (Table 5.1, run 8) [51]. The molecular weights and yields of the poly(vinylene-arsine) slightly decreased when the copolymerization was carried out at 0 °C or 50 °C (Table 5.1, runs 9 and 10). Various solvents were found usable for affording copolymers with number-average molecular weight of more than ten thousand in moderate yields at 25 °C (Table 5.1, runs 11–13). The copolymerization of *cyclo*-$(MeAs)_5$ with **1a** was conducted in chloroform at 25 °C in various monomer feed ratios (Table 5.1, runs 14–16). In all cases the copolymerization resulted in the corresponding alternating copolymer, poly(vinylene-arsine). The 1 : 1 alternating structure of MeAs and **1a** is controlled very strictly under a wide variety of copolymerization conditions. This copolymerization system showed a behavior of stepwise reaction polymerization in the conversion-molecular weight plot. The steep increase of the molecular weight of the polymer with the increase of the conversion rate of **1a** was observed in the range of high conversion. The present copolymerization can be regarded as a polyaddition of MeAs having biradical structure with phenylacetylene. Alternatively, an adduct of MeAs with acetylene works as AB monomer.

5.5.2 Properties of Poly(vinylene-arsine)s

Poly(vinylene-arsine)s showed remarkable air oxidation stability. The arsenic in **2a** existed in a trivalent state and no arsenic–arsenic bond or no oxidized arsenic was present even after isolation under air. No decrease of the molecular weight or no change of the structure was observed even after exposing them to air for several months. In order to examine the stability toward air moisture and oxidation more precisely, the chloroform solution of methyl-substituted poly(vinylene-arsine) was stirred in air atmosphere at (50 °C), and the molecular weight change was monitored by GPC. The GPC analysis showed no decrease of the molecular weight after stirring for 24 h. The ^{1}H NMR spectrum of the recovered copolymer indicated no change of the polymer structure. After **3a** was stirred vigorously with 30% H_2O_2 at 60 °C for 3 h, the ^{1}H-NMR spectrum and the GPC curve of **3a** were the same as those for the starting polymer. These results suggest that the trivalent state of arsenic in the main chain was insensitive to the oxidation.

The methyl-substituted poly(vinylene-arsine) showed 10% weight losses at 265 °C (under N_2) and 205 °C (under air), and the phenyl-substituted one showed them at

Table 5.2 Optical properties of polymers.

Run	Arsenic monomer	1	Polymer	Emission λ_{max} (nm)[a]	Excitation λ_{max} (nm)[a]
1	*cyclo*-$(PhMe)_6$	**1a**	**3a**	437	375
2	*cyclo*-$(PhMe)_6$	**1b**	**3b**	443, 466	394
3	*cyclo*-$(PhMe)_6$	**1c**	**3c**	402	341
4	*cyclo*-$(PhMe)_6$	**1d**	**3d**	350	315
5	*cyclo*-$(AsMe)_5$	**1a**	**2a**	485	396
6	*cyclo*-$(AsMe)_5$	**1e**	**2e**	450	350
7	*cyclo*-$(MeP)_5$	**1a**	**4a**	470, 500	400, 430
8	*cyclo*-$(PhSb)_6$	**1f**	**5f**	no	no

a. Recorded in dilute $CHCl_3$ solutions at room temperature.

284 °C (under N_2) and 250 °C (under air) (Figure 5.3). The poly(vinylene-arsine) with phenyl-substitution was thermally more stable than that with methyl-substitution. The glass transition temperatures (T_g) of the methyl- and phenyl-substituted polymers were 58 and 93 °C, respectively, determined by DSC analysis.

In UV-vis analyses, not only strong absorption in the UV region derived from the π–π* transition of the benzene ring but also small absorption in the visible region was observed, which is due to the *n*–π* transition in the main chain. Each emission peak maximum of **3a–c** was red-shifted in the order of **3c** < **3a** < **3b** (Table 5.2), which coincides with the order of strength of electron-withdrawing properties of the substituents. This indicates that the donor-acceptor (arsenic atom with a lone pair and vinylene unit with electron-accepting group) repeating units made the band gap narrower and resulted in the lower energy of the emission. The poly(vinylene-arsine)s with no aromatic substitution (**2e**) also showed the emission observed in the visible blue-violet region with a peak at 450 nm by an irradiation at 350 nm. In the case of **3d**, the naphthyl group itself has a fluorescent property that is stronger than that derived from the *n*–π* transition in the main chain of the polymer, and therefore the emission peak due to the main chain was hidden.

The polymer metal complexes were prepared by adding an acetone solution of **2a** or **3a** to an acetone solution of cobalt(II) chloride hexahydrate [52]. The methyl-substituted one **2a** formed the **2a**-Co complexes due to its less bulky structure. On the other hand, the phenyl-substituted one **3a** formed no cobalt complex due to its bulky structure. The methyl-substituted one **2a** also formed soluble **2a**-Ru complex as red solid by adding half equimolar amount of *cis*-$(bpy)_2RuCl_2{\cdot}2H_2O$ against the vinylene-arsine structure and **2a** in a refluxing deaerated $CHCl_3$ / ethanol (1/1). Polymer **3a** resulted in no or little complexation with ruthenium ions. The UV-vis absorption spectra of **2a**-Ru complex showed absorption peaks at around 470 nm, which is derived from metal-to-ligand charge transfer (MLCT) as well as a strong absorption due to the aromatic groups in the UV region, indicating the incorporation of the ruthenium complexes in the polymer.

The reduction of Ag ions in DMF took place spontaneously in the presence and the absence of **2a** or **3a** at 100 °C [53]. When no polymer was present in the system, the reduction led to silver deposition. On the other hand, in the presence of **2a** or **3a** in 1:1 ratio with $AgNO_3$, colloidal solutions were obtained. This means that poly(vinylene-arsine)s stabilized the silver nanoparticles. The UV-vis absorption spectra of the colloidal solutions indicated the formation of the silver nanoparticles with a surface plasmon absorption band at 400 nm. Polymer **2a** is a better stabilizer for the silver nanoparticles in DMF owing to its less bulky structure than **3a**.

5.5.3 Poly(vinylene-phosphine)s and Poly(vinylene-stibine)s

Phosphine and stibine analogues of poly(vinylene-arsine)s, poly(vinylene-phosphine) (**4a**) and poly(vinylene-stibine) (**5f**), were also prepared by RCRAC of pentamethylcyclopentaphosphine, *cyclo*-$(MeP)_5$, and hexaphenylcyclohexastibine, *cyclo*-$(PhSb)_6$, with **1a** and dimethyl acetylenedicarboxylate (**1f**), respectively (Table 5.1, runs 17 and 18) (Scheme 5.9) [54, 55]. The radical reaction of *cyclo*-$(MeP)_5$ with 4-cyanophenylacetylene (**1b**) and dimethyl acetylenedicarboxylate (**1f**) gave only oligomeric products. Radicals of lighter elements are expected to have less capture ability of vinyl radicals. The capture of the vinyl radicals is hardly achieved in RCRAC of *cyclo*-$(MeP)_5$ with electron-deficient acetylenes, which provide less reactive vinyl radicals. The phenyl-substituted cyclooligophosphine, *cyclo*-$(PhP)_5$, did not provide poly(vinylene-phosphine) due to low reactivity of phenyl-substituted phosphorous radicals toward the ethynyl group of **1a**.

Although RCRAC of *cyclo*-$(PhSb)_6$ and **1f** gave the polymer **5f**, no polymerization proceeded when *cyclo*-$(PhSb)_6$ was used in RCRAC system with **1f**. The phenylacetylene derivatives (**1a–1c**) gave only low molecular weight poly(vinylene-stibine)s. The effects of conjugation or electron-withdrawing of the substituent of acetylenic monomer promoted RCRAC. This result suggested that the arsenic radical preferred reacting with electron-accepting monomers due to the lone pair on the arsine atom. However, an excessive conjugative or electrostatic effect of the substitution of an acetylenic monomer such as **1f** decreased the reactivity of the vinyl radicals formed from the acetylenic compounds toward the arsenic radical due to increase stability

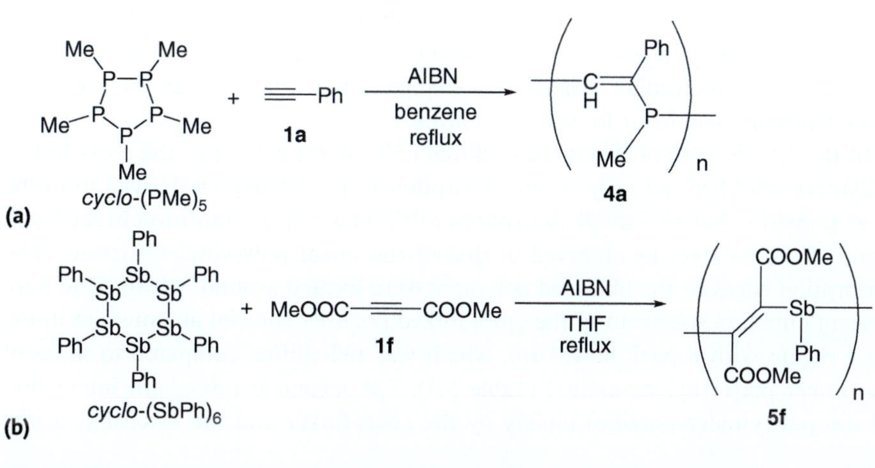

Scheme 5.9 RCRAC of (a) *cyclo*-$(MeP)_5$ with **1a** and (b) *cyclo*-$(PhSb)_6$ with **1f**.

of the vinyl radicals. The arsenic radicals have more reactivity to the ethynyl groups than the antimony radicals, and the antimony radicals have more reactivity toward the vinyl radicals. These differences between the cases of *cyclo*-$(PhSb)_6$ and *cyclo*-$(PhSb)_6$ are derived from the reactivity of the corresponding radicals.

Poly(vinylene-phosphine) (**4a**) is air- and moisture-stable in the solid state at room temperature. After being exposed to air in the solution state for several days, the polymer decomposed to show no polymeric peaks in a GPC trace. The fluorescence spectrum of a dilute chloroform solution of **4a** showed an emission peak at 470 nm, which derived from the n–π^* transition along the polymer backbone (Table 5.2). The excitation spectrum of **4a** showed peaks at 400 and 430 nm, monitored at 470 nm. The bathochromic shift observed in the excitation spectrum of **4a** compared with that of **2a** might be attributed to the conformation of phosphorus or arsenic atom in the main chain. The s-character of the lone pair on a phosphorous atom is weaker than that of an arsenic atom. This might lead to increase delocalization between the lone pair electron and the p electron of the vinylene unit.

The polymer **5f** was air-stable in the solid state at room temperature. No changes in the molecular weight and the structure were observed after exposing **5f** to air for several days. No emission was observed when **5f** was excited at 350 or 400 nm, while the poly(vinylene-arsine)s showed the emission. As the atomic number of pnictogen increases, the s character of the lone pair enhances. In polymer **5f,** the lone pairs are much localized on antimony element because of the stronger s-character. We assume interaction between the lone pair electron and the π electron of the vinylene unit decreases compared to the case of the poly(vinylene-arsine)s.

5.5.4 Cross-Linked Poly(vinylene-arsine)s

The radical terpolymerization of *cyclo*-$(PhAs)_6$, **1a**, and p- or m-diethynylbenzene (DVB) gave soluble conjugated cross-linked poly(vinylene-arsine)s by controlling the molar ratio of **1a** and DVB (Scheme 5.10) [56]. The glass transition temperatures (T_g) value increases with the content of the cross-linker (Table 5.3). When the feed ratio of the p- and m-DVB was the same, the polymers from p-DVB showed higher T_g than those from m-DVB. This might be due to increased average molecular weights of the cross-linked polymers by increasing efficiency of the cross-linking agents. All of the resulting polymers were stable in the solid state at room temperature. No decrease of the molecular weight or no change of the structure was observed even after exposing them to air for several months.

In the UV-vis spectra measured in dilute chloroform solutions, the cross-linked polymers exhibited not only strong absorption in the UV region derived from the π–π^* transition but also small absorptions attributed to n–π^* transition in the backbone of the polymer as observed in that of the linear poly(vinylene-arsine). The absorption edges of the obtained polymers were located around 500 nm. The fluorescent emission spectrum of the cross-linked polymer showed an emission in the blue region with a peak at 473 nm, which was red-shifted compared to those of the linear poly(vinylene-arsine) (Table 5.3). The favorable interchain interaction of the poly(vinylene-arsine) moiety by the cross-linker and the extension of the

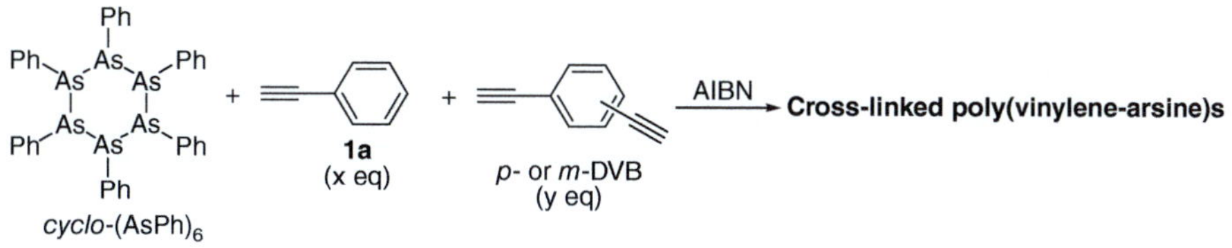

Scheme 5.10 Preparation of cross-linked poly(vinylene-arsine)s by the radical terpolymerization of *cyclo*-$(PhAs)_6$, **1a**, and *p*- or *m*-DVB.

Table 5.3 Results of radical terpolymerization of *cyclo*-$(PhMe)_6$, **1a**, and *p*- or *m*-diethynylbenzene.

Run	DVB	Feed ratio 1a/DVB	Yield (%)[a]	M_n[b]	M_w[b]	M_w/M_n[b]	T_g[c] (°C)	Emission λ_{max} (nm)[d]	Excitation λ_{max} (nm)[d]
1	no	6.0/0	27	1700	3100	1.8	93	437	375
2	*p*-DVB	5.0/0.5	29	1900	3700	1.9	93	470	392
3	*p*-DVB	3.0/1.5	51	2800	11,800	4.2	102	473	394
4	*p*-DVB	2.0/2.0	62	ND	ND	ND	105	ND	ND
5	*m*-DVB	3.0/1.5	48	2700	7800	2.9	99	465	392
6	*m*-DVB	2.0/2.0	54	2600	8000	3.1	104	464	391

a. Isolated yields after reprecipitation into *n*-hexane.
b. Determined by GPC analysis (DMF, polystyrene standards).
c. Recorded in dilute $CHCl_3$ solutions at room temperature.
d. Determined by DSC measurement.
ND = not determined.

effective conjugation length through the conjugated cross-linking bond resulted in the narrower band gap. The longer extension of the conjugation path by the *p*-substituted cross-linker than by the *m*-substituted one might be derived from the more efficient resonance effect of the conjugated *p*-substitution.

5.6 Arsole Polymers

Since the first report of poly(vinylene-arsine)s in 2002 [44], research on arsenic-containing conjugated polymers was largely dormant for over a decade. This changed in 2016 when dithienoarsole polymers were independently reported by Heeney and coworkers [28], as well as by the authors [29]. Heeney's group synthesized a dithienoarsole polymer ($M_n = 12{,}300$, $M_w/M_n = 1.9$) using a microwave-assisted Stille coupling reaction between dithienoarsole dibromide and *trans*-1,2-bis(tributylstannyl) ethane (Scheme 5.11a). This polymer, tested in top-gate/bottom-contact transistor configurations, exhibited a hole mobility of 0.08 $cm^2/V{\cdot}s$ and excellent air stability. The promising stability of the trivalent arsenic center suggests potential for developing

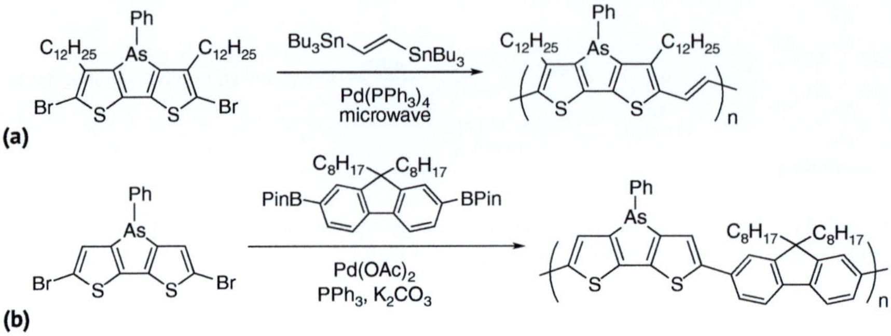

Scheme 5.11 Synthesis of dithienoarsole polymers.

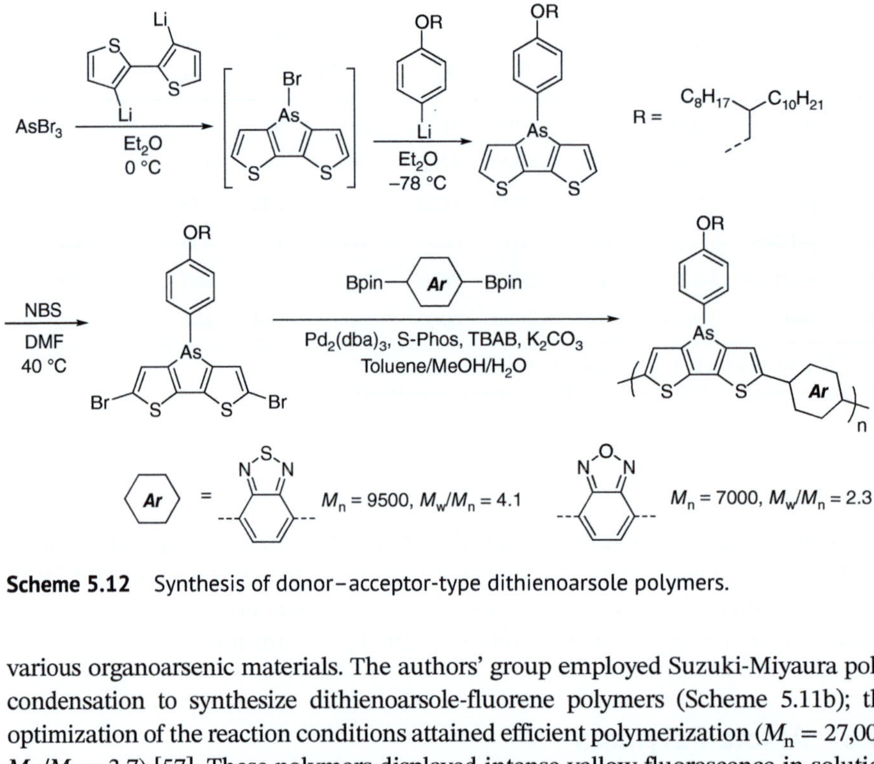

Scheme 5.12 Synthesis of donor–acceptor-type dithienoarsole polymers.

various organoarsenic materials. The authors' group employed Suzuki-Miyaura polycondensation to synthesize dithienoarsole-fluorene polymers (Scheme 5.11b); the optimization of the reaction conditions attained efficient polymerization ($M_n = 27{,}000$, $M_w/M_n = 3.7$) [57]. These polymers displayed intense yellow fluorescence in solution ($\lambda = 524$ nm, $\Phi = 0.52$) but experienced aggregation-caused quenching (ACQ) in the solid state ($\lambda = 575$ nm, $\Phi = 0.01$). Remarkably, despite arsenic's ability to coordinate to transition metals, residual catalysts were effectively removed due to arsenic's weaker coordination ability compared to phosphorus. Additionally, arsenic-containing polymers demonstrated reduced catalytic poisoning effects during the Suzuki-Miyaura reaction, unlike their phosphorus-based counterparts.

Functionalization of dithienoarsole monomers with long alkyl chains improved the solubility and planarity of the resulting polymers (Scheme 5.12) [58].

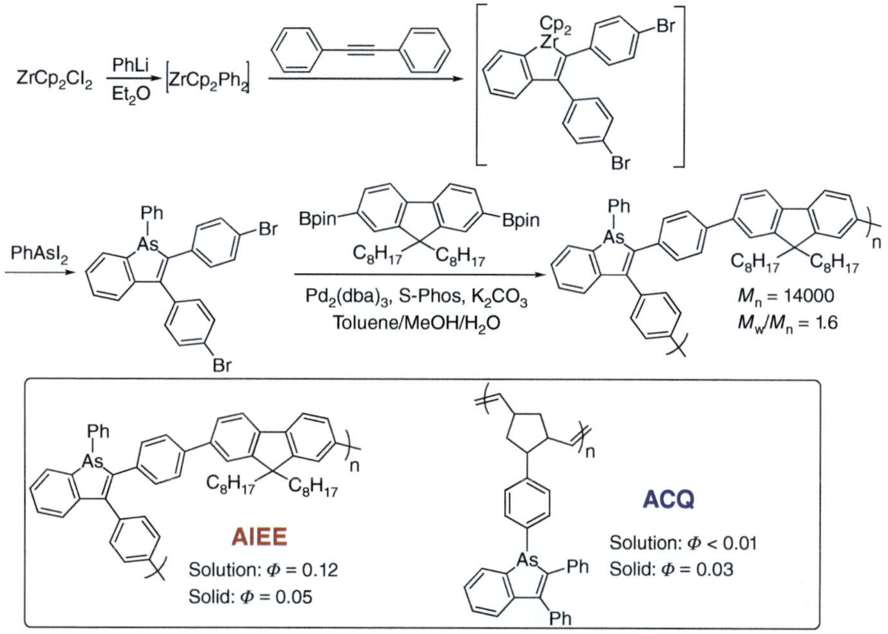

Scheme 5.13 Synthesis of 2,3-diarylbenzo[*b*]arsole polymer. The chemical structure should be regiorandom, but only the head-to-tail structure is shown here for clarity.

Copolymerization with benzothiadiazole and benzoxadiazole produced donor–acceptor-type polymers ($M_n = 9{,}500$ and 7,000, respectively), which exhibited near-infrared (NIR) emission. The dithienoarsole-benzoxadiazole copolymer reached an emission maximum of 886 nm, with moderate quantum yield ($\Phi = 0.01$).

A different monomer, 2,3-diarylbenzo[*b*]arsole dibromide, was synthesized from bis(*p*-bromophenyl)acetylene via zirconacycle intermediates (Scheme 5.13) [59]. Suzuki-Miyaura polycondensation with fluorene diboronic ester yielded polymers with highly bent structures ($M_n = 14{,}000$, $M_w/M_n = 1.6$). These polymers exhibited strong solution-phase fluorescence ($\Phi = 0.12$), though emission was quenched in the solid state ($\Phi = 0.05$). In contrast, side-chain-type polymers demonstrated aggregation-induced emission enhancement (AIEE), with quantum yields increasing in the solid state ($\Phi < 0.01$ in solution, $\Phi = 0.03$ in solid).

Heeney and coworkers also synthesized dithienoarsole polymers with stannylated benzodithiophene derivatives ($M_n = 14{,}000$–$20{,}000$, $M_w/M_n = 1.86$–1.93) for organic photovoltaic devices (Scheme 5.14) [60]. Blends of these polymers with $PC_{71}BM$ achieved moderate power conversion efficiencies (~4%) when alkyl chain lengths were optimized. The polymers were stable under ambient conditions, highlighting their practical utility.

Arsole polymers were further applied as photosensitizers for singlet oxygen ($^1O_2$) generation, driven by intersystem crossing from singlet to triplet states facilitated by arsenic's heavy atom effect [61]. The dithienoarsole polymer exhibited a $^1O_2$ generation quantum yield of 0.54, significantly higher than that of a bithiophene polymer

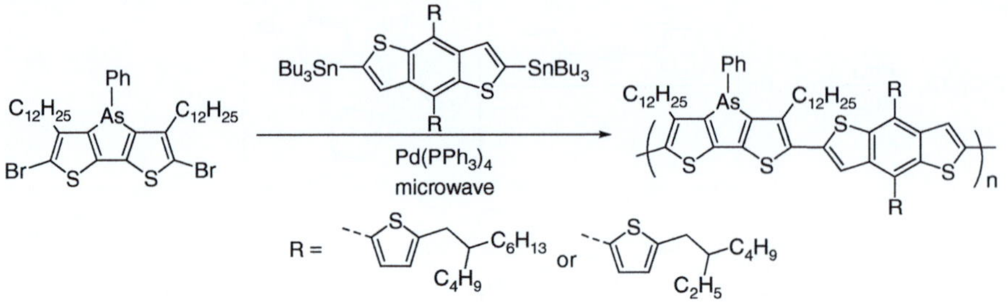

Scheme 5.14 Synthesis of dithienoarsole polymers for organic photovoltaic devices.

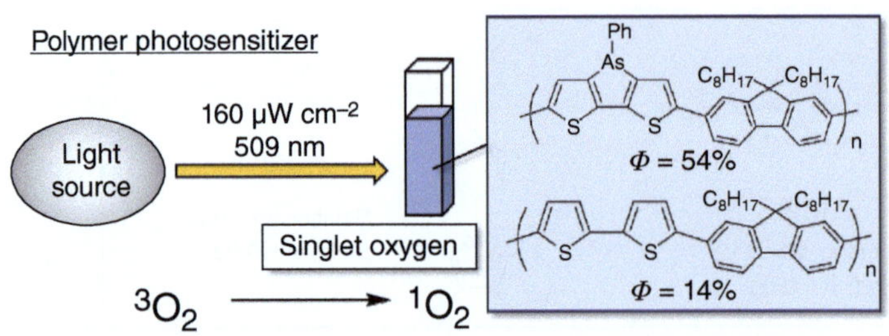

Figure 5.7 Singlet oxygen generation by polymer photosensitizer.

without arsenic bridging ($\Phi = 0.14$) (Figure 5.7). The dithienoarsole polymer also demonstrated excellent resistance to degradation by reactive $^1O_2$. The stability of arsole polymers was further showcased in amplified spontaneous emission (ASE) applications [62]. Dithienoarsole copolymers with bithiophene and diethynylbenzene exhibited ASE with exceptional durability under laser excitation, outperforming widely used laser dyes like rhodaminc 6G. In contrast, low-molecular-weight arsole derivatives and non-arsenic-containing polymers did not exhibit ASE, underscoring the importance of the polymeric structure.

Tomita and coworkers developed a "post-element transformation technique" to efficiently synthesize heterole-containing polymers [63]. For instance, phosphole-containing polymers were obtained by reacting titanacyclopentadiene polymers (formed in situ from aromatic diynes and low-valent titanium complexes) with dichlorophenylphosphine [64]. Similarly, reaction with diiodophenylarsine yielded arsole-containing polymers ($M_n = 5{,}500$, $M_w/M_n = 2.5$) (Scheme 5.15) [65]. Cyclic voltammetry (CV) revealed that arsole-containing polymers exhibited quasi-reversible oxidation, with consistent oxidation peaks across multiple sweeps. In contrast, phosphole-containing polymers underwent irreversible structural changes after the first oxidation sweep, reflecting the superior oxidative resistance of arsenic.

A film of dithienoarsole polymer was fabricated via electropolymerization (Figure 5.8) [66]. Using tetraethylammonium tetrafluoroborate (Et_4NBF_4) as the electrolyte, the high oxidation potential of the dithienoarsole monomer (1.22–1.39 V) initially hindered polymerization. However, the addition of a BF_3-Et_2O complex lowered the oxidation potential (0.49 V), likely due to π-complex

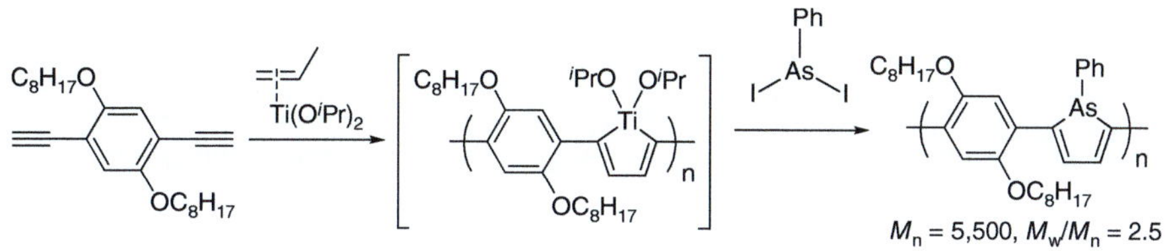

Scheme 5.15 Synthesis of arsole polymer by post-element transformation technique.

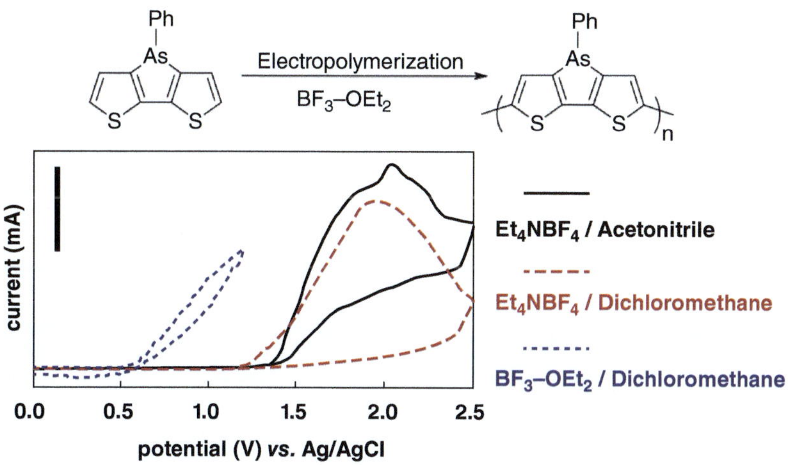

Figure 5.8 Cyclic voltammograms of dithienoarsole for electropolymerization.

formation with the monomer. This enabled successful fabrication of a polymer film on an ITO electrode. UV-Vis absorption spectra and X-ray photoelectron spectroscopy confirmed the formation of the polymer.

5.7 π-Conjugated Polymers with Other Arsenic Units

Orthaber and coworkers reported the synthesis of an arsaalkene-containing polymer [67]. Fluorene dibromide was lithiated, and subsequent reaction with $MesAsCl_2$ produced an arsaalkene dibromide (Scheme 5.16). Microwave-assisted Stille coupling with 2-(tributylstannyl)thiophene yielded a thienyl-substituted arsaalkene monomer. Electropolymerization of this monomer produced a polymer film with electrochromic properties. The film's color reversibly transitioned between neutral and oxidized states, with the oxidized state exhibiting a stable dark blue color for several days under inert conditions, demonstrating the stability of the doped polymer species.

Triphenylarsine, one of the simplest arsenic-containing conjugated molecules, was also utilized in polymer synthesis. For the preparation of the dibromide monomer, bis(*p*-bromophenyl)phenylarsine, $PhAsI_2$ was reacted with *p*-bromolithiobenzene (Scheme 5.17) [68]. Suzuki-Miyaura polycondensation of this monomer with

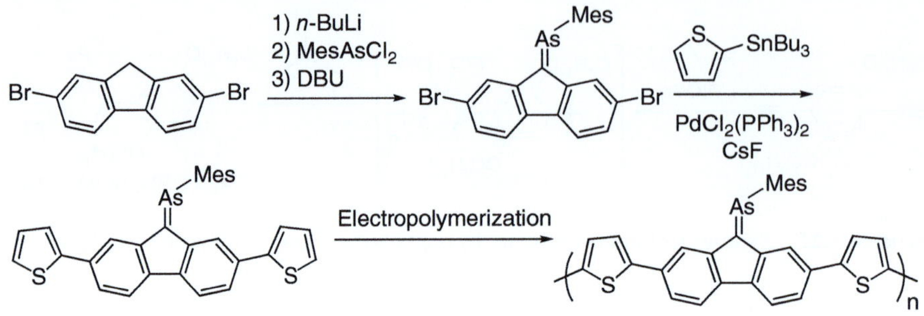

Scheme 5.16 Synthesis of arsaalkene polymer.

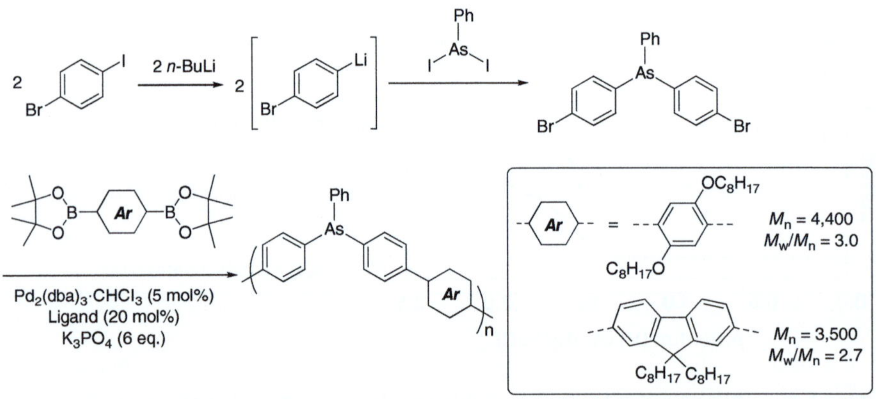

Scheme 5.17 Synthesis of triphenylarsine polymer.

benzene and fluorene diboronic ester comonomers resulted in conjugated polymers ($M_n = 3{,}500$, $M_w/M_n = 2.7$) exhibiting fluorescence ($\lambda = 405$ nm, $\Phi = 0.65$). Notably, the conjugation system extended through the arsenic atom, enhancing the overall electronic properties.

5.8 Conclusion

Since the discovery of the first organometallic compound, "cacodyl," in the seventeenth century, organoarsenic chemistry has significantly expanded, yielding a wide variety of arsenic derivatives. Despite this progress, many researchers remain reluctant to conduct experimental studies involving arsenic compounds due to the need for volatile and toxic precursors. However, the development of safe and efficient methods for As—C bond formation has paved the way for further advancements in functional organoarsenic chemistry. This chapter has summarized the fundamentals of organoarsenic chemistry and highlighted several practical methods for As—C bond formation. Notably, organoarsenic radicals, electrophiles, and nucleophiles can now be generated from non-volatile precursors such as arsenic homocycles and As_2O_3. These reactive species have enabled the synthesis of functional organoarsenic compounds, including conjugated molecules.

One particularly useful method is radical-mediated bismetallation of carbon–carbon triple bonds via the homolytic cleavage of element–element bonds. This atom-economical approach has been applied to produce poly(vinylene-arsine)s, marking a significant milestone in arsenic-based polymer chemistry. However, after the initial report on poly(vinylene-arsine)s in 2002, research on other arsenic-containing conjugated polymers remained stagnant for over a decade. The breakthrough came in 2016 with the development of dithienoarsole polymers, reigniting interest in this field. Since then, various arsole- and arsenic-containing polymers have been reported, showcasing unique properties such as luminescence, photosensitization, hole transport, and electrochromism, as well as high oxidative resistance. These emerging classes of conjugated polymers hold great promise for advancing the field of highly functional organic materials.

References

1 Krannich, L.K. and Watkins, C.L. (1994). *Encyclopedia of Inorganic Chemistry* (ed. R.B. King), Vol. 1, p. 200. Chichester, England: John Wiley & Sons.

2 Ehrlich, P. and Bertheim, A. (1912). Über das salzsaure 3,3′-Diamino-4,4′-dioxyarsenobenzol und seine nächsten Verwandten. *Berichte der Deutschen Chemischen Gesellschaft* 45: 756.

3 Lloyd, N.C., Morgan, H.W., Nicholson, B.K. et al. (2005). The composition of Ehrlich's salvarsan: resolution of a century-old debate. *Angewandte Chemie International Edition* 44: 941.

4 Leavitt, F.C., Manuel, T.A., and Johnson, F. (1959). Novel heterocyclo pentadienes. *Journal of the American Chemical Society* 81: 3163.

5 Märkl, G., Hauptmann, H., and Merz, A. (1983). Synthese von 1-phenyl-2, 5-diaryl (dialkyl)-arsolen; umsetzung der arsole mit alkalimetallen und lithiumorganylen. *Journal of Organometallic Chemistry* 249: 335.

6 Downing, J.H. and Smith, M.B. (2004). Phosphorus ligands. In: *Comprehensive Coordination Chemistry II* (ed. J.A. McCleverty and T.J. Meyer), Vol. 1, p. 253. Amsterdam, The Netherlands: Elsevier Science.

7 Campbell, I.G.M. and Poller, R.C. (1956). The synthesis and stereochemistry of some tervalent arsenic compounds. *Journal of the Chemical Society* 1956: 1195.

8 Horner, L. and Hofer, W. (1966). Photoracemisierung von optisch aktiven tertiären arsinen. *Tetrahedron Letters* 1966: 3323.

9 Kerr, P.G., Leung, P.-H., and Wild, S.B. (1987). Optically active arsenic macrocycles. Stereospecific syntheses of enantiomers and diastereomers of 14-membered trans-As2S2 chelating macrocycles containing resolved asymmetric tertiary arsine donors. *Journal of the American Chemical Society* 109: 4321.

10 Bosnich, B., Jackson, W.G., and Wild, S.B. (1973). Metal complexes of dissymmetric arsines. Stereochemistry, topological stability, and spectra of cobalt (III) complexes containing a linear quadridenate tetra-tert-arsine ligand. *Journal of the American Chemical Society* 95: 8269.

11 Bosnich, B., Jackson, W.G., and Wild, S.B. (1974). Dissymmetric arsine complexes. I. Preparation and properties of a series of cobalt (III) complexes containing a linear quadridentate arsine. *Inorganic Chemistry* 13: 1121.

12 Bonaterra, M., Martín, S.E., and Rossi, R.A. (2003). One-pot palladium-catalyzed cross-coupling reaction of aryl iodides with stannylarsanes and stannylstibanes. *Organic Letters* 5: 2731.

13 Montimer, C.T. and Skinner, H.A. (1952). The thermochemistry of organo-arsenic compounds. *Journal of the Chemical Society* 1952: 4331.

14 Kauffmann, T., Berghus, K., and Ennen, J. (1985). Vielelektronenliganden, XII. Übergangsmetall-Liganden durch Umsetzung von Spiro [2.4] hepta-4, 6-dien und Spiro [4.4] nona-1, 3-dien mit Lithium-arseniden. *Chemische Berichte* 118: 3724.

15 Tzschach, A. and Baensch, S. (1971). Organoarsen-Verbindungen. XVII. Zur Umsetzung von Biphosphinen und Biarsinen mit Phenylacetylen. *Journal für Praktische Chemie* 313: 254.

16 Kato, T., Tanaka, S., and Naka, K. (2015). In-situ iodination of organoarsenic homocycles: facile synthesis of 9-arsafluorene. *Chemistry Letters* 44: 1476.

17 Tanaka, S., Imoto, H., Kato, T. et al. (2016). A practical method for the generation of organoarsenic nucleophiles towards the construction of a versatile arsenic library. *Dalton Transactions* 45: 7937.

18 Irving, S.N. (1968). *Dangerous Properties of Industrial Materials*, 3e, p. 437. New York: Reinhold Book Corporation.

19 Cubicciotti, D. and Eding, H. (1965). The vapor pressure and enthalpy of arsenic triiodide and its absolute entropy at 298°K. *Journal of Physical Chemistry* 69: 2743.

20 Tanaka, S., Konishi, M., Imoto, H. et al. (2020). Fundamental study on arsenic(III) halides (AsX_3; X = Br, I) toward the construction of C_3-symmetrical monodentate arsenic ligands. *Inorganic Chemistry* 59: 9587.

21 Sumida, A., Imoto, H., and Naka, K. (2022). Synthetic strategy for AB_2-type arsines via bidentate dithiolate leaving groups. *Inorganic Chemistry* 61: 17419.

22 Sumida, A., Ogawa, K., Imoto, H. et al. (2023). Steric and electronic effects of arsa-Buchwald ligands on Suzuki–Miyaura coupling reaction. *Dalton Transactions* 52: 2838.

23 Aeschlimann, J.A., Lees, N.D., McCleland, N.P. et al. (1925). X.—Organic compounds of arsenic. Part II. Derivatives of the arsenic analogue of carbazole. *Journal of the Chemical Society, Transactions* 127: 66.

24 Raciszewski, Z. and Braye, E.H. (1970). Fluorescence of phenyl-substituted heterocyclopentadienes: comparison of the effects of N, P, As, P→O and of the number of phenyl substituents. *Photochemistry and Photobiology* 12: 429.

25 Muranaka, A., Yasuike, S., Liu, C.Y. et al. (2009). Effect of periodic replacement of the heteroatom on the spectroscopic properties of indole and benzofuran derivatives. *Journal of Physical Chemistry A* 113: 464.

26 Ishidoshiro, M., Matsumura, Y., Imoto, H. et al. (2015). Practical synthesis and properties of 2, 5-diarylarsoles. *Organic Letters* 17: 4854.

27 Ishidoshiro, M., Imoto, H., Tanaka, S. et al. (2016). An experimental study on arsoles: structural variation, optical and electronic properties, and emission behavior. *Dalton Transactions* 45: 8717.

28 Green, J.P., Han, Y., Kilmurray, R. et al. (2016). An air-stable semiconducting polymer containing dithieno[3,2-b:2′,3′-d]arsole. *Angewandte Chemie International Edition* 55: 7148.

29 Kato, T., Imoto, H., Tanaka, S. et al. (2016). Facile synthesis and properties of dithieno [3, 2-b: 2′, 3′-d] arsoles. *Dalton Transactions* 45: 11338.

30 Imoto, H., Kawashima, I., Yamazawa, C. et al. (2017). Multi-mode emission color tuning of dithieno [3, 2-b: 2′, 3′-d] arsoles. *Journal of Materials Chemistry C* 5: 6697.

31 Takahara, C., Iwasaki, S., Kihara, H. et al. (2022). 4-Aryldithieno [3, 2-b: 2′, 3′-d] arsoles: effects of the As-substituent on the structure, photophysical properties, and stability. *Dalton Transactions* 51: 13734.

32 Urushizaki, A., Yumura, T., Kitagawa, Y. et al. (2020). Dithieno[3,4-b:3′,4′-d]arsole: a novel class of hetero[5]radialenes. *European Journal of Organic Chemistry* 26: 3965–3970.

33 Imoto, H., Urushizaki, A., Kawashima. I. et al. (2018). Peraryl Arsoles: practical synthesis, electronic structures, and solid-state emission behaviors. *Chemistry: A European Journal* 23: 8058.

34 Iwasaki, S., Miyake, Y., Imoto, H. et al. (2022). 1,2,5-Triarylcycloalka[c]arsoles: structural effects of fused-cycloalkanes on stability and photophysical properties. *Chemistry: A European Journal* 28: e202202084.

35 Iwasaki, S., Imoto, H., and Naka, K. (2023). Tuning emission color and efficiency of 1,2,5-Triaryl-3,4-cycloalka[c]arsoles. *Bulletin of the Chemical Society of Japan* 96: 208.

36 Takahara, C., Iwasaki, S., Miyake, Y. et al. (2024). Element recycling in arsoles. *Bulletin of the Chemical Society of Japan* 97: uoad028.

37 Fujii, T., Tanaka, S., Hayashi, S. et al. (2020). Dipyridinoarsole: a new class of stable and modifiable heteroatom-bridged bipyridines. *Chemical Communications* 56: 6035.

38 Kikuchi, K., Sei, H., Okubo, K. et al. (2024). Breathing metal–organic frameworks supported by an arsenic-bridged 4, 4′-bipyridine ligand. *Inorganic Chemistry* 63: 4337.

39 Kawashima, I., Imoto, H., Ishida, M. et al. (2019). Dibenzoarsepins: planarization of 8π-electron system in the lowest singlet excited state. *Angewandte Chemie International Edition* 58: 11686.

40 Baird, N.C. (1972). Quantum organic photochemistry. II. Resonance and aromaticity in the lowest 3. pi.. pi.* state of cyclic hydrocarbons. *Journal of the American Chemical Society* 94: 4941.

41 Fujii, T., Kusukawa, T., Imoto, H. et al. (2023). Pnictogen-bridged diphenyl sulfones as photoinduced pnictogen bond forming emission motifs. *Chemistry: A European Journal* 29: e202202572.

42 Matsuo, K., Okumura, R., Hayashi, H. et al. (2022). Phosphaacene as a structural analogue of thienoacenes for organic semiconductors. *Chemical Communications* 58: 13576.

43 Sumida, A., Saeki, A., Matsuo, K. et al. (2025). Dithienoarsinines: stable and planar π-extended arsabenzenes. *Chemical Science* 16: 1126.

44 Naka, K., Umeyama, T., and Chujo, Y. (2002). Synthesis of poly (vinylene-arsine) s: alternating radical copolymerization of arsenic atomic biradical equivalent and phenylacetylene. *Journal of the American Chemical Society* 124: 6600.

45 Kawaguchi, S., Nagata, S., Shirai, T. et al. (2006). Photochemical behaviors of tetraphenyldiphosphine in the presence of alkynes. *Tetrahedron Letters* 47: 3919.

46 Ogawa, A., Ogawa, I., Obayashi, R. et al. (1999). Highly selective thioselenation of vinylcyclopropanes with a $(PhS)_2$–$(PhSe)_2$ binary system and its application to thiotelluration. *Journal of Organic Chemistry* 64: 86.

47 Sato, A., Yorimitsu, H., and Oshima, K. (2005). Synthesis of (E)-1,2-diphosphanylethene derivatives from alkynes by radical addition of tetraorganodiphosphane generated in situ. *Angewandte Chemie International Edition* 44: 1694.

48 Dodds, D.L., Haddow, M.F., Orpen, A.G. et al. (2006). Stereospecific diphosphination of activated acetylenes: a general route to backbone-functionalized, chelating 1, 2-diphosphinoethenes. *Organometallics* 25: 5937.

49 Umeyama, T., Naka, K., Nakahashi, A. et al. (2004). Radical copolymerization of acetylenic compounds with phenyl-substituted cyclooligoarsine: substituent effect and optical properties. *Macromolecules* 37: 1271.

50 Umeyama, T., Naka, K., and Chujo, Y. (2004). Synthesis of poly (vinylene arsine) s through the ring-collapsed radical alternating copolymerization of an organoarsenic homocycle with aliphatic acetylenes and their properties. *Journal of Polymer Science, Part A: Polymer Chemistry* 42: 3604.

51 Umeyama, T., Naka, K., and Chujo, Y. (2004). Spontaneous ring-collapsed alternating copolymerization of a homocyclic arsenic compound and phenylacetylene. *Macromolecules* 37: 5952.

52 Nakahashi, A., Bravo, M., Naka, K. et al. (2009). Coordination ability of poly (vinylene-arsine) s towards transition metal ions. *Journal of Physics: Conference Series* 184: 012021.

53 Naka, K., Nakahashi, A., Bravo, M. et al. (2010). Synthesis of poly (vinylene-arsine) s-stabilized silver nanoparticles. *Applied Organometallic Chemistry* 24: 573.

54 Naka, K., Umeyama, T., Nakahashi, A. et al. (2007). Synthesis of poly (vinylene–phosphine) s: ring-collapsed radical alternating copolymerization of methyl-substituted cyclooligophosphine with acetylenic compounds. *Macromolecules* 40: 4854.

55 Naka, K., Nakahashi, A., and Chujo, Y. (2006). Ring-collapsed radical alternating copolymerization of phenyl-substituted cyclooligostibine and acetylenic compounds. *Macromolecules* 39: 8257.

56 Umeyama, T., Naka, K., Bravo, M. et al. (2004). Synthesis and properties of cross-linked poly (vinylene-arsine). *Polymer Bulletin* 52: 191.

57 Yamazawa, C., Imoto, H., and Naka, K. (2018). Syntheses of dithienoarsole-containing polymers via Suzuki-Miyaura and Sonogashira-Hagihara coupling reactions. *Chemistry Letters* 47: 887.

58 Takahara, C., Nakamura, M., Aoyama, Y. et al. (2023). Versatile and practical design of dithieno[3,2-b:2′,3′-d]arsole polymers. *Macromolecules* 56: 6758.

59 Ishijima, K., Tanaka, S., Imoto, H. et al. (2021). 2,3-Diarylbenzo[b]arsole: structural modification and polymerization for tuning of photophysical properties. *Chemistry: A European Journal* 27: 4676.

60 Green, J.P., Cha, H., Shahid, M. et al. (2019). Dithieno [3, 2-b: 2′, 3′-d] arsole-containing conjugated polymers in organic photovoltaic devices. *Dalton Transactions* 48: 6676.

61 Tanaka, S., Enoki, T., Imoto, H. et al. (2020). Highly efficient singlet oxygen generation and high oxidation resistance enhanced by arsole-polymer-based photosensitizer: application as a recyclable photooxidation catalyst. *Macromolecules* 53: 2006.

62 Yamazawa, C., Hirano, Y., Imoto, H. et al. (2021). Superior light-resistant dithieno [3, 2-b: 2′, 3′-d] arsole-based polymers exhibiting ultrastable amplified spontaneous emission. *Chemical Communications* 57: 1595.

63 Tomita, I. (2018). Element-blocks π-conjugated polymers by post-element-transformation technique. In: *New Polymeric Materials Based on Element-Blocks* (ed. Y. Chujo), Vol. 1, p. 49. Singapore: Springer.

64 Matsumura, Y., Ueda, M., Fukuda, K. et al. (2015). Synthesis of π-conjugated polymers containing phosphole units in the main chain by reaction of an organometallic polymer having a titanacyclopentadiene unit. *ACS Macro Letters* 4: 124.

65 Matsumura, Y., Ishidoshiro, M., Irie, Y. et al. (2016). Arsole-containing π-conjugated polymer by the post-element-transformation technique. *Angewandte Chemie International Edition* 55: 15040.

66 Imoto, H., Yamazawa, C., Hayashi, S. et al. (2018). Electropolymerization of dithieno[3,2-b:2′,3′-d]arsole. *ChemElectroChem* 5: 3357.

67 Morales Salazar, D., Mijangos, E., Pullen, S. et al. (2017). Functional small-molecules & polymers containing PQC and AsQC bonds as hybrid p-conjugated materials. *Chemical Communications* 53: 1120.

68 Kihara, H., Imoto, H., and Naka, K. (2023). Synthesis of main-chain-type triphenylarsine polymers. *Polymer Journal* 55: 555.

6

Carborane-containing Conjugated Polymers and Their Applications for Optoelectronic Materials

Kazuo Tanaka[1,2]

[1]*Department of Polymer Chemistry, Graduate School of Engineering, Kyoto University, Katsura, Nishikyo-ku, Kyoto, Japan*
[2]*Graduate School of Global Environmental Studies, Kyoto University, Katsura, Nishikyo-ku, Kyoto, Japan*

6.1 Introduction

Because of superior optical and electronical properties, conjugated polymers have attracted attention as a key material for constructing optoelectronic devices. One of the valid strategies for obtaining advanced properties is to employ heteroatoms. By forming conjugated systems involving heteroatoms, complexes, and cluster compounds, unique functions that cannot be obtained only with organic compounds are often expressed. Therefore, many researchers have devoted their efforts to establishing introduction manners of inorganic elements and discovering unexpected functions. Moreover, these findings often contribute to establishing new principles for material designs as well as fundamental elemental chemistry.

In this chapter, unique properties and applications of carborane-containing π-conjugated polymers are described. Carboranes have an icosahedral cluster consisting of two carbon and ten boron atoms. Boron atoms can form characteristic three-center two-electron (3c2e) bonds derived from their electron-deficient properties. Owing to its 3D electron delocalization involving the 3c2e bonds, unique functions have been discovered. As the carbon atoms get closer to each other, their electron acceptance increases, reaching a maximum in the *ortho* form (Figure 6.1). In particular, *o*-carborane-based materials have been paid attention as a platform for constructing advanced materials in photochemistry. In the chapter, initially, various applications, such as hydrogen storage substances, thermally stable materials, and electronic materials, including carborane-containing π-conjugated polymers, are surveyed. Next, optical properties of mainly *o*-carborane-containing materials will be described. In particular, from the first discovery of aggregation-induced emission

Conjugated Polymer Synthesis and Materials Applications, First Edition. Edited by Yoshiki Chujo.
© 2026 Wiley-VCH GmbH. All rights reserved, including rights for text and data mining and training of artificial intelligence technologies or similar technologies. Published 2026 by Wiley-VCH GmbH.

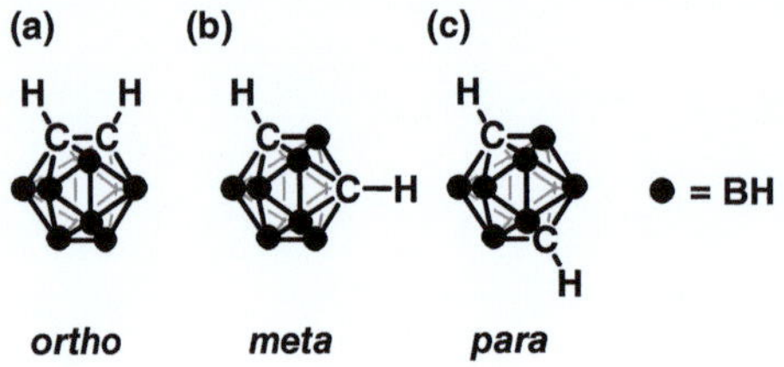

Figure 6.1 Chemical structures of carborane isomers.

(AIE) from *o*-carborane-containing polymers, various studies on mechanisms and applications to solid-state luminescent materials have been actively conducted. These studies will be reviewed.

6.2 Advanced Functional Materials

6.2.1 Hydrogen Storage

Because of extraordinarily high surface areas, porous organic polymers (POPs) are considered very promising candidates for hydrogen storage materials. Therefore, POPs have been prepared through various coupling and condensation reactions. However, the heat of hydrogen adsorption of pure hydrocarbon-based POPs is often too low. Currently, cryogenic cooling such as a liquid nitrogen bath is generally required to realize sufficient hydrogen absorption capacity, which increases the weight and cost of the storage system. Three kinds of POPs involving *m*-carborane were reported to work as adsorbents for gas storage applications, especially hydrogen storage (Figure 6.2) [1]. A series of POPs containing highly electron-deficient carborane components were prepared to enhance the interaction between the adsorbent and hydrogen. These polymers have narrow pore size distributions, with the majority of the dimensions in the range of 0.7–1.0 nm. High Brunauer–Emmett–Teller (BET)=specific surface areas up to 1023 m^2/g were obtained. The hydrogen adsorption capacities at 77, 195, and 298 °K were measured using a Sievert isotherm apparatus. The initial heat of adsorption of the carborane-containing polymer was calculated to be 8–10 kJ/mol, which is at the same level as the highly conjugated polyphenylene (8–9 kJ/mol).

6.2.2 Enhancement of Thermal Stability of Composites

Carbon materials, such as graphite and carbon/carbon composites, have been widely used due to their extremely high heat resistance, high strength, low thermal expansion, and low density. However, the use of carbon materials in oxidizing environments is limited because they are oxidized even at temperatures as low as 400 °C. To reinforce the oxidation resistance of carbon materials, ceramics such as SiC, B_4C, $MoSi_2$, BN, and TiC are usually applied. Carbon/ceramic composites like C/SiC—B_4C have attracted much attention because of their excellent oxidation resistance at temperatures up to 1400 °C in air. The new oxidation-resistant SiC/B_4C/C nanocomposite was prepared by pyrolyzing carborane-containing poly(silyleneethynylene-phenyleneethynylene) (CB-PSEPE) at 1450 °C in argon (Figure 6.3) [2]. The phase composition and microstructure of the carbon-rich ceramics were investigated

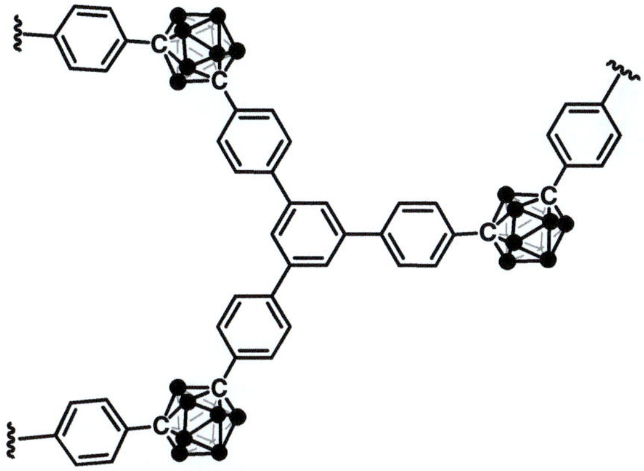

Figure 6.2 One of the POPs reported in ref. [1] can work as a hydrogen storage material.

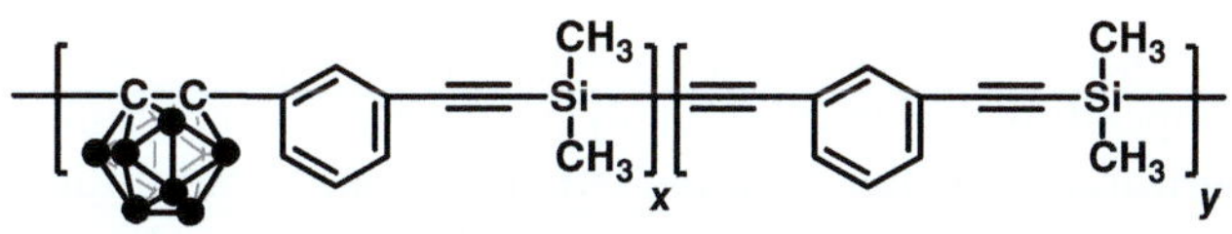

Figure 6.3 Chemical structure of CB-PSEPE.

by X-ray diffraction, Raman spectroscopy, and transmission electron microscopy. Accordingly, it was found that nanocrystals containing β-SiC and B_4C with sizes of about 10–20 nm are formed during the pyrolysis process. The boron element in CB-PSEPE promotes the growth of β-SiC nanocrystals in the nanocomposite. The resulting nanocomposite exhibited excellent thermo-oxidative stability with a residual yield of 105.7% at 1000 °C in air. It is speculated that the borosilicate layer on the surface of the ceramic could prevent oxidation. This carborane-containing polymer could be a potential precursor for advanced carbon-based materials with high oxidation resistance.

6.3 Electronic Properties of Carborane-containing Conjugated Polymers

6.3.1 Electrochemical Polymerization

Early studies focused mainly on the polymer synthesis and characterization of the products prepared by electrochemical polymerization with carborane-containing thiophenes and pyrroles. For example, the conducting polypyrrole films functionalized with neutral *o*-carboranes were prepared [3]. The synthesized redox-active polymers exhibited significantly higher electrochemical stability than unsubstituted polypyrroles. These functions are of great interest for some applications such as electrochromic and energy storage devices. Moreover, the synthesis and electrochemical studies of novel pyrrole derivatives covalently linked to neutral *o*- (**1C** and **2C**) or anionic *nido*-carboranes (**1N** and **2N**) via methylene or ethylene spacers

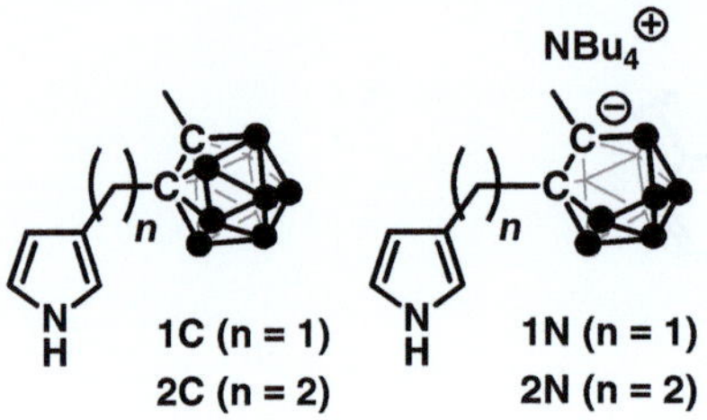

Figure 6.4 Chemical structures of carborane-modified pyrroles.

were reported (Figure 6.4) [4]. The influence of the spacer length and the charge of the boron cluster on the electropolymerization efficiency and the properties of the resulting conducting polymer films were evaluated. Only the neutral derivative **2C**, containing the ethyl spacer between the pyrrole ring and the boron cluster, produced better electroactive and conducting polypyrrole films. Additionally, it was proposed that owing to the existence of carboranes, the polymer film from **2C** can be more thermally and electrochemically stable than unsubstituted polypyrrole. It is suggested that these organic–inorganic hybrid polymers might be applicable as a building block for high-temperature devices or as a polymer precursor for ceramics.

Electrochemical polymerization with the carborane-tethered 3- and 3,4-substituted pyrroles and thiophenes (**3–8**) was reported (Figure 6.5) [5]. Owing to the enhanced stability of the carborane cage, the electroactivity of the polythiophenes hardly decreased even upon application of high oxidation potentials (up to 2.0 V). In contrast, polypyrrole derivatives became completely electrically inactive under similar conditions. According to UV–vis absorption measurements with the electrogenerated polymers, it was suggested that the introduction of the carborane cage should be responsible for stabilizing the average conjugation length of the resulting polymers compared to the reference phenyl-substituted polypyrroles and polythiophenes. Furthermore, these functionalized materials exhibited less than 10% mass loss up to 500 °C under inert atmosphere. These results also demonstrated superior thermal stability of the synthesized polymer because of the carborane modification.

Vicente et al. synthesized polymers by electrochemical polymerization using thiophene derivatives **9–13** containing each carborane isomer as a monomer (Figure 6.6) [6]. The conductivity profiles of poly(**10**) and poly(**11**) showed maxima at 12 $S{\cdot}cm^{-1}$ at 0.9 and 1.2 V versus Ag/Ag^{+}, while poly(**9**) exhibited higher conductivity of about 20 $S{\cdot}cm^{-1}$ at lower voltage (0.6 V) and reached a quasi-plateau. Furthermore, poly(**9**) showed high electrochemical stability, with only a slight decrease in conductivity up to 1000 cycles. The conductivity data also confirmed that of the three carborane isomers studied, the *o*-carborane cluster has the highest conductivity.

Similarly, in electrochemical polymerization in dichloromethane, the increase in voltage from about 0.50 to 1.15 V versus Ag/Ag^{+} in the order of *ortho* (**9**) < *meta* (**10**) < *para* (**11**) (Figure 6.6) [7]. Furthermore, from UV–vis absorption measurements, it was shown that the optical band gaps of *ortho* (**9**), *meta* (**10**), and *para* (**11**) were detected to be 1.8, 2.0, and 2.2 eV, respectively. These data suggested that the higher electrical conductivity of *ortho* (**9**) should be because of the planar conformation of the conjugated backbone. Molecular modeling calculations using density functional theory (DFT) methods also supported this speculation.

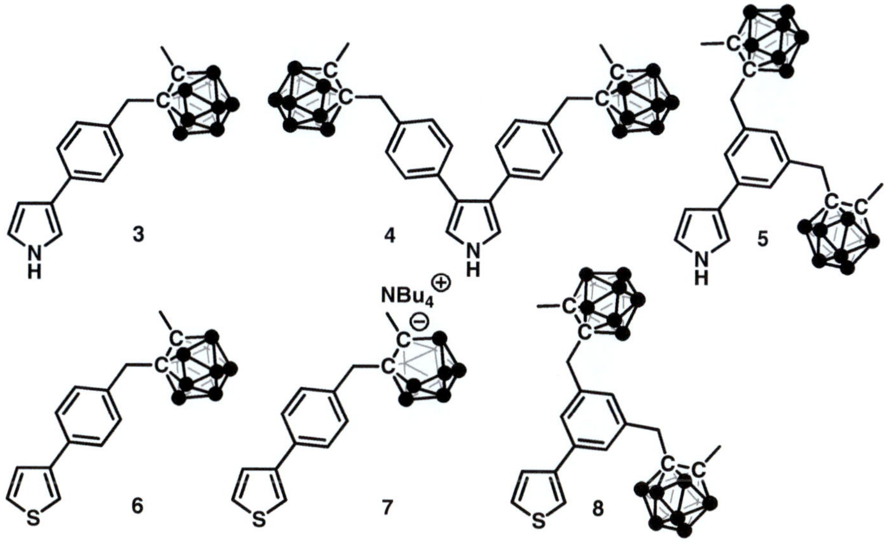

Figure 6.5 Chemical structures of modified pyrroles and thiophenes.

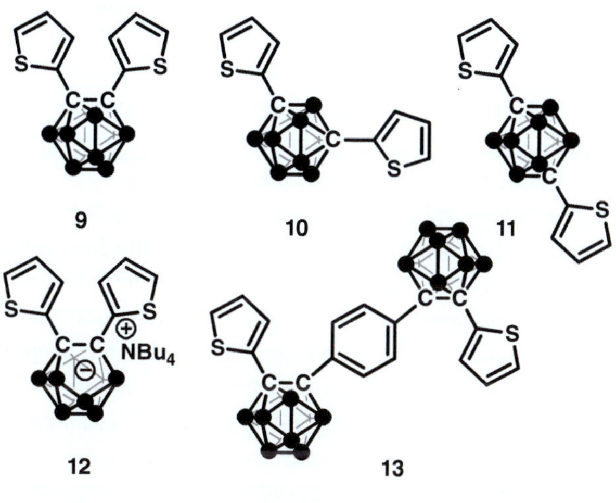

Figure 6.6 Chemical structures of thiophene derivatives.

6.3.2 Electric Devices

The *m*-carborane-containing conjugated polymer with **10** was synthesized, and the electrochromic behaviors were reported (Figure 6.7) [8]. The corresponding polymer exhibited ambipolar properties and had an optical band gap of 1.95 eV. The polymer film showed the multicolor behavior from orange, yellow-green, pale green, greenish blue, and blue during the transition from the neutral to the oxidized state under an applied external potential. The poly(**10**)/PEDOT-based electrochromic device showed a transparent yellow color at −2.0 V and a blue color at +2.0 V. Since the carborane units often improve stability, it was proposed that these devices might be applicable for practical usages.

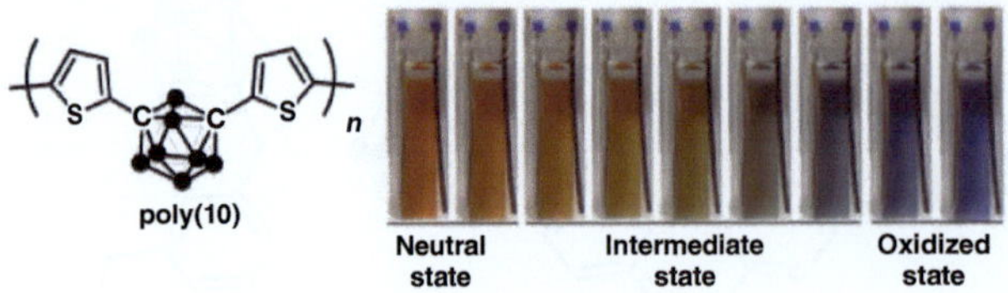

Figure 6.7 Electrochromic properties of poly(**10**). *Source*: Adapted from [8].

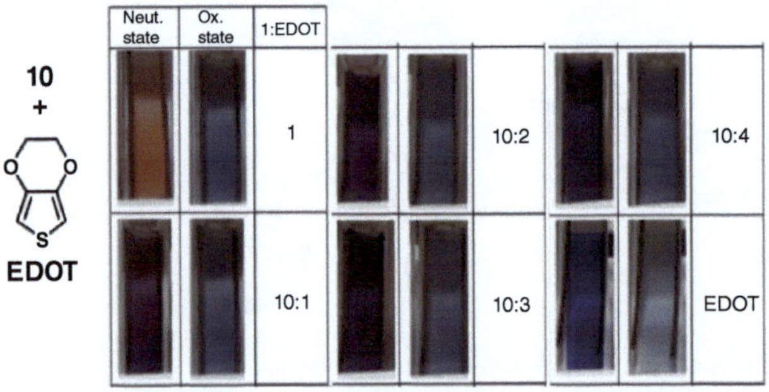

Figure 6.8 Electrochromism of the copolymers with **10** and EDOT. *Source*: Adapted from [9].

The copolymers were synthesized through electrochemical polymerization with **10** and 3,4-ethylenedioxythiophene (EDOT) units (Figure 6.8) [9]. The resulting films exhibited different electrochemical and electrochromic behaviors compared to the homopolymers. The band gap (1.69–1.82 eV) and energy levels of each frontier molecular orbital of the copolymer films were tunable by changing the feed ratio. The copolymers also maintained their charge/discharge properties at both low and high scan rates, indicating that these polymers can be used as good alternatives for supercapacitor devices. The copolymer films also exhibited electrochromic properties, suggesting that these films could be applicable for electrochromic devices.

The main-chain-type conjugated polymers having 1,4-difunctionalized benzocarborane **pBZ-CDT** and **pBZ-DPP** were synthesized, and their optoelectronic properties were examined (Figure 6.9) [10]. The copolymers containing cyclopentadithiophene and diketopyrrolopyrrole derivatives were prepared with good molecular weights through the Stille polymerization. From the measurements, it was indicated that the introduction of the carborane units into conjugated polymers affects electronic properties in organic field-effect transistors. It was shown that the steric hindrances of the benzocarborane units and the backbone twisting should prevent close overlap of the conjugated chains, resulting in the suppression of charge carrier mobility.

6.3.3 Applications for OPV

The organic photovoltaics (OPV) properties of carborane-containing conjugated polymers have been reported (Figure 6.10a) [11]. Heeney et al. synthesized a series of modified polythiophenes with the pendant vinylene-linked carborane side chains.

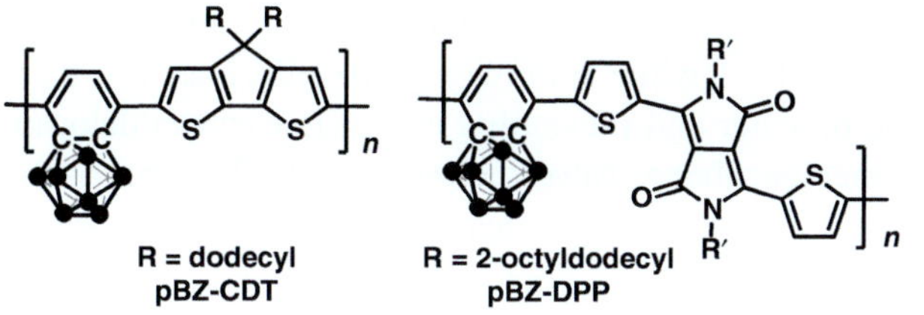

Figure 6.9 Chemical structures of the benzocarborane-fused copolymers.

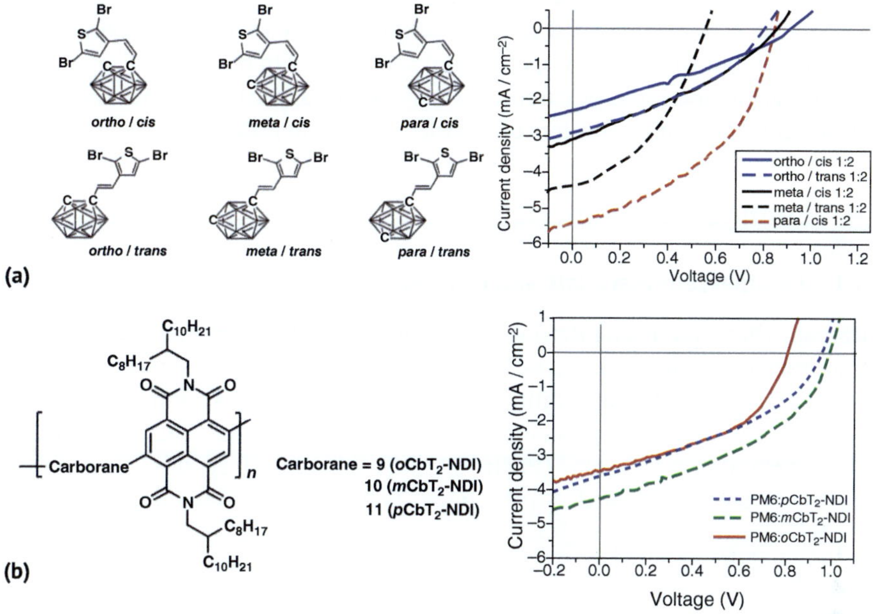

Figure 6.10 OPV properties of (a) the pendant-type polymers. *Source*: [11]/with permission of Royal Society of Chemistry and (b) NDI-containing copolymers. *Source*: Adapted from [12].

The polymers containing *o*-, *m*-, and *p*-carboranes attached to *cis*- or *trans*-double bonds were prepared. It was demonstrated that the inclusion of carborane side chains can improve solubility without the need for additional alkyl groups. It was indicated that the structures of carboranes and double bonds can affect the optoelectronic, photovoltaic, and charge transport properties. In particular, the incorporation of *o*-carborane exhibited the pronounced electron-withdrawing effect, followed by the significant increase in the ionization potential of the polymer. The *para*/*cis* combination showed V_{oc} of 0.85 V, J_{sc} of 5.44 mA cm^{-2}, and *FF* of 0.44, and consequently, the PCE value was determined as 2.03%. It is believed that backbone distortion could influence these physical properties.

Furthermore, the copolymers involving naphthalimide (NDI) and carborane in the main chain were reported for realizing n-type polymers (Figure 6.10b) [12]. The fundamental optical and electronic properties of the synthesized polymers were found to be similar. In addition, DFT calculation results suggested that the conjugated system should be localized in the aromatic segment, and the carborane acted as a "conjugation breaker." It was proposed that the carborane exerts an inductive electron-withdrawing effect while minimizing the effect on the frontier molecular orbitals of the polymer and preventing complete electron delocalization throughout the polymer backbone. Nevertheless, in thin-film transistors (TFTs) and OPVs, the nature of the carborane isomer had a significant impact on the device performance. Both ***p*****CbT$_2$-NDI** and ***m*****CbT$_2$-NDI** exhibited n-type semiconducting behaviors in the TFT devices, with reasonable mobility values of up to $1.22 \times 10^{-3}\ cm^2 \cdot V^{-1} \cdot s^{-1}$ in fully amorphous films. Furthermore, all three polymers can be used as an electron-acceptor material in all-polymer organic solar cells with power conversion efficiencies up to 1.8%. All three carborane isomers were directly incorporated into a conjugated polymer backbone and were firstly compared in terms of their optoelectronic properties and applications in this study. These results suggest that the inclusion of carboranes may be useful for tuning the mechanical and/or semiconducting behaviors of optoelectronic materials.

6.4 Luminescent Materials

6.4.1 Luminescent Materials and Devices

Carboranes have also been introduced into conjugated polymers and their optoelectronic properties have been investigated. The polymers having silylcarboranes at the side chains [13], the benzodithiophene-fused oligomers [14], and the oligomers modified with the energy-transfer pair at both ends through the bridged conjugated system containing *p*-carborane (Figure 6.11a) [15] have been synthesized and their properties were investigated. Furthermore, polyfluorenes containing carboranes

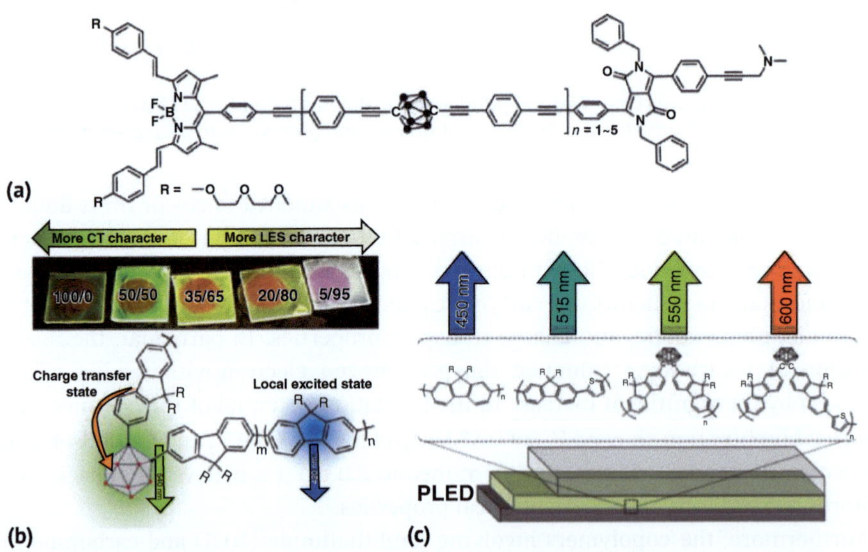

Figure 6.11 (a) Energy transfer system through *p*-carborane. (b) Emission color tuning in the polyfluorene copolymers including *o*-carborane. *Source*: [19]/with permission of American Chemical Society. (c) Organic light-emitting devices with various modified polyfluorenes. *Source*: [20]/with permission of American Chemical Society.

have been prepared [16], and it has been demonstrated that modified polyfluorenes can be a platform as an optically functional materials.

The synthesis of carborane-containing poly(dihexylfluorene)s was accomplished, and it was shown that these polymers exhibit solvatochromism in both solution and solid states because of the formation of the CT state between the fluorene units and the carborane molecules [17]. Furthermore, the dual emission property consisting of the locally excited (LE) emission from the fluorene moiety and the CT emission between the fluorene and carborane was also reported in the latter paper [18]. Diverse colors were observed by changing solvents in solution (Figure 6.11b) [19]. In addition, by exposing the polymer films to solvent vapors, significant changes in emission color were detected. These data suggest that the polymers could be applicable for chemical sensing. Moreover, carborane-based polyfluorene derivatives were used in light-emitting diodes and transistors (Figure 6.11c) [20]. The data from optical measurements revealed that although carborane does not participate in π-conjugated networks, its presence causes a large red shift in the electroluminescence and TFT devices. In field-effect transistors, charge carrier mobility is enhanced by an order of magnitude despite disrupting π-conjugation. Carborane-containing conjugated polymers is promising as a new material in optoelectronic applications.

6.4.2 Aggregation-induced Emission

Next, the solid-state luminescence properties of *o*-carborane-containing materials are explained. Chujo et al. synthesized the polymers containing *m*-carborane in the main chain [21]. Similarly to the previous examples, optical data revealed that main-chain conjugation was broken at the *m*-carborane unit. On the other hand, when *o*-carborane was introduced, luminescence was observed in the longer wavelength region, meaning that electronic communications should be developed through the polymer main chains (Figure 6.12) [22]. In particular, strong luminescence was obtained only in the film state, clearly indicating that these polymers should have AIE properties [22]. Furthermore, it was shown from the mechanistic study that this luminescence band should have the CT character [23, 24]. Additionally, it has been demonstrated that the color of AIE can be tuned before and after complexation in coordination polymers [25]. In these studies, it was suggested that the vibration at the carbon–carbon bond in the *o*-carborane unit might be the main factor in the expression of the AIE property. This mechanism was supported by the fact that AIE was not observed in the polymers with condensed benzene rings [26]. It was also reported that *o*-carborane acts as an electron acceptor when bonded to aromatics at the carbon site. Interestingly, *o*-carborane can be an electron donor toward aromatics when bonded to the boron site, and the light-emitting material was obtained based on this electron-donating ability [27]. From these reports, researches on the mechanism of solid-state luminescent properties and applications for luminochromic materials have been conducted mainly with small molecular systems. Recent progress is described here.

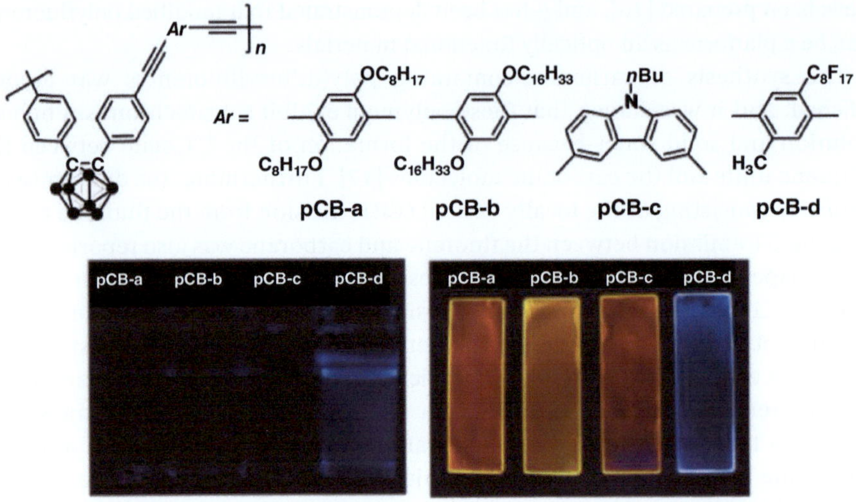

Figure 6.12 Representative aggregation-induced emission (AIE) materials of *o*-carborane-containing conjugated polymers. *Source*: [22]/with permission of American Chemical Society.

6.5 Photochemistry of *o*-carboranes

6.5.1 Demands for Solid-state Luminescent Materials

As next-generation technology on organic light-emitting devices, sensing has attracted attention. To fulfill this purpose, one of the strategies is to utilize optically functional molecules as a probe. In detecting target phenomena or substances using molecular probes, there are several advantages to using luminescence over color change. First, since the background is dark, it is easy to improve detection sensitivity by eliminating noise. In addition, by using a microscope or a charge-coupled device camera, it is possible to obtain information from tiny areas ranging from nanometers to molecular scales. Furthermore, spectroscopic analyses enable us to gather detailed information. By monitoring real-time and time-course changes in emission wavelength and intensity, various parameters can be obtained regarding the targets. Therefore, it is important to develop molecules that change their luminescence properties in response to external stimuli or environmental changes.

In order to achieve more accurate detection, it is essential to improve sensitivity. However, there is a critical issue to be overcome in organic luminescent dyes. Most of emission properties of organic dyes are commonly spoiled in the condensed state due to aggregation-caused quenching (ACQ) caused by nonspecific intermolecular interactions. Although to avoid ACQ, introduction of bulky substituents in the molecular skeleton is effective, another problem is rising. Since bulky substituents contribute to isolation of each molecule, the resulting materials often show insensitivity toward molecular environmental changes. In other words, sensitivity toward external stimuli is lowered by employing bulky substituents. Therefore, it can be said that there is a tradeoff between solid-state luminescence and stimuli-responsiveness. To solve this problem, a new strategy is required.

Heteroatoms possess a wide variety of unique properties derived from their inherent characteristics, and therefore, "element-blocks," which are minimum functional units containing heteroatoms, are promising platforms for developing functional materials [28–32]. Therefore, the exploration of new element-blocks and understanding of the origins of these properties are of great importance. From this standpoint, we also regard *o*-carboranes as attractive element-blocks (Figure 6.1). As mentioned above, unique functions have been discovered, and *o*-carborane-based materials have been paid attention as a platform for constructing advanced materials, especially in photochemistry.

Most luminescent organic dyes exhibit poor emission in the solid state due to ACQ caused by nonspecific intermolecular interactions in the condensed state, whereas various types of aryl-modified *o*-carboranes exhibit emission even in the solid state. Moreover, some aryl-modified *o*-carboranes exhibit environment-sensitive intensity changes. In the aggregated state, more intense emission can be obtained than in dilute solution, meaning AIE behaviors [33]. Furthermore, since degrees of intermolecular interactions would be significantly dependent on molecular distributions, solid-state emission properties, such as luminescent color, are drastically influenced by morphology changes. These results suggest that the *o*-carborane unit is a promising "element-block" for developing environmentally sensitive functional materials with solid-state luminescent properties.

From the first report on AIE properties from the *o*-carborane-containing conjugated polymers [22], fundamental studies and applications have been performed with aryl-modified *o*-carboranes. In this part, the relevant issues on AIE property of *o*-carboranes are described mainly from our studies. Recently, we proved that the extension of the carbon–carbon bond caused after photo-excitation plays a critical role in rapid decay of the excited state (Figure 6.13) [34]. From this study, the mechanism for emission annihilation in solution will be illustrated. Next, unique solid-state luminescent properties of a series of aryl-modified *o*-carboranes are explained. The *o*-carborane unit can work as an electron acceptor when bonded at the carbon atom. Therefore, intense CT emission can be induced by the combination with aromatic units. In particular, owing to the steric structure of *o*-carborane, ACQ can be suppressed, and consequently, solid-state emission with the CT character is obtained. Based on this principle, various types of functional solid-state luminescent materials have been developed. The design strategies for these materials will be surveyed. More recently, excimer emission has been observed from some of the aryl-modified *o*-carboranes in the solid state [35]. Additionally, the crystal structure–property relationship for obtaining excimer emission has been clarified. These results will also be explained.

6.5.2 Experimental Proofs for Emission Annihilation Through the Bond Elongation at the Carbon–Carbon Bond in *o*-carborane

Because of the superior solid-state luminescent properties, *o*-carborane has attracted much attention as a building block for constructing stimuli-responsive solid-state luminescent materials. However, the detailed mechanism of emission quenching in solution of aryl-substituted *o*-carboranes was still veiled. So far, aryl-modified *o*-carboranes with AIE properties exhibit intense emission not in solution but in solid. At first, the vibration at the carbon–carbon bond was presumed to be responsible for emission annihilation in solution. After that, according to quantum calculations, the elongation at the carbon–carbon bond in *o*-carborane in the excited state has been proposed as a main path for emission annihilation (Figure 6.13) [36]. However, due to intramolecular rotation processes, the critical mechanism was not determined yet. In this research, we can firstly prove that emission annihilation can proceed only through the bond elongation without rotation. The details are illustrated.

We designed and synthesized two types of biphenyl-substituted *o*-carboranes, **B(3)** and **B(4)**, having the fused structures at the neighbor carbon and boron atoms for fixing molecular conformation (Figure 6.14) [34]. In these molecules, the bond elongation is allowed, while rotation would be prohibited. Significantly, their fluorescence quantum yields were extremely low ($\Phi_{PL} < 0.01$) in solution, while the values increased to 0.04 (**B(3)**) and 0.08 (**B(4)**) in solid, clearly indicating that both compounds have AIE properties. From the emission measurements with various solvents, typical peak shifts of the emission bands were detected from both compounds, meaning that the emission bands should be derived from electronic transition from the CT state, as observed in the previous aryl-modified *o*-carboranes.

To elucidate the origin of these photophysical properties, we also performed quantum calculations. It was shown that the highest occupied molecular orbitals (HOMOs) are delocalized through the whole molecules in the S_1 excited state, meanwhile the lowest unoccupied molecular orbitals (LUMOs) were largely

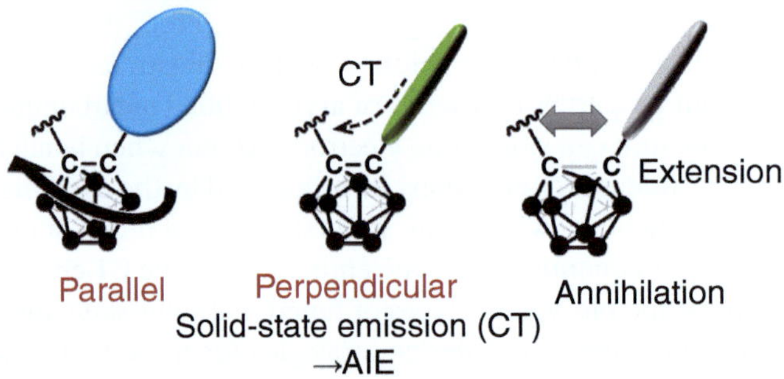

Figure 6.13 Summary of the aggregation-induced emission (AIE) mechanism in aryl-modified *o*-carboranes.

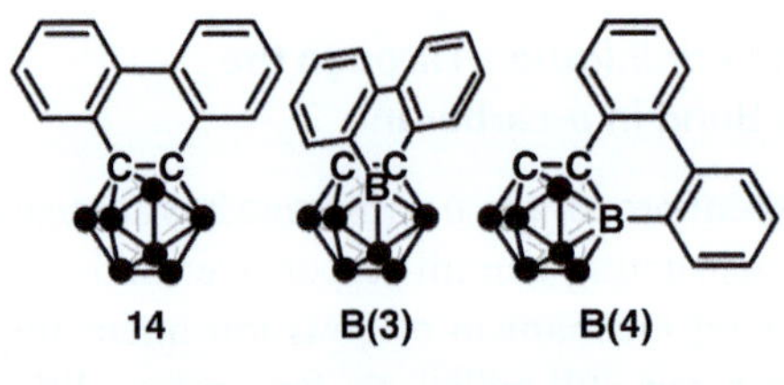

Figure 6.14 Chemical structures of biphenyl-fused *o*-carboranes.

distributed on the *o*-carborane unit, suggesting the fact that the intramolecular CT should occur from the *o*-carborane unit to the biphenyl moiety in the S_1–S_0 transition. These results support the CT character of the emission bands. Notably, in the S_1 excited states, the characteristic carbon–carbon bond elongation was detected. The estimated carbon–carbon bond length of **B(3)** was 1.64 Å in the ground state. Interestingly, the bond extension to 2.40 Å was proposed in the excited state. The elongation process was also observed in **B(4)**, 1.64 Å in the ground state and 2.38 Å in the excited state. These results clearly indicate that bond elongation should be induced in the excited state.

We then performed quantum mechanics and molecular mechanics (QM/MM) calculations to estimate the S_1-optimized structures in the crystalline state, where structural relaxation would be restricted. In the calculations, the molecular conformations of **B(3)** and **B(4)** were referred to those in the single crystal packing. One of the central molecules was treated as a QM molecule using a time-dependent-density functional theory (TD-DFT), and the surrounding molecules were modeled using a universal force field. Under this condition, after structural optimizations with both **B(3)** and **B(4)** in the S_1 state, shorter carbon–carbon bond lengths (2.28 Å for **B(3)** and 2.32 Å for **B(4)**) were obtained than those in the isolated state. It is because the surrounding molecules interfere with the structural relaxation. These results mean that incomplete structural relaxation did not lead to quenching, and solid-state emission could be observed. From these optical measurements and theoretical investigations, we confirmed that emission annihilation in solution can be induced not by rotation but by the bond elongation in the excited state.

6.5.3 CT Emission of Aryl-modified *o*-carboranes

Although the CT emission from aryl-modified *o*-carboranes, the detailed mechanism is described here. *o*-Carborane has a σ^* orbital on the carbon–carbon bond. Therefore, in the molecules with a π-conjugated unit bonded to the carbon, CT interaction can be formed through the $\sigma^* - \pi^*$ conjugation in the excited state. Through this mechanism, the CT emission can be obtained. This process can be realized in quantum calculations. In the cases of **B(3)** and **B(4)**, both HOMO and LUMO are located on the biphenyl moiety in the ground state. These results correspond to the finding that the peak wavelength of the absorption spectrum does not change even in various solvents.

In contrast, in the S_1 excited state, the HOMO is distributed throughout the molecule, whereas the LUMO is mainly localized on the *o*-carborane unit, indicating that an intramolecular CT occurs from the *o*-carborane unit to the biphenyl moiety in the S_1–S_0 transition. These behaviors represent typical CT emission. From the emission measurements, the CT character can be confirmed. The peak position of the emission band of **B(3)** was obtained at around 580 nm (*n*-hexane), and peak shifts to longer wavelength regions were observed in the polar solvents such as 628 nm (MeCN). This tendency was also observed from **B(4)**. These behaviors are typical of CT emission, and since the emission bands can be observed at the similar positions even in solid, it can be said that AIE is attributable to the CT emission.

According to the above mechanism, it was presumed that the expression of AIE property should be controllable by regulating the extension of the carbon–carbon bond in the *o*-carborane unit. Herein, the representative example, where the ACQ-presenting molecule can be transformed to the AIE-active one by introducing the single chemical bond, is explained. For obtaining intense CT emission, bis(benzo[*b*]thiophene)-bonded *o*-carborane (**BC**) was designed (Figure 6.15a) [36, 37]. In this study, DFT and TD-DFT calculations were used for evaluating structural relaxation in the excited state. Similarly to the above data, the carbon–carbon bond length in the *o*-carborane unit was elongated in the excited state. In the most stable conformation in the excited state estimated by TD-DFT calculation, the bond length increased by 30%. In contrast, such a significant change was hardly observed in the ground state. These data suggest that structural relaxation proceeds after photoexcitation to form a thermally stable conformation in the excited state, and emission annihilation should occur. In addition, emission enhancement can be observed in solid because of suppression of bond elongation. Therefore, the AIE behavior was expectable. We also theoretically examined the conformational changes in the excited state using bi(benzo[*b*]thiophene)-fused *o*-carborane (**FC**), in which the carbon–carbon bond was immobilized by condensing both benzo[*b*]thiophene

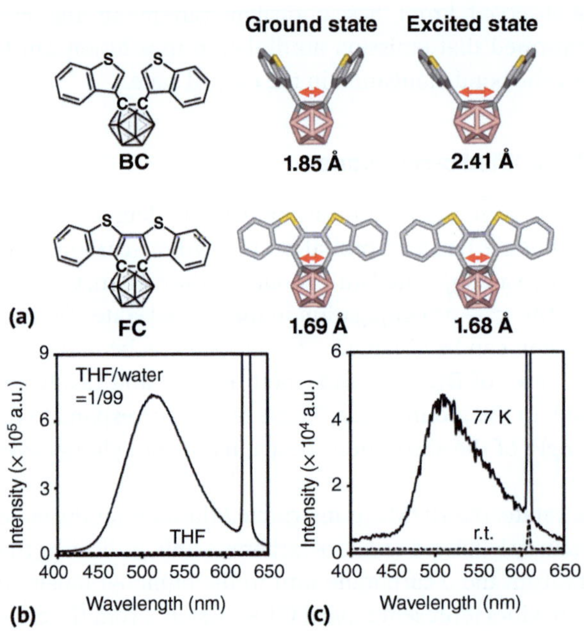

Figure 6.15 (a) Chemical structures of *o*-carboranes and their thermally stable conformations obtained from theoretical calculations. DFT and TD-DFT were performed at B3LYP/6-31G(d,p)//B3LYP/6-31G(d,p) and TD-B3LYP/6-31+G(d,p)//TD-B3LYP/6-31+G(d,p) levels, respectively. (b) Emission spectra of bis(benzo[*b*]thiophene)-bonded *o*-carborane (**BC**) in THF (dashed line) and THF/water = 1/99 vol% (solid line) with the excitation light at 304 nm and (c) in 2-MeTHF with variable temperature. *Source*: Reproduced from [36]/with permission of Royal Society of Chemistry.

moieties [36]. As we expected, the TD-DFT calculation gave almost the same structure obtained from the DFT calculation.

We actually synthesized both modified *o*-carboranes and evaluated their optical properties [37]. First, we monitored the emission change during aggregate formation. Typical AIE behavior was obtained only from **BC**, as we expected. Almost no emission band was obtained from the THF solution of **BC** ($\Phi_{PL} < 0.01$) (Figure 6.15b). By increasing the water content to 99 wt% to induce aggregate formation, the intense emission band with the peak at 514 nm was observed. The powder sample of **BC** also showed bright emission ($\Phi_{PL} = 0.61$, $\lambda_{em} = 507$ nm). Furthermore, in 2-methyltetrahydrofuran (2-MeTHF) at 77 K, where almost all molecular motions would be suppressed, significant increase in emission intensity was detected (Figure 6.15c). These results clearly indicate that **BC** should be an AIE-active molecule. Conversely, for FC, almost no significant emission increase in the solid state was observed ($\Phi_{PL,solution} < 0.01$, $\Phi_{PL,solid} = 0.05$). These data represent that the expression of AIE can be controlled by the presence or absence of a single chemical bond.

6.5.4 Angular Dependency of Electron Acceptor Properties of the *o*-carborane Unit

As mentioned above and in the previous studies, it was described that CT emission can be induced from aryl-modified *o*-carboranes, where the *o*-carborane unit acts as an electron acceptor [38, 39]. Next, it is explained that the magnitude of electronic interaction can vary depending on the angle between the π-conjugated plane and the carbon–carbon bond in the *o*-carborane unit. The above-mentioned **B(3)** and **B(4)** are fixed at different angles toward the π-conjugated plane. In addition, the biphenyl-modified *o*-carborane **14** having the parallel conformation, where the π-conjugated system of the biphenyl moiety and the carbon–carbon bond are included in the same plane, was reported from the previous work [40]. Therefore, they are representative examples for discussing the angle dependency of electron-accepting properties of the *o*-carborane unit.

The UV–vis absorption spectra revealed that **B(3)**, **B(4)**, and **14** exhibited almost identical band shapes with similar vibrational structures attributable to $\pi - \pi^*$ transitions at the biphenyl moiety (Figure 6.14). The absorption maxima of both carbon–boron fused *o*-carboranes were observed around 275 nm, which is identical to that of biphenyl (248 nm), indicating that the same electronic structure as biphenyl is realized in the ground state.

In the emission spectra, significant differences were observed. Compound **14** showed the emission band with a vibrational structure, with the peak at around 325 nm in chloroform. This can be attributed to emission from the LE state at the biphenyl moiety. On the other hand, **B(3)** and **B(4)** showed the broad emission band with the peak at around 580 nm in chloroform, as mentioned above, and further showed solvatochromism. These data mean that in the case of the parallel conformation, electronic interaction is hardly formed between the π-conjugated system and the *o*-carborane unit, but as the angle changes, electronic interaction is generated and subsequently the *o*-carborane unit can act as an electron acceptor.

6.5.5 Twisted intramolecular Charge Transfer in the Crystalline State

Another unique property of the aryl-modified *o*-carboranes is the twisted intramolecular charge transfer (TICT) process in the crystalline state (Figure 6.13). TICT is a class of CT emission that occurs via a structural change in the excited state. The significant point in TICT from aryl-modified *o*-carboranes is that the *o*-carborane units can rotate even in solid such as crystal, and the emission band originating from the transition from the TICT state can be observed. In aryl-modified *o*-carboranes, the π plane and the carbon–carbon bond are generally on the same plane (parallel conformation) to extend the π-conjugated system in the ground state. As mentioned above, the electron-accepting property of *o*-carborane changes with rotation. Therefore, in the excited state, CT states can be formed after rotation, and subsequently thermally stable structure (vertical conformation) is formed. After that, the emission band via the TICT process is obtained. The discovery of the TICT process in crystal is described.

We synthesized the anthracene-connected *o*-carborane **Ant-H** and investigated optical properties (Figure 6.16) [41]. The dual emission properties were observed from the solution samples in all solvents. The emission band around 450 nm was barely affected by changing the solvent type, meaning that emission should be caused from the LE state. In contrast, the emission band around 600 nm was bathochromically shifted with increasing solvent polarity, which is the typical CT character. Although LE emission can generally only be observed under frozen or structurally restricted environmental conditions, **Ant-H** exhibited emission from the LE state, suggesting that the energy barrier for intramolecular rotation of the

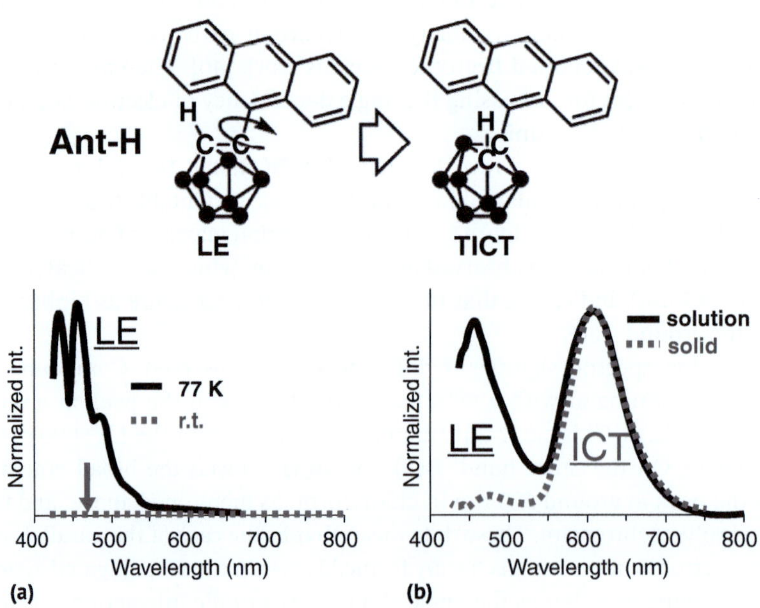

Figure 6.16 Emission spectra (a) with variable temperature and (b) in solution and solid states of **Ant-H**. *Source*: Reproduced from [41]/with permission of John Wiley & Sons.

anthracene moiety may be small. Indeed, from the variable-temperature emission spectra, the difference of energy levels between LE and CT states was estimated as 0.72 kcal/mol. This result suggests that the transition from the CT state to the LE one is allowed to occur at room temperature in the excited state.

It was shown from the measurements with the aggregation and crystalline samples that the intensity of the CT emission band increased significantly in the solid states. In the THF solution, weak emission was observed ($\Phi_{PL} = 0.02$), while **Ant-H** exhibited intense emission in both the aggregated and crystalline states ($\Phi_{PL} = 0.18$ and 0.38, respectively), clearly indicating that the **Ant-H** is an AIE- and CIE-active molecule. The intense CT emission observed in the crystalline state is of significance. According to the results of X-ray crystallography, it was observed that the anthracene moiety should have a parallel conformation along the carbon–carbon bond in the *o*-carborane unit, which is unfavorable for the formation of the CT state. This fact represents that the rotation of *o*-carborane is capable in the crystal packing, and consequently the CT emission observed around 600 nm can be induced. In other words, the TICT process is permitted in **Ant-H** even in the crystalline state. Owing to the spherical structure of the *o*-carborane unit and less intermolecular interactions, intramolecular rotation can occur in crystal.

To support the solid-state TICT process, the emission spectrum of **Ant-H** was measured in the frozen matrix of 2-MeTHF at 77 K. As a result, only LE emission originating from the anthracene moiety was observed at around 450 nm with a clear vibrational structure. In particular, significant CT emission at around 600 nm was hardly detected. Moreover, in the crystalline state at 77 °K, the relative intensity of CT emission to LE was lower than that at room temperature. It is likely that the rotation of the *o*-carborane unit is suppressed in the frozen medium and in the crystalline state at low temperatures. Consequently, the enhancement of LE emission can be detected. Conversely, the LE emission gradually decreased at high temperatures. It is because the rotation is thermally activated by heating. These data also clearly support that the emission band at around 600 nm is derived from the TICT state. In general, organic compounds cannot show TICT emission in the crystalline state due to structural restriction in the condensed state. Since *o*-carborane has a compact spherical shape, rotation proceeds after photoexcitation. Moreover, ACQ is effectively avoided, and therefore TICT emission can be obtained from the well-ordered packing states.

The TICT process was able to be theoretically reproduced according to the potential energy curves calculated by varying the dihedral angle φ [42, 43]. Figure 6.17 demonstrates the height of the energy level of each conformation [41]. The potential energy curve in the ground state shows that the most stable conformation is the parallel conformation ($\varphi = -15°$), which is in good agreement with the angle determined by X-ray structure analysis ($\varphi = -13.6°$). The second stable point was a perpendicular conformation ($\varphi = -90°$) with an energy 1.5 kcal/mol higher. Furthermore, the rotation barrier was also obtained as 12 kcal/mol. From these data, it is shown that the rotation process can occur (from −15° to 90°) upon excitation in the parallel conformation (process 1).

In summary, we demonstrated solid-state luminescence from the TICT state. After this finding, we also observed TICT emission by replacing the anthracene moiety with other aromatic compounds. The results and applications will be described below.

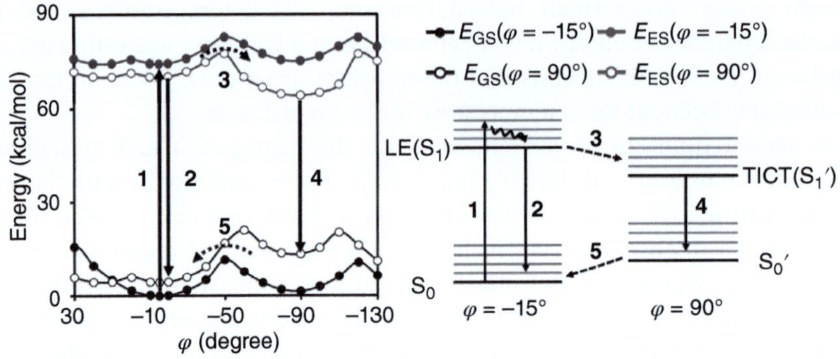

Figure 6.17 Calculation results with the anthracene-*o*-carborane **Ant-H**. E_{GS} and E_{ES} represent energy in the ground and the excited states, respectively. The φ value means the angle between the C—C bond and the π-plane ($\varphi = -15°$: parallel, $\varphi = 90°$: perpendicular). *Source*: Reproduced from [41]/with permission of John Wiley & Sons.

6.6 Applications for Solid-state Luminescent Materials

6.6.1 Development of Highly Efficient Solid-state Luminescent Materials

According to the numerous numbers of studies on solid-state luminescent materials consisting of aryl-modified *o*-carboranes, it is suggested that the sphere structure is advantageous for inhibiting intermolecular interactions that lead to ACQ in the solid state. Therefore, it has been expected that highly efficient solid-state luminescent materials can be obtained based on the structural feature. In this part, several strategies for obtaining high emission efficiencies ($\Phi_{PL} > 0.90$) in the solid state.

We examined the emission efficiencies of anthracene- and pyrene-modified *o*-carboranes in the solid state (Figure 6.18). Although crystallization-induced emission (CIE) properties were observed from both compounds, there was much room for improvement of emission efficiency. According to the AIE mechanism of aryl-modified *o*-carboranes, the extension of the carbon–carbon bond in the *o*-carborane units should be responsible for emission annihilation. It was assumed that the bond extension is not sufficiently prohibited even in the crystalline state. To completely suppress molecular motions in the condensed state, bulky substituents were introduced into the adjacent carbon atom [44]. Consequently, by introducing the substituent group, such as methyl or trimethylsilyl (TMS) groups, into pyrene-modified *o*-carborane ($\Phi_{PL} = 0.80$), drastic improvements of emission efficiencies were observed in the solid state. Interestingly, the emission color was varied by introducing the substituents on the adjacent carbons. The pristine pyrene-modified *o*-carborane showed orange emission, whereas the methyl- and TMS-modified derivatives exhibited green and yellow emission, respectively. In the absence of the bulky substituent on the adjacent carbon, intramolecular rotation at the *o*-carborane unit is allowed in the excited state. Therefore, the excited molecule can form the most energetically stable twist structure, and the emission

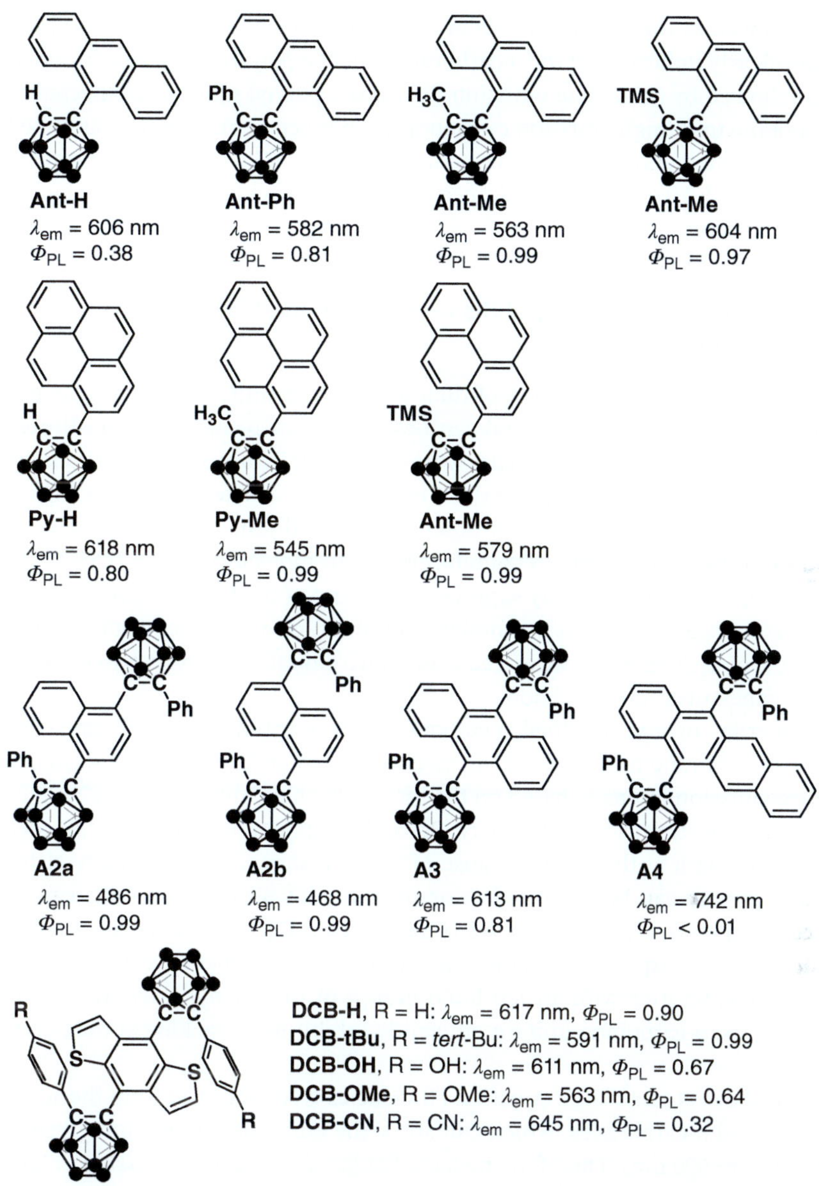

Figure 6.18 Chemical structures and solid-state luminescent properties of aryl-modified *o*-carboranes.

band should appear in the longer wavelength region. In the substituent-modified *o*-carboranes, rotation and the bond extension at the carbon–carbon bond in the *o*-carborane unit are suppressed because of structural restriction. Thus, emission color should be determined by the conformation in the crystalline packing.

Similar results were also obtained from the anthracene-modified *o*-carboranes. The pristine anthracene-modified *o*-carborane showed orange emission ($\Phi_{PL} = 0.38$), while the introduction of methyl and TMS groups induced shorter wavelength emissions as well as the improvements of emission efficiencies (Figure 6.18) [45]. These data clearly indicate that the *o*-carborane unit can play a critical role in avoiding ACQ followed by solid-state emission. When the phenyl group is connected to the *o*-carborane unit, high emission efficiency has not been obtained. It is assumed that electronic interaction might be formed, followed by ACQ, although the phenyl group has larger degree of steric hindrance. Alkyl groups are favorable for inducing solid-state emission.

6.6.2 High Emission Efficiency of Bis-*o*-carborane-modified Aromatic Systems

Another strategy is the introduction of dual *o*-carborane units into the single aromatic rings. Chujo et al. designed and synthesized the bis-*o*-carborane substituted anthracene **A3**, and its optical properties were examined (Figure 6.18) [46]. From the aggregation of **A3** in THF/water (1/99 vol%) mixing solvent system, relatively higher emission efficiency ($\Phi_{PL} = 0.77$) was observed. In addition, similar values were observed from the crystal samples containing solvent molecules ($\Phi_{PL} = 0.66$~0.81). Although ACQ seems to be largely suppressed because of the bis-*o*-carborane units, ACQ was still observed due to incomplete inhibition of π-stacking at the central anthracene moiety. Therefore, we aimed to enhance emission intensity based on the bis-*o*-carborane-substituted skeleton.

AIE properties of aryl-modified *o*-carboranes are attributable to CT emission, where the *o*-carborane unit works as an electron acceptor. Therefore, we designed to enhance emission intensity based on the bis-*o*-carborane-substituted skeleton by employing the electron-rich aryl moiety. Benzodithiophene is able to be regarded as an analogue with anthracene. Because of thienyl rings, electron density at the central aryl moiety can be enhanced. Moreover, the bis-*o*-carborane units can be introduced at the same positions of **A3** (Figure 6.18) [47]. Thus, intense emission with CT character is expected from the benzodithiophene analogue. We synthesized two compounds with or without *tert*-butyl groups at the phenyl rings, which are connected to the *o*-carborane unit for efficiently inhibiting π-stacking at the centra aryl unit.

From the above studies, high emission efficiencies were obtained from the series of aryl-modified *o*-carboranes, which can show the emission bands from green to orange (500 nm~600 nm). Therefore, next challenges are to realize emission in blue ($\lambda_{em} < 500$ nm) and NIR ($\lambda_{em} > 700$ nm) regions. By modulating the size of aryl moieties, we accomplished to luminescent color changes based on the bis-*o*-carborane-substituted systems.

We presumed that the peak wavelength of emission bands could be shifted to the shorter wavelength region by shortening the π-conjugated system at the central aryl moiety [47]. To evaluate the validity of this idea, we designed and synthesized the benzene- (**A1**) and naphthalene-containing triads (**A2a** and **A2b**). In **A2b**, we proposed that π-conjugation should be distorted, and consequently an emission band in much shorter wavelength region could be induced. According to the optical measurements, significant solid-state emission was not confirmed from **A1**. On the other hand, we observed the emission bands around 480 nm and especially around

460 nm from **A2a** and **A2b** with high emission efficiencies, respectively. These data represent that our design strategies based on bis-*o*-carborane substitution should be valid for obtaining intense emission in the visible region.

Next, we also aimed to apply this molecular design for obtaining NIR-emitting materials (Figure 6.18) [47]. Instead of anthracene in **A3**, tetracene was employed at the central aryl moiety. Unfortunately, although slight AIE and CIE were detected in the NIR region (THF/water=1/99 vol%: λ_{em} = 759 nm, crystal: λ_{em} = 742 nm), emission efficiency in the crystalline state was very small (Φ_{PL} = 0.01). In the series of acenes, it is known that oxidation followed by generation of quinone structures readily occurs at the bridging ring positions. We assume that bis-*o*-carboranes are insufficient for completely covering the tetracene moiety. So far, TMS-acetylene groups have been conventionally used for enhancing oxidation resistance. The *o*-carborane substitution might be a potential functional group that can contribute to enhancing stability as well as emission efficiency.

We prepared bis-*o*-carborane-modified benzodithiophenes **DCB-H** and **DCB-tBu** through the decaborane insertion at the ethynyl moiety in the corresponded precursors (Figure 6.18) [48]. From the structural analysis of the products, it was found that **DCB-tBu** has the designed structure, meanwhile **DCB-H** has the *cis* form, where both phenyl groups located at the same side. However, desired emission properties were obtained from both molecules. **DCB-H** and **DCB-tBu** showed AIE properties in yellow region and efficient CIE as we expected (**DCB-H**: Φ_{PL} = 0.90, **DCB-tBu**: Φ_{PL} = 0.94). In addition to the steric hindrance of the *o*-carborane units, the smaller size of benzodithiophene could be favorable for avoiding ACQ.

6.6.3 Solid-state Luminescence of *Nido*-carborane

Next, the electronic and luminescent properties of the helicene-like bibenzothiophene between *o*-carborane and 5,6-dicarba-*nido*-decaborane are illustrated (Figure 6.19) [49]. We found that the *o*-carborane and *nido*-decaborane-fused bibenzothiophenes are simultaneously produced in the decaborane insertion at the ethynyl moiety and successfully isolated both compounds. Then, the helicene skeleton was constructed through the coupling reaction. Interestingly, the single crystal X-ray analysis revealed that one has *o*-carborane, meanwhile the other has *nido*-decaborane, which can be obtained by subtraction of two boron atoms from *o*-carborane. In particular, although the boron cluster units work as an anchor, the helicene-like distorted structure was realized in the *nido*-decaborane-fused bibenzothiophene. From the optical measurements with the solutions, different properties depending on the type of anchors were observed. In the *o*-carborane-fused compound, it was suggested from the absorption and emission spectra that weak electronic interactions should be formed between *o*-carborane and the bibenzothiophene moiety. The weak emission band (Φ_{PL} = 0.01) was observed only from the hexane solution with the peak at 376 nm. These data can be explained by the theoretical results that π-conjugation was restrictedly developed within the bibenzothiophene moiety in frontier orbitals of the *o*-carborane-fused compound. On the other hand, it was revealed that the *nido*-decaborane unit can play significant roles not only in maintaining the helical structure but also in extending

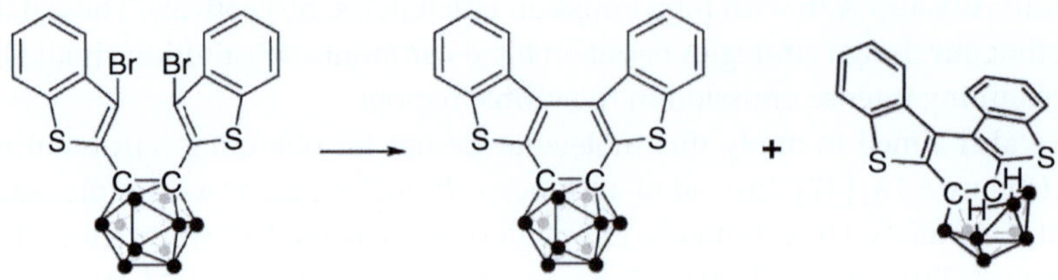

	hexane		crystal		scratched	
	λ_{PL} (nm)	Φ_{PL}	λ_{PL} (nm)	Φ_{PL}	λ_{PL} (nm)	Φ_{PL}
ortho	376	<0.01	–	–	–	–
nido	521	0.64	556	0.49	547	0.57

Figure 6.19 Chemical structures and optical properties of *ortho*- and *nido*-carboranes.

π-conjugation involving the boron cluster even through the distorted bibenzothiophene because of the *nido*-decaborane unit. Intense luminescence with solvatochromism in the solution state was obtained from the induced charge transfer state where the *nido*-carborane unit worked as an electron acceptor and the CT state was formed in the excited state. Finally, it was also demonstrated that the *nido*-decaborane unit is responsible for presenting the mechanochromic luminescent behavior as well as solid-state emission property. In this case, the π-conjugation on the distorted structure supported by the *nido*-decaborane anchor plays a critical role in suppressing ACQ followed by presenting solid-state emission with stimuli responsiveness.

6.6.4 Enhancement to Luminescent Efficiencies by Thermal Rearrangement from *Ortho*- to *Meta*-carborane

Because of its steric molecular shape and 3D conjugation, *o*-carborane has worked as a key unit for exhibiting solid-state luminescence by suppressing ACQ. On the other hand, so far, *m*-carboranes were simply used as an electron-accepting unit in commodity luminescent dyes. In this section, it is described that by thermally induced isomerization from *o*- to *m*-carborane, emission properties are enhanced in both solution and solid states (Figure 6.20) [50]. We eventually found the thermal reaction of **A3** proceeded near the melting point and subsequently generated the luminescent product **A3′**. To understand this unexpected reaction, we initially investigated the structure of the luminescent product from **A3** and estimated the mechanism. According to the data from the single crystal X-ray analysis with the product, it was clarified that transformation of the molecular structure from *o*- to *m*-carborane proceeds. Moreover, the rearrangement occurs at only one of two *o*-carborane units by heating at 300 °C, which is the relatively lower compared to the temperature ranges used in the previous studies. According to

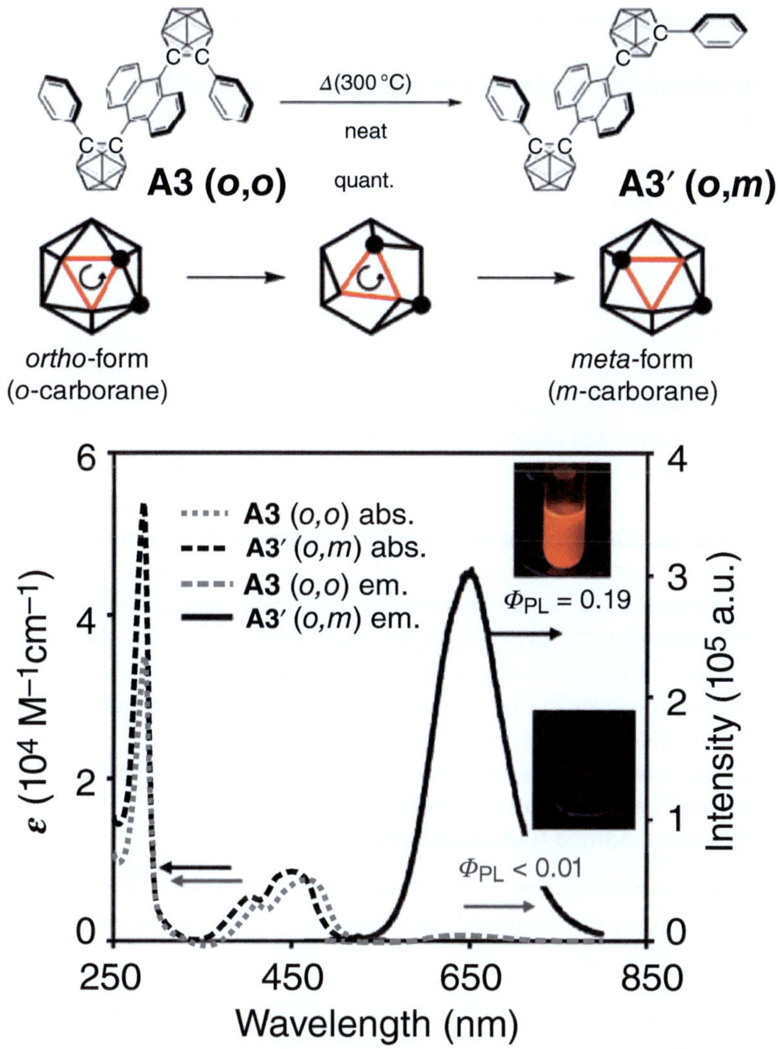

Figure 6.20 Thermal rearrangement in **A3** and optical spectra of the product **A3′**. *Source*: Reproduced from [50]/with permission of John Wiley & Sons.

theoretical investigations with **A3** for transition states with quantum calculation, the mechanism was well explained. By connecting to anthracene, the energy barrier in the first rearrangement step should be lowered. Therefore, isomerization can proceed efficiently. Finally, it was demonstrated from optical measurements that this isomerization can contribute to enhancing emission efficiencies in solution and solid states of the compound. Furthermore, the product showed deep-red emission in crystal (λ_{em} = 670 nm, Φ_{PL} = 0.07). These findings would be valid for designing functional luminescent materials based on thermally induced rearrangement in *o*-carborane.

6.7 Stimuli-responsive Luminochromic Materials

6.7.1 Multi-stimuli-responsiveness of Bis-*o*-carborane-modified Anthracene

In Section 6.6.4, the solid-state luminescent properties of **A3** was described. Furthermore, it has been reported that some solvents are able to be incorporated into the crystal when **A3** is recrystallized by changing the solvent, and each crystal having different types of solvent molecules shows diverse luminescent properties (Figure 6.21) [46]. **A3** forms the dimer structure in the crystalline state. It is likely that changes in crystal packing can vary the degree of π-stacking in the dimer followed by molecular environment. As a consequence, different optical properties can be induced. Interestingly, it is possible to dynamically change the incorporated solvent molecules to other types by fuming solvent vapor fuming in the crystalline state. Indeed, the crystals incorporating chloroform, dichloromethane, and benzene molecules exhibited different luminescence colors. Their optical properties were able to be switched by exposing the crystals to the other types of solvent vapors. Moreover, by heating the crystals and applying physical force to disrupt the crystalline packing, the incorporated solvent molecules were released, and additional changes in optical properties were observed. These results mean that the crystal samples incorporating solvents should have thermochromic and mechanochromic luminescent properties. It is generally difficult to induce structural changes in the crystalline state. Hence, it is intriguing that the bis-*o*-carborane anthracene can readily undergo such diverse changes.

6.7.2 Real-time Thermochromic Luminescent Properties of Bis-*o*-carborane-modified Benzodithiophenes

In the presence of *tert*-Bu groups at the phenyl groups in **DCB-H**, the *trans* conformation was obtained in the crystalline sample of **DCB-tBu**. By changing the

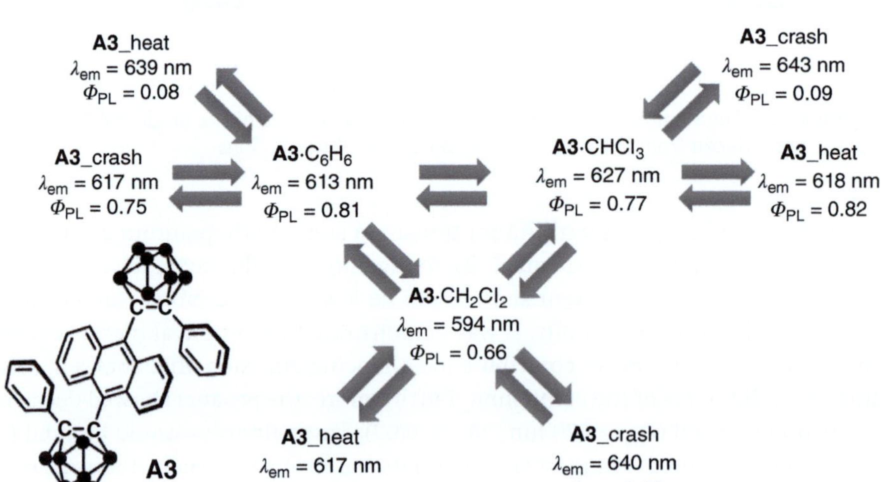

Figure 6.21 Multichromic behaviors of **A3** by heating, crashing, and vapor fuming.

conformation, sensitivity toward environmental changes might be varied because intermolecular interactions, mainly through π-stacking, are greatly changed. From both conformations, solid-state luminescent properties were obtained. Next, we examined influence of the substituent effect on optical properties, especially stimuli-responsive luminochromic characters. We further synthesized a series of **DCB-H** derivatives bearing various substituents on the phenyl ring, such as hydroxyl, methoxy, and cyano groups, and evaluated their optical properties (Figure 6.22) [51].

All products showed not only AIE but also CIE properties. From the ^{1}H NMR spectroscopic analysis, it was suggested that **DCB-OMe**, **DCB-CN**, and **DCB-tBu** have the *trans* conformation in the crystalline state, while **DCB-H** and **DCB-OH** form the *cis* conformation. From the optical measurements with each crystalline sample, thermochromic luminescent behaviors were observed from the derivatives except for **DCB-OH**. In particular, these molecules showed rapid and reversible luminochromic behaviors in the crystalline state. Degree of molecular motions should be responsible for stimuli-responsiveness in emission properties. Due to hydrogen-bond formation, **DCB-OH** hardly showed luminochromic behaviors.

6.7.3 Constant Luminescent Properties During Phase Changes and Luminochromic Behaviors

Two types of multi-functional solid-state emissive *o*-carborane derivatives with or without substituents on the *o*-carborane units have been reported (Figure 6.23) [52]. From the emission spectra, it was found that the pristine bis-*o*-carborane-substituted 1,4-bis(phenylethynyl)benzene (**CBH**) simultaneously exhibited dual emission from the LE and TICT states in the solution. By changing intensity ratios between these emission bands with solvent polarity and heating, clear luminochromic behaviors were obtained. These properties can be explained by the fact that TICT emission has higher sensitivity in environmental changes than LE emission. Interestingly,

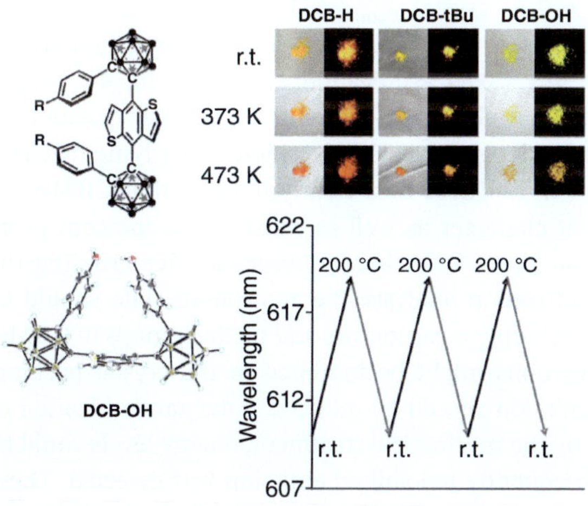

Figure 6.22 Rapid thermochromic luminescent properties of bis-*o*-carborane-substituted dithienylbenzenes. *Source*: [51]/with permission of John Wiley & Sons.

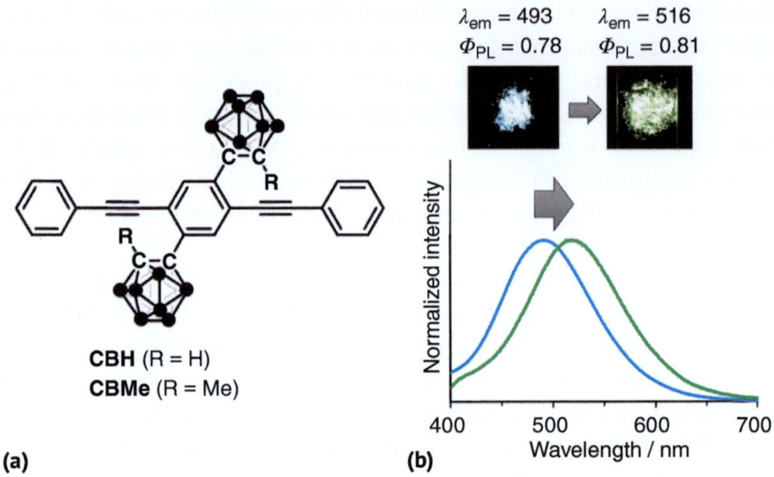

Figure 6.23 (a) Chemical structures of **CBH** and **CBMe** and (b) mechanochromic luminescent behaviors of **CBH.** *Source*: [52]/with permission of Royal Society of Chemistry.

TICT emission was observed even in the solid state. Similarly to pyrene- and anthracene-modified *o*-carboranes, intramolecular rotations at the *o*-carborane units are acceptable even in the condensed state. Furthermore, AIE and CIE were also observed. These solid-state emission enhancements can be rationalized in a manner similar to the conventional mechanism in which the two *o*-carborane units effectively inhibit nonspecific intermolecular interactions of the central conjugated moiety, thereby avoiding ACQ.

Next, we also synthesized the methyl-substituted derivative (**CBMe**) and discovered its constant emission properties in the solution, aggregation, and solid states, although luminescent color was drastically changed. In the case of **CBH**, the emission bands were observed in the similar wavelength regions (solution: $\lambda_{em} = 566$ nm, $\Phi_{PL} = 0.36$, aggregation: $\lambda_{em} = 538$ nm, $\Phi_{PL} = 0.30$, crystal: $\lambda_{em} = 537$ nm, $\Phi_{PL} = 0.57$). In contrast, **CBMe** showed drastic luminescent color changes with higher emission efficiencies (solution: $\lambda_{em} = 540$ nm, $\Phi_{PL} = 0.84$, aggregation: $\lambda_{em} = 542$ nm, $\Phi_{PL} = 0.81$, crystal: $\lambda_{em} = 493$ nm, $\Phi_{PL} = 0.78$). It should be noted that the peak position of the crystalline sample was shifted to longer wavelength region by applying mechanical forces ($\lambda_{em} = 516$ nm, $\Phi_{PL} = 0.61$). These data clearly indicate that **CBMe** has the mechanochromic luminescent character as well as constant luminescent properties. Since significant diffraction peaks were hardly observed after grinding the crystalline sample in the X-ray diffraction analysis, the amorphous state should be realized. Because of bulky units, including *o*-carboranes and methyl groups in **CBMe**, nonspecific intermolecular interactions might be inhibited in the crystal packing, and subsequently blue-shifted emission should be induced. After randomization of molecular distribution triggered by the mechanical treatment, energy levels could be stabilized by π-stacking, and subsequently red-shifted emission was detected. These data from **CBH** and **CBMe** suggest another important issue relating to emission efficiency.

6.7.4 Aggregation-caused Luminochromic Properties in the Polymer Film for Constructing Film-type Sensors

In general, since structural constraints are weak in the amorphous state, critical thermal motions that cause the deactivation of the excited state occur compared to in the crystalline state, resulting in a decrease in luminescence efficiency. On the other hand, since the main channel for nonradiative decay in *o*-carborane opens after the carbon–carbon bond is extended, it is suggested that constant emission efficiency was observed even in the amorphous state. This result suggests that **CBH** and **CBMe** are a versatile platform for developing luminescent materials that can maintain their luminescence efficiency even in various matrices. Therefore, we next designed polymer composite films possessing thermochromic luminescent properties based on the concept of **CBH**.

Stimuli-responsive luminochromic materials have attracted tremendous attention as a key component for developing next generation of film-type optical sensors. To meet a demand in industrial fields, such as for the detection of tiny stimuli and slight environmental changes, luminochromic organic dyes are regarded as a potential candidate because of tunability of material properties and processability. So far, although a wide variety of molecular skeletons that show intriguing optical responses toward external stimuli have been reported, most of materials can show the expected properties in the crystalline state. Therefore, there are difficulties in the applications of these conventional small-molecular based materials to practical film-type sensors. Since molecules would be randomly distributed in amorphous, nonspecific intermolecular interaction occurs, followed by critical ACQ. Moreover, external stimuli and environmental factors tend to be canceled in the amorphous state. In order to realize stimuli-responsive luminochromism with amorphous films, a new design was needed. In this situation, we designed the system based on dye-loaded polymer films consisting of conventional polymer, polystyrene (PS) as a matrix, and the **CBH** derivative (Figure 6.24) [53].

The bis-*o*-carborane-modified conjugated molecule is utilized as a key chromophore because of effective emission efficiencies, especially in the amorphous state, and potential environment-sensitive luminochromic properties. Based on **CBH**, we designed bis-*o*-carborane-substituted bis(thienylethynyl)benzene, **thio-CBH** in which bis-thienyl units were incorporated as electron-donating functional groups for enhancing environmental sensitivity by increasing electron density in the π-conjugation unit. From the optical measurements, it was shown that **thio-CBH** has suitable properties, including luminochromism in solid as well as solid-state emission, for constructing polymer-based luminochromic film materials. In the PS spin-coating film, dual-emission properties consisting of LE and CT emission bands were observed at around 400 and 600 nm, respectively. By increasing the concentration of **thio-CBH** in the PS matrix, the alteration of intensity ratios proceeded, followed by luminescent color change. Since aggregation formation was confirmed and enhanced in the films with increasing concentrations of **thio-CBH** in films, it can be said that **thio-CBH** can present aggregation-induced luminochromism. Finally, on the basis of this mechanism, we realized thermochromic luminescence from red to orange with the **thio-CBH**-loaded PS films. According to emission spectra and

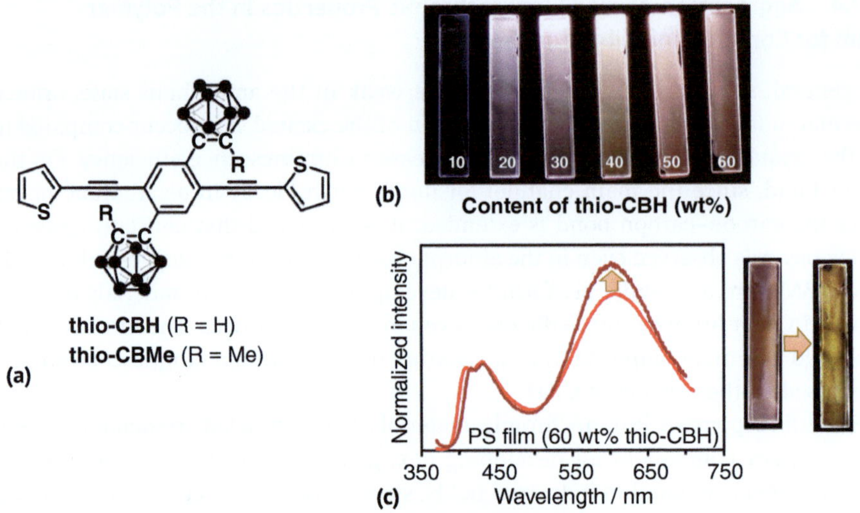

Figure 6.24 (a) Chemical structures of **thio-CBH** and **tho-CBMe**. (b) Emission properties of PS composite films containing variable concentrations of **thio-CBH**. (c) Thermochromic luminescence of the PS film containing 60 wt% **thio-CBH.** *Source*: [53]/with permission of John Wiley & Sons.

microscopic observations, it was shown that thermochromic luminescence should be originated from aggregation of **thio-CBH** in the PS matrix, followed by significant enhancement of intensity of CT emission. These irreversible film sensors are applicable for preparing temperature-history memory devices.

6.7.5 Thermochromic Luminescence by the Regulation of Rotation of *o*-carborane

In this section, thermochromic luminescent behaviors and regulation of their sensitivity are described based on the triphenylamine-modified-*o*-carborane system. In the previous parts, it was illustrated that electron-accepting ability of *o*-carborane could be varied by rotation at the connecting bond between aryl and *o*-carborane units. Based on this fact, we presumed that control of thermochromic luminescent behaviors might be possible by modulating the height of energetic barriers of intramolecular rotation. In other words, we expected to observe both sensitive and clear luminochromic responses from aryl-modified-*o*-carboranes. To confirm validity of this idea, we synthesized a series of triphenylamine-*o*-carborane derivatives with various numbers of substituents for tuning rotation barriers (Figure 6.25a) [54].

From the emission measurements, dual-emission properties consisting of the emission bands from the LE and TICT states were observed with the crystalline samples. Especially, from variable-temperature measurements, diverse responses were obtained. By cooling, the pristine and mono-methylated *o*-carboranes exhibited thermally independent emission during temperature changes, while thermochromic luminescence with drastic color change from red to blue was observed from bis-methylated *o*-carborane **Me2CB**. From the mechanistic study, it was

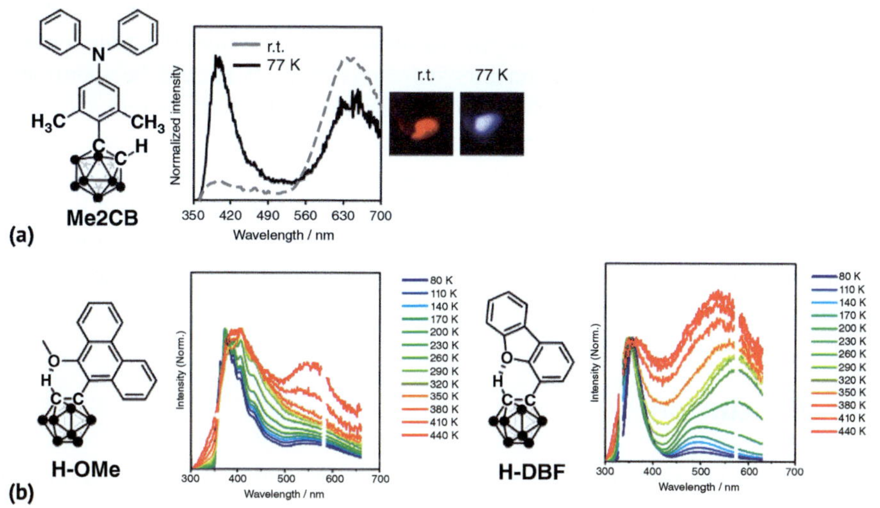

Figure 6.25 Thermochromic luminescent properties of (a) **Me2CB**, (b) **H-OMe** and **H-DBF**. *Source*: [54]/with permission of John Wiley & Sons.

revealed that alteration of intensity ratios between LE and TICT emission bands should be responsible for thermochromic luminescence. Furthermore, from the experimental and theoretical investigations, it was clearly indicated that the height of the rotation barrier of the *o*-carborane can be tuned to the substituent effect, and the degree of intramolecular rotations at the *o*-carborane units should play an important role in variable sensitivity in luminochromic behaviors in temperature change. In summary, we demonstrated not only thermally resistant solid-state emission but also clear thermochromic luminescence from the modified *o*-carboranes by tuning molecular motions in the excited state. These data mean that the new strategy for modulating solid-state emissive properties can be established utilizing the *o*-carborane dyad structure.

In the above study, steric hindrances were utilized for regulating the degree of intramolecular rotations. Next, by employing hydrogen bonds, we achieved modulated rotations (Figure 6.25b) [55]. It is known that the hydrogen atom on the cage-carbon atom ($C_{cage}H$) has weak acidity and can form noncovalent interactions with electron-rich atoms. We introduced larger π-planes, such as phenanthrene and dibenzofuran (DBF), possessing intramolecular interaction sites into the *o*-carborane scaffold and evaluated optical properties. From all compounds, dual emission properties consisting of LE and CT emission bands were observed. With the crystalline samples, at 80 K, intensity of the LE emission band was much larger than that of the CT emission band. Because of hydrogen bonding between $C_{cage}H$ and π-plane moieties, molecular structures should be fixed at the parallel conformation in which electronic accepting ability of *o*-carborane is minimized toward the conjugated system on the aryl moiety. By increasing temperature, relative intensity of the CT emission band increased. These results can be explained by the fact that thermal motion is enhanced as the temperature increases, and more molecules are released from fixation through the intramolecular $C_{cage}H \cdots O$ interaction. Finally, molecular rotations should be allowed, followed by the formation of the TICT state. Finally, thermochromic luminescent behaviors were observed. It should be mentioned that significant enhancement of the CT emission was observed around 400 °K in phenanthrene-modified *o*-carborane, meanwhile at 230 °K, the CT emission clearly appeared in dibenzofuran-modified *o*-carborane. By tuning bond energy, these data were in good agreement with the value of binding energy calculated by quantum chemical calculations. Thus, we can demonstrate that the temperature range, where the thermochromic luminescence begins to change color, can be adjusted by intermolecular interactions.

6.8 Conclusion

In this chapter, recent examples of research on the properties of carborane-containing polymers and conjugated molecules are described. In addition, since the progress on optical properties has been remarkable, details on luminescent properties were surveyed in the latter half. More recently, solid-state excimer emission has been observed from some kinds of aryl-modified *o*-carboranes [56–64], and other clusters also show unique behaviors [65–68]. Therefore, there is much room to explore new luminescent materials based on carboranes. For example, in the past, the main purpose of aryl-modified *o*-carborane research was to obtain solid-state luminescence based on the strategy utilizing the isolation effect of the aromatic ring moiety by *o*-carborane and to realize stimuli-responsive luminochromism. In many cases, molecular design has been carried out with the aim of avoiding intermolecular interactions, which cause ACQ. Furthermore, solid-state excimer luminescence has been discovered recently, as mentioned above. Since the interaction between two molecules is necessary for excimer formation, it can be said that the new strategy has been established to actively use intermolecular interactions. As a perspective, by using exciplexes, it is expected that even more diverse luminescence properties can be derived. On the other hand, since control of crystal structures has not yet been achieved, and a molecular design method from scratch has not yet been established, the synthesis of various molecules is still needed.

In order to control the solid-state luminescence of *o*-carboranes, it is essential to comprehend the molecular mobility in the excited states. We call molecules that undergo large structural relaxation in the excited state "excitation-driven molecules" and have focused as a platform for constructing stimuli-responsive luminochromic materials. In particular, although these molecular motions in the excited state generally cause emission annihilation as observed in conventional AIE-active dyes, if we can keep suppressing the channel for nonradiative decay processes, it is possible to obtain unique luminochromic behaviors toward tiny external stimuli and slight environmental changes as well as intense emission. It is relatively easy to predict the molecular motion modes in aryl-modified *o*-carboranes and estimate the energy states of each stage using quantum chemical calculations. Therefore, we can expect to continue developing luminescent materials with sensitive environmental responsiveness based on regulating molecular motions in excited states.

References

1 Yuan, S.W., White, D., Mason, A. et al. (2011). Porous organic polymers containing carborane for hydrogen storage. *International Journal of Energy Research* 37: 732–740. https://doi.org/10.1002/er.1886.

2 Wang, C.F., Huang, F.R., Jiang, Y. et al. (2012). A novel oxidation resistant $SiC/B_4C/C$ nanocomposite derived from a carborane-containing conjugated polycarbosilane. *Journal of the American Ceramic Society* 95: 71–74. https://doi.org/10.1111/j.1551-2916.2011.04972.x.

3 Fabre, B., Chayer, S., and Vicente, M.G.H. (2003). First conducting polymer functionalized with covalently linked carborane units. *Electrochemistry Communications* 5: 431–434. https://doi.org/10.1016/s1388-2481(03)00095-x.

4 Fabre, B., Clark, J.C., Vicente, M.G.H. (2006). Synthesis and electrochemistry of carboranylpyrroles. Toward the preparation of electrochemically and thermally resistant conjugated polymers. *Macromolecules* 39: 112–119. https://doi.org/10.1021/ma051508v.

5 Hao, E., Fabre, B., Fronczek, F.R. et al. (2007). Syntheses and electropolymerization of carboranyl-functionalized pyrroles and thiophenes. *Chemistry of Materials* 19: 6195–6205. https://doi.org/10.1021/cm701935n.

6 Hao, E., Fabre, B., Fronczek, F.R. et al. (2007). Poly[di(2-thiophenyl) carborane]s: conducting polymers with high electrochemical and thermal resistance. *Chemical Communications* 42: 4387–4389. https://doi.org/10.1039/b710375a.

7 Barrière, F., Fabre, B., Hao, E.H. et al. (2009). Electropolymerizable 2,2′-carboranyldithiophenes. Structure-property investigations of the corresponding conducting polymer films by electrochemistry, UV-visible spectroscopy and conducting probe atomic force microscopy. *Macromolecules* 42: 2981–2987. https://doi.org/10.1021/ma802369z.

8 Cansu-Ergun, E.G. and Cihaner, A. (2013). Electro-optical properties of poly[di(2-thiophenyl)carborane] and its opto-electronic application. *Materials Chemistry and Physics* 143: 387–392. https://doi.org/10.1016/j.matchemphys.2013.09.015.

9 Cansu-Ergun, E.G. and Cihaner, A. (2013). A new carborane based polymeric electrochrome. *Journal of Electroanalytical Chemistry* 707: 78–84. https://doi.org/10.1016/j.jelechem.2013.08.026.

10 Marshall, J., Fei, Z.P., Yau, C.P. et al. (2014). Incorporation of benzocarborane into conjugated polymer systems: synthesis, characterisation and optoelectronic properties. *Journal of Materials Chemistry C* 2: 232–239. https://doi.org/10.1039/c3tc31663g.

11 Marshall, J., Hooton, J., Han, Y. et al. (2014). Polythiophenes with vinylene linked *ortho, meta* and *para*-carborane sidechains. *Polymer Chemistry* 5: 6190–6199. https://doi.org/10.1039/c4py00767k.

12 Aniés, F., Qiao, Z.R., Nugraha, M.I. et al. (2022). *N*-type polymer semiconductors incorporating *para, meta*, and *ortho*-carborane in the conjugated backbone. *Polymer* 240: 124481. https://doi.org/10.1016/j.polymer.2021.124481.

13 Simon, Y.C., Peterson, J.J., Mangold, C. et al. (2009). Synthesis of polyfluorenes with pendant silylcarboranes. *Macromolecules* 42: 512–516. https://doi.org/10.1021/ma8018816.

14 Marshall, J., Schroeder, B.C., Bronstein, H. et al. (2014). Benzocarborano[2,1-b:3,4-b′]dithiophene containing conjugated polymers: synthesis, characterization, and optoelectronic properties. *Macromolecules* 47: 89–96. https://doi.org/10.1021/ma402273w.

15 Hablot, D., Ziessel, R., Alamiry, M.A.H. et al. (2013). Nanomechanical properties of molecular-scale bridges as visualised by intramolecular electronic energy transfer. *Chemical Science* 4: 444–453. https://doi.org/10.1039/C2SC21505E.

16 Peterson, J.J., Simon, Y.C., Coughlin, E.B. et al. (2009). Polyfluorene with p-carborane in the backbone. *Chemical Communications* 2009: 4950–4952. https://doi.org/10.1039/b908131c.

17 Peterson, J.J., Davis, A.R., Werre, M. et al. (2011). Carborane-containing poly(fluorene): response to solvent vapors and amines. *ACS Applied Materials & Interfaces* 3: 1796–1799. https://doi.org/10.1021/am2000856.

18 Martin, K.L., Krishnamurthy, A., Strahan, J. et al. (2019). Excited state characterization of carborane-containing poly(dihexyl fluorene)s. *The Journal of Physical Chemistry A* 123: 1701–1709. https://doi.org/10.1021/acs.jpca.8b07955.

19 Martin, K.L., Smith, J.N., Young, E.R. et al. (2019). Synthetic emission tuning of carborane-containing poly(dihexylfluorene)s. *Macromolecules* 52: 7951–7960. https://doi.org/10.1021/acs.macromol.9b01325.

20 Davis, A.R., Peterson, J.J., and Carter, K.R. (2012). Effect of *o*-carborane on the optoelectronic and device-level properties of poly(fluorene)s. *ACS Macro Letters* 1: 469–472. https://doi.org/10.1021/mz300037u.

21 Kokado, K., Tokoro, Y., and Chujo, Y. (2009). Luminescent *m*-carborane-based π-conjugated polymer. *Macromolecules* 42: 2925–2930. https://doi.org/10.1021/ma900174j.

22 Kokado, K. and Chujo, Y. (2009). Emission via aggregation of alternating polymers with *o*-carborane and *p*-phenylene–ethynylene sequences. *Macromolecules* 42: 1418–1420. https://doi.org/10.1021/ma8027358.

23 Kokado, K., Tokoro, Y., and Chujo, Y. (2009). Luminescent and axially chiral π-conjugated polymers linked by carboranes in the main chain. *Macromolecules* 42: 9238–9242. https://doi.org/10.1021/ma902094u.

24 Kokado, K. and Chujo, Y. (2010). Polymer reaction of poly(*p*-phenylene-ethynylene) by addition of decaborane: modulation of luminescence and heat resistance. *Polymer Journal* 42: 363–367. https://doi.org/10.1038/pj.2010.13.

25 Kokado, K. and Chujo, Y. (2011). A luminescent coordination polymer based on bisterpyridyl ligand containing *o*-carborane: two tunable emission modes. *Dalton Transactions* 40: 1919–1923. https://doi.org/10.1039/C1DT90011K.

26 Kokado, K., Tominaga, M., and Chujo, Y. (2010). Aromatic ring-fused carborane-based luminescent π-conjugated polymers. *Macromolecular Rapid Communications* 31: 1389–1394. https://doi.org/10.1002/marc.201000160.

27 Tominaga, M., Morisaki, Y., and Chujo, Y. (2013). Luminescent polymer consisting of 9,12-linked *o*-carborane. *Macromolecular Rapid Communications* 34: 1357–1362. https://doi.org/10.1002/marc.201300368.

28 Chujo, Y. and Tanaka, K. (2015). New polymeric materials based on element-blocks. *Bulletin of the Chemical Society of Japan* 88: 633–643. https://doi.org/10.1246/bcsj.20150081.

29 Gon, M., Tanaka, K., and Chujo, Y. (2018). Recent progress in the development of advanced element-block materials. *Polymer Journal* 50: 109–126. https://doi.org/10.1038/pj.2017.56.

30 Tanaka, K. and Chujo, Y. (2020). Modulation of the solid-state luminescent properties of conjugated polymers by changing the connecting points of flexible boron element-blocks. *Polymer Journal* 52: 555–566. https://doi.org/10.1038/s41428-020-0316-y.

31 Tanaka, K. and Chujo, Y. (2023). Frustrated element-blocks: a new platform for constructing unique stimuli-responsive luminescent materials. *Polymer Journal* 55: 353–364. https://doi.org/10.1038/s41428-022-00709-w.

32 Ito, S., Gon, M., Tanaka, K. et al. (2021). Recent developments in stimuli-responsive luminescent polymers composed of boron compounds. *Polymer Chemistry* 12: 6372–6380. https://doi.org/10.1039/D1PY01170G.

33 Ochi, J, Tanaka, K., and Chujo, Y. (2020). Recent progresses in the development of solid-state luminescent *o*-carboranes with stimuli responsivity. *Angewandte Chemie International Edition* 132: 9841–9855. https://doi.org/10.1002/anie.201916666.

34 Ochi, J., Tanaka, K., and Chujo, Y. (2021). Experimental proofs for emission annihilation through bond elongation at the carbon–carbon bond in *o*-carborane with fused biphenyl-substituted compounds. *Dalton Transactions* 50: 1025–1033. https://doi.org/10.1039/D0DT03618H.

35 Tanaka, K. (2024). Solid-state excimer emission of *o*-carborane derivatives and applications for stimuli-responsive luminescent materials. *Dalton Transactions* 53: 9240–9247. https://doi.org/10.1039/D4DT01072H.

36 Tanaka, K., Nishino, K., Ito, S. et al. (2017). Development of solid-state emissive *o*-carborane and theoretical investigation of the mechanism of the aggregation-induced emission behaviors of organoboron "element-blocks". *Faraday Discussions* 196: 31–42. https://doi.org/10.1039/C6FD00155F.

37 Nishino, K., Hashimoto, K., Tanaka, K. et al. (2016). Synthesis and properties of highly-rigid conjugation system based on Bi(benzo[*b*]thiophene)-fused *o*-carborane. *Tetrahedron Letters* 57: 2025–2028. https://doi.org/10.1016/j.tetlet.2016.03.069.

38 Nishino, K., Uemura, K., Tanaka, K. et al. (2018). Dual emission via remote control of molecular rotation of *o*-carborane in the excited state by the distant substituents in tolane-modified dyads. *New Journal of Chemistry* 16: 4210–4214. https://doi.org/10.1039/C7NJ04283C.

39 Nishino, K., Uemura, K., Tanaka, K. et al. (2017). Enhancement of aggregation-induced emission by introducing multiple *o*-carborane substitutions into triphenylamine. *Molecules* 22: 2009. https://doi.org/10.3390/molecules22112009.

40 Morisaki, Y., Tominaga, M., Ochiai, T. et al. (2014). *o*-Carborane-based biphenyl and *p*-terphenyl derivatives. *Chemistry: An Asian Journal* 9: 1247–1251. https://doi.org/10.1002/asia.201400067.

41 Naito, H., Nishino, K., Morisaki, Y. et al. (2017). Solid-state emission of the anthracene–*o*-carborane dyad from the twisted-intramolecular charge transfer in the crystalline state. *Angewandte Chemie International Edition* 56: 254–259. https://doi.org/10.1002/ange.201609656.

42 Ochi, J., Yanagihara, T., Tanaka, K. et al. (2023). Tuning of the height of energy barrier between locally-excited and charge transfer states by altering the fusing position of *o*-carborane in phenylnaphthalene. *Physical Chemistry Chemical Physics* 25: 11839–11844. https://doi.org/10.1039/D3CP00334E.

43 Ochi, J., Tanaka, K., and Chujo, Y. (2023). Investigation of the substitution site effect on *o*-carborane-based chromophores by anthracene introduction at the B(3) position. *Bulletin of the Chemical Society of Japan* 96: 98–102. https://doi.org/10.1246/bcsj.20220310.

44 Nishino, K., Yamamoto, H., Tanaka, K. et al. (2016). Development of solid-state emissive materials based on multi-functional *o*-carborane-pyrene dyads. *Organic Letters* 18: 4064–4067. https://doi.org/10.1021/acs.orglett.6b01920.

45 Naito, H., Nishino, K., Morisaki, Y. et al. (2017). Highly-efficient solid-state emissions of anthracene–*o*-carborane dyads with various substituents and their thermochromic luminescence properties. *Journal of Materials Chemistry C* 4: 10047–10054. https://doi.org/10.1039/C7TC02682J.

46 Naito, H., Morisaki, Y., and Chujo, Y. (2015). *o*-Carborane-based anthracene: a variety of emission behaviors. *Angewandte Chemie International Edition* 54: 5084–5087. https://doi.org/10.1002/anie.201500129.

47 Naito, H., Nishino, K., Morisaki, Y. et al. (2017). Luminescence color tuning from blue to near infrared of stable luminescent solid materials based on bis-*o*-carborane-substituted oligoacenes. *Chemistry: An Asian Journal* 12: 2134–2138. https://doi.org/10.1002/asia.201700815.

48 Nishino, K., Uemura, K., Tanaka, K. et al. (2018). Modulation of the *cis*- and *trans*-conformations in the bis-*o*-carborane substituted benzodithiophenes and emission enhancement effect on luminescent efficiency by solidification. *European Journal of Organic Chemistry* 2018: 1507–1512. https://doi.org/10.1002/ejoc.201701641.

49 Nishino, K., Hashimoto, K., Tanaka, K. et al. (2018). Comparison of luminescent properties of helicene-like bibenzothiophenes with *o*-carborane and 5,6-dicarba-*nido*-decaborane. *Science China Chemistry* 61: 940–946. https://doi.org/10.1007/s11426-018-9258-y.

50 Naito, H., Uemura, K., Morisaki, Y. et al. (2018). Enhancement of luminescence efficiencies by thermal rearrangement from *ortho*- to *meta*-carborane in bis-carborane-substituted acenes. *European Journal of Organic Chemistry* 2018: 1885–1890. https://doi.org/10.1002/ejoc.201800151.

51 Nishino, K., Tanaka, K., Morisaki, Y. et al. (2019). Design of thermochromic luminescent dyes based on the bis(*ortho*-carborane)-substituted benzodithiophene structure. *Chemistry: An Asian Journal* 14: 789–795. https://doi.org/10.1002/asia.201801529.

52 Mori, H, Nishino, K., Wada, K. et al. (2018). Modulation of luminescence chromic behaviors and environment-responsive intensity changes by substituents in bis-*o*-carborane-substituted conjugated molecules. *Materials Chemistry Frontiers* 2: 573–579. https://doi.org/10.1039/C7QM00486A.

53 Wada, K., Hashimoto, K., Ochi, J. et al. (2021). Rational design for thermochromic luminescence in amorphous polystyrene films with bis-*o*-carborane-substituted enhanced-conjugated molecule having aggregation-induced luminochromism. *Aggregate* 2: e93. https://doi.org/10.1002/agt2.93.

54 Nishino, K., Tanaka, K., and Chujo, Y. (2019). Tuning of sensitivity in thermochromic luminescence by regulating molecular rotation based on triphenylamine-substituted *o*-carboranes. *Asian Journal of Organic Chemistry* 8: 2228–2232. https://doi.org/10.1002/ajoc.201900537.

55 Ochi, J., Yuhara, K., Tanaka, K. et al. (2022). Controlling the dual-emission character of aryl-modified *o*-carboranes by intramolecular CH···O interaction sites. *Chemistry: A European Journal* 28: e202200155. https://doi.org/10.1002/chem.202200155.

56 Nishino, K., Yamamoto, H., Tanaka, K. et al. (2017). Solid-state themochromic luminescence through twisted intramolecular charge transfer and excimer formation of a carborane–pyrene dyad with an ethynyl spacer. *Asian Journal of Organic Chemistry* 6: 1818–1822. https://doi.org/10.1002/ajoc.201700390.

57 Nishino, K., Yamamoto, H., Ochi, J. et al. (2019). Time-dependent emission enhancement of the ethynylpyrene–*o*-carborane dyad and its application as a luminescent color sensor for evaluating water contents in organic solvents. *Chemistry: An Asian Journal* 14: 1577–1581. https://doi.org/10.1002/asia.201900396.

58 Yamamoto, H., Ochi, J., Yuhara, K. et al. (2022). Switching between intramolecular charge transfer and excimer emissions in solids based on aryl-modified ethynyl–*o*-carboranes. *Cell Reports Physical Science* 3: 100758. https://doi.org/10.1016/j.xcrp.2022.100758.

59 Ochi, J., Tanaka, K., and Chujo, Y. (2019). Improvement of solid-state excimer emission of the aryl–ethynyl–*o*-carborane skeleton by acridine introduction. *European Journal of Organic Chemistry* 2019: 2984–2988. https://doi.org/10.1002/ejoc.201900212.

60 Ochi, J., Tanaka, K., and Chujo, Y. (2021). Dimerization-induced solid-state excimer emission showing consecutive thermochromic luminescence based on acridine-modified *o*-carboranes. *Inorganic Chemistry* 60: 8990–8997. https://doi.org/10.1021/acs.inorgchem.1c00901.

61 Ochi, J., Tanaka, K., and Chujo, Y. (2023). Alternately π-stacked systems assisted by *o*-carborane: dual excimer emission and color modulation by B_{cage}–methylation. *Angewandte Chemie International Edition* 62: e202214397. https://doi.org/10.1002/anie.202214397.

62 Yuhara, K. and Tanaka, K. (2023). Direction switching and self-recovering mechanochromic luminescence of anthracene-modified *o*-carboranes. *Chemistry: A European Journal* 29: e202301189. https://doi.org/10.1002/chem.202301189.

63 Yuhara, K., Tanaka, K., and Chujo, Y. (2022). Regulation of solid-state dual-emission properties by switching luminescence processes based on a bis-*o*-carborane-modified anthracene triad. *Materials Chemistry Frontiers* 6: 1414–1420. https://doi.org/10.1039/D2QM00211F.

64 Yuhara, K. and Tanaka, K. (2024). The photosalient effect and thermochromic luminescence based on *o*-carborane-assisted π-stacking in the crystalline state. *Angewandte Chemie International Edition* 63: e202319712. https://doi.org/10.1002/anie.202319712.

65 Nishino, K., Morisaki, Y., Tanaka, K. et al. (2017). Electron-donating abilities and luminescence properties of tolane-substituted *nido*-carboranes. *New Journal of Chemistry* 15: 10550–10554. https://doi.org/10.1039/C7NJ02438J.

66 Uemura, K., Tanaka, K., and Chujo, Y. (2022). Conformation-dependent electron donation of *nido*-carborane substituents and its influence on phosphorescence of tris(2,2′-bipyridyl)ruthenium(II) complex. *Crystals* 12: 688. https://doi.org/10.3390/cryst12050688.

67 Yanagihara, T. and Tanaka, K. (2023). Photophysical properties of pyrene-substituted *nido*-carborate anion and the change of luminescence species by counter cations in solution and solid states. *Advanced Optical Materials* 11: 2300492. https://doi.org/10.1002/adom.202300492.

68 Ochi, J., Yanagihara, T., Tanaka, K. et al. (2024). Synthesis, cation exchange, and optical properties of pyrene-based chromophores containing $[CB_9H_9]^-$ and $[CB_{11}H_{11}]^-$ cluster units. *Bulletin of the Chemical Society of Japan* 97: uoae005. https://doi.org/10.1093/bulcsj/uoae005.

7

Low-bandgap Conjugated Polymers

Yi Zhang, Chunchen Liu, and Fei Huang

State Key Laboratory of Luminescent Materials and Devices, Guangdong Basic Research Center of Excellence for Energy & Information Polymer Materials, Institute of Polymer Optoelectronic Materials and Devices, School of Materials Science and Engineering, South China University of Technology, Guangzhou 510640, P. R. China

7.1 Introduction

Conjugated polymers (CPs) are a class of polymers characterized by a backbone of alternating single and double bonds, which facilitates the delocalization of π-electrons [1]. One of their most critical properties is the bandgap, which correlates to their photophysical properties. Greater π-conjugation along the polymer backbone and stronger intramolecular charge transfer typically lead to a reduced bandgap, defined as the energy difference between the highest occupied molecular orbital (HOMO) and the lowest unoccupied molecular orbital (LUMO), as illustrated in Figure 7.1a [2]. Generally, the bandgap (E_g) of CPs is commonly estimated from the absorption edge of their thin films, as shown in Figure 7.1b, using a specific calculation formula ($E_g = \frac{1240}{\lambda_{onset}}$ eV) [3]. Alternatively, it can also be determined electrochemically by measuring the difference between the first oxidation potential and the first reduction potential (Figure 7.1c), which is calculated using the given formula ($E_g = |E_{ox} - E_{red}|$ eV) [4]. Unless otherwise specified in the chapter, the term "bandgap" generally refers to the value derived from the absorption edge of thin films. Based on their E_g, CPs can be categorized into wide-, medium-, and low-bandgap groups. Historically, polymers with an E_g below 1.5 eV have been classified as low-bandgap conjugated polymers (LBG-CPs) [1, 2, 4]. Extensive research efforts have led to the development of numerous polymers with E_g values below this threshold, many of which have been thoroughly reviewed in previous literatures. In the chapter, we specifically define LBG-CPs as those with an E_g smaller than the silicon bandgap of 1.12 eV [5–8].

Since the pioneer work reporting LBG-CP, poly(isothianaphthene) (PITN), with a bandgap of approximately 1.0 eV in 1984, the synthesis of such polymers has attracted significant research interest [9, 10]. Various synthetic methods, including

Conjugated Polymer Synthesis and Materials Applications, First Edition. Edited by Yoshiki Chujo.
© 2026 Wiley-VCH GmbH. All rights reserved, including rights for text and data mining and training of artificial intelligence technologies or similar technologies. Published 2026 by Wiley-VCH GmbH.

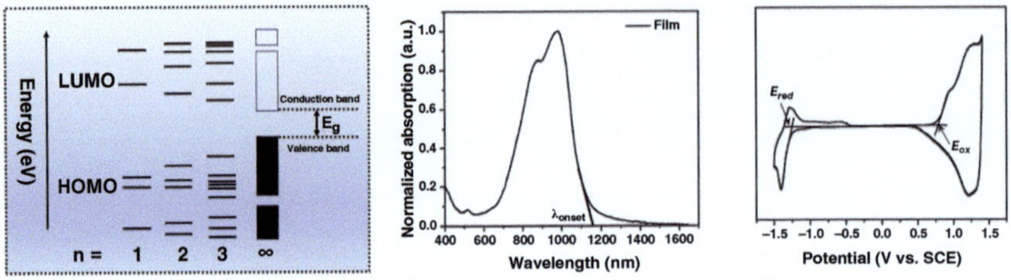

Figure 7.1 (a) The illustration of bandgap (E_g) formation during the polymerization of a conjugated monomer into a conjugated polymer (CP), (b) The illustration of E_g calculated by the onset wavelength (λ_{onset}) of thin-film absorption, and (c) illustration of E_g calculated by electrochemical method.

Stille coupling, Suzuki coupling, and Aldol condensation, have been developed to produce a diverse range of LBG-CPs [11–14]. The reduction in E_g of LBG-CPs usually affords them some unique properties. For instance, lowering the bandgap can facilitate the generation of free carriers within the polymer, enabling metallic-like conductivity and strong absorption in the near-infrared (NIR) region of LBG-CPs. Additionally, LBG-CPs offer several advantages, such as tunable energy levels, solution processability, and narrow bandgaps, making them promising candidates for applications in organic photodetectors (OPDs), organic field-effect transistors (OFETs), organic thermoelectrics (OTEs), organic electrochemical transistors (OECTs), NIR fluorescence imaging (FLI), and photothermal therapy (PTT) [5, 8, 15–20]. The chapter provides a systematic review of the representative design principles, synthetic strategies, and applications of LBG-CPs.

7.2 Design Principles of LBG-CPs

Over the past four decades since the discovery of the first LBG-CP, PITN, chemists have developed a wide range of LBG-CPs using various design strategies. Despite their structural diversity, the approaches to bandgap reduction can generally be classified into two main categories: (1) utilizing donor–acceptor (D–A) interactions along the polymer backbone to lower the bandgap and (2) incorporating stabilized quinoidal resonance structures into the main chain [5, 10, 18].

7.2.1 D–A Interaction for Bandgap Reduction

Incorporating alternating donor (D) and acceptor (A) units within the polymer backbone is an effective strategy for reducing the bandgap, giving rise to polymers collectively known as "D–A conjugated polymers" (D–A CPs) (Figure 7.2a) [1]. This approach gained prominence after Havinga et al. demonstrated that a polysquaraine polymer with a bandgap below 0.5 eV could be synthesized via condensation reactions [21]. In D–A copolymers, the molecular orbitals of the D and A units interact, leading to hybridized HOMO and LUMO levels. Specifically, the

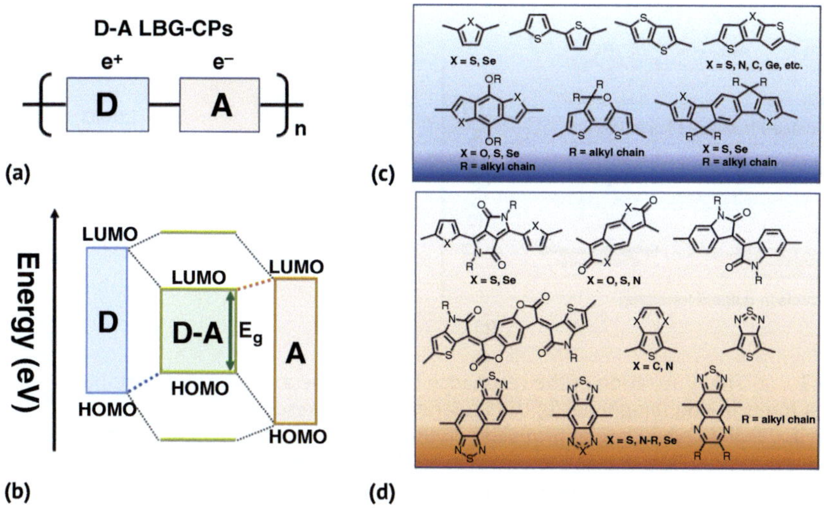

Figure 7.2 (a) The illustration of D–A low-bandgap conjugated polymers (LBG-CPs), (b) the simplified schematic of bandgap reduction through D–A orbital hybridization, and representative donor (D) (c) and acceptor (A) (d) units used for constructing LBG-CPs.

HOMO is primarily governed by the D unit, while the LUMO is dominated by the A unit, resulting in a narrower optical bandgap (E_g) compared to the respective homopolymers of the D or A units (illustrated in Figure 7.2b) [4]. By enhancing the electron-donating strength of the D unit (raising the HOMO level) and/or increasing the electron-withdrawing ability of the A unit (lowering the LUMO level), orbital hybridization can be further intensified, strengthening intramolecular charge transfer and further narrowing E_g. Thus, the rational selection of D and A units enables precise modulation of the HOMO and LUMO levels in CPs. A wide range of D and A units with strong electronic properties have been developed, facilitating the design of LBG-CPs with absorption onset wavelengths exceeding 1100 nm (corresponding to bandgaps below 1.12 eV). Figure 7.2c, d illustrate commonly used D and A units for constructing LBG-CPs.

7.2.2 Stabilized Quinoidal Resonance for Bandgap Reduction

Incorporating quinoidal resonance structures into the polymer backbone is another well-established strategy for significantly reducing the bandgap of CPs. It is widely recognized that all CPs exhibit both aromatic and quinoidal resonance forms, as illustrated in Figure 7.3a [2]. While the quinoidal form has a lower bandgap compared to the aromatic form, it disrupts aromaticity and is energetically less stable, thereby limiting its natural contribution to bandgap reduction. For instance, polythiophene, which predominantly adopts the aromatic form, has a relatively wide E_g of approximately 2.0 eV. However, when an aromatic ring is fused to the 3,4-positions of the thiophene unit, as shown in Figure 7.3b, the polymer backbone stabilizes in the quinoidal form to restore aromaticity [22, 23]. This stabilization reduces bond length alternation, thereby increasing the quinoidal character and

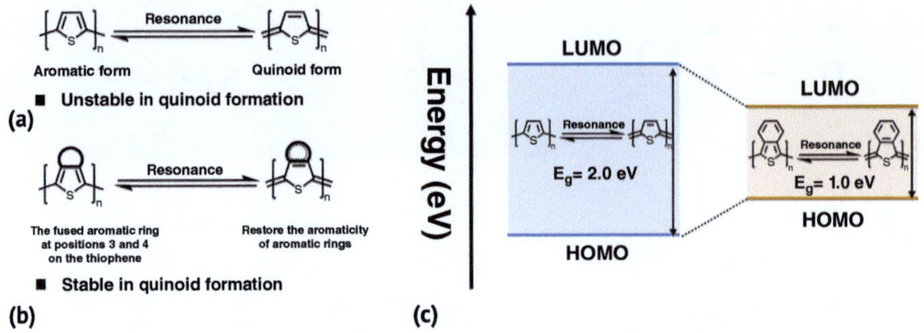

Figure 7.3 (a) The illustration of the resonance between the aromatic and unstable quinoidal forms in polythiophene, (b) the illustration of the resonance between the aromatic and stabilized quinoidal forms in polythiophene with an aromatic ring fused at the 3,4-positions of the thiophene unit, and (c) the schematic presentation of E_g and resonance structures in polythiophene and poly(thieno[3,4-b]pyrazine) (PTIN).

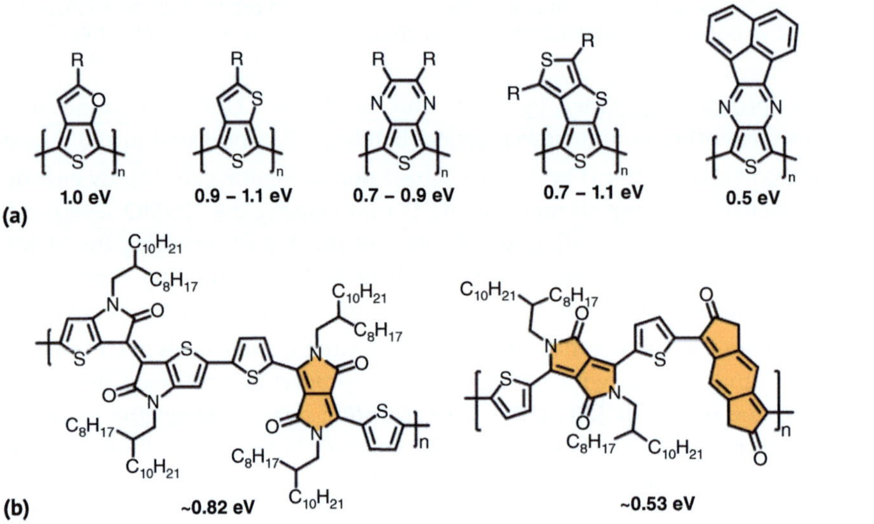

Figure 7.4 (a) The representative structures of quinoidal CPs based on polythiophene derivative backbones, (b) the molecular structures of alternating quinoidal resonance units with D or A units.

effectively lowering E_g. For example, in poly(thieno[3,4-b]pyrazine) (PTIN), where a benzene ring is fused to the β-position of thiophene, the bandgap is reduced to approximately 1.0 eV, making it the first recognized LBG-CP (Figure 7.3c) [22]. This breakthrough inspired the development of numerous LBG-CPs using similar quinoidal stabilization strategies, with representative examples shown in Figure 7.4a.

Notably, incorporating alternating quinoidal resonance units with donor (D) or acceptor (A) units allows for precise bandgap (E_g) tuning, leading to polymers with further reduced E_g for various applications. Representative molecular structures of such LBG-CPs are shown in Figure 7.4b.

7.3 Synthesis of LBG-CPs

The research and development of LBG-CPs have flourished, leading to the synthesis of numerous polymers based on the aforementioned design principles. Among these, the Stille coupling reaction has played a pivotal role in the design, synthesis, and advancement of CPs, particularly those with low bandgaps. In addition to Stille coupling, several other polymerization methods have been employed for the synthesis of LBG-CPs. This section introduces these polymerization techniques and provides representative examples to enhance understanding of their underlying mechanisms. The widely used Stille coupling method is discussed in detail, categorized by common structural units, while other polymerization approaches are briefly introduced.

7.3.1 Stille Coupling for Constructing LBG-CPs

Stille coupling, a crucial carbon–carbon coupling reaction, has become an indispensable tool in organic synthesis since its formal introduction in 1978. Over the past four decades, this method has garnered significant attention in the synthesis of CPs for organic optoelectronics, thanks to its mild reaction conditions, straightforward preparation of polymerization monomers, and high monomer conversion efficiency. As illustrated in Figure 7.5, the Stille coupling reaction involves conjugated dibromo monomers and distannyl compounds dissolved in appropriate solvents, catalyzed by palladium (Pd) and conducted under suitable reaction temperatures to form CPs. In addition to traditional binary copolymers, ternary and quaternary copolymers can also be synthesized by introducing additional dibromo or distannyl monomers into the reaction system (Figure 7.5). Moreover, the ratio of specific segments in

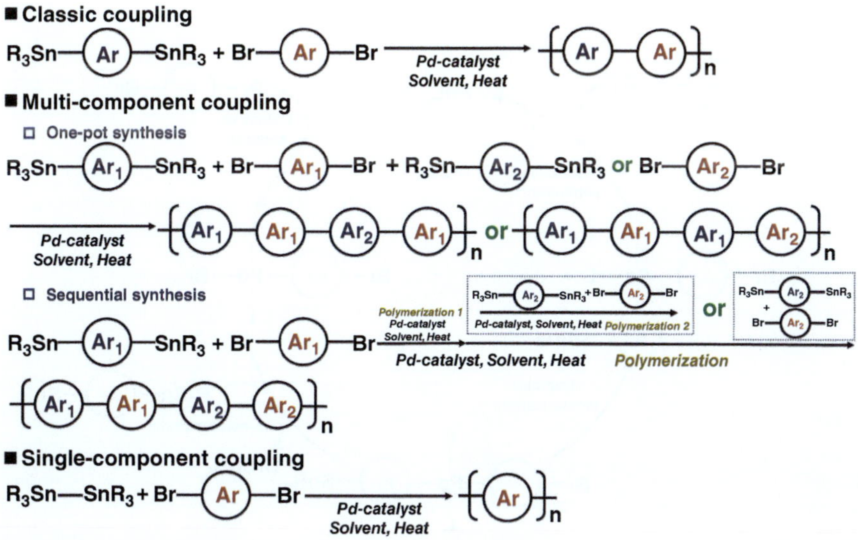

Figure 7.5 The simplified schematic of three categories of Stille coupling for preparing conjugated polymers (CPs).

multicomponent copolymers can be controlled by adjusting the sequence and quantity of monomer additions. Beyond copolymers, Stille coupling can also be used to synthesize homopolymers through reactions between dibromo monomers and hexabutyldistannane or hexamethyldistannane (Figure 7.5). This versatility highlights the significant advantages of Stille coupling in constructing CPs. By incorporating donor, acceptor, or quinoidal units into the reaction, researchers can prepare a wide variety of LBG-CPs with tunable energy levels and absorption properties.

The Stille coupling mechanism, essential for understanding its underlying process, consists of four stages: oxidative addition, transmetalation, isomerization, and reductive elimination. The active zero-valent palladium catalyst first undergoes oxidative addition with halogenated compounds, forming a cis intermediate, which then rapidly isomerizes into a more stable trans intermediate. The trans intermediate subsequently reacts with the organotin compound in a transmetalation step, followed by reductive elimination, which regenerates the zero-valent palladium catalyst and produces the coupled product, completing one catalytic cycle (Figure 7.6). In polymerization reactions, this cycle repeats until the reaction is terminated.

For Stille coupling monomers, electron-withdrawing units often have high electron affinity, stabilizing the carbon—halogen bonds (e.g. C—Br), particularly in dibromo-substituted compounds, which facilitates the reaction. In contrast, electron-donating units possess high electron density and nucleophilicity, enhancing their interaction with organotin groups (e.g. trimethylstannyl or tributylstannyl) and ensuring high selectivity with minimal side reactions. Electron-withdrawing dibromo monomers are typically synthesized using N-bromosuccinimide (NBS), while electron-donating stannyl monomers are prepared via organolithium reagents such as n-butyllithium (n-BuLi) or lithium diisopropylamide (LDA) and trimethyltin chloride. Below, representative reactions are presented based on the classification of electron-accepting units, providing readers with insights into specific methods for synthesizing LBG-CPs from these units.

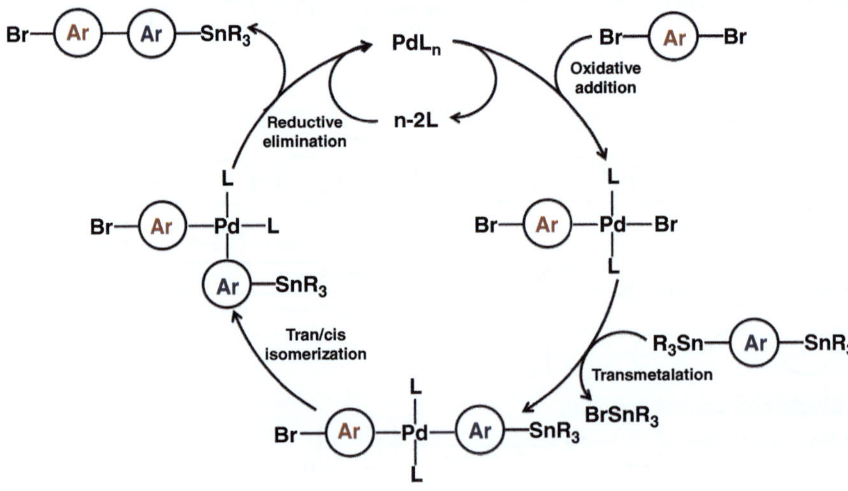

Figure 7.6 The illustration of mechanism of Stille coupling.

7.3.1.1 Synthesis of LBG-CPs Based on Thieno[3,4-b]Thiadiazole (TT) and Its Derivatives

Thieno[3,4-b]thiadiazole (TT), known for its strong electron-withdrawing properties and quinoidal resonance characteristics, has been widely used in the design of LBG-CPs. In 2009, Hegger et al. synthesized a binary CP, PDDTT, through the Stille coupling of a dibromo-substituted TT derivative with 2,5-bis(tri-n-butyltin) thiophene under palladium (Pd) catalysis. The resulting polymer exhibited an absorption onset (λ_{onset}) at 1450 nm and an optical bandgap of 0.85 eV (Figure 7.7a) [24]. In 2014, Wang et al. modified PDDTT by introducing a third component-a dibutyltin-substituted thiophene with a long alkyl chain at the 3,4-positions of the thiophene unit-into the copolymerization. By adjusting the ratios of the three monomers and using a catalyst system consisting of Tris(dibenzylideneacetone) dipalladium ($Pd_2(dba)_3$) and tri(o-tolyl)phosphine ($P(o\text{-}tol)_3$), they successfully synthesized ternary copolymers with tunable optical bandgaps. UV-visible absorption spectra showed that varying the proportion of the third component initially lowered and then increased the E_g of polymers (Figure 7.7b) [25].

7.3.1.2 Synthesis of LBG-CPs Based on Benzothiadiazole (BT) and Its Derivatives

Benzothiadiazole (BT) and its derivatives, such as benzoselenadiazole (BSe), are widely used as building blocks in the construction of LBG-CPs. Ng et al. synthesized a series of ultra-low bandgap CPs ($E_g < 1.12$ eV) by performing Stille coupling of dibromo-substituted BT derivatives with strongly electron-donating, exocyclic-olefin-substituted cyclopentadithiophene distannyl compounds, using Tetrakis(triphenylphosphine)palladium ($Pd(PPh_3)_4$) as the catalyst in toluene (Figure 7.8a) [26]. Tang and co-workers further advanced this approach by replacing the sulfur atom in pyridyl[2,1,3]thiadiazole (PT) with selenium and coupling it with cyclopentadithiophene D unit through Stille coupling, yielding a low bandgap polymer (PCDTPTSe) with an E_g of 1.0 eV (Figure 7.8b) [27, 28].

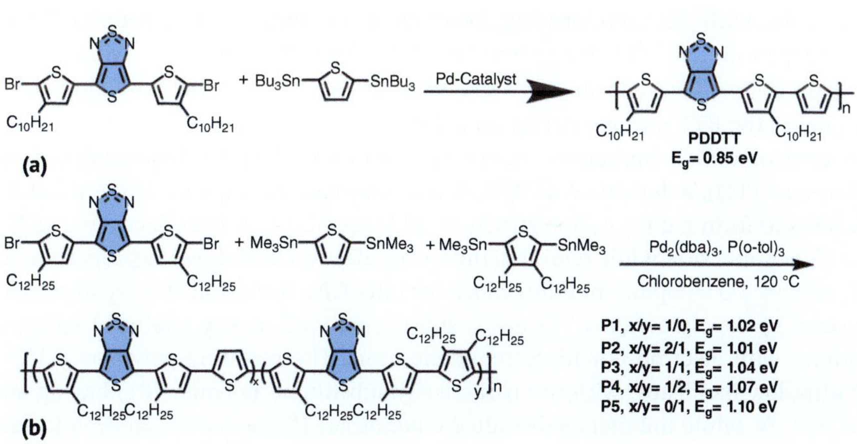

Figure 7.7 The representative Stille coupling alternating copolymers (a) and random copolymers (b) based on thieno[3,4-b]thiadia.

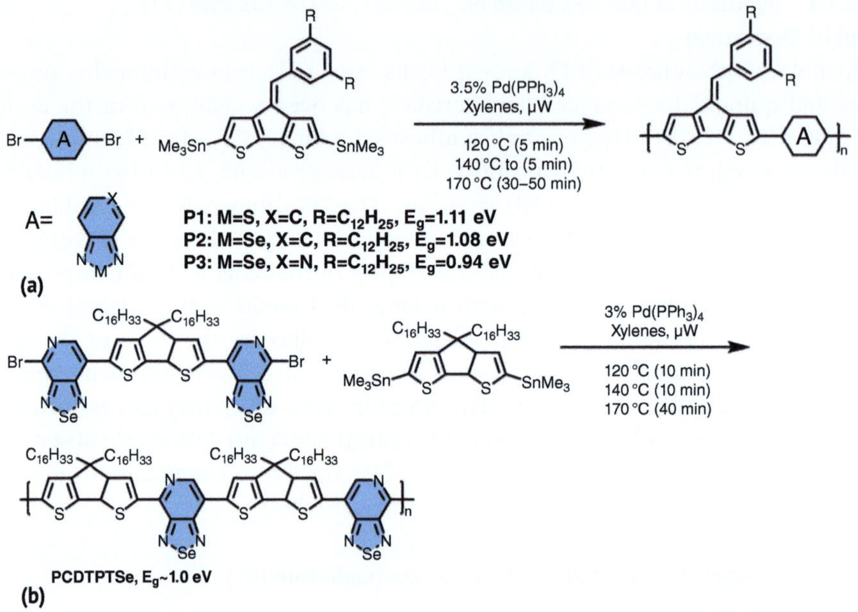

Figure 7.8 The representative Stille coupling alternating copolymers based on benzothiadiazole (BT) and its derivatives: (a) with a vinylphenyl-substituted thiophene–cyclopentadiene donor; (b) with an alkyl-substituted thiophene–cyclopentadiene donor.

7.3.1.3 Synthesis of LBG-CPs Based on Benzobisthiadiazole (BTT) and Its Derivatives

Benzobisthiadiazole (BTT) units, in contrast to BT units, exhibit stronger electron-withdrawing properties. Polymers that use BTT as the acceptor unit typically display pronounced intramolecular push-pull effects and exceptionally narrow optical bandgaps. As shown in Figure 7.9a, Huang et al. synthesized CPs by coupling dibromo-substituted BTT derivatives, bearing long alkyl chains on the thiophene units, with electron-donating distannyl compounds via Stille coupling [29]. The strong intramolecular charge transfer (ICT) effects resulted in CPs with optical bandgaps (E_g) as low as 0.6 and 0.55 eV, respectively. However, the limited modifiable sites in the BTT unit restrict its versatility.

To overcome this limitation, researchers introduced [1,2,5]thiadiazolo[3,4-g]quinoxaline (TQ), a derivative of BTT, which involves opening one of the thiadiazole units to form a quinoxaline structure. This modification introduces two additional functional sites while retaining the strong electron-withdrawing properties of BTT, making TQ a popular building block for LBG-CPs. Anderson et al. synthesized polymers by copolymerizing TQ units, substituted with alkoxy chains at various positions, with electron-donating thiophene units. The resulting polymers exhibited ultra-low bandgaps, with the para-alkoxy-substituted polymer (P1) having an E_g of 0.71 eV, while the meta-substituted counterpart (P2) achieved an even lower bandgap of 0.64 eV (Figure 7.9b) [30].

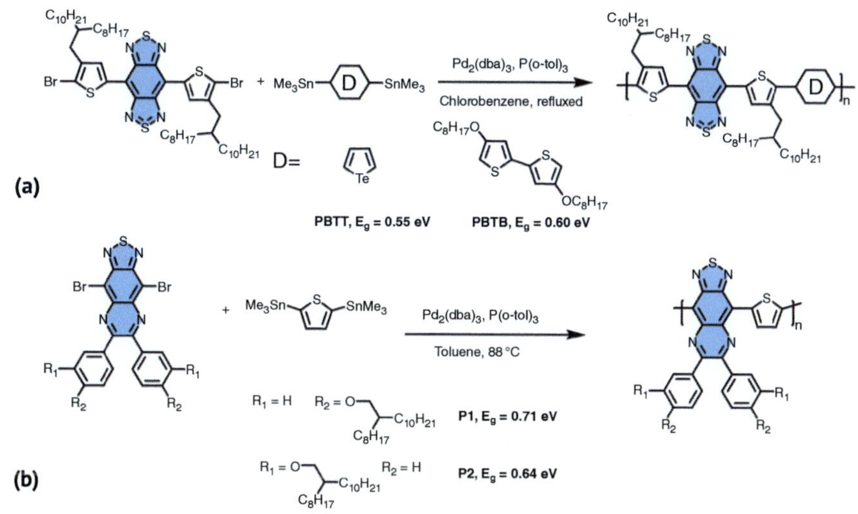

Figure 7.9 The representative Stille coupling block copolymers: (a) polymers with a benzobisthiadiazole (BBT) core; (b) polymers with BBT derivatives.

7.3.1.4 Synthesis of LBG-CPs Based on Diketopyrrolopyrrole (DPP) and Its Derivatives

Diketopyrrolopyrrole (DPP) is a widely used structural unit for constructing CPs with high charge mobility. Its ability to stabilize in a quinoidal resonance structure also makes DPP ideal for designing extremely LBG-CPs. Lei et al. synthesized "A–A" type CPs by coupling DPP units, flanked by thiophene side groups, with TQ and BTT units via Pd-catalyzed Stille coupling. The resulting polymers, P(TPDD-TQ) and P(TPDD-BBT), exhibited high HOMO levels and low LUMO levels, as shown in Figure 7.10a [31]. Similarly, Geng et al. synthesized LBG-CPs by coupling DPP-based dibromo monomers, end-capped with indole oxides, to cyanide-substituted bithiophene distannyl compounds. By substituting halogens at different positions on the dibromo monomers, they were able to fine-tune the polymer bandgaps. The synthetic routes, polymer structures, and corresponding bandgaps are presented in Figure 7.10b [32].

7.3.1.5 Synthesis of LBG-CPs Based on Isoindigo (IID) and Its Derivatives

Isoindigo (IID) and its derivatives exhibit strong electron-withdrawing properties, which facilitate the construction of CPs with broad absorption ranges and high hole mobilities. Wang et al. employed thienoisoindigo (TII) units as the A unit and copolymerized them with benzotristhiophene (BTT) and dithienobenzotristhiophene (DTBTT) donor units through Pd-catalyzed Stille coupling. The resulting polymers, PBT and PDT, demonstrated exceptionally low bandgaps of 1.00 and 0.92 eV, respectively (Figure 7.11a) [33]. Additionally, Maes et al. introduced a modified IID derivative, bay-annulated indigo (BAI), as the A unit. BAI, with its strong electron-withdrawing properties and relatively high LUMO levels, proved to be an ideal candidate for low-bandgap polymers. They synthesized polymers by coupling BAI units with donor units

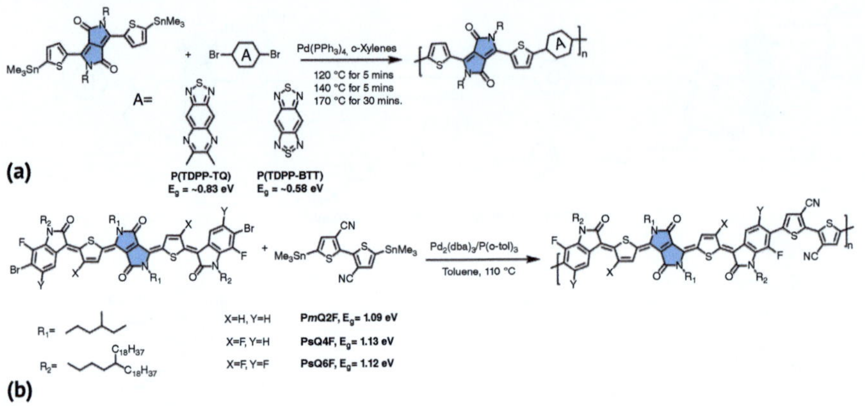

Figure 7.10 The representative Stille coupling polymers: (a) diketopyrrolopyrrole (DPP)-based block copolymers; (b) quinoid-resonance DPP-based block copolymers.

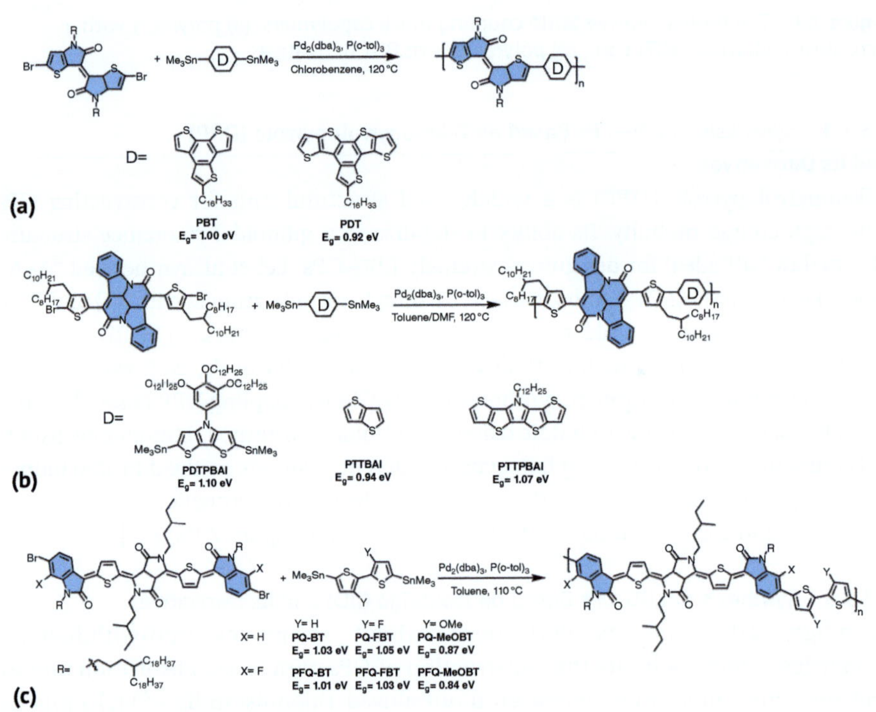

Figure 7.11 The representative Stille coupling polymers: (a) thienoisoindigo (IDD)-based block copolymers; (b) bay-annulated IDD-based block copolymers; (c) quinoid-resonance DPP-incorporated IDD-based block copolymers.

such as dithieno[3,2-b:2′,3′-d]pyrrole (DTP), thieno[3,2-b]thiophene (TT), and thienothieno[3,2-b]thienopyrrole (TTP) via Stille coupling, yielding CPs PDTPBAI, PTTBAI, and PTTPBAI, with bandgaps (Eg) of 0.94, 1.10, and 1.07 eV, respectively (Figure 7.11b) [34]. Furthermore, Geng et al. synthesized a series of IID-based LBG-CPs with strong second near-infrared (NIR-II) absorption by introducing various substituents on the conjugated backbone. These polymers and their corresponding bandgaps are shown in Figure 7.11c [35].

7.3.1.6 Synthesis of LBG-CPs Based on Benzofurandione (BFDO) and Its Derivatives

Benzofurandione (BFDO) units, known for their high planarity and excellent electron-withdrawing properties, have been widely used in the design of LBG-CPs. Polymers incorporating BFDO units typically exhibit ultra-low bandgaps due to strong intramolecular charge transfer (ICT) effects. Pei et al. synthesized CPs TBDPPV, TBDOPV-T, and TBDOPV-2T, with bandgaps (E_g) of 0.78, 0.92, and 1.04 eV, respectively, by coupling BFDO-containing dibromo monomers with common distannyl units such as ethylene, thiophene, and bithiophene (Figure 7.12a) [36]. Later, Huang and co-workers developed a method to brominate BFDO units, producing highly electron-withdrawing and quinoidal resonance-stabilized BFDO-2Br units. Using these units, along with BFDO derivatives containing thiophene side groups, they synthesized a series of ultra-low bandgap polymers via Stille coupling with donor units like DPP, thiophene, and selenophene, under Pd-catalysis. The synthetic routes, polymer structures, and corresponding bandgaps are shown in Figure 7.12b, c [37, 38].

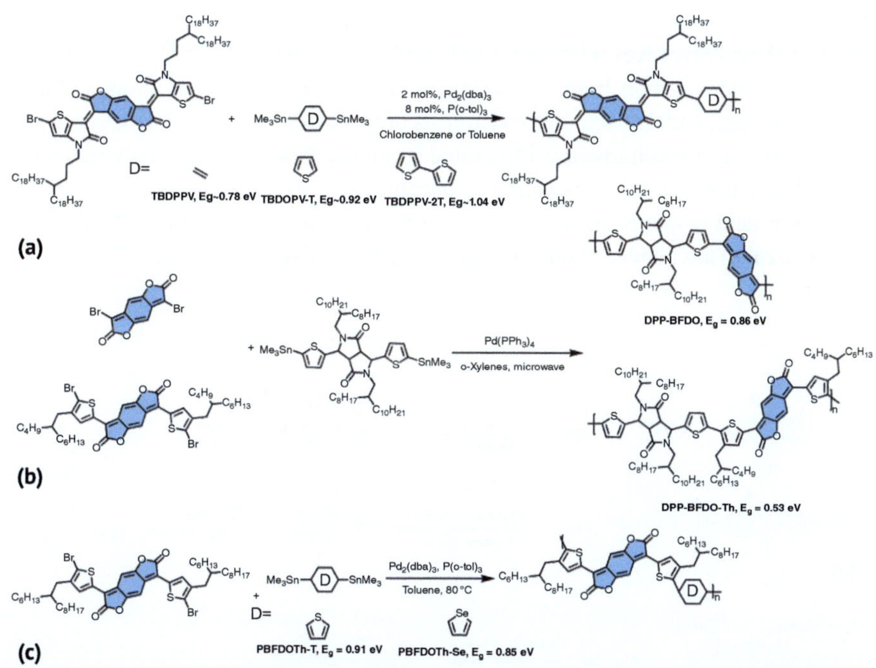

Figure 7.12 The representative Stille coupling polymers: (a) benzo[1,2-b:4,5-b']difuran-2,6(3H,7H)-dione (BFDO)-based block copolymers; (b) and (c) quinoid-resonance BFDO-based block copolymers.

7.3.1.7 Synthesis of LBG-CPs Based on 4,4′-difluoro-4-bora-3a, 4a-diaza-s-indacene (BODIPY) and Its Derivatives

Since its first report in 1968, BODIPY and its derivatives have become widely used as building blocks in CPs due to their tunable optoelectronic properties. The BODIPY unit, characterized by its two-equivalent boron—nitrogen bonds, can significantly reduce polymer bandgaps through resonance stabilization. Recent studies by Liu et al. demonstrated that the resonance-stabilized $N - B \leftarrow N$ structure in BODIPY could be utilized to design ultra-narrow bandgap CPs [39–41]. For instance, Liu et al. synthesized two polymers by coupling dibromo-substituted BODIPY derivatives with different bromination positions and 1,2-bis(tributylstannyl)ethylene via Stille coupling. The resulting polymers, P–α–V and P–β–V, exhibited bandgaps of 0.80 eV and 1.39 eV, respectively, depending on the position of bromination (Figure 7.13a) [42]. Expanding on this approach, the same group developed thiophene-fused BODIPY derivatives to promote a planar conformation, which enhanced conjugation and further reduced the bandgap. The resulting polymer, IP1, synthesized via Stille coupling with hexabutylditin-substituted thiophene, exhibited an ultra-low bandgap of 0.83 eV (Figure 7.13b) [41].

7.3.2 Other Polymerization Methods for Constructing LBG-CPs

7.3.2.1 Suzuki Coupling

The Suzuki coupling reaction, developed by Japanese chemist Akira Suzuki, is a significant organic reaction that involves the cross-coupling of aryl or alkenyl boronic acids and their derivatives with aryl or alkenyl halides under palladium catalysis to form new carbon–carbon bonds. Known for its mild reaction conditions and high yields, it has been widely used in the synthesis of functional small molecules and polymers. The reaction mechanism is illustrated in Figure 7.14, using the polymerization of typical aryl dibromo monomers and bisboronic ester aryl monomers as an example. The boronic acid reagents used in Suzuki coupling are less toxic, and the by-products (e.g. boron salts) are generally nontoxic and easy to handle. In contrast, Stille coupling

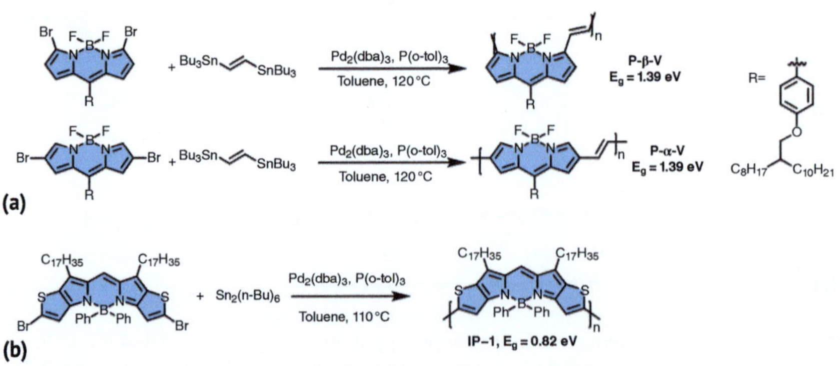

Figure 7.13 The representative Stille coupling polymers: (a) block copolymers coupled at different positions of 4,4′-difluoro-4-bora-3a,4a-diaza-s-indacene (BODIPY); (b) homopolymers based on conjugation-extended BODIPY derivatives.

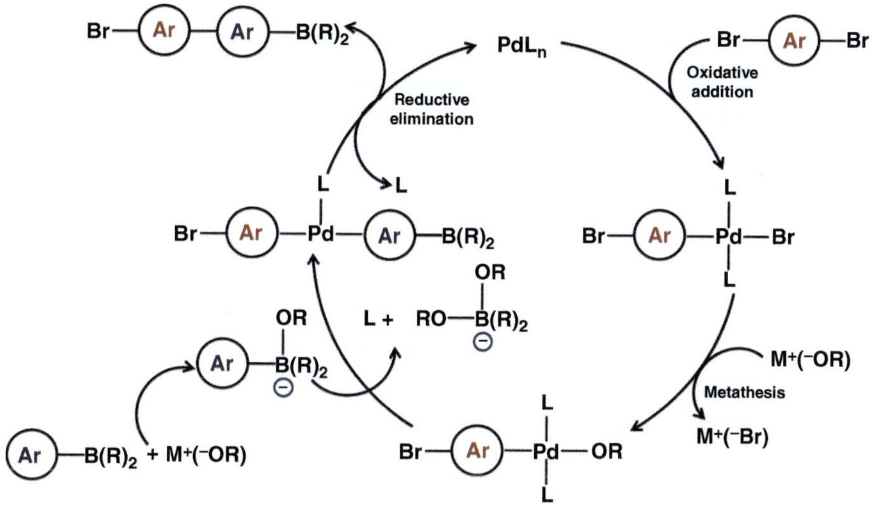

Figure 7.14 The illustration of mechanism of Suzuki coupling.

involves the use of organotin reagents (e.g. tributyl tin), which are highly toxic, and the resulting tin-based waste is both environmentally hazardous and costly to dispose of.

However, the preparation of boronic acid reagents mainly involves nucleophilic addition and dehydration-condensation reactions, which are more straightforward for aryl systems with evenly distributed electron density and some nucleophilicity, such as benzene. In contrast, thiophene, with its sulfur atom exhibiting higher electron density and soft nucleophilicity, tends to react more efficiently with softer reagents like organotin compounds rather than harder boron reagents. Furthermore, reactions between thiophene and tin reagents typically result in stable thiophene-tin derivatives, which are highly active and reactive in coupling reactions. In Stille coupling, thiophene-tin compounds readily undergo tin elimination to form C—C bonds, with low reaction barriers and minimal side reactions. As a result, conjugated systems based on thiophene generally favor the preparation of organotin reagents and utilize Stille coupling.

It is important to note that the sulfur atoms in thiophene contribute additional p-orbitals, which strengthen the π-conjugation system with the carbon backbone. This enhanced conjugation improves the material's light absorption and electronic delocalization, which helps reduce the bandgap (E_g). In contrast, the larger E_g of benzene makes it less suitable for constructing LBG-CPs. Additionally, compared to benzene-based conjugated systems, which often exhibit weaker conjugation and poor molecular packing, thiophene enhances π–π stacking interactions, thereby improving charge carrier mobility in thiophene-based conjugated materials. Moreover, the electronic structure, energy levels, and optoelectronic properties of thiophene materials can be further fine-tuned by introducing substituents or extending the conjugation length (e.g. by incorporating bithiophene, benzothiophene, etc.). These advantages make thiophene-based five-membered ring structures the preferred choice for constructing LBG-CPs and small molecules, while benzene is typically used as an auxiliary or structural modifier rather than as the core functional unit.

Given these factors, the synthesis and development of LBG-CPs are primarily focused on Stille coupling, with fewer examples utilizing Suzuki coupling. A brief overview of the application of Suzuki coupling will be presented as follows.

Mase et al. utilized a dibromo TQ unit with strong electron-withdrawing and quinoidal characteristics, along with 1,4-benzenediboronic acid bis(pinacol) ester, in the presence of a $Pd_2(dba)_3$ catalyst, triphenylphosphine (PPh_3) ligand, and potassium carbonate (K_2CO_3) as the base [43]. The polymerization was carried out in a mixed solvent of toluene and water, resulting in the polymer PBTQ(OD). Compared to a series of polymers using thiophene and its derivatives as donor units (D), the bandgap (E_g) of PBTQ(OD) was significantly wider at 1.14 eV, which does not meet the definition of LBG-CPs outlined in the chapter. This further highlights that benzene, in contrast to thiophene and its derivatives, is not an ideal core functional unit for constructing LBG-CPs. However, it is important to emphasize that Suzuki coupling offers considerable advantages, particularly its environmental friendliness. As such, the development of Suzuki coupling methodologies for constructing LBG-CPs remains essential. Given the limitations of benzene, researchers have increasingly turned to benzene derivatives as main-chain functional units. For example, BT derivatives not only improve the stacking of benzene units but also enhance their electron-withdrawing properties.

Lim et al. synthesized the polymer PDBT by using DPP and BT as strong electron-accepting units, with thiophene spacers acting as electron donors. This polymer, with a E_g of 1.1 eV, was prepared by polymerizing DPP–thiophene dibromo monomers and BT diboronic acid bis(pinacol) esters under $Pd(PPh_3)_4$ catalysis, with sodium carbonate (Na_2CO_3) as the base and a toluene-water solvent system [44]. Despite these advancements, thiophene and its derivatives remain the most commonly used building blocks for constructing LBG-CPs. As a result, developing boronate ester derivatives of thiophene and its derivatives is crucial for synthesizing LBG-CPs. For example, Wudl et al. prepared DPP–thiophene-connected units as bisboronate esters and reacted them with dibromo BTT units-which possess stronger electron-withdrawing properties than BT-via Suzuki coupling. The resulting polymer, PBBTDPP, exhibited an ultra-narrow E_g of 0.65 eV, as determined by electrochemical measurements [45] (Figure 7.15).

These examples show the potential of Suzuki coupling in constructing LBG-CPs. However, the range of building blocks available for Suzuki coupling to prepare LBG-CPs is considerably more limited compared to Stille coupling. Given its excellent environmental friendliness, which aligns with the green chemistry principles essential for the future development of chemical sciences, researchers are encouraged to develop more modular components that are compatible with Suzuki coupling for synthesizing LBG-CPs.

7.3.2.2 Aldol Condensation

Since 1860, when the French chemist Wurtz first discovered that aldehyde compounds (such as acetaldehyde) undergo self-addition under basic conditions to produce β-hydroxy aldehydes-later known as the aldol condensation reaction-the theory and mechanism of aldol condensation have been continuously refined. It is now well-understood as an organic reaction in which an aldehyde or ketone containing active α-hydrogens reacts with another molecule of aldehyde or ketone under

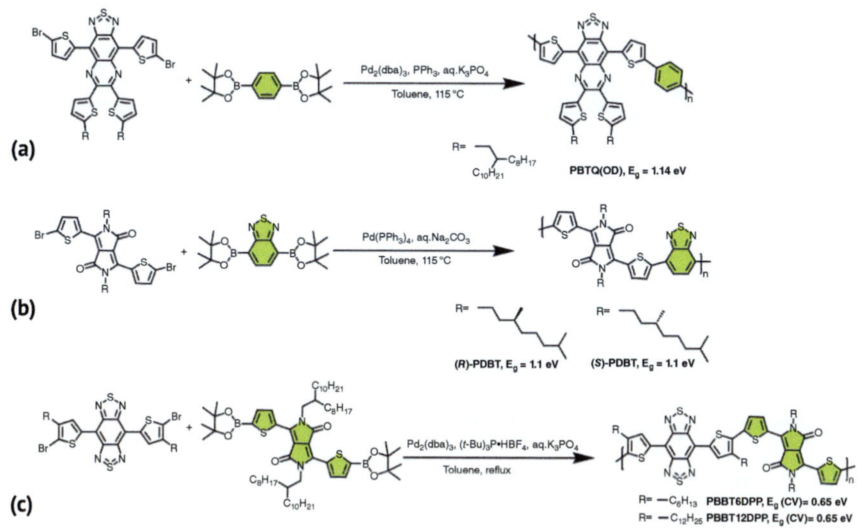

Figure 7.15 The representative low-bandgap conjugated polymers (LBG-CPs) synthesized via Suzuki coupling: (a) block copolymers containing benzothiadiazole–quinoxaline (BTQ) and benzene; (b) block copolymers containing diketopyrrolopyrrole (DPP) and benzothiadiazole (BT); (c) block copolymers containing benzobisthiadiazole (BBT) and DPP.

basic or acidic catalytic conditions. This process involves the deprotonation of the α-hydrogen and the formation of a new C—C bond, yielding β-hydroxy aldehydes or β-hydroxy ketones as products. These products can undergo subsequent dehydration to form α, β-unsaturated carbonyl compounds. The reaction mechanism is illustrated in Figure 7.16. Compared to Stille and Suzuki couplings, which require expensive transition metal catalysts, multi-step reaction conditions, and costly purification processes, aldol condensation has attracted significant research interest due to its environmentally friendly nature. It does not require costly metals and generates water as a by-product [46]. This reaction has been utilized to synthesize various novel organic materials, including small molecules and polymers. In recent years, several researchers have employed aldol condensation for the synthesis of LBG-CPs. A brief overview of these efforts is provided below [47, 48].

McCulloch et al. used a traditional Lewis acid catalyst, p-toluenesulfonic acid (PTSA), for aldol condensation polymerization, with toluene as the solvent at 120 °C. The monomers used were enolizable bis-oxindole and nucleophilic bis-isatin. By varying the conjugated segments and alkyl chain lengths of the copolymer units, they synthesized five LBG-CPs, labeled P1–P5, with their structures and E_g values shown in Figure 7.17 [49].

Pei et al. synthesized two monomers, bis(isatin) and benzodifurandione, following the literature and carried out polymerization using 20 mol% TsOH as the catalyst in toluene at 120 °C for 48 hours [50]. The resulting LBG-CP, LPPV, was further modified by adjusting the alkyl chain lengths, producing polymers LPPV-1 and LPPV-2, both exhibiting low E_g values of 0.89 and 0.92 eV, respectively (Figure 7.18a). The McCulloch group also synthesized a series of LBG-CPs by

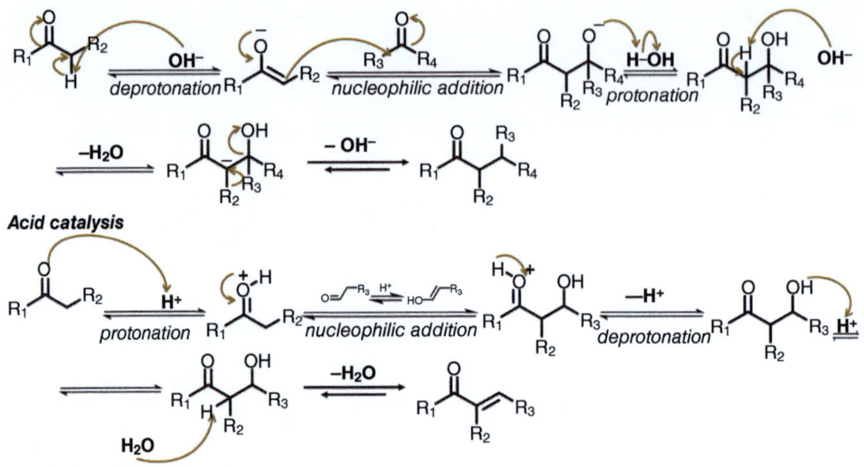

Figure 7.16 The illustration of mechanism of aldol condensation.

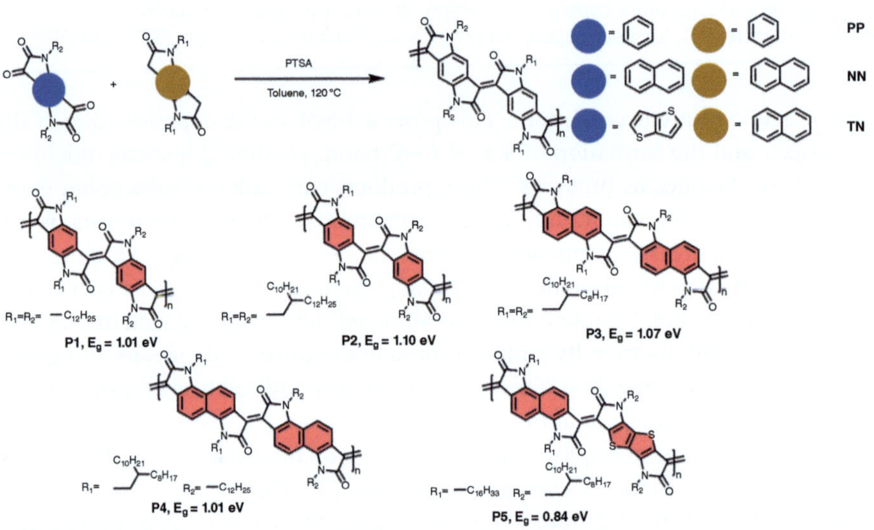

Figure 7.17 Synthetic methods, structures, and E_g of LBG-CPs P1–P5 synthesized via Aldol condensation.

varying the alkyl side chain lengths of bis-oxindole monomers. These monomers were polymerized with bis-isatin monomers containing ethylene glycol (EG) side chains under PTSA catalysis in toluene at 130 °C. All polymers exhibited low Eg values, with their structures and E_g values detailed in Figure 7.18b [51]. Meanwhile, Yue and Zeglio constructed a lactone structure containing linked EG chains using IID-capped bithiophene units with EG side chains, polymerizing them with BFDO units under PTSA catalysis. The resulting LBG-CP, p(C-g2T), exhibited an E_g of 0.92 eV (Figure 7.18c) [52].

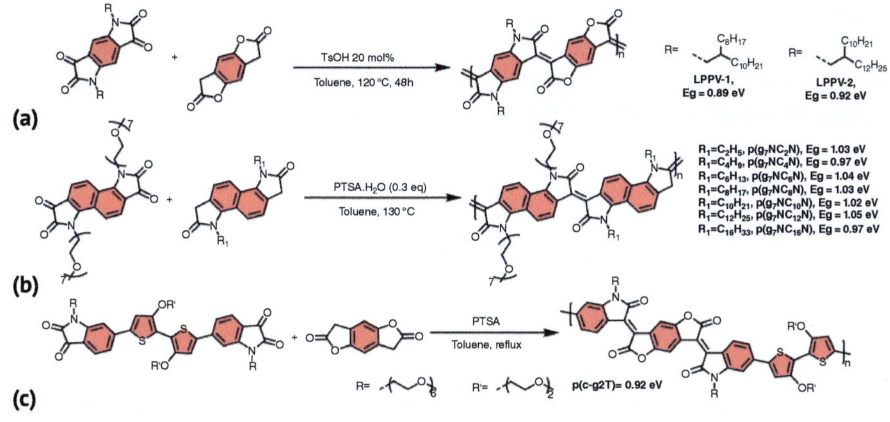

Figure 7.18 Representative low-bandgap conjugated polymers (LBG-CPs) synthesized via Aldol condensation: (a) rigid coplanar poly(p-phenylene vinylene) (PPV) derivatives; (b) acid-catalyzed aldol polycondensation between the enolizable carbonyl units of bis-oxindole monomers and the electrophilic carbonyl units of the corresponding bis-isatin comonomers; and (c) a series of novel n-type donor–acceptor conjugated polymers based on glycolated lactone and bis-isatin units.

These examples demonstrate that aldol condensation, as a green, nontoxic, and metal-free polymerization method, is a viable approach for constructing LBG-CPs. Moreover, the double-bond linkage between copolymer units effectively restricts torsional angles, enhancing interchain interactions. This feature significantly advances the development and application of these polymers.

In addition to aldol condensation, Stille and Suzuki couplings are among the most commonly used polymerization methods for synthesizing LBG-CPs. Other methods, such as C—H direct arylation and oxidative polymerization, are also frequently employed. With continued research and innovation, more synthesis methods and suitable monomers are expected to emerge, further enriching the diverse family of LBG-CPs and expanding their scope and potential applications.

7.4 Application of LBG-CPs

7.4.1 Application in OPDs

OPDs have gained significant attention in recent years due to their unique properties, including flexibility, lightweight, and compatibility with large-scale manufacturing processes [18, 53, 54]. Owing to the rapid development of novel photoactive materials and device configurations, OPDs have achieved comparable performance with commercial silicon-based photodetectors in the NIR region ($\leq$ 1000 nm). As mentioned in last section, further decreasing the bandgap of CPs will redshift absorption over 1000 nm but also will lower the LUMOs or lift the HOMOs, resulting in unmatched energy levels and thus limiting the efficiency of OSCs. However, polymers with absorption over 1000 nm can be promising candidates for short-wave infrared OPDs that show application in image sensing, artificial intelligence, quality inspection, machine vision, optical communication, and night surveillance [5, 15, 16, 19, 55]. The important parameters determining the performance of OPDs include dark current density (J_d), responsibility (R), and detectivity (D^*). These parameters are measured from Eqs. (7.1) and (7.2):

$$R = \frac{\text{EQE} \times \lambda}{1240} \tag{7.1}$$

where λ is the wavelength of the incident light.

$$D^* = \frac{R}{\sqrt{2qJ_d}} \tag{7.2}$$

where q is 1.602×10^{-19}, J_d is dark current density.

It can be found that the R and D^* are closely relevant with the absorption wavelength of OPDs, the higher R and D^* in the longer wavelength indicate the higher detective ability of OPDs in the long wavelength region.

In 2007, Yang et al. reported a LBG-CP, namely PTT, with alternating thieno[3,4-b]thiophene and thiophene backbones (Figure 7.19) [56]. The absorption edge of PTT could reach 970 nm, making it a promising candidate for NIR OPDs. In 2009, Heeger et al. copolymerized thiophene flanked thieno[3,4-b] thiadiazole with thiophene to afford a novel LBG-CP, PDDTT. PDDTT exhibits a bandgap of ~0.85 eV, enabling absorption to ~1500 nm. The PDDTT: $PC_{61}BM$ based OPDs achieved a D^* over 10^{13} Jones in the range of 300–1150 nm and over 10^{12} Jones in 1150–1450 nm [57]. These remarkable results strongly validate the great potential of LBG-CP-based OPDs to achieve comparable performance with commercial inorganic photodetectors, such as Si and InGaAs photodetectors. Furthermore, the solubility and molecular packing of PDDTT have been tried to optimized by terpolymerization. A series of thieno[3,4-b]thiadiazole based polymers, W1–W5, have been synthesized by Wang et al. for OPDs [25]. All fabricated OPDs exhibited a photo-response in the range of 400–1200 nm, and the WP4-based device achieved a D^* over 10^{12} Jones in the range of 330–950 nm. In addition, the thieno[3,4-b]pyrazine can also utilized to construct LBG-CPs for OPDs. In 2011, Chen et al. reported a LBG polymer with alternating thieno[3,4-b]pyrazine and bithiophene, namely PDTTP, and the resulted OPDs realized a spectra response reaching 1100 nm [58].

Constructing polymers with a strong electron-donating unit and a strong electron-withdrawing unit has been recognized as a widely used approach to design LBG-CPs for OPDs. The [1,2,5]thiadiazolo[3,4-g]quinoxaline (TQ) unit has been widely used as a strong electron-withdrawing unit to construct various LBG-CPs for OPDs (Figure 7.19). In 2007, Andersson et al. reported the LBG polymer, LBPP-1, based on the TQ unit with a bandgap of ~1.0 eV. The OPD based on LBPP-1: $PC_{61}BM$ has realized a spectra response to 1200 nm and a peak responsibility at 950 nm [59]. Copolymerizing the TQ unit with a stronger electron-donating unit CPDT, a lower bandgap polymer CPDT-TQ has been synthesized. The CPDT-TQ: $PC_{71}BM$-based OPD has achieved a spectra response to 1550 nm, and the maximum D^* of 5×10^{10} Jones can be obtained [60]. Later, regulation on the side chain of CPDT unit and the structure of bridging unit has been conducted to afford two new LBG polymers,

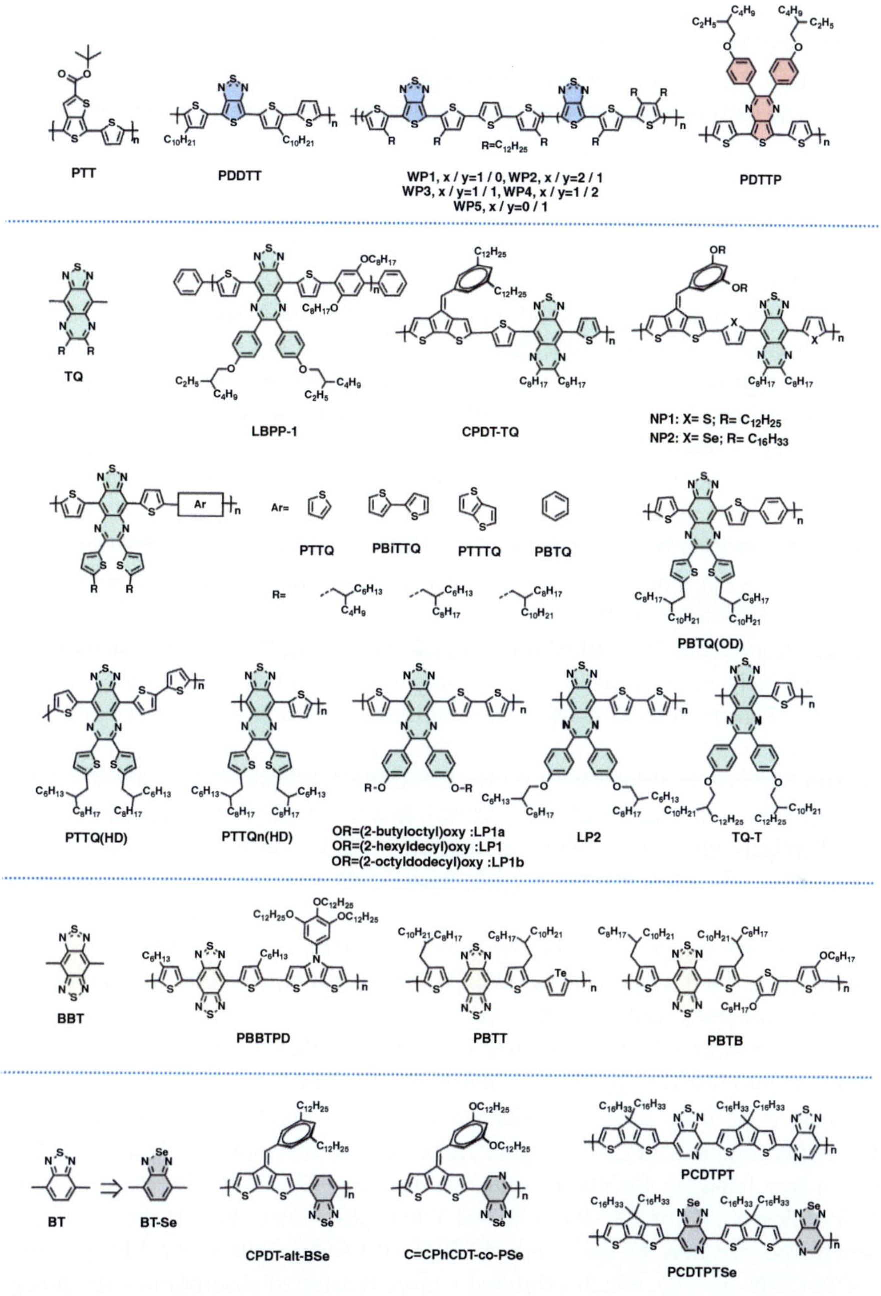

Figure 7.19 The illustration of representative low-bandgap conjugated polymers (LBG-CPs) for organic photodetector (OPDs).

NP1 and NP2 [61]. Comparing the NP1 with the thiophene as the bridge, the NP2 with selenophene bridge exhibited a lower bandgap, which is benefit to achieving a longer wavelength response in OPDs. It was found that the NP1-based OPD could achieve a peak EQE of 26% and a D^* about 10^{11} Jones at 1100 nm under

0 bias. In 2020, Maes et al. synthesized a series of TQ-based LBG polymers, PTTQ, PBiTTQ, PTTTQ, and PBTQ [43]. Through regulating the structure of copolymerizing unit, the bandgap of resulting polymers could be controlled from 0.87 to 1.14 eV. The PTTQ(BO)-based OPD could realize a response to 1500 nm, and a EQE of 18% was obtained at 1130 nm under −2 V bias. In the same year, Vandewal et al. reported a series of TQ-based LBG polymers, PTTBAI, PBTQ(OD), PTTQ(HD), and PTTQn(HD), with bandgaps of 0.81–1.12 eV for OPDs. And a response to 2000 nm with D^* of 10^{10} Jones has been achieved. In 2022, Leem et al. reported two LBG polymers for OPDs, LP1 and LP2, with alternating TQ as the A unit and tri-thiophene unit or di-thiophene unit as the D unit [62]. Polymer LP1 based on tri-thiophene exhibited a higher absorption coefficient and preferable "face-on" stacking. Further, the alkoxy chain length of phenyl groups on the TQ unit was regulated for another two LP1-analogs (LP1a and LP1b) with varied length of alkoxy chains. With Y7 as the acceptor materials, the LP1: Y7-based OPD achieved a EQE of 19.5% at 1200 nm under −1 V bias and a D^* of 2.96×10^{10} Jones for spectrum region over 1000 nm. Additionally, Gasparini et al. reported a TQ-based D–A CP, TQ-T, with a bandgap lower than 0.7 eV, and the TQ-T: IEICO-4F-based device achieved a J_d of 8.4×10^{-3} mA cm^{-2} under −2 V bias and a D^* of 10^9–10^{10} Jones in the wavelength range of 350–1800 nm [63].

In addition to the TQ unit, the benzo[1,2-c;4,5-c′]bis[1,2,5]thiadiazole (BBT) is another widely used electron-withdrawing unit for constructing LBG polymers for OPDs (Figure 7.19). Notably, it was found that OPDs based on BBT-containing polymers usually exhibited relatively inferior performance due to the increased non-irradiation recombination caused by the strong diradicaloid character of the BBT unit along the conjugated backbone [64, 65]. In 2018, Wudl et al. prepared a CP with a low bandgap, PBBTPD, containing alternated BBT and dithieno[3,2-b:2′,3′-d]pyrrole unit [66] The PBBTPD: Tri-PC$_{61}$BM-based OPD achieved a wide response from 350 to 2500 nm, and a D^* surpassing 10^{11} Jones was obtained. In 2020, Huang et al. used BBT unit to construct two LBG polymers, PBTT and PBTB, for OPDs [67]. Notably, the flexible OPD based on PBTB has been prepared and considerable device performance could be achieved.

As aforementioned, the selenium (Se) atom replacement has been demonstrated as an effective approach to decrease the bandgap of CPs. Therefore, the Se-replaced BT unit, 2,1,3-benzoselenadiazole (BT–Se), was also utilized to construct LBG polymers for OPDs (Figure 7.19). Ng et al. developed a CP, CPDT-alt-BSe, with a low bandgap for OPDs, and the resulting CPDT-alt-BSe: PC$_{71}$BM-based OPD achieved a D^* of 10^{12} Jones at −12 V bias [26]. Later, the nitrogen (N) atoms were introduced into the backbone of CPDT-alt-BSe to obtain a new LBG polymer, C=CPhCDT—co—PSe, which exhibited a more redshifted absorption with an edge of 1300 nm [68]. The resulting OPDs achieved an optimal D^* of 3×10^{10} Jones at 0 bias. In 2021, Zheng et al. developed a LBG polymer, PCDTPTSe, with alternating [1,2,5]selenadiazolo[3,4-c]pyridine unit and cyclopenta[2,1-b:3,4-b′]dithiophene unit [27]. Compared with the PCDTPT with S atoms, the bandgap of PCDTPTSe was decreased by 0.15 eV. In the cavity-enhanced PCDTPTSe: PC$_{71}$BM-based OPDs, spectral response full-width at half-maximum (FWHM) values as low as 22 nm were achieved, and the calculated D^* was over 10^{11} Jones.

7.4.2 Application in OTEs

Organic thermoelectric (OTE) devices are recognized as an ideal option for wearable sensors because of the inherent advantages of organic semiconductors, such as low thermal conductivity, molecular diversity, nontoxicity, and solution processability. Through the development of new materials and the improvement of device fabrication processes, the performance of n-type and p-type OTE devices has been significantly improved [69–71]. The thermoelectric figure-of-merit (ZT) is the key parameter to evaluate the properties of thermoelectric materials, which can be measured from Eq. (7.3):

$$ZT = S^2\sigma T/\kappa \tag{7.3}$$

where S is Seebeck coefficient, σ is conductivity, κ is thermal conductivity, T is absolute temperature. Compared to inorganic semiconductors, organic materials are less crystalline and tend to have lower thermal conductivity, which is usually below $0.5\ \mathrm{W\,m^{-1}\,K^{-1}}$ [70, 72]. Therefore, in the field of OTEs, the power factor (PF) is often used as an indicator to evaluate the thermoelectric performance, which is defined as the product of squared Seebeck coefficient and conductivity.

In general, the charge carrier concentration in organic materials is very low, resulting in low conductivity that does not meet the requirements of thermoelectric applications. Chemical doping is the most widely used method to improve the thermoelectric properties of organic materials. Doping of organic materials is usually achieved by the intermolecular charge carrier transfer between the dopant and the host molecules, which is essentially a redox chemical reaction. The doping efficiency is largely determined by the HOMO and the LUMO energy levels of polymers [73–75]. To achieve efficient p-type doping, a high HOMO energy level is usually required, while efficient n-type doping is usually required to have a lower LUMO energy level.

Silicon is the most heavily used inorganic semiconductor, the n- and p-type doping process of silicon can be achieved by implanting phosphorus atoms or boron atoms to replace the silicon atoms. By controlling the implanted ratio of dopant atoms, the doping efficiency can be even reach 100% [76, 77]. The successful applications of n-type and p-type silicon inspired the researchers to explore the similar doping methods of CPs. LBG-CPs can be designed and adjusted to achieve both high HOMO and low LUMO energy levels, it's a suitable property to realize both p- and n-doping in a single host polymer (Figure 7.20).

Lei et al. synthesized three LBG-CPs through copolymerizing the trimethyltin product of TDPP and the dibromo compounds of BT, TQ, and BBT [31]. The HOMO/LUMO energy levels of p(TDPP-BT), p(TDPP-TQ), and p(TDPP-BBT) were −5.34/−3.55, −5.23/−3.94, and −5.20/−4.17 eV, respectively. The decrease of the bandgap is probably because of the enhanced planarity, better conjugation, and more readily aromatic-to-quinoidal transformation. These polymers showed high charge carrier mobilities, p(TDPP-TQ) displayed the highest mobilities with electron mobility of $7.76\ \mathrm{cm^2\,V^{-1}\,s^{-1}}$ and hole mobility of $6.16\ \mathrm{cm^2\,V^{-1}\,s^{-1}}$, and could also be n-doped and p-doped by N-DMBI and $FeCl_3$ effectively, resulting in n-type and p-type electrical conductivities of 16.1 and $348.3\ \mathrm{S\,cm^{-1}}$. In 2023, Deng et al. designed and synthesized a new quinoidal compound TQDPP, with the ThDPP quinoidal core and thienoisatin end groups [78]. Two TQDPP-based CPs,

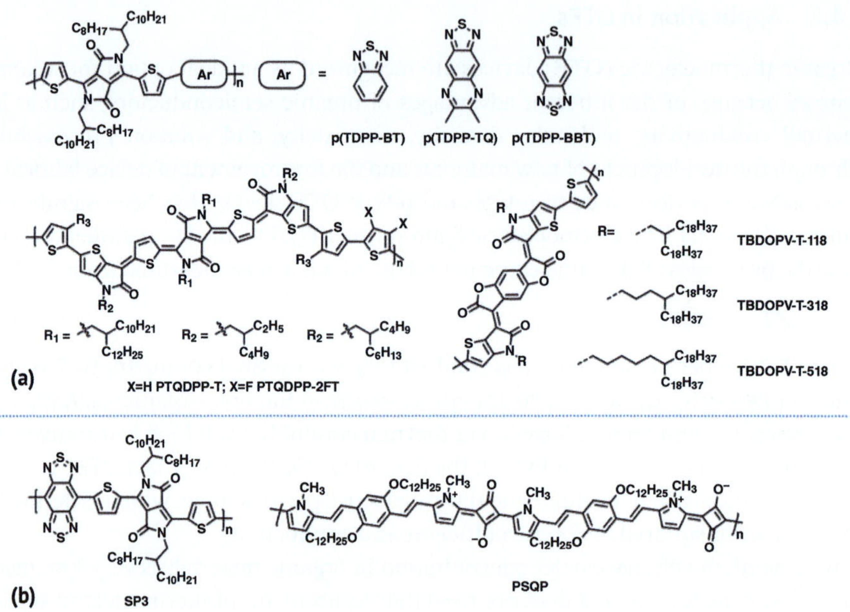

Figure 7.20 The structure illustration of some representative LBG-CPs for (a) organic thermoelectrics (OTEs) and (b) bioimaging and therapy.

PTQDPP-T and PTQDPP-2FT, displayed strong absorption in the NIR-II region, with the thin-film absorption onset above 1500 nm and optical band gaps lower than 0.80 eV. The HOMO/LUMO energy levels were −4.96/−3.85 eV for PTQDPP-T and −5.04/−3.90 eV for PTQDPP-2FT. The two polymers could be both n-doped and p-doped. Benefiting from the better packing order in thin-film, PTQDPP-2FT showed a better performance in OTE devices, achieving a $FeCl_3$ doped p-type PF of 278.2 $\mu W\ m^{-1}\ K^{-2}$ and a N-DMBI doped n-type PF of 2.37 $\mu W\ m^{-1}\ K^{-2}$, respectively.

The energetic disorder leads to a large negative effect on the charge transport properties of CPs. Designing rigid rod–CPs with ideally planar chain conformation with minimum variations in torsion angles along the polymer backbone is an effective method to decrease energetic disorder. In this regard, Pei et al. synthesized three thiophene-based TBDOPV copolymers (TBDOPV-T) with different branched side chains, TBDOPV-T-118, TBDOPV-T-318, and TBDOPV-T-518 [79] and studied the impact of the crystallization rate of the alkyl side chains on the performance of the polymers. Through reducing the crystallization rate of side chains, the π–π contacts could be formed during solution-deposition and annealing processes, which promoted the polymers to achieve excellent conductivity and thermoelectric performance. TBDOPV-T-518 showed the highest n-type and p-type conductivities, the n-type conductivity of 114 S cm^{-1} doped by N-DMBI and the p-type conductivity of 452 S cm^{-1} doped by $FeCl_3$ could be realized. Due to the relatively high Seebeck coefficient, TBDOPVV-T-518 showed high PF values for both types (more than 200 $\mu W\ m^{-1}\ K^{-2}$ for n-type and more than 100 $\mu W\ m^{-1}\ K^{-2}$ for p-type).

7.4.3 Application in Bioimaging and PTT

Fluorescence imaging (FLI) and photoacoustic imaging (PAI), coupled with PTT, have emerged as excellent phototheranostic systems in preclinical studies [80, 81]. However, it is a big challenge to achieve high-quality bioimaging (FLI and PAI) and therapy in deep tissue because of the limited penetration ability of commonly used visible light (380–780 nm). By contrast, the near-infrared (NIR, 700–900 nm) and the second near-infrared (NIR-II, 900–1700 nm) light can achieve an outstanding penetration depth due to the decreased light absorption and scattering in the longer wavelength region. Benefiting from the strong absorption in NIR region and good biocompatibility, LBG-CPs have recently showcased remarkable potential in NIR FL and PA imaging, as well as NIR PTT domains [82]. Notably, owing to their extensive π-conjugated structures, a majority of LBG-CPs tend to be hydrophobic, which is less conducive to applications in biological scenarios. Consequently, these LBG-CPs are often encapsulated within amphiphilic polymers or self-assembled into water-soluble conjugated polymer nanoparticles (CPNs), which usually exhibit strong absorption in the NIR region, even in the NIR-II region (900–1700 nm) and can be used as the photosensitive agents for being injected into the living body. Generally, CPNs designed to convert light into fluorescence are commonly harnessed for FL imaging, those adept at converting light into photoacoustic signals find application in PA imaging, and CPNs with the ability to transform light into heat serve the purpose of PTT (Figure 7.21). In certain instances, the prepared CPs showcased multifunctionality, adeptly converting incident light into fluorescence, photoacoustic signals, and heat simultaneously. This versatile capability makes them as suitable candidates for conducting FLI or PAI-guided PTT within living organisms.

As aforementioned, BBT-based LBG-CPs hold significant promise for applications in OPDs and OTEs. Besides, the BTT-based LBG-CPs usually exhibit a strong absorption in the NIR-II region (1000–1700 nm) and can effectively transform the photo-irradiation into heat or photoacoustic signal. Therefore, the BBT-based LBG-CPs have been widely used as an attractive PTT or PAI agents. In addition, Huang et al. constructed a A–A structural LBG-CP, SP3, by copolymerizing BBT unit with the DPP unit [83]. The obtained CPNs exhibited high photothermal conversion efficiency and strong PA signals at 1280 nm that were detected using commercial NIR-II PAI systems, and the PAI-imaging-guided photothermal tumor therapy was realized in live animals under NIR-II light excitation. In addition, the squaraine-based LBG-CPs demonstrated significant potential as biocompatible agents for NIR-II FLI and NIR-II PTT treatment. (Figure 7.20) Huang et al. constructed a squaraine-based CP, PSQP, which could be encapsulated by an amphiphilic polymer to afford CPNs [84]. The PSQP CPNs exhibited strong absorption at 1064 nm and good photothermal effects and a redshifted NIR-II emission peak (1290 nm) after excitation by a 1064 nm laser. As a result, combined with specific biorthogonal labeling, the 1064 nm excited NIR-II imaging precision and NIR-II PTT efficiency of PSQP CPNs could be improved in live mice. Up to date, increasing number of LBG-CPs have been developed for NIR-II photoacoustic/FLI and PTT, and the recent progress has been summarized in several reviews [20, 85].

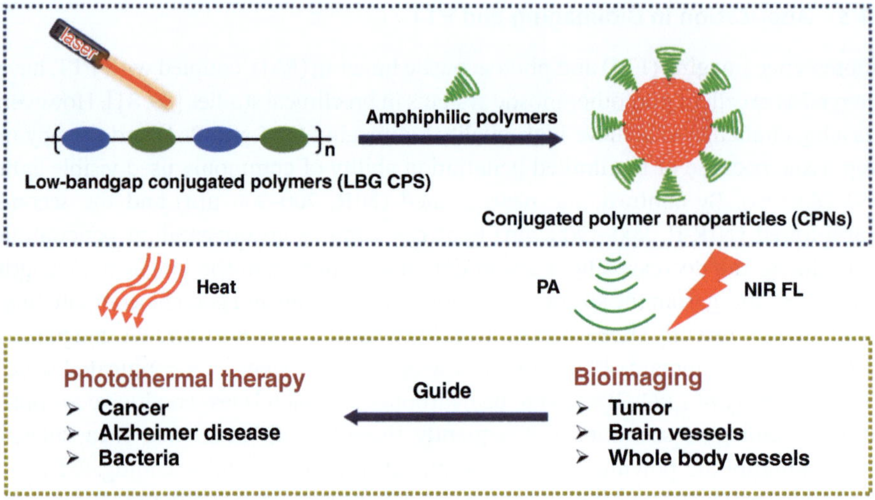

Figure 7.21 The illustration of low-bandgap conjugated polymers (LBG-CPs) for bioimaging and photothermal therapy (PTT).

7.5 Conclusions and Perspective

Over the years, a growing global community of researchers has made significant strides in the development of LBG-CPs, unlocking their potential for a wide array of applications, including OPDs, OTEs, NIR-FLI, PAI, and PTT. The chapter has provided a comprehensive overview of the molecular design strategies behind LBG-CPs, elucidating the relationship between structure and photophysical properties. Furthermore, recent advances in their applications across OSCs, OPDs, OTEs, NIR-II FLI, PAI, and PTT have been discussed in detail.

However, despite these breakthroughs, several challenges remain in the further development and widespread application of LBG-CPs. For OPDs, LBG-CPs have shown promise, but a significant gap remains in achieving high detectivity in the short-wave infrared region (1000–2500 nm). While LBG-CPs have been successful in NIR OPDs, there are few reports of materials with high detectivity in the 1200–1700 nm region, which limits their competition with commercial InGaAs detectors. Synthesis strategies focused on enhancing the NIR absorption and carrier mobility of LBG-CPs will be essential to address this gap. In the realm of OTEs, significant progress has been made in p-type materials, but n-type LBG-CPs with high conductivity are still scarce. Although many LBG-CPs exhibit notable n-type conductivity in doped states, their limited stability under ambient conditions restricts their practical use in thermoelectric applications. The synthesis of stable, high-conductivity n-type LBG-CPs will be crucial for advancing this field. For applications in FLI, PAI, and PTT, LBG-CPs offer excellent biocompatibility and performance, but there is still a need for more research into their metabolic pathways and long-term biocompatibility in vivo. The development of LBG-CPs with controlled degradation and clearance rates will be essential for their successful translation into clinical practice.

Synthesis approaches for LBG-CPs remain a key area for improvement. Current polymerization techniques, such as Stille, Suzuki, and Aldol condensation, have enabled the development of a variety of LBG-CPs with controlled molecular weight and narrow dispersity. However, the synthesis of LBG-CPs with precise control over both the electronic structure and morphology remains challenging. Innovations in polymerization methods, including new strategies for incorporating heteroatoms and modulating the polymer backbone, will be essential for improving the performance of LBG-CPs in high-performance devices.

In conclusion, while significant progress has been made in the design, synthesis, and application of LBG-CPs, there are still many challenges to overcome. By addressing the issues related to stability, conductivity, absorption properties, and biocompatibility, LBG-CPs can realize their full potential in a wide range of applications. Continued research into both the molecular design and synthesis of these materials will be key to unlocking their full capabilities across diverse fields.

Acknowledgments

The work was financially supported by the Fundamental Research Funds for the Central Universities (No. 2024ZYGXZR082).

References

1 Ajayaghosh, A. (2003). Donor–acceptor type low band gap polymers: polysquaraines and related systems. *Chemical Society Reviews* 32 (4): 181–191.

2 Mikie, T. and Osaka, I. (2020). Small-bandgap quinoid-based π-conjugated polymers. *Journal of Materials Chemistry C* 8 (41): 14262–14288.

3 Zhang, Y., Chen, J., Yang, J. et al. (2024). Sensitive SWIR organic photodetectors with spectral response reaching 1.5 μm. *Advanced Materials* 36 (41): 2406950.

4 Liu, C., Shao, L., Chen, S. et al. (2023). Recent progress in π-conjugated polymers for organic photovoltaics: solar cells and photodetectors. *Progress in Polymer Science* 143: 101711.

5 Huang, X., Lan, N., Chen, W. et al. (2021). Low-bandgap conjugated polymers with photocurrent response over 1000 nm. *Journal of Materials Science* 56 (14): 8334–8357.

6 Jia, T., Zhang, J., Zhang, G. et al. (2022). Rationally regulating the terminal unit and copolymerization spacer of polymerized small-molecule acceptors for all-polymer solar cells with high open-circuit voltage over 1.10 V. *Journal of Materials Chemistry A* 10 (30): 15932–15940.

7 Zhuo, H., Qiu, B., Li, X. et al. (2024). Giant molecule acceptors for high performance polymer solar cells. *Giant* 20: 100336.

8 Xiao, L., Chen, X., Yang, X. et al. (2020). Recent advances in polymer-based photothermal materials for biological applications. *ACS Applied Polymer Materials* 2 (10): 4273–4288.

9 Roncali, J. (1997). Synthetic principles for bandgap control in linear π-conjugated systems. *Chemical Reviews* 97 (1): 173–206.

10 Rasmussen, S.C., Schwiderski, R.L., and Mulholland, M.E. (2011). Thieno[3,4-b]pyrazines and their applications to low band gap organic materials. *Chemical Communications* 47 (41): 11394–11410.

11 Zhang, X., Shi, Y., Deng, Y. et al. (2023). Direct arylation polycondensation of thiophene-based C—H monomers. *Chinese Journal of Chemistry* 41 (21): 2908–2924.

12 Cordovilla, C., Bartolomé, C., Martínez-Ilarduya, J.M. et al. (2015). The stille reaction, 38 years later. *ACS Catalysis* 5 (5): 3040–3053.

13 Farhang, M., Akbarzadeh, A.R., Rabbani, M. et al. (2022). A retrospective-prospective review of Suzuki–Miyaura reaction: from cross-coupling reaction to pharmaceutical industry applications. *Polyhedron* 227: 116124.

14 Zhu, X., Duan, J., Chen, J. et al. (2024). Aldol condensation for the construction of organic functional materials. *Angewandte Chemie International Edition* 63 (2): e202311879.

15 Lim, D.H., Ha, J.W., Choi, H. et al. (2021). Recent progress of ultra-narrow-bandgap polymer donors for NIR-absorbing organic solar cells. *Nanoscale Advances* 3 (15): 4306–4320.

16 Scharber, M.C. and Sariciftci, N.S. (2021). Low band gap conjugated semiconducting polymers. *Advanced Materials Technologies* 6 (4): 2000857.

17 Liu, Y., Wu, Y., Geng, Y. et al. (2022). Managing challenges in organic photovoltaics: properties and roles of donor/acceptor interfaces. *Advanced Functional Materials* 32 (43): 2206707.

18 Xie, B., Chen, Z., Ying, L. et al. (2019). Near-infrared organic photoelectric materials for light-harvesting systems: organic photovoltaics and organic photodiodes. *InfoMat* 2 (1): 57–91.

19 Meng, D., Zheng, R., Zhao, Y. et al. (2022). Near-infrared materials: the turning point of organic photovoltaics. *Advanced Materials* 34 (10): 2107330.

20 Wang, Y., Zhang, H., Wang, Z. et al. (2020). Photothermal conjugated polymers and their biological applications in imaging and therapy. *ACS Applied Polymer Materials* 2 (10): 4222–4240.

21 Pomfret, S.J., Monkman, A.P., and Havinga, E.E. (1996). Electroabsorption measurements of polysquaraine. *Synthetic Metals* 78 (3): 285–288.

22 van Mullekom, H.A.M., Vekemans, J.A.J.M., Havinga, E.E. et al. (2001). Developments in the chemistry and band gap engineering of donor–acceptor substituted conjugated polymers. *Materials Science and Engineering: R: Reports* 32 (1): 1–40.

23 Bundgaard, E. and Krebs, F.C. (2007). Low band gap polymers for organic photovoltaics. *Solar Energy Materials and Solar Cells* 91 (11): 954–985.

24 Xia, Y., Wang, L., Deng, X. et al. (2006). Photocurrent response wavelength up to 1.1 μm from photovoltaic cells based on narrow-band-gap conjugated polymer and fullerene derivative. *Applied Physics Letters* 89 (8): 081106.

25 Qi, J., Zhou, X., Yang, D. et al. (2014). Optimization of solubility, film morphology and photodetector performance by molecular side-chain engineering of low-bandgap thienothiadiazole-based polymers. *Advanced Functional Materials* 24 (48): 7605–7612.

26 Wu, Z., Yao, W., London, A.E. et al. (2017). Temperature-dependent detectivity of near-infrared organic bulk heterojunction photodiodes. *ACS Applied Materials & Interfaces* 9 (2): 1654–1660.

27 Yang, J., Huang, J., Li, R. et al. (2021). Cavity-enhanced near-infrared organic photodetectors based on a conjugated polymer containing [1,2,5]Selenadiazolo[3,4-c]Pyridine. *Chemistry of Materials* 33 (13): 5147–5155.

28 Ying, L., Hsu, B.B.Y., Zhan, H. et al. (2011). Regioregular Pyridal[2,1,3]thiadiazole π-conjugated copolymers. *Journal of the American Chemical Society* 133 (46): 18538–18541.

29 Lv, L., Dang, W., Wu, X. et al. (2020). Flexible short-wave infrared image sensors enabled by high-performance polymeric photodetectors. *Macromolecules* 53 (23): 10636–10643.

30 Steckler, T.T., Henriksson, P., Mollinger, S. et al. (2014). Very low band gap thiadiazoloquinoxaline donor–acceptor polymers as multi-tool conjugated polymers. *Journal of the American Chemical Society* 136 (4): 1190–1193.

31 Chen, X.X., Li, J.T., Fang, Y.H. et al. (2022). High-mobility semiconducting polymers with different spin ground states. *Nature Communications* 13 (1): 2258.

32 Wang, T., Xu, C., Deng, Y. et al. (2025). Conjugated polymers with an oxindole-terminated quinoidal unit: energy level modulation toward air-stable n-doped conductors. *Advanced Functional Materials* 35 (2): 2412647.

33 Han, J., Qi, J., Zheng, X. et al. (2017). Low-bandgap donor–acceptor polymers for photodetectors with photoresponsivity from 300 nm to 1600 nm. *Journal of Materials Chemistry C* 5 (1): 159–165.

34 Verstraeten, F., Gielen, S., Verstappen, P. et al. (2018). Near-infrared organic photodetectors based on bay-annulated indigo showing broadband absorption and high detectivities up to 1.1 μm. *Journal of Materials Chemistry C* 6 (43): 11645–11650.

35 Zhao, X., Cai, H., Deng, Y. et al. (2021). Low-band gap conjugated polymers with strong absorption in the second near-infrared region based on diketopyrrolopyrrole-containing quinoidal units. *Macromolecules* 54 (7): 3498–3506.

36 Lu, Y., Yu, Z.D., Un, H.I. et al. (2021). Persistent conjugated backbone and disordered lamellar packing impart polymers with efficient n-doping and high conductivities. *Advanced Materials* 33 (2): 2005946.

37 Cai, H., Tang, H., Wang, T. et al. (2024). An n-type open-shell conjugated Polymer with high-spin ground-state and high intrinsic electrical conductivity. *Angewandte Chemie International Edition* 63 (25): e202402375.

38 Luo, X., Xiong, W., Tang, H. et al. (2024). Electron-withdrawing quinone polymers with enhanced conjugated planarity for n-type organic thermoelectrics. *Journal of Materials Chemistry C* 12 (44): 17943–17951.

39 Shao, X., Liu, M., Liu, J. et al. (2022). A resonating B, N covalent bond and coordination bond in aromatic compounds and conjugated polymers. *Angewandte Chemie International Edition* 61 (34): e202205893.

40 Xu, J., Zhang, Y., Liu, J. et al. (2023). NIR-II absorbing monodispersed oligomers based on $N-B \leftarrow N$ Unit. *Angewandte Chemie International Edition* 62 (42): e202310838.

41 Xu, J., Liu, J., and Wang, L. (2023). Ultra-small-bandgap conjugated polymers based on an N − B ← N unit. *Angewandte Chemie International Edition* 62 (24): e202303870.

42 Shao, X., Fan, X., Cao, X. et al. (2023). Electron-transporting properties and ultrasmall band gap of BODIPY-based conjugated polymers: teach an old dog new tricks. *Macromolecules* 56 (23): 9489–9497.

43 Verstraeten, F., Gielen, S., Verstappen, P. et al. (2020). Efficient and readily tuneable near-infrared photodetection up to 1500 nm enabled by thiadiazoloquinoxaline-based push–pull type conjugated polymers. *Journal of Materials Chemistry C* 8 (29): 10098–10103.

44 Han, H., Choi, J.H., Ahn, J. et al. (2023). Chiral diketopyrrolopyrrole-based conjugated polymers with intramolecular rotation–isomeric conformation asymmetry for near-infrared circularly polarized light-sensing organic phototransistors. *ACS Applied Materials & Interfaces* 15 (49): 57447–57460.

45 Yuen, J.D., Fan, J., Seifter, J. et al. (2011). High performance weak donor–acceptor polymers in thin film transistors: effect of the acceptor on electronic properties, ambipolar conductivity, mobility, and thermal stability. *Journal of the American Chemical Society* 133 (51): 20799–20807.

46 Chen, J., Yang, J., Guo, Y. et al. (2022). Acceptor modulation strategies for improving the electron transport in high-performance organic field-effect transistors. *Advanced Materials* 34 (22): 2104325.

47 Zhang, G., Dai, Y., Liu, Y. et al. (2017). Facile green synthesis of isoindigo-based conjugated polymers using aldol polycondensation. *Polymer Chemistry* 8 (22): 3448–3456.

48 Xiao, M., Carey, R.L., Chen, H. et al. (2021). Charge transport physics of a unique class of rigid-rod conjugated polymers with fused-ring conjugated units linked by double carbon-carbon bonds. *Science Advances* 7 (18): eabe5280.

49 Onwubiko, A., Yue, W., Jellett, C. et al. (2018). Fused electron deficient semiconducting polymers for air stable electron transport. *Nature Communications* 9 (1): 416.

50 Lu, Y., Yu, Z.D., Zhang, R.Z. et al. (2019). Rigid coplanar polymers for stable n-type polymer thermoelectrics. *Angewandte Chemie International Edition* 58 (33): 11390–11394.

51 Chen, H., Moser, M., Wang, S. et al. (2021). Acene ring size optimization in fused Lactam polymers enabling high n-type organic thermoelectric performance. *Journal of the American Chemical Society* 143 (1): 260–268.

52 Wang, Y., Zeglio, E., Wang, L. et al. (2022). Green synthesis of lactone-based conjugated polymers for n-type organic electrochemical transistors. *Advanced Functional Materials* 32 (16): 2111439.

53 Ren, H., Chen, J.D., Li, Y.Q. et al. (2020). Recent progress in organic photodetectors and their applications. *Advanced Science* 8 (1): 2002418.

54 Wang, C., Zhang, X., and Hu, W. Organic photodiodes and phototransistors toward infrared detection: materials, devices, and applications. *Chemical Society Reviews* 49 (3): 653–670.

55 Shao, L., Hong, L., Cao, Y. et al. (2023). Spiro-conjugation in narrow-bandgap nonfullerene acceptors enables broader spectral response and higher detectivity for near-infrared organic photodetectors. *Advanced Optical Materials* 11 (13): 2202823.

56 Yao, Y., Liang, Y., Shrotriya, V. et al. (2007). Plastic near-infrared photodetectors utilizing low band gap polymer. *Advanced Materials* 19 (22): 3979–3983.

57 Gong, X., Tong, M., Xia, Y. et al. (2009). High-detectivity polymer photodetectors with spectral response from 300 nm to 1450 nm. *Science* 325 (5948): 1665–1667.

58 Chen, E.C., Tseng, S.R., Chao, Y.C. et al. (2011). Polymer infrared photo-detector with high sensitivity up to 1100 nm. *Synthetic Metals* 161 (15–16): 1618–1622.

59 Perzon, E., Zhang, F., Andersson, M. et al. (2007). A conjugated polymer for near infrared optoelectronic applications. *Advanced Materials* 19 (20): 3308–3311.

60 Wu, Z., Yao, W., London, A. E. et al. (2018). Elucidating the detectivity limits in shortwave infrared organic photodiodes. *Advanced Functional Materials* 28 (18): 1800391.

61 Wu, Z., Zhai, Y., Yao, W. et al. (2018). The role of dielectric screening in organic shortwave infrared photodiodes for spectroscopic image sensing. *Advanced Functional Materials* 28 (50): 1805738.

62 Park, I., Kim, C., Kim, R. et al. (2022). High performance shortwave infrared organic photodetectors adopting thiadiazole quinoxaline-based copolymers. *Advanced Optical Materials* 10 (19): 2200747.

63 Jacoutot, P., Scaccabarozzi, A.D., Zhang, T. et al. (2022). Infrared organic photodetectors employing ultralow bandgap polymer and non-fullerene acceptors for biometric monitoring. *Small* 18 (15): 2200580.

64 Yuen, J.D., Kumar, R., Zakhidov, D. et al. (2011). Ambipolarity in benzobisthiadiazole-based donor-acceptor conjugated polymers. *Advanced Materials* 23: 3780–3785.

65 Yuen, J.D., Wang, M., Fan, J. et al. (2015). Importance of unpaired electrons in organic electronics. *Journal of Polymer Science Part A: Polymer Chemistry* 53 (2): 287–293.

66 Zheng, L., Zhu, T., Xu, W. et al. (2018). Solution-processed broadband polymer photodetectors with a spectral response of up to 2.5 μm by a low bandgap donor–acceptor conjugated copolymer. *Journal of Materials Chemistry C* 6 (14): 3634–3641.

67 Lei Lv, W.D., Xiaoxi W., Hao, C. et al. (2020). Flexible short-wave infrared image sensors enabled by high performance polymeric photodetectors. *Macromolecules* 53: 10636–10643.

68 Yao, W., Wu, Z., Huang, E. et al. (2019). Organic bulk heterojunction infrared photodiodes for imaging out to 1300 nm. *ACS Applied Electronic Materials* 1 (5): 660–666.

69 Goel, M. and Thelakkat, M. (2020). Polymer thermoelectrics: opportunities and challenges. *Macromolecules* 53 (10): 3632–3642.

70 Shao, S. and Liu, J. (2021). Molecular insights into n-type organic thermoelectrics. *CCS Chemistry* 3 (10): 2702–2716.

71 Stefan Wilhelm, A.J.T., Qin D., Seiichi O. et al. (2016). Analysis of nanoparticle delivery to tumours. *Nature Reviews Materials* 1 (5): 16014.

72 Yang, C.Y., Jin, W.L., Wang, J. et al. (2018). Enhancing the n-Type conductivity and thermoelectric performance of donor-acceptor copolymers through donor engineering. *Advanced Materials* 30 (43): 1802850.

73 Lüssem, B., Keum, C.M., Kasemann, D. et al. (2016). Doped organic transistors. *Chemical Reviews* 116 (22): 13714–13751.

74 Zhao, W., Ding, J., Zou, Y. et al. (2020). Chemical doping of organic semiconductors for thermoelectric applications. *Chemical Society Reviews* 49 (20): 7210–7228.

75 Tripathi, A., Lee, Y., Lee, S. et al. (2022). Recent advances in n-type organic thermoelectric materials, dopants, and doping strategies. *Journal of Materials Chemistry C* 10 (16): 6114–6140.

76 Yanfeng Wang, K.K.L., Tsung-Ta, H., Ling P. et al. (2005). Use of phosphine as an n-type dopant source for vapor—liquid—solid growth of silicon nanowires. *Nano Letters* 5: 2139.

77 Chen, J., Li, D., Zhang, Y. et al. (2020). Comparative study on P and B doped nano-crystalline Si multilayers. *Applied Surface Science* 529: 146971.

78 Geng, X., Du, T., Xu, C. et al. (2023). Realizing p-type and n-type doping of a single conjugated polymer via incorporation of a thienoisatin-terminated quinoidal unit. *Advanced Functional Materials* 33 (28): 2300809.

79 Yu, Z.D., Lu, Y., Wang, Z.Y. et al. (2023). High n-type and p-type conductivities and power factors achieved in a single conjugated polymer. *Science Advances* 9 (8): eadf3495.

80 Wang, Y., Feng, L., and Wang, S. (2019). Conjugated polymer nanoparticles for imaging, cell activity regulation, and therapy. *Advanced Functional Materials* 29 (5): 1806818.

81 Kenry., Duan, Y., and Liu, B. (2018). Recent advances of optical imaging in the second near-infrared window. *Advanced Materials* 30 (47): 1802394.

82 Miao, Q. and Pu, K. (2018). Organic semiconducting agents for deep-tissue molecular imaging: second near-infrared fluorescence, self-luminescence, and photoacoustics. *Advanced Materials* 30 (49): 1801778.

83 Zhang, W., Sun, X., Huang, T. et al. (2019). 1300 nm absorption two-acceptor semiconducting polymer nanoparticles for NIR-II photoacoustic imaging system guided NIR-II photothermal therapy. *Chemical Communications* 55 (64): 9487–9490.

84 Zhang, W., Deng, W., Zhang, H. et al. (2020). Bioorthogonal-targeted 1064 nm excitation theranostic nanoplatform for precise NIR-IIa fluorescence imaging guided efficient NIR-II photothermal therapy. *Biomaterials* 243: 119934.

85 Li, J., Rao, J., and Pu, K. (2018). Recent progress on semiconducting polymer nanoparticles for molecular imaging and cancer phototherapy. *Biomaterials* 155: 217–235.

8

Polymers with Three-dimensional Conjugated Molecules

Satoru Hiroto

Graduate School of Human and Environmental Studies, Kyoto University, Sakyo-ku, Kyoto, Japan

8.1 Introduction

Three-dimensional (3D) π-conjugated systems, represented by buckybowls, have developed tremendously owing to the establishment of facile synthetic methods for construction of curved π-conjugated structures [1–3]. The curved structure has large strain energy that has hampered easy access to the molecules. The representative molecules in chemistry are fragments of carbon materials such as fullerenes and carbon nanotubes, so-called buckybowls, and carbon nanorings. In addition, helicene, which is the most conventional model of curved π-conjugated molecules by M.S. Newman and D. Lednicer in 1955, has helical fused π-conjugated structure (Figure 8.1) [4]. In the past decades, the chemistry of curved π-conjugated molecules has been focused on the development of their efficient synthesis. Recently, curved π-conjugated molecules have become accessible molecules that are ready for application in molecular materials. In the course of these studies, various unique properties originating from their curved structures were elucidated. The curved structure provides mainly three kinds of properties: (1) increase of solubility in organic solvent, (2) chirality other than quaternary carbon due to their 3D structure, and (3) multidirectional intermolecular interactions in solid states. These properties can be included by using curved π-conjugated molecules either as a branch or as a main-chain component of polymers. A polymeric curved π-conjugated system was initially discovered in helicene chemistry. Katz reported the synthesis of helical conjugated ladder polymer by metal coordination with helicene-based ligands in 1996 [5]. After that, there have been several reports about helicene-containing polymers. After remarkable development of synthetic strategy of corannulene, reports on polymers with corannulenes and other buckybowls have been emerged [6]. In the chapter, we would like to focus on polymers with buckybowls and helicenes, including supramolecular polymers.

Conjugated Polymer Synthesis and Materials Applications, First Edition. Edited by Yoshiki Chujo.
© 2026 Wiley-VCH GmbH. All rights reserved, including rights for text and data mining and training of artificial intelligence technologies or similar technologies. Published 2026 by Wiley-VCH GmbH.

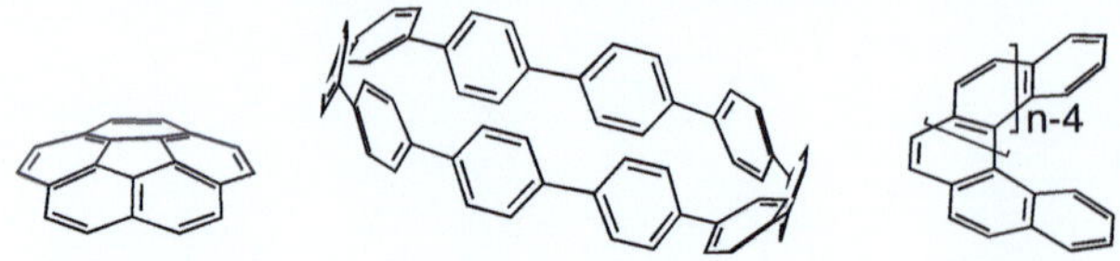

Figure 8.1 Representative curved π-conjugated molecules.

8.2 Corannulene-containing Polymer

Buckybowl, a bowl-shaped π-conjugated molecule, has been regarded as a fragment of fullerenes. The bowl-shaped conformation provides concave and convex surfaces on the π-conjugated planes, enabling different intermolecular interactions with other π-conjugated molecules. In particular, the concave surface of buckybowls is effective for binding fullerene molecules via concave–convex interactions. The concave and convex surfaces are interconvertible bowl-to-bowl inversions in the case of low energy barrier (Figure 8.2) [7]. Therefore, incorporation of buckybowl units into polymers enhances the ability to capture fullerenes. The most common families of buckybowls are corannulene and sumanene. Corannulene was firstly synthesized in 1996 [8], and after enthusiastic studies to develop their synthetic methods, corannulene is now prepared on a kilogram scale according to the Siegels strategy [9, 10]. Currently, corannulene is a commercially available compound. On the other hand, the history of sumanene is rather short, and an efficient synthetic method was reported by Sakurai, Hirao et al. in 2003 [11]. In the case of sumanene, hetero-sumanene, which contains heteroatoms in sumanene skeleton, has been the main component in polymeric compounds [12, 13]. First, covalently bonded polymers based on corannulenes will be presented.

The first report of the corannulene-based polymer was reported by Stuparu in 2012 [14]. Corannulene is regarded as a useful building block for C_5-symmetric star-shaped polymeric materials. The star-shaped polymers have been synthesized by two protocols (Scheme 8.1). The poly(lactic acid) substituted corannulene was prepared by penta-substituted corannulene as an initiator. In this report, the author also investigated the association with C_{60} and discovered the concentration dependent association with C_{60}. During the study, the author also found the solvent and alkyl chain length dependency of the association. To gain in-depth information, the author synthesized new corannulene-branched polymers using a free base polymerization technique. These polymers were found to interact with fullerene C_{60} through "ball-and-socket" type π–π stacking interactions. This "ball-and-socket" interactions are dependent on alkyl-chain length and solvents. Considering the parent corannulene does not bind C_{60} in solution, this phenomenon is one of the unique properties of corannulene-containing polymers [15].

In 2013, Stuparu reported the synthesis of corannulene-branched polymers (Figure 8.3a) [16]. The author prepared polymers with alkyl chains of different lengths between the polymer main chain and corannulene units. The interactions of the synthesized homopolymers with fullerene C_{60} were evaluated by ^{13}C NMR

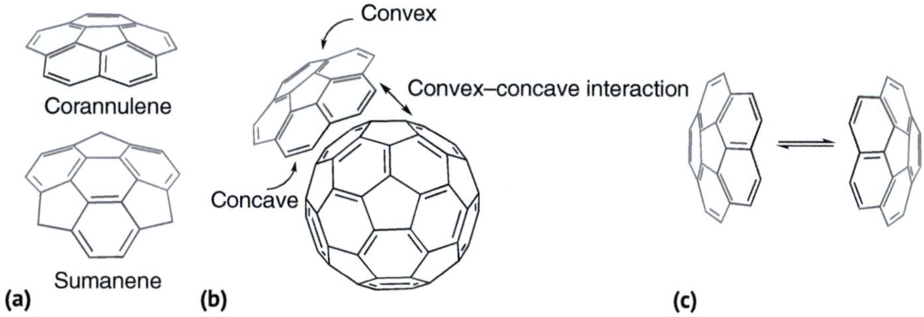

Figure 8.2 (a) Representative buckybowls, (b) concave–convex interaction with C_{60}, and (c) bowl-to-bowl inversion.

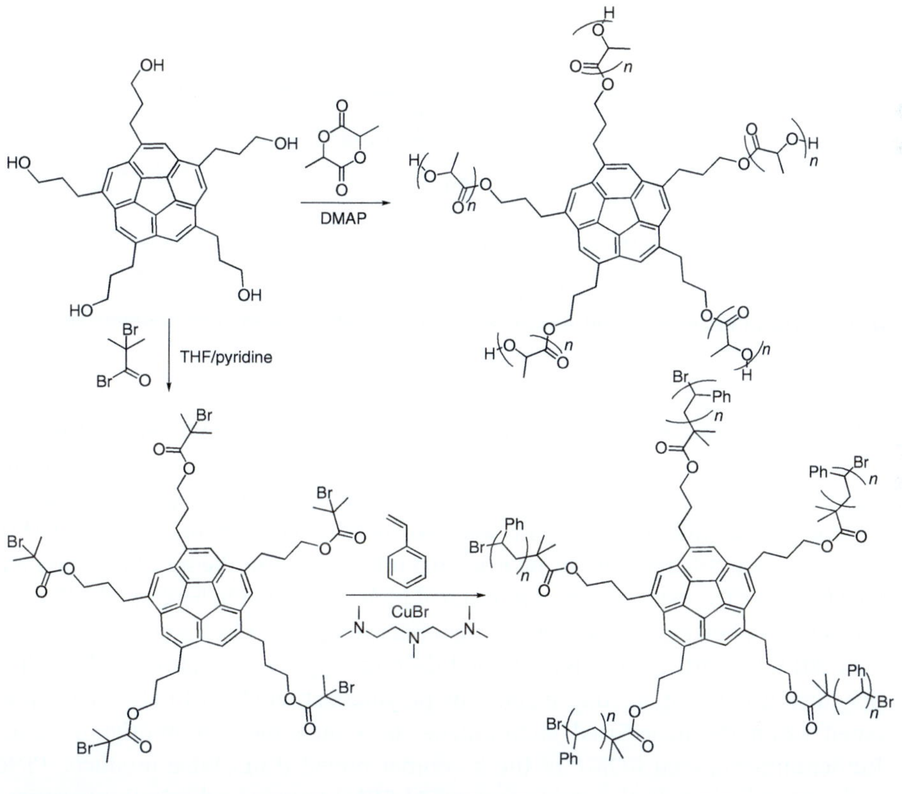

Scheme 8.1 Synthesis of the star-shaped polymers.

spectroscopy, and it was confirmed that the shorter the linker length, the stronger the interaction with fullerene. The nature of the solvent was also shown to affect the interaction with fullerenes, and experiments using chlorobenzene and dichloromethane revealed that the choice of solvent could tune the strength of the interaction. Furthermore, it was shown that the host–guest interaction can be tuned by changing the temperature, with the interaction strengthening as the temperature

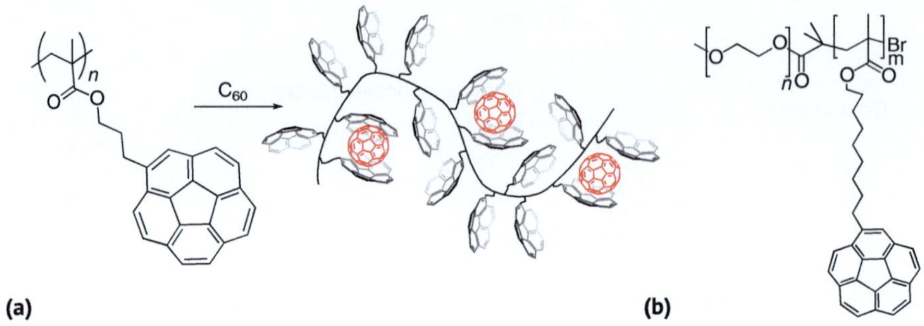

Figure 8.3 (a) Corannulene-branched polymer receptors and "bowl-and-socket" interaction with C_{60} and (b) amphiphilic host copolymer.

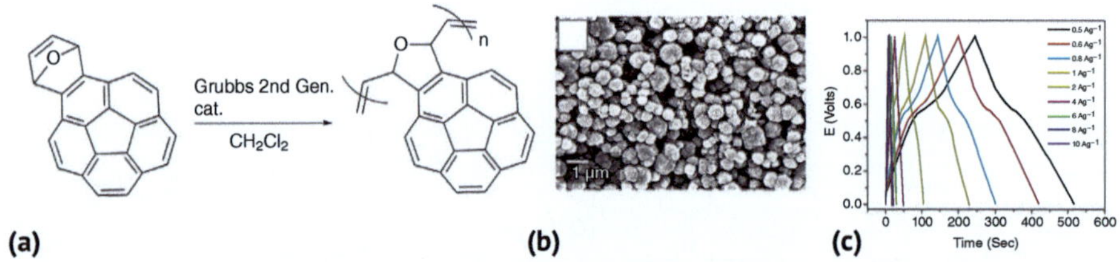

Figure 8.4 (a) Synthesis of polymers by ring-opening metathesis polymerization (ROMP) of corannulene-encoded oxanorbornadiene monomer, (b) FE-SEM images of polymer nanoparticles, and (c) galvanostatic charge-discharge (GCD) profiles of polymeric nanostructures in 1M H_2SO_4 aq. electrolyte. *Source*: [17]/with permission of American Chemical Society.

is lowered and weakened as the temperature is raised. The author also synthesized a block copolymer with polyethylene glycol (PEG) and found that it formed micelle structures in water (Figure 8.3b). Micelles containing fullerene C_{60} were observed by atomic force microscopy (AFM) and transmission electron microscopy (TEM), indicating that fullerenes were incorporated into the core of the micelles. This enabled fullerene C_{60} to be dispersed in water, which is expected to have applications in the fields of photoelectric conversion materials and plastic electronics.

In 2017, Stuparu group also reported different types of corannulene-branched polymers using ring-opening metathesis polymerization (ROMP) [17]. They prepared oxa-norbornene-fused corannulene as a monomer on multigram scale. Ruthenium-catalyzed ROMP of the monomer provided insoluble products. Field emission scanning electron microscopy (FM-SEM) revealed spherically structured polymer aggregates in the size range of 400–980 nm were formed (Figure 8.4). Isothermal N_2 adsorption/desorption measurements showed the porous features of the nanostructures. Electrochemical studies further disclosed its electrochemical supercapacitor properties, exhibiting a high capacitance value of approximately 134 $F \cdot g^{-1}$ at 0.5 $A \cdot g^{-1}$, as determined by galvanostatic charge-discharge (GCD) spectroscopy. In addition, these corannulene-based nanomaterials exhibited exceptional long-term stability through charge and discharge cycles, showing 90% capacitance retention even after undergoing 10,000 cycles.

After these reports, polymers containing corannulene in the backbone have been reported by several researchers. In 2014, Cao's group reported corannulene-containing polymer by cross-coupling of dibromocorannulene with bithiophene by Stille cross-coupling reactions (Figure 8.5) [18]. The corannulene units were fused with a diimide skeleton to construct a donor–acceptor type polymer. They also investigated the organic field-effect transfer performance of the obtained polymer, showing moderate ambipolar transporting properties with an electron mobility of 7.45×10^{-5} cm^2 V^{-1} s^{-1} and hole mobility of 0.025 cm^2 V^{-1} s^{-1}. AFM and X-ray diffraction analyses revealed that the film was almost amorphous.

Smaldone's group reported the corannulene-containing porous polymers [19]. They prepared three types of corannulene-based POPs (**BB-POPs**) by Sonogashira copolymerization of 1,2,5,6-tetrabromocorannulene (TBC) with 1,4-diethynylbenzene (**BB-POP-1**), 1,3,5-trisethynylbenzene (**BB-POP-2**), and tetrakis(4-ethynyl)-tetraphenylmethane (**BB-POP-3**) (Figure 8.6). Surface area and porosity measurements revealed that **BB-POP-3** exhibited the highest surface area (BET = 560 m^2/g) and CO_2 adsorption capacity (11.7 wt%) among the three polymers. The polymers were found to be microporous with pore sizes predominantly below 2 nm. **BB-POP-3** exhibited the highest nitrogen uptake

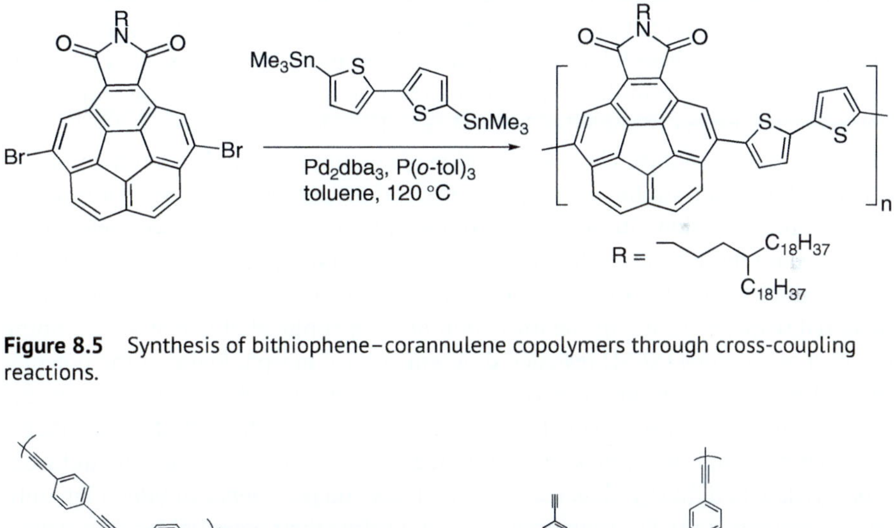

Figure 8.5 Synthesis of bithiophene–corannulene copolymers through cross-coupling reactions.

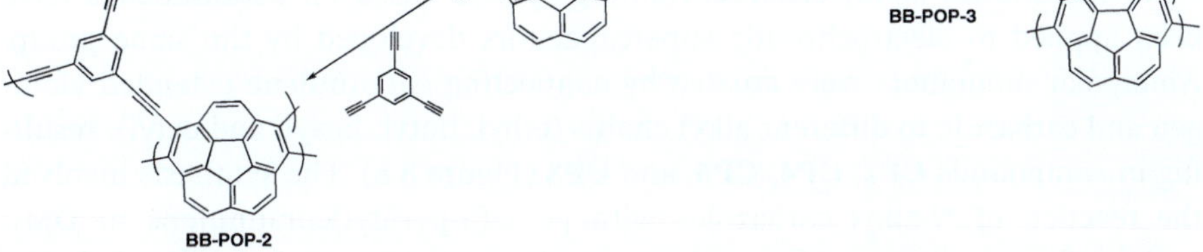

Figure 8.6 Synthesis of **BB-POPs** by Sonogashira couplings with tetrabromocorannulene.

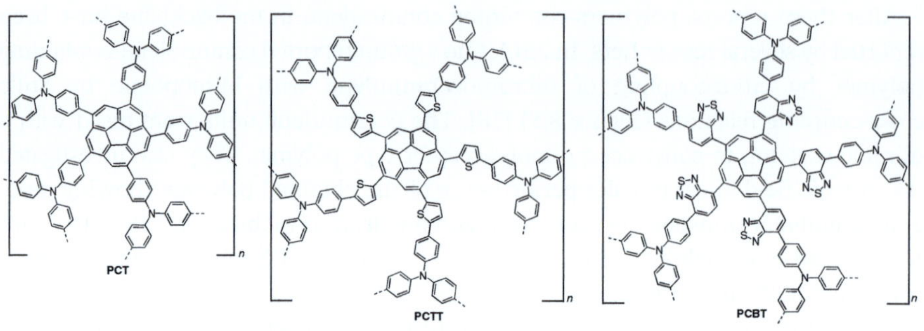

Figure 8.7 Structures of **PCT**, **PCTT**, and **PCBT** as electrofluorochromic materials.

capacity (284 cm^3/g) and CO_2 uptake at 273 K (57 cm^3/g or 11.7 wt%), which were significantly higher than those of **BB-POP-1** and **BB-POP-2**. The redox properties of corannulene were retained in the POPs, as evidenced by cyclic voltammetry (CV) studies. **BB-POP-2** and **BB-POP-3** showed distinct reduction events, indicating their potential for electrochemical applications such as supercapacitors and batteries.

Another study on the electrochemical applications of corannulene-based polymers was conducted by Luo et al. [20–22]. They investigated the electrochromic properties of corannulene-based donor–acceptor-type conjugated polymers. They prepared three star-shaped corannulene-based monomers bearing acceptors based on triphenylamine, thiophene-triphenylamine, or benzothiazole-triphenylamines by Suzuki–Miyaura cross-couplings. The incorporation of corannulenes and electron-accepting groups allowed the tuning of the photophysical properties, resulting in a range of fluorescence emissions from green to orange-red. The monomers were electropolymerized on indium tin oxide (ITO) glass to form conjugated polymer films (**PCT**, **PCTT**, and **PCBT**) (Figure 8.7). The electropolymerization process involved cyclic potential scans, resulting in the formation of 3D π-conjugated networks. Scanning electron microscopy (SEM) revealed different surface morphologies of the polymer films. The electrochromic properties were analyzed by applying potentials to the polymer films and observing color changes. The color of **PCT**, **PCTT**, and **PCBT** was changed from pale yellow to azure blue, from mimosa to ivy green, and from moon yellow to light sky blue, respectively. Thus, the polymers exhibited high optical contrast ratios and fast switching times, indicating their potential for electrochromic applications. The polymer films retained their fluorescence in the solid state and demonstrated reversible electrofluorescence switching.

Further studies about electrochromic polymers based on corannulenes have been applied to electrochromic supercapacitors developed by the same group. Ambipolar monomers were created by connecting corannulene-extended viologen and carbazole to different alkyl chains (ethyl, butyl, hexyl, and octyl), resulting in compounds **CP2**, **CP4**, **CP6**, and **CP8** (Figure 8.8). The synthesis involved the reaction *of N*-alkyl carbazoles with penta(4-pyridyl)corannulene in DMF under reflux conditions, followed by ion exchange to improve solubility. Polymer

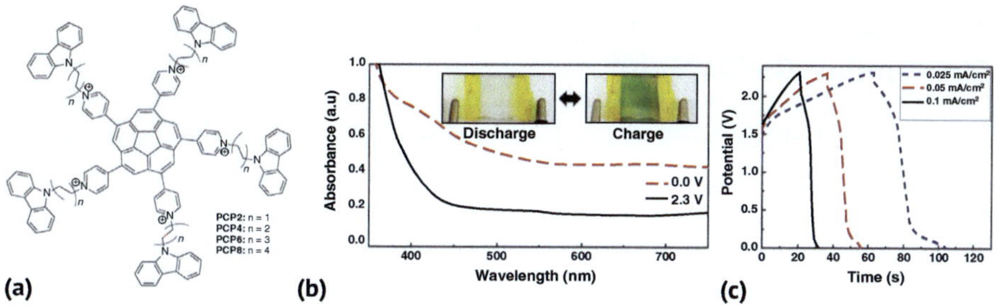

Figure 8.8 (a) Schematic views of **PCPns**; (b) UV/vis absorption spectra on charge–discharge conversion in the visible region; and (c) Galvanostatic charge-discharge (GCD) profiles of the device using **PCP4.** *Source*: [21]/with permission of American Chemical Society.

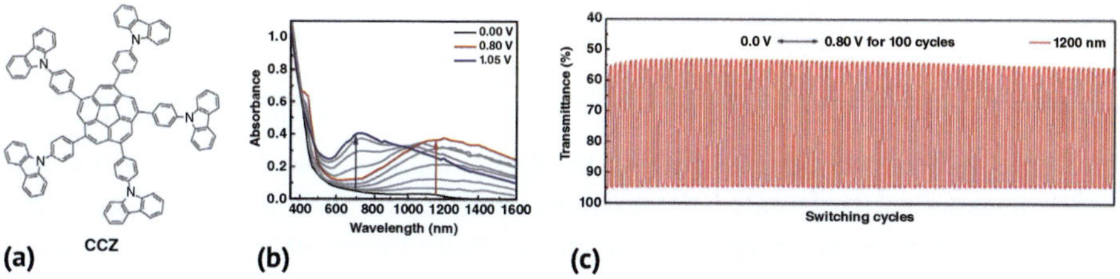

Figure 8.9 (a) Schematic views of corannulene–carbazole (**CCZ**); (b) UV/vis absorption spectra on charge–discharge conversion in the visible region; and (c) GCD profiles of the device using **PCP4.** *Source*: [22]/with permission of American Chemical Society.

films were prepared on ITO glass via electrochemical polymerization. The process involved cycling the potential scans in a mixed solvent of dichloromethane and acetonitrile containing the monomers and an electrolyte. The resulting polymers were named **PCP2**, **PCP4**, **PCP6**, and **PCP8**. The spectroelectrochemical spectra showed that the polymers were transparent in the neutral state and exhibited color changes upon doping. Optical contrast and switching times were measured, with **PCP4** exhibiting the highest optical contrast and fastest switching times. A proof-of-concept electrochromic ambipolar supercapacitor device was fabricated using **PCP4** as the electrochemically active layer. The device demonstrated a wide operating voltage window and transparent to green color switching, making it a promising material for electrochromic applications. In 2023, the Luo's group developed copolymer films made from corannulene–carbazole (**CCZ**) and 3,4-ethylenedioxythiophene (EDOT) for near-infrared (NIR) electrochromic applications (Figure 8.9). In this report, **P0**, which is a polymer film formed by electropolymerization of **CCZ**, showed NIR electrochromic behavior upon application of voltage, faster electrochromic switching, and higher reversibility than the planar polymer film. In these cases, they concluded that the bowl-shaped conformation of corannulene would work effectively for high capacitance capability owing to its ion-accepting ability into its concave capability.

8.3 Supramolecular Polymers Based on Bowl-shaped π-conjugated Molecules

As mentioned above, buckybowls have attracted attentions for building unit for supramolecular assembly with convex–concave interaction. In the case of corannulene, investigations for regular packing in crystal have been enthusiastically investigated by numerous researchers [23]. The pristine corannulene shows herringbone packing pattern due to the presence of CH–π interactions rather than convex–concave interactions [24]. To gain columnar packing in a cup stacking mode for the sake of improvement of charge transfer mobility, substituent effect on crystal packing has been investigated. Afterward, introduction of electron-withdrawing groups such as trifluoromethyl substituents at their periphery proved to show a perpendicular columnar stacking [25]. Another method is to introduce long-chain alkyl groups at the periphery of corannulene. In this case, the liquid crystalline (LC) mesophase appears. In 2009, Aida et al. reported the first LC corannulene derivative responsive to an electrical field (Figure 8.10) [26]. They synthesized two new corannulene derivatives, **Cor1** and **Cor2**, with peripheral amide functionalities expected to form multiple hydrogen bonds, assisting in the formation of stable columnar structures. DSC measurements revealed the existence of the LC mesophase of **Cor2** upon second heating. POM analysis on second cooling process showed a fan texture at

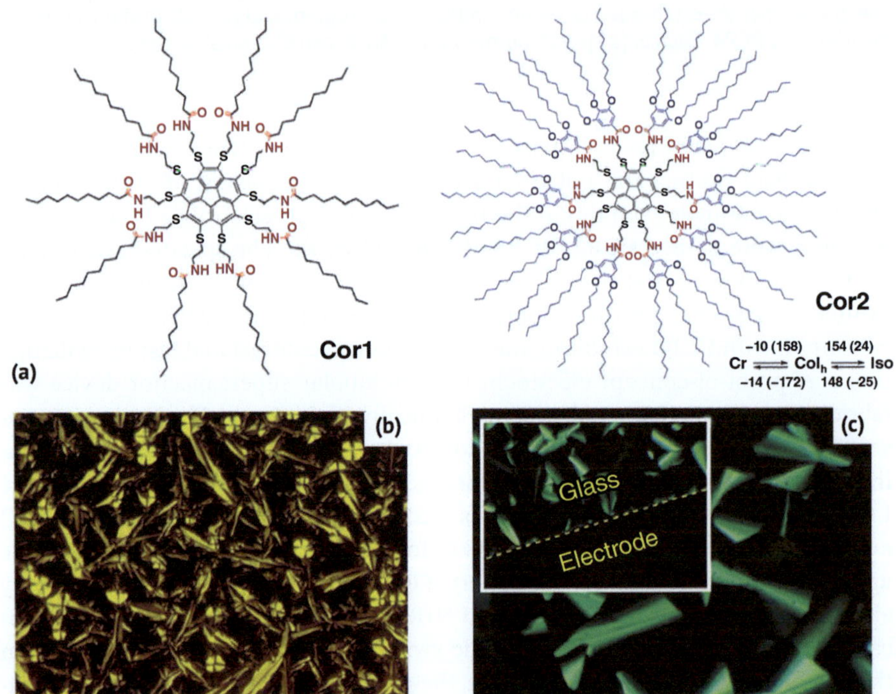

Figure 8.10 (a) Structures of **Cor1** and **Cor2**, (b) POM image of **Cor1** at 160 °C, and (c) POM image of **Cor2** at 115 °C. *Source*: [26]/with permission of American Chemical Society.

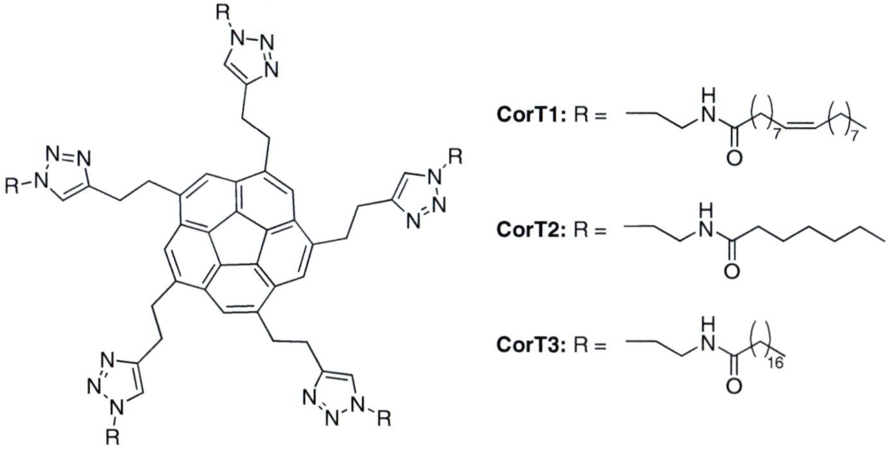

Figure 8.11 Structures of **CorT1**, **CorT2**, and **CorT3**.

115 °C, indicating its hexagonal columnar LC assemblies. Compound **Cor1** showed a narrow LC phase range at higher temperatures. Interestingly, the LC columns of **Cor2** aligned homeotropically to an ITO electrode surface under an applied electric field. This alignment was maintained even after the electric field was switched off but could be reset by heating above the clearing temperature. This phenomenon is attributed to the nonplanar geometry of corannulene, which induces a dipole moment.

Seigel's group independently developed LC corannulenes in 2013 [27]. The study demonstrates that the nature of hydrocarbon chains influences the phase behavior and gelation properties of C5-symmetric pentakislipido-corannulene derivatives. They prepared three types of fivefold symmetric penta-substituted corannulenes (Figure 8.11). Three derivatives, labeled **CorT1**, **CorT2**, and **CorT3**, were synthesized and their phase behaviors were studied using DSC and POM. **CorT1** exhibited thermal transitions and birefringence, indicative of a liquid-crystalline phase, whereas **CorT2** showed no such properties. **CorT3** displayed phase transitions and birefringence only beyond its melting point. Small- and wide-angle X-ray scattering (SAXS and WAXS) experiments revealed a lamellar phase for **CorT1** with a layer-to-layer distance of 4.4 nm, as confirmed by TEM. The gelation properties of these compounds were tested in various solvents. **CorT1** formed thermoreversible gels in cyclohexane and methylcyclohexane (MCH), while **CorT2** did not exhibit gelation behavior. **CorT3** formed gels in cyclohexane, similar to **CorT1**. UV-vis and emission spectroscopy, along with FTIR analysis, suggested the absence of ordered supramolecular organization or hydrogen bonding in the gel phase for both **CorT1** and **CorT3**.

In 2017, Stuparu and Mahadevegowda developed thermoresponsive corannulene supramolecular assembly [28, 29]. Specifically, they focused on corannulenes substituted with the triethylene glycol unit **CorTeg** (Figure 8.12). In solvents such as methanol and acetonitrile, **CorTeg** exhibited two major absorption bands at 250 and 335 nm. The emission spectra showed moderate intensities in these solvents,

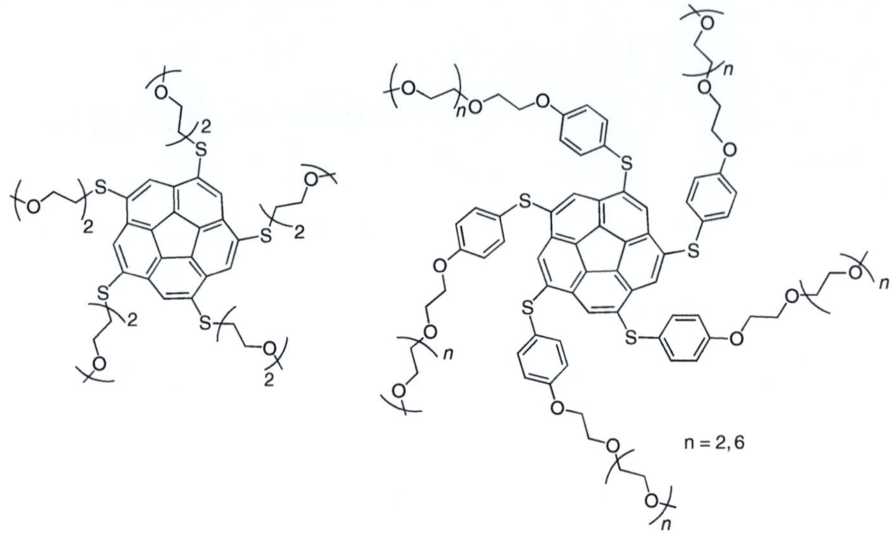

Figure 8.12 Structures of thermoresponsive corannulene polymers.

whereas in water, both the absorption and emission intensities were significantly reduced. Heating its aqueous solution led to a loss of optical clarity above 40 °C, indicating the formation of larger structures. This process was reversible because cooling restored the clarity of the solution. In addition, increasing the temperature resulted in enhanced emission intensity. The quantum yields for fluorescence at 20 and 70 °C were 0.4 and 1.2%, respectively, suggesting that thermal aggregation restricted molecular motion and enhanced emission. AFM and SEM revealed that the thermally triggered assembly produced large fibrillar structures with heights ranging from 10 to 100 nm and lengths in the micrometer range. SEM images showed uninterrupted fibers hundreds of micrometers long. The authors also prepared eight new corannulene derivatives combining a hydrophobic corannulene core with hydrophilic PEG arms. The water solubility of the derivatives was assessed, revealing that only pentasubstituted compounds with five PEG arms were highly soluble in water. The thermoresponsive behavior was studied by measuring the transmittance of the aqueous solutions of the derivatives at different temperatures. The results show that the lower critical solution temperature (LCST) can be modulated between 30 and 50 °C by adjusting the molecular structure. Compounds with phenyl rings exhibited the highest LCST, whereas those with shorter PEG segments showed a lower LCST.

One of the landmarks about supramolecular chemistry is living supramolecular polymerization [30]. This phenomenon has been firstly found in the chemistry of corannulene. In 2015, Aida's group demonstrated chain-growth polymerization using corannulene-based metastable monomers (Figure 8.13) [31, 32]. They synthesized monomers with a corannulene core and amide-appended thioalkyl side chains, which were conformationally restricted from spontaneous polymerization at ambient temperatures. These monomers, identified as **M**, adopt a cage-like closed conformation due to intramolecular hydrogen bonding. The synthesis involved the

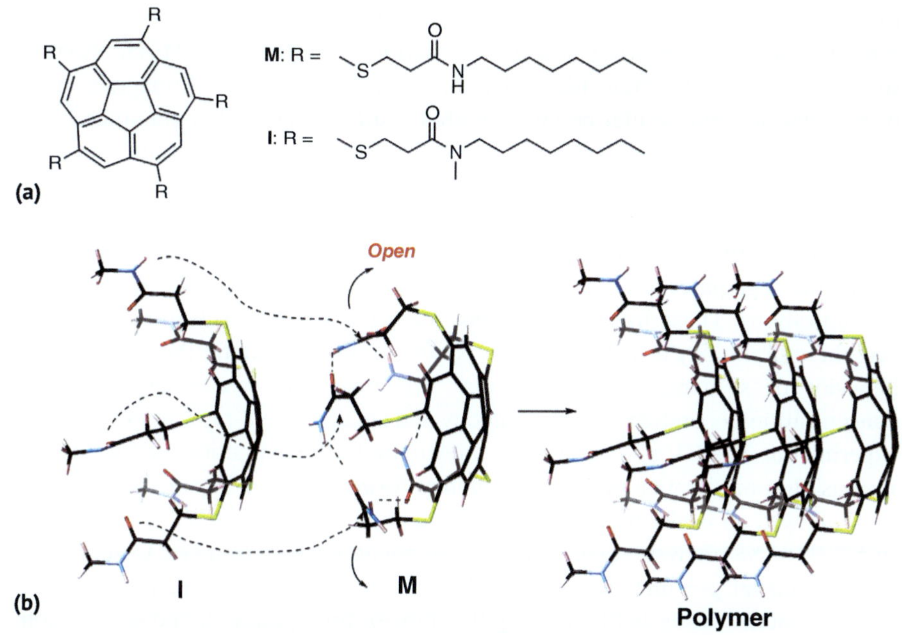

Figure 8.13 (a) The key components of living supramolecular polymerization with corannulenes monomer (**M**) and initiator (**I**); (b) schematic description of supramolecular polymerization.

creation of *N*-methylated derivatives of these monomers, referred to as initiators (**I**), which lack the capacity for intramolecular hydrogen bonding and thus remain in an open conformation. The polymerization process was initiated by mixing the monomers with the tailored initiators in a low-polarity solvent, MCH, at 25 °C. The chain-growth polymerization proceeded with characteristics typical of a living mechanism, where the polymer chains grew exclusively by sequential monomer addition to the reactive terminus. This process was monitored by DLS, DOSY NMR, AFM, and SEC. The results showed that the formed polymer chains had a small polydispersity index (PDI), and their lengths were proportional to the monomer-to-initiator ratio. The physical properties of the polymers were further analyzed using infrared spectroscopy, which indicated that the polymerization was accompanied by a shift in the hydrogen-bonded amide C═O vibration, suggesting robust intermolecular hydrogen bonding in the polymer chain. The stability of the polymers was tested by heating and cooling cycles, which demonstrated that the polymers could depolymerize and repolymerize under specific conditions. Additionally, this study explored the stereoselective polymerization of chiral monomers using chiral initiators, achieving high enantioselectivity, and enabling the optical resolution of racemic mixtures.

Sumanene-based supramolecular polymers were firstly reported in 2023 by Sukurai and Fukuhara groups [33]. The researchers synthesized sumanene and investigated its ability to form supramolecular polymers. Sumanene, with a deeper bowl depth of 1.11 Å compared to corannulene, exhibits slower bowl-to-bowl inversion, which is crucial for its spontaneous supramolecular polymerization. The synthesis involved preparing a saturated solution of sumanene in MCH, observing the formation of supramolecular polymers using AFM. The AFM images revealed bundle-like supramolecular polymers with a width of 74.0 ± 28.6 nm and a height of 2.7 ± 1.6 nm. This study further explored the effects of the solvent and concentration on the morphology of supramolecular polymers. When the concentration of sumanene in MCH was reduced, the polymers transitioned from bundles to globule aggregates. Similarly, changing the solvent to dichloromethane (CH_2Cl_2) resulted in globule aggregates at lower concentrations, but bundle-like polymers at higher concentrations. These findings highlight the critical role of solvation in controlling the morphology of the sumanene polymers. To understand the solution-state behavior of sumanene supramolecular polymers, the researchers conducted concentration-dependent UV spectroscopy and small-angle X-ray scattering (SAXS) experiments. The UV spectra showed a decrease in the extinction coefficient with increasing concentration, indicating polymer formation. The degree of polymerization followed an isodesmic model, with an equilibrium constant (K) of 18600 M^{-1} in MCH. SAXS analysis suggested that sumanene forms tetrameric to septameric supramolecular polymers in solution. The dynamic control of the supramolecular polymerization was tested by varying the temperature, pressure, and solvent composition. Temperature changes from 0 to 60 °C had negligible effects on polymer formation, indicating the thermodynamic stability of the polymers. Hydrostatic pressure also had little impact, with a slightly positive volume change (ΔV_1) of 2.7 ± 0.7 cm^3. However, the solvent composition significantly influenced polymerization, with the equilibrium constant varying 30-fold between MCH and CH_2Cl_2. A threshold solvent composition of 25% MCH was identified, beyond which polymer formation was favored.

One of the important works related with supramolecular assemblies of sumanenes is its ferroelectric assembly reported in 2021 by Akutagawa's group (Figure 8.14) [34]. The researchers focused on trithiasumanene (**SS**) as the core molecule due to its low bowl–inversion barrier [35]. The bowl depth and inversion barrier of **SS** are smaller than those of other bowl-shaped molecules, such as sumanene and corannulene, making it suitable for ferroelectric applications. The **SS** core was modified by introducing six long alkoxy chains ($-OC_nH_{2n+1}$) to enhance the motional

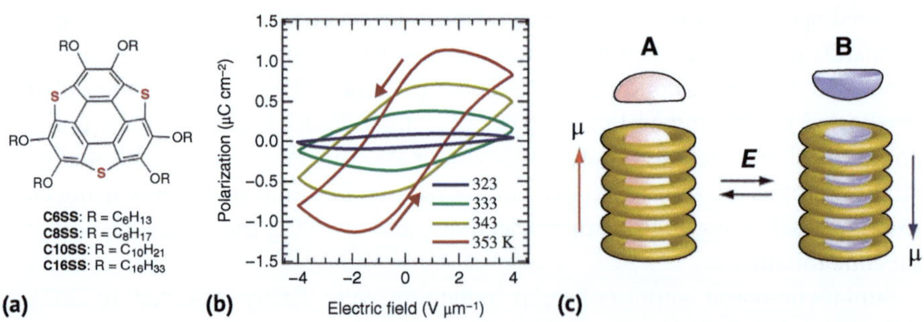

Figure 8.14 (a) Structures of **CnSS**, (b) temperature-dependent *P-E* hysteresis curves of **C10SS**, and (c) schematic model of the bowl-to-bowl inversion of sumanene molecules on application of electric fields. *Source*: [36]/with permission of Royal Society of Chemistry.

freedom and facilitate bowl inversion. The synthesis involved the dealkylation of **C2SS** to obtain the hexahydroxy intermediate **C0SS**, followed by re-alkylation with alkyl bromides (C_nH_{2n+1}—Br) to produce **CnSS** derivatives ($n = 6, 8, 10$, or 16). X-ray diffraction analyses provided insights into the molecular assembly, revealing one-dimensional stacking of bowl-shaped cores in a columnar fashion. The polar space group *P*3c1 was identified for **C2SS**, indicating a ferroelectric ground state. Dielectric measurements were conducted to evaluate the ferroelectric properties of **the CnSS** derivatives. The real (ε^1) and imaginary (ε^2) components of the dielectric constants were measured as functions of the temperature and frequency. The results showed dielectric anomalies and ferroelectric polarization–electric field (P–E) hysteresis in the high-temperature S_2 phase. The bowl-to-bowl inversion mechanism was confirmed by observing the retention of polarization after applying a pulse voltage. The ferroelectric response was attributed to the thermally activated structural inversion of the π-stacking columns. This chapter presents a novel approach for designing organic ferroelectric materials using the bowl-to-bowl inversion of π-aromatic cores. The **CnSS** derivatives exhibit unique ferroelectric properties due to their structural dynamics and thermal behavior. These findings suggest that the wide variation in molecular design for organic π-molecules can lead to the development of flexible and high-density organic memory devices.

Supramolecular polymers with other bowl-shaped molecules have been reported in 2018 [36]. Hiroto and Shinokubo group succeeded in elucidation of supramolecular polymeric behavior with nitrogen-embedded buckybowl, so-called "azabuckybowl" (Figure 8.15). The parent corannulene showed very weak convex–concave interactions with fullerenes. This interaction was observed only in the solid state. On the other hand, azabuckybowl showed strong binding with C_{60} even in solution [37]. In particular, the azabuckybowl constructed segregate stack in co-crystal with C_{60}, enhancing charge-carrier mobility compared to that of azabuckybowl. In this crystal, azabuckybowl formed 3D π-stacking mode, convex–concave interaction with C60 and π–π interaction with adjacent azabuckybowls. Such 3D stacking fashion and strong association of azabuckybowl enable to construct supramolecular assembly with C_{60} in solution processible process [38]. They prepared directly

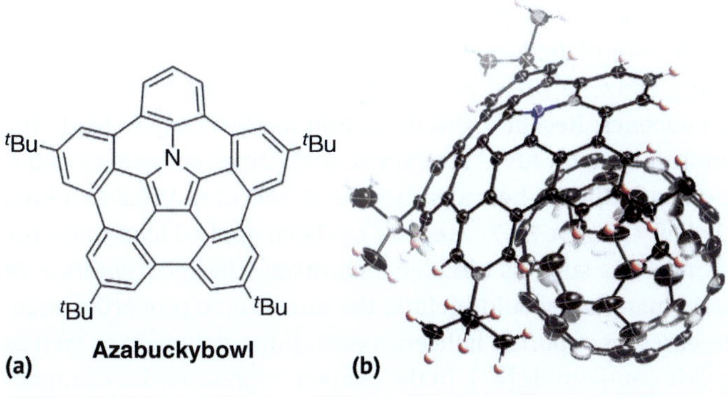

Figure 8.15 (a) Structure of azabuckybowl and (b) binding structure of azabuckybowl with C_{60} in co-crystal. *Source*: Satoru Hiroto.

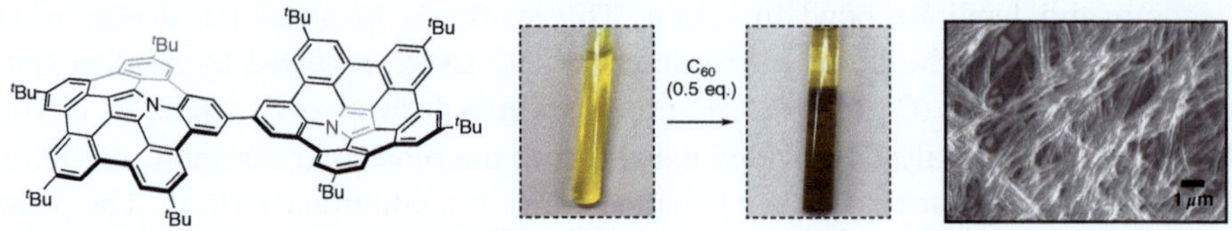

(a) (b) (c)

Figure 8.16 (a) Structure of azabuckybowl dimer, (b) views on addition of C_{60} into the solution of azabuckybowl dimer, and (c) SEM image of azabuckybowl dimer and C_{60} precipitates on plate. *Source*: Satoru Hiroto.

linked azabuckybowl dimer by palladium-catalyzed C-H/C-Br coupling reaction from a tribrominated precursor. Titration experiments in 1,2-dichlorobenzene and toluene revealed that the dimer forms 1:1 complexes with C_{60} in solution, with an association constant that is significantly higher than that of the monomer. The binding constant was solvent-dependent, which was attributed to the different solvophobicities of fullerenes in each solvent. The study also explored the supramolecular assembly of the dimer with C_{60}. Titration experiments monitored by ^{1}H NMR spectroscopy showed that the dimer formed larger structures in the presence of C_{60}, leading to the formation of fiber-like aggregates (Figure 8.16). SEM confirmed the fibrous morphology, and MALDI-TOF mass spectrometry indicated a 1:1 ratio of the dimer to C_{60} in the fibers. Powder X-ray diffraction analysis suggested a different packing structure for the fibers compared to that of single crystals. The research demonstrates that the nitrogen-embedded buckybowl dimer can serve as a strong binding motif for C_{60}, forming supramolecular assemblies with potential applications in materials science. This study highlights the importance of the electron-donating properties and the role of non-covalent interactions in the formation of higher-order structures.

8.4 Helicene-based Polymers

Helicenes are representative curved π-conjugated molecules with helical and fused conformations [39]. As a narrow definition, "helicene" is categorized as an *ortho*-fused benzene with a helical shape due to steric repulsion at the edge benzene rings. The nomenclature of helicenes is [*n*]helicene, in which *n* represents the number of consisting benzene rings. According to IUPAC definition, [5]helicenes are the smallest helicenes. Recently, this definition has been expanded to the helical π-conjugated molecules with fused structures, sometimes containing heteroatoms into their skeletons [40]. The helical structure of helicenes induces a kind of chirality, so-called helicity (Figure 8.17). Helicity has been applied to asymmetric catalysis, chiroptical properties such as circular dichroism. The incorporation of helicenes into polymeric materials would include the anisotropic properties mentioned above. Some researchers reported helicenes containing polymers as well as supramolecular polymeric compounds [41]. In the chapter, representative examples are described to gain insight about this study.

In 1998, one of the first reports on polymeric, oligomeric, and cyclophane materials was based on poly(arylene ethynylene)-containing [6]helicene as an arylene

Figure 8.17 Enantiomers of [6]helicenes.

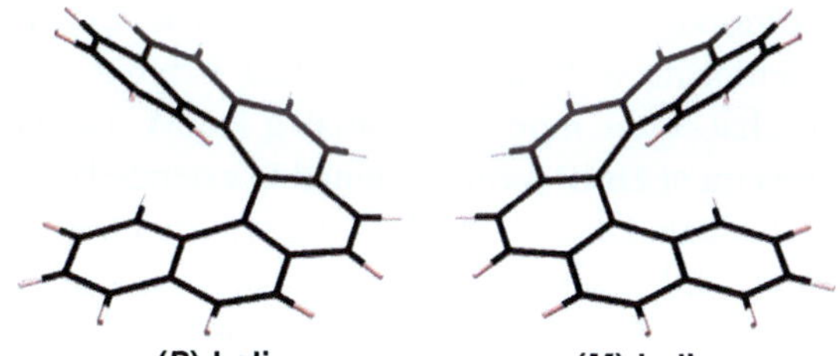

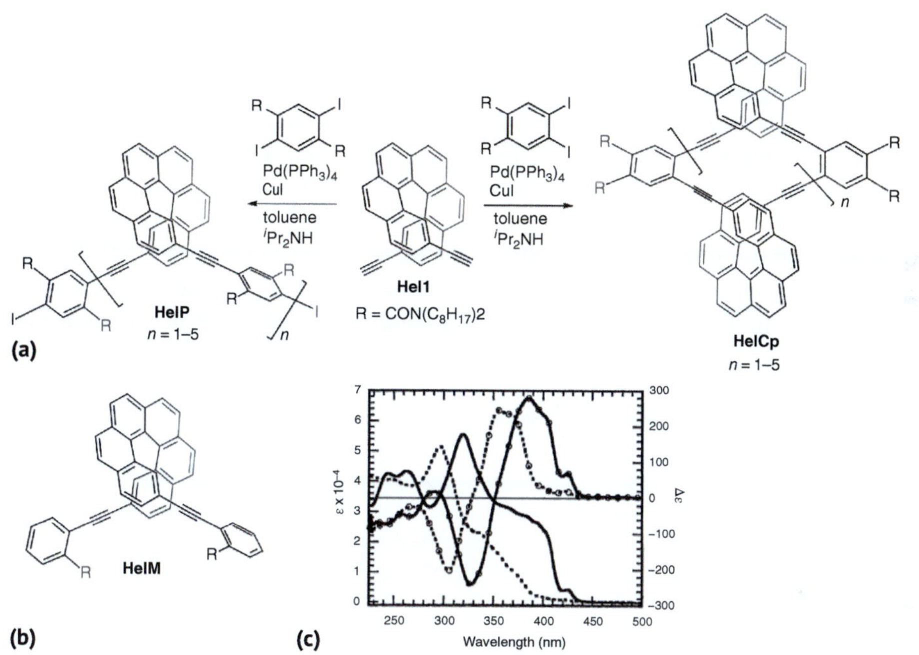

Figure 8.18 (a) The synthesis of **HelP** and **HelCp** from **Hel1**; (b) the structure of **HelM**; (c) UV and CD spectra of **HelP** (solid line) and **HelM** (dashed line). *Source*: [42]/with permission of American Chemical Society.

unit [42]. The synthesis begins with the preparation of 2,15-diethynyl[6]helicene through a series of steps starting from bis-stilbenes (Figure 8.18). The synthesis begins with the preparation of nonracemic 2,15-diethynyl[6]helicene (**Hel1**) through a series of steps involving the photocyclization of bis-stilbenes, followed by various chemical transformations, including DIBAL-H reduction and Dess–Martin oxidation. The helicene was then reacted with the *p*-diiodobenzene derivative to form the polymer **HelP** and with the *o*-diiodobenzene derivative to form cyclophane **HelCp**. The resulting polymers, **HelP** and cyclophane **HelCp**, were highly soluble in common organic solvents. The molar rotation of the polymer **HelP** was greater than that of the monomeric analog **HelM**, indicating extended conjugation. The UV and CD spectra of the polymer **HelCp** showed peaks shifted to longer wavelengths compared to those of **HelM**, further supporting the presence of extended conjugation. The UV spectra of **HelP** and cyclophane **HelCp** showed characteristic absorptions corresponding to π–π* transitions. The CD spectra revealed significant circular dichroism, with the polymer **HelP** exhibiting enhanced optical activity compared to that of the monomeric analog **HelM**. The bathochromic shifts observed in the spectra of **HelP** were attributed to extended conjugation along the polymer chain.

In 2016, Lu and Chen's group reported benzo[5]helicene-based conjugated polymers [43]. A series of conjugated polymers based on benzo[5]helicene were synthesized. The synthesis began with 2,15-dimethoxybenzo[5]helicene, which underwent demethylation, etherification, and bromination to yield the key intermediates. These intermediates were then polymerized to form polymers **P1–P5**, which exhibited good solubility in common organic solvents and high thermal stability with decomposition temperatures ranging from 268 to 330 °C (Figure 8.19). This study focused on the photophysical properties of these polymers and their application as fluorescent chemosensors for detecting nitroaromatic explosives such as 2,4,6-trinitrotoluene (TNT), 2,4-dinitrotoluene (DNT), picric acid (PA), and 4-nitrotoluene (NT). The photophysical properties were characterized in both solution and solid states, revealing that the polymers had similar absorption and emission spectra in both states, with significant red shifts due to the large conjugated π-system. The fluorescence lifetimes of the polymers were in the nanosecond range, and their fluorescence quantum yields varied between 0.14 and 0.71 in chloroform. The polymers

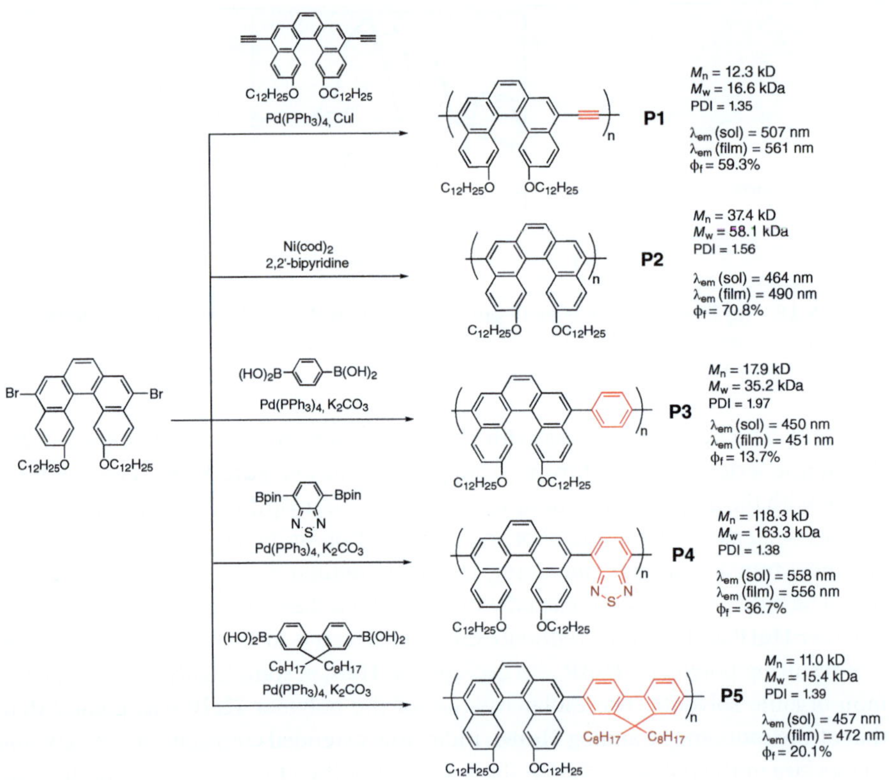

Figure 8.19 Synthesis and optical properties of [5]helicene polymer and copolymers **P1–P5**.

demonstrated high selectivity and sensitivity toward nitroaromatic explosives, with fluorescence quenching efficiencies in the order PA > TNT > DNT > NT, correlating with the electron-deficient nature of the analytes. The study also highlighted the rapid and reversible fluorescence response of polymer **P2** films to nitroaromatic vapors, achieving up to 87% quenching efficiency for DNT vapor within 10 seconds. The results suggest that these benzo[5]helicene-based conjugated polymers are promising materials for practical applications in explosive detection due to their excellent photophysical properties and effective sensing capabilities.

The Crassous group reported in 2020 regarding the synthesis and chiroptical properties of novel π-conjugated helical polymers based on chiral diketopyrrolopyrrole (DPP) and helicene units (Figure 8.20) [44]. This study aimed to develop efficient red circularly polarized (CP) light emitters, which have potential applications in optoelectronics and bioimaging. In previous report, the authors reported the synthesis of DPP appended [6]helicene oligomers, **H6(DPP)$_2$** and **DPP(H6DPP)$_2$**, showing red emission due to the presence of exciton couplings (Figure 8.20a) [45, 46]. The synthesis involved the Sonogashira coupling reaction between helicene derivatives and DPP cores, resulting in polymers with high solubility in common organic solvents and significant thermal stability (Figure 8.20b). The polymers, specifically *rac*-, (*P*)-, and (*M*)-(**H6DPP**)$_n$, exhibited intense electron circular dichroism (ECD) and CP luminescence (CPL) in the far-red spectral region due to exciton coupling between achiral DPPs within the chiral polymeric structure. The UV-Vis and ECD spectra revealed strong absorption and chiroptical responses, with the polymers displaying mirror-image ECD spectra for the *P*- and *M*-enantiomers. The CPL measurements showed that the polymers emitted intense red light with high fluorescence

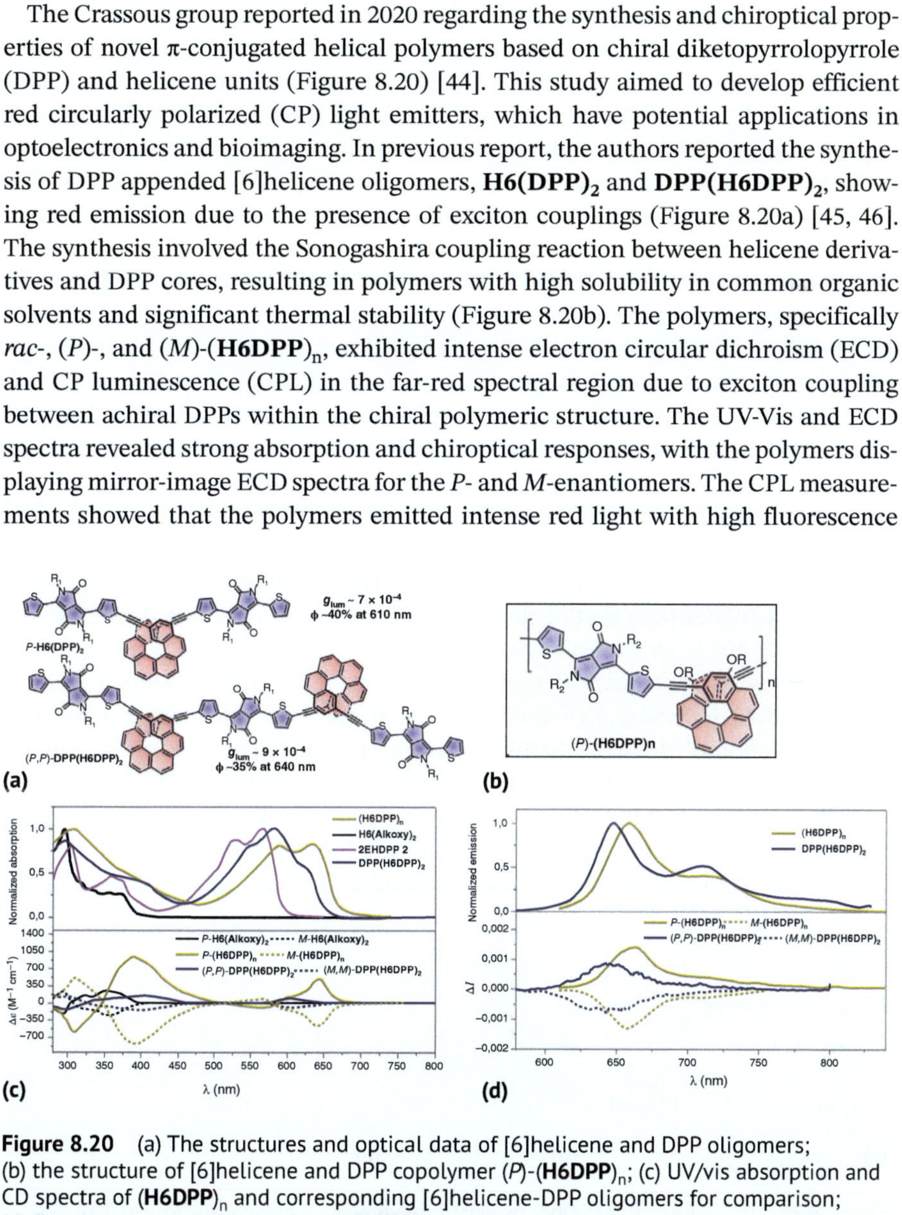

Figure 8.20 (a) The structures and optical data of [6]helicene and DPP oligomers; (b) the structure of [6]helicene and DPP copolymer (*P*)-(**H6DPP**)$_n$; (c) UV/vis absorption and CD spectra of (**H6DPP**)$_n$ and corresponding [6]helicene-DPP oligomers for comparison; (d) fluorescence and CPL spectra of (**H6DPP**)$_n$ and the corresponding [6]helicene-DPP oligomers. *Source*: [44]/Frontiers Media/CC BY 4.0.

quantum yields ($\Phi \approx 35\%$) and dissymmetry factors ($g_{lum} \approx 1.3 \times 10^{-3}$), indicating efficient chiral excited-state properties. The study also highlighted the electrochemical behavior of the polymers, with CV revealing reversible redox processes attributed to the DPP units. The results suggest that these helical DPP-helicene polymers are promising candidates for CP light-emitting applications, offering an alternative to metal-based systems and demonstrating the potential of chiral organic materials in advanced optoelectronic devices. The chapter concludes by emphasizing the importance of molecular design in achieving high CPL efficiency and the potential for further optimization of these materials for practical applications.

Kamikawa and Chujo et al. detail the synthesis and characterization of novel optically active polymers incorporating [7]helicene-fused oxanorbornene units [47]. The authors synthesized monomer **2** through the [4+2] cycloaddition of in situ-generated [7]helicenyl aryne and furan, followed by optical resolution using a chiral stationary phase to obtain the optically active **(*P*)**-[7]**hel** and **(*M*)**-[7]**hel** isomers. The polymerization was carried out using various Grubbs catalysts, with the first-generation Grubbs catalyst proving optimal for the ring-opening metathesis polymerization (ROMP) of monomer [7]**hel**. The resulting polymers, **P**[7]**hel**, exhibited significant chiroptical properties, which were studied using circular dichroism (CD) and circularly polarized luminescence (CPL) spectroscopy (Figure 8.21). The synthesis began with the preparation of 9,10-dibromo[7]helicene, which was treated with n-BuLi to generate the [7]helicenyl aryne intermediate. This intermediate underwent cycloaddition with furan to yield the desired monomer, which was then polymerized. The polymers displayed high solubility in common organic solvents and were characterized by UV-Vis absorption and fluorescence spectroscopy, revealing a bathochromic shift in the fluorescence maximum compared to that of the monomer. The fluorescence quantum yield of the polymer was slightly higher than that of the monomer, which was attributed to enhanced fluorescence due to superposition and extended elongation. The CD spectra of the polymers showed significant Cotton effects, with the polymer exhibiting approximately 10-fold increased CD intensities compared to the monomer, indicating the preservation of the helical structure in the polymer. The CPL measurements revealed that the sign of the CPL was reversed for the monomer and the polymer, with the polymer showing a larger absolute value of the dissymmetry factor (g_{lum}) than the monomer. This reversal was attributed to the structural changes in the core unit during polymerization. The study concluded that the polymerization of the

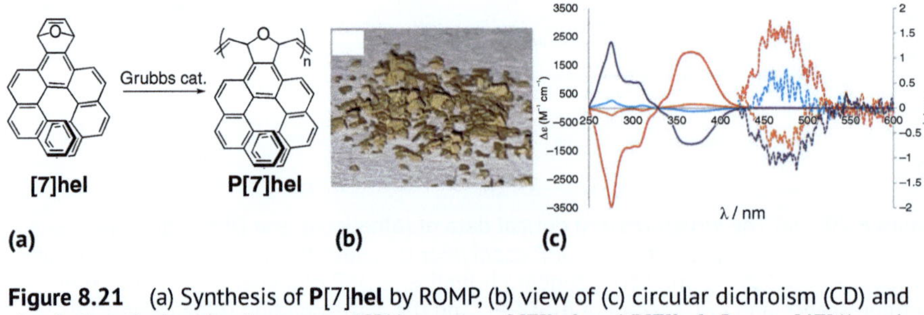

Figure 8.21 (a) Synthesis of **P**[7]**hel** by ROMP, (b) view of (c) circular dichroism (CD) and circularly polarized luminescence (CPL) spectra of [7]**hel** and **P**[7]**hel**. *Source:* [47]/American Chemical Society.

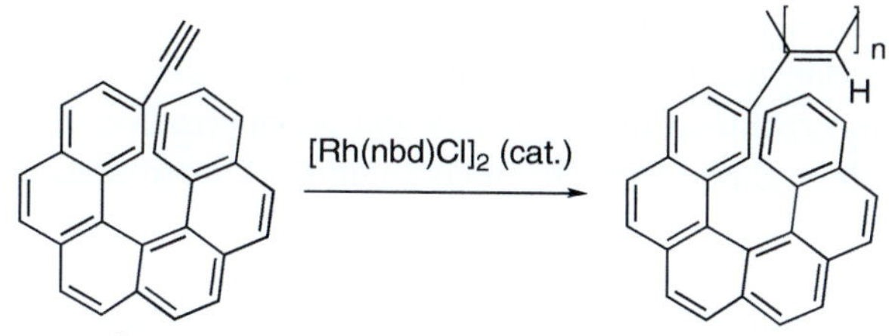

Figure 8.22 Synthesis of [6]helicene-branched polymers.

[7]helicene-fused oxanorbornene monomer resulted in new polymers that retained the helicene properties and exhibited promising chiroptical properties, making them potential candidates for applications in chiral materials, asymmetric catalysis, chiral separation techniques, and CP light-emitting devices. The authors emphasized the importance of further research to explore the full potential of helicene-based polymers in various applications, including the creation of supramolecular assemblies and self-assembled structures. The paper provided a comprehensive analysis of the synthesis, characterization, and potential applications of these novel optically active polymers, highlighting their unique properties and the challenges associated with their synthesis and polymerization.

Other than polymers containing helicenes as main chains, Yashima's group explored the synthesis of novel helical polyacetylenes with optically active [6]helicene units as pendant groups (Figure 8.22) [48]. They synthesized polyacetylenes by polymerizing acetylenes (**1-*rac***, **1-*P***, and **1-*M***) using a rhodium catalyst. The monomers included racemic (**1-*rac***) and optically active (**1-*P*** and **1-*M***) [6]helicene units. The polymerization was conducted under various conditions, including different solvents, concentrations, and rhodium catalysts. Despite challenges with polymer precipitation and low molecular weights, the resulting polymers were stereoregular and exhibited *cis*-transoidal configurations. The optically active polymers (**P1-*P*** and **P1-*M***) formed preferred-handed helical conformations, as evidenced by induced circular dichroism (ICD) in their π-conjugated backbones. These polymers demonstrated high chiral recognition abilities, selectively adsorbing one enantiomer of racemates, such as 1,1′-binaphthyl analogs. The chiral recognition was attributed to the helical arrangement of helicene pendants along the polymer backbone, which facilitated enantioselective interactions.

8.5 Conclusion and Outlook

Since the chemistry of curved π-conjugated molecules is recently developing field in chemistry, their application in various research areas is emerging. Their polymerization enhanced their ability such as binding fullerenes for buckybowls and CPL for helicenes. In the case of the fluid LC phase, the dynamic behavior of curved π-conjugated molecules has led to the development of a novel approach to electronic devices. These developments are owing to the improvement of accessibility to curved π-conjugated molecules without special apparatus such as photochemical reactors. In particular, corannulene is currently commercially available, and recently, the cycloparaphenylene, which is a fragment of carbon nanotube, can be purchased from commercial sources. Currently, a limited number of curved π-conjugated molecules composed of only carbon and hydrogen have been studied for incorporation into polymers and polymerization. Therefore, as an extension of the physical properties, the incorporation of guest molecules based on 3D structures and the utilization of vacancies has been studied. Helicenes have shown the benefit of polymerization for the amplification of chiral optical properties based on their helicity, but there is still a lack of research on the application of their photochemical and electrochemical properties to polymeric materials. On the other hand, synthetic methods for curved π-conjugated molecules incorporating heteroatoms have been improved one after another, and their excellent photochemical and electrochemical properties have been attracting attention. It is expected that polymer materials incorporating such curved molecules with excellent photochemical properties will be developed one after another in the future. For example, Diederich group reported butadiyne linked carbo [6] helicene oligomers, which showed reduction of HOMO-LUMO gap on oligomerization [49, 50]. On the other hand, recently Hiroto's group developed butadiyne-bridged azahelicene oligomers, which exhibited enhancement of emission efficiency on oligomerization as well as bathochromic shifts of optical spectra (Figure 8.23) [51]. The development of curved π-conjugated molecules into polymers has just begun, and we can expect to see further expansion as new synthetic methods for various curved π-conjugated molecules are developed in the future.

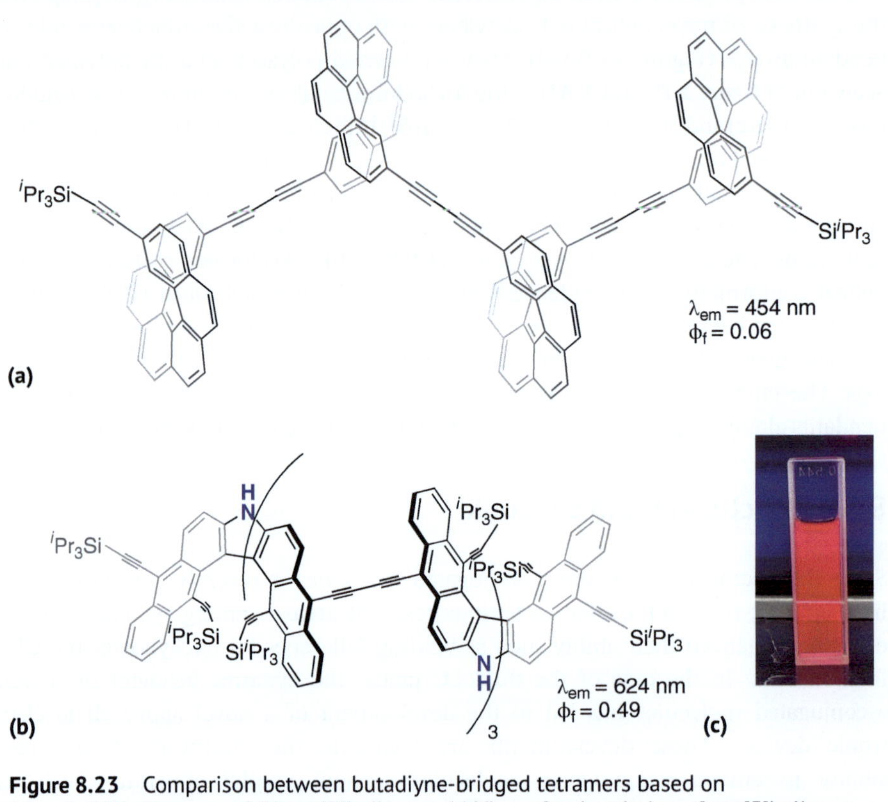

Figure 8.23 Comparison between butadiyne-bridged tetramers based on (a) carbo[6]helicenes and (b) aza[5]helicenes. (c) View of red emission of aza[5]helicene tetramer in CH_2 Cl_2. *Source*: [5]/John Wiley & Sons.

Acknowledgment

The contribution from our group described in the chapter has been supported by KAKENHI Grant Number JP 22K05080 for financial support.

References

1 Petrukhina, M.A., Scott, L.T., and Kroto H.W. (eds.) (2011). *Fragments of Fullerenes and Carbon Nanotubes: Designed Synthesis, Unusual Reactions, and Coordination Chemistry*. Wiley-VCH.

2 Saito, M., Shinokubo, H., and Sakurai H. (2018). Figuration of bowl-shaped π-conjugated molecules: properties and functions. *Materials Chemistry Frontiers* 2: 635.

3 Stuparu, M.C. (2023). Macromolecular architectures of corannulene: synthesis, properties, and applications of polymers containing a molecular bowl of carbon. *Chemistry of Materials* 35: 1836.

4 Newman, M.S., Lutz, W.B., and Lednicer, D. (1955). A new reagent for resolution by complex formation; the resolution of phenanthro-[3,4-c]phenanthrene. *Journal of the American Chemical Society* 77: 3420.

5 Dai, Y., Katz, T.J., and Nichols, D.A. (1996). Synthesis of a helical conjugated ladder polymer. *Angewandte Chemie International Edition in English* 35: 2109.

6 Sygula, A. (2011). Chemistry on a half-shell: synthesis and derivatization of buckybowls. *European Journal of Organic Chemistry* 2011: 1611.

7 Wu, Y.T. and Siegel, J.S. (2006). Aromatic molecular-bowl hydrocarbons: synthetic derivatives, their structures, and physical Properties. *Chemical Reviews* 106: 4843.

8 Barth, W.E. and Lawton, R.G. (1966). Dibenzo[ghi,mno]fluoranthene. *Journal of the American Chemical Society* 88: 380.

9 Butterfield, A.M., Gilomen, B., and Siegel, J.S. (2012). Kilogram-scale production of corannulene. *Organic Process Research & Development* 16: 664.

10 Seiders, T.J., Elliot, E.L., Grube, G.H. et al. (1999). Synthesis of corannulene and alkyl derivatives of corannulene. *Journal of the American Chemical Society* 121: 7804.

11 Sakurai, H., Daiko, T., and Hirao, T. (2003). A synthesis of sumanene, a fullerene fragment. *Science* 301: 5641.

12 Saito, M., Furukawa, S., Kobayashi, J. et al. (2016). The chemistry of heterasumanenes. *The Chemical Record* 16: 64.

13 Higashibayashi, S. and Sakurai, H. (2011). Synthesis of sumanene and related buckybowls. *Chemistry Letter* 40: 122.

14 Stuparu, M.C. (2012). Synthesis and properties of star polymers with a C5-symmetric bowl-shaped aromatic core. *Journal of Polymer Science Part A: Polymer Chemistry* 50: 2641.

15 Stuparu, M.C. (2021). Corannulene: a curved polyarene building block for the construction of functional materials. *Accounts of Chemical Research* 54: 2858.

16 Stuparu, M.C. (2013). Rationally designed polymer hosts of fullerene. *Angewandte Chemie International Edition* 52: 7786.

17 Mishra, A., Ulaganathan, M., Edison, E. et al. (2017). Polymeric nanomaterials based on the buckybowl motif: synthesis through ring-opening metathesis polymerization and energy storage applications. *ACS Macro Letters* 6: 1212.

18 Lu, R.Q., Xuan, W., Zheng, Y.Q. et al. (2014). A corannulene-based donor–acceptor polymer for organic field-effect transistors. *RSC Advances* 4: 56749.

19 Karunathilake, A.A.K., Thompson, C.M., Perananthan, S. et al. (2016). Electrochemically active porous organic polymers based on corannulene. *Chemical Communications* 52: 12881.

20 Huang, J.J., Lin, H.A., Chen, C. et al. (2021). Corannulene-based donor–acceptor-type conjugated polymers with electrochromic properties. *Journal of Materials Chemistry C* 9: 7919.

21 Huang, J.J., Liang, H.I., Lin, H.A. et al. (2022). Corannulene extended viologen-based ambipolar polymers for transparent-to-color electrochromic supercapacitors. *ACS Applied Electronic Materials* 4: 4996.

22 Huang, J.J., Lin, H.A., and Luo, SC. (2023). Enhancing NIR electrochromism with twisted copolymer films of corannulene-carbazole and 3,4-ethylenedioxythiophene. *ACS Applied Polymer Materials* 5: 5727.

23 Leith, G.A. and Shustova, N.B. (2021). Graphitic supramolecular architectures based on corannulene, fullerene, and beyond. *Chemical Communications* 57: 10125.

24 Li, X. and Inagaki, M. (2016). Buckybowls: corannulene and its derivatives. *Small* 12: 3206.

25 Nestoros, E. and Stuparu, M.C. (2018). Corannulene: a molecular bowl of carbon with multifaceted properties and diverse applications. *Chemical Communications* 54: 6503.

26 Miyajima, D., Tashiro, K., Araoka, F. et al. (2009). Liquid crystalline corannulene responsive to electric field. *Journal of the American Chemical Society* 131: 44.

27 Mattarella, M., Haberl, J.M., Ruokolainen, J. et al. (2013). Five-fold symmetric penta-substituted corannulene with gelation properties and a liquid-crystalline phase. *Chemical Communications* 49: 7204.

28 Mahadevegowda, S.H. and Stuparu, M.C. (2017). Thermoresponsive corannulene. *European Journal of Organic Chemistry* 2017: 570.

29 Mahadevegowda, S.H. and Stuparu, M.C. (2017). Amphiphilic corannulene derivatives: synthetic access and development of a structure/property relationship in thermoresponsive buckybowl amphiphiles. *ACS Omega* 2: 4964.

30 Ogi, S., Sugiyasu, K., Mannna, S. et al. (2014). Living supramolecular polymerization realized through a biomimetic approach. *Nature Chemistry* 6: 188.

31 Kang, J., Miyajima, D., Mori, T. et al. (2015). A rational strategy for the realization of chain-growth supramolecular polymerization. *Science* 347: 6222.

32 Kang, J., Miyajima, D., Itoh, Y. et al. (2014). C5-symmetric chiral corannulenes: desymmetrization of bowl inversion equilibrium via "intramolecular" hydrogen-bonding network. *Journal of the American Chemical Society* 136: 10640.

33 Mizuno, H., Nakazawa, H., Harada, M. et al. (2023). Sumanene-stacked supramolecular polymers. Dynamic, solvation-directed control. *Chemical Communications* 59: 9595.

34 Furukawa, S., Wu, J., Koyama, M. et al. (2021). Ferroelectric columnar assemblies from the bowl-to-bowl inversion of aromatic cores. *Nature Communications* 12: 768.

35 Priyakumar, U.D. and Sastry, G.N. (2001). Heterobuckybowls: a theoretical study on the structure, bowl-to-bowl inversion barrier, bond length alternation, structure-inversion barrier relationship, stability, and synthetic feasibility. *The Journal of Organic Chemistry* 66: 6523.

36 Yokoi, H., Hiroto, S., Sakamaki, D. et al. (2018). Supramolecular assemblies of a nitrogen-embedded buckybowl dimer with C60. *Chemical Science* 9: 819.

37 Yokoi, H., Hiraoka, Y., Hiroto, S. et al. (2015). Nitrogen-embedded buckybowl and its assembly with C60. *Nature Communications* 6: 8215.

38 Hiroto, S. (2018). Innovative synthesis and functions of curved-π conjugated molecules. *Bulletin of the Chemical Society of Japan* 91: 829.

39 Shen, Y. and Chen, C.F. (2012). Helicenes: synthesis and applications. *Chemical Reviews* 112: 1463.

40 Pop, F., Zigon, N., and Avarvari, N. (2019). Main-group-based electro- and photoactive chiral materials. *Chemical Reviews* 119: 8435.

41 Crassous, J., Stará, I.G., and Star′y, I. (eds.) (2022). *Helicenes*. Germany: Wiley-VCH.

42 Fox, J.M. and Lin, D. (1998). Synthesis of conjugated helical acetylene-bridged polymers and cyclophanes. *The Journal of Organic Chemistry* 63: 2031.

43 Zhou, L.L., Li, M., Lu, H.Y. et al. (2016). Benzo[5]helicene-based conjugated polymers: synthesis, photophysical properties, and application for the detection of nitroaromatic explosives. *Polymer Chemistry* 7: 310.

44 Dhbaibi, K., Shen, C., Jean, M. et al. (2020). Chiral diketopyrrolopyrrole-helicene polymer with efficient red circularly polarized luminescence. *Frontiers in Chemistry* 8: 237.

45 Dhbaibi, K., Favereau, L., Srebro-Hooper, M. et al. (2018). Exciton coupling in diketopyrrolopyrrole–helicene derivatives leads to red and near-infrared circularly polarized luminescence. *Chemical Science* 9: 735.

46 Dhbaibi, K., Quinton, C., Favereau, L. et al. (2020). Modulation of circularly polarized luminescence through excited-state symmetry breaking and interbranched exciton coupling in helical push–pull organic systems. *Chemical Science* 11: 567.

47 Fujikata, K., Gon, M., Tanaka, K. et al. (2023). Synthesis and properties of optically active [7]Helicene-Fused oxanorbornene polymers by ring-opening metathesis. *Macromolecules* 56: 4550.

48 Anger, E., Iida, H., Yamaguchi, T. et al. (2014). Synthesis and chiral recognition ability of helical polyacetylenes bearing helicene pendants. *Polymer Chemistry* 5: 4909.

49 Schaack, C., Sidler, E., Trapp, N. et al. (2017). Helical threads: enantiomerically pure carbo[6]Helicene oligomers. *Chemistry–A European Journal* 23: 14153.

50 Scaack, C., Arrico, L., Sidler, E. et al. (2019). Helicene monomers and dimers: chiral chromophores featuring strong circularly polarized luminescence. *Chemistry–A European Journal* 25: 8003.

51 Hiroto, S. and Chujo, M. (2025). Donor-acceptor-donor DADs with electron-rich π-extended azahelicenes to panchromatic absorbing dyes. *Chemistry–An Asian Journal* 20 (1): e202400830.

9

Conjugated Polymers for Energy Storage Devices

Bharat S. Mantripragada, Sameer N. Mishra, Kottisa S. Patnaik, Uday S. Reddi, Pavithra Kasthurirangan, Bipashna Pradhan, Saibrata Punyasloka, and Noriyoshi Matsumi

Graduate School of Advanced Science and Technology, Japan Advanced Institute of Science and Technology, Nomi, Ishikawa, Japan

9.1 Introduction

A π-conjugated molecule is a molecule in which π-conjugation prevails through entire polymer backbone, leading to delocalization of electrons in molecules with single and multiple bonds [1]. In 1977, Shirakawa et al. discovered that polyacetylene exhibits metallic conductivity [1]. Since this discovery, the interest in conjugated polymers has increased significantly, leading to the exploration of their properties in various fields.

In this context, in recent years, there has been an exponential increase in the energy demands in the world. To this end, lithium-ion batteries (LIBs) emerged as promising devices owing to the possibility of achieving high energy density and power density [2]. In this respect, there is a tremendous interest in preparing novel materials to enhance the performance of LIBs. Typical LIBs consist of two electrodes (cathode and anode), electrolyte, and a separator. Electrolyte and separator are electrically insulating materials [3]. Cathodic and anodic materials reversibly store lithium ions during the charge–discharge process. In the context of LIBs, polymers play a crucial role in enabling the energy storage process; for instance, typical LIBs consist of graphite- or silicon (Si)-based anodic active materials [3, 4]. Such polymers have been designed to maintain adhesion between active materials and current collector and thereby to maintain the integrity of the electrode [4]. In this respect, polymers such as poly(vinylidene fluoride) (PVDF), carboxymethyl cellulose (CMC), styrene-butadiene rubber (SBR), etc., are employed. These polymers rely on van der Waals interactions to maintain mechanical stability [5, 6]. Conjugated polymers, on the other hand, can enhance the adhesive properties between binder, active materials, and current collector through various other

Conjugated Polymer Synthesis and Materials Applications, First Edition. Edited by Yoshiki Chujo.
© 2026 Wiley-VCH GmbH. All rights reserved, including rights for text and data mining and training of artificial intelligence technologies or similar technologies. Published 2026 by Wiley-VCH GmbH.

interactions such as π–π interactions, H-bonding [6], etc. Apart from maintaining adhesion, conjugated polymers also decrease the electrical resistance in the anode compared to the non-conjugated counterparts. Besides adhesion, another interesting property of conjugated polymers (CPs) is their ability of accommodating electronegative heteroatoms like N, O, and S in the conjugation framework [7]. This property can be exploited to enable p- or n-doping of the polymers. On anodic side, n-doping of the polymers helps in lowering the lowest unoccupied molecular orbital (LUMO) energy, which in turn restricts the excessive reduction of electrolyte when used as anodic binder. Conversely, on cathodic side, p-doping of polymers helps in restricting excessive oxidation of the electrolyte. Figure 9.1 shows the role of n- or p-type of polymers in protecting the electrolyte. In the case of n-doping (where

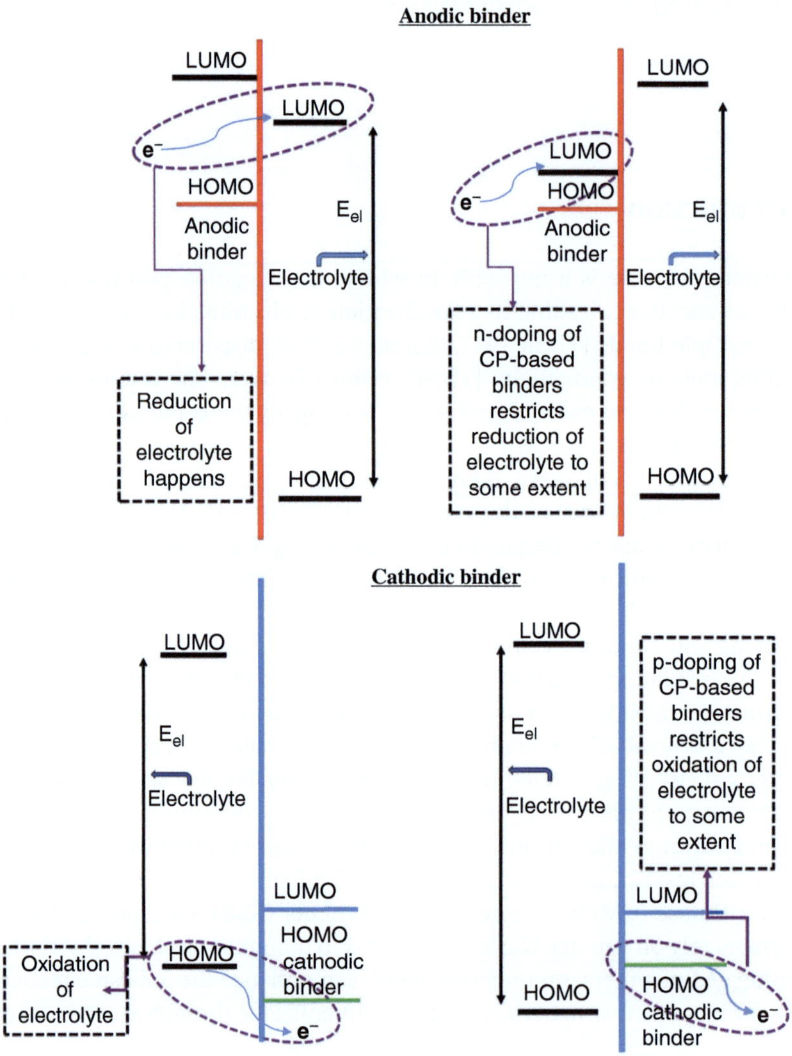

Figure 9.1 Schematic representation of n- and p-doping of CP binders.

LUMO of binder is lower than the LUMO of the electrolyte), during the reduction step, electrons will be able to access the LUMO of the binder instead of accessing the LUMO of the electrolyte to restrict, therefore restricting the extent of electrolyte reduction [8]. Similarly, on cathode side, if the highest occupied molecular orbital (HOMO) of cathodic binder is higher than that of electrolyte, excessive oxidation of the electrolyte can be restricted [9]. Hence, a careful design of binder is essential for optimal performance of a battery.

Besides binder applications, the versatility of organic reactions can also be exploited to enable ion storage properties as active material. An example of conjugated polymer with ion storage ability is shown in Figure 9.2b [10]. Further, based on the structure of reactants, the geometry of the CPs can be predetermined (demonstrated in Figure 9.2a). A detailed explanation of such structure-charge storage ability relation will be presented in the later sections.

As discussed previously, the polarity induced in organic molecules due to the electronegative atoms can also be exploited to make CPs as sites to adsorb oxygen and catalyze oxygen reduction reaction (ORR) and oxygen evolution reaction (OER) processes. Through appropriate design, CPs can also behave as ligands to transition metal ions [13]. These properties of CPs are beneficial in devices like Li-O_2 batteries and fuel cells. Hence, in this chapter a detailed explanation on the applications of CPs as (1) binders for active materials, (2) active materials for Li/Na storage, and (3) catalysts for ORR/OER reactions are presented with case studies.

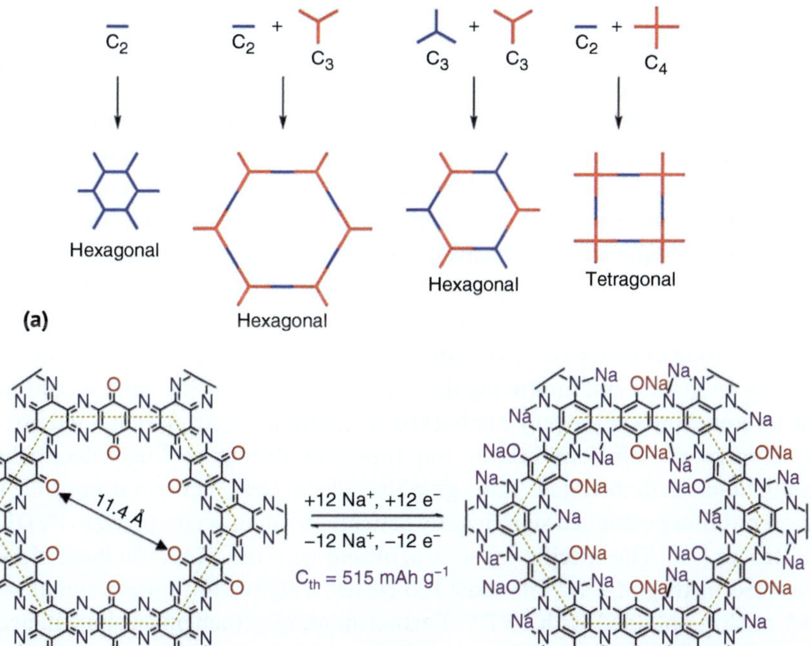

Figure 9.2 (a) Relation between geometry of reactants and CPs [11] / Royal Society of Chemistry. (b) Ion storage in CPs [12] / Springer Nature / CC BY 4.0.

9.2 Conjugated Polymers as Binder Materials in LIBs

The development of binders for LIBs has undergone a substantial transformation, transitioning from simple, passive components to active, multifunctional materials capable of greatly enhancing battery performance. Traditional binders like poly(vinylidene fluoride) (PVDF) were primarily used to provide structural integrity and adhesion [14]. As the demand for higher-performance batteries has grown, especially for fast-charging, high-energy-density, and long-cycle-life applications, the need for more advanced binder systems has emerged.

In response to these demands, conjugated polymer binders have gained attention due to their unique ability to offer mechanical robustness while also providing electronic and ionic conductivity. In this section, the effect of conjugated structures on the battery performance with bis-imino-acenapthenequinone (BIAN)-based n-type conjugated polymers will be presented as a specific case study for graphite and Si anodes [15, 16].

9.2.1 Graphite Anodes

Graphite, the most used anode material in LIBs, faces a series of challenges that limit its performance, particularly at high-rate cycling conditions. These challenges include poor solid electrolyte interface (SEI) formation, weak adhesion to the current collector, and limited electronic conductivity. Conventional binders, such as PVDF, fail to fully address these issues due to their limited interaction with both the graphite framework and the current collector, resulting in suboptimal electrochemical performance. It is also found that PVDF possesses safety issues such as formation of LiF and unsaturated $>C{=}CF^-$ bonds with resultant heat release and dead lithium deposition in graphite anodes, especially at high C rates [16].

To overcome these limitations, innovative binder systems have been developed, such as the bis-imino-acenaphthenequinone (BIAN)–fluorene copolymer binder (Figure 9.3a), as reported by Gourang et al. at our group [17, 18]. This binder exploits the unique structural features of its components to enhance performance. The planar naphthalene group of the BIAN unit enables π–π stacking interactions with the graphite surface, improving the mechanical integrity of the anode. The diimine group of BIAN enhances adhesion to the copper current collector, reducing contact resistance and improving the overall efficiency of the battery. Additionally, the fluorene units in the polymer backbone maintain electronic conductivity, which is essential for efficient lithium-ion transport during cycling. Electrochemical evaluations demonstrated that graphite electrodes incorporating the BIAN–fluorene binder exhibited significantly better performance compared to PVDF-based electrodes [16]. The specific capacity achieved with the BIAN–fluorene binder was over 250 mAh/g at a 1C rate after 100 cycles, a significant improvement over the 165 mAh/g obtained with PVDF. Furthermore, electrochemical impedance spectroscopy and dynamic impedance analysis revealed the formation of a more stable and conductive electrode–electrolyte interface, which is crucial for improving cycle life and high-rate performance [17].

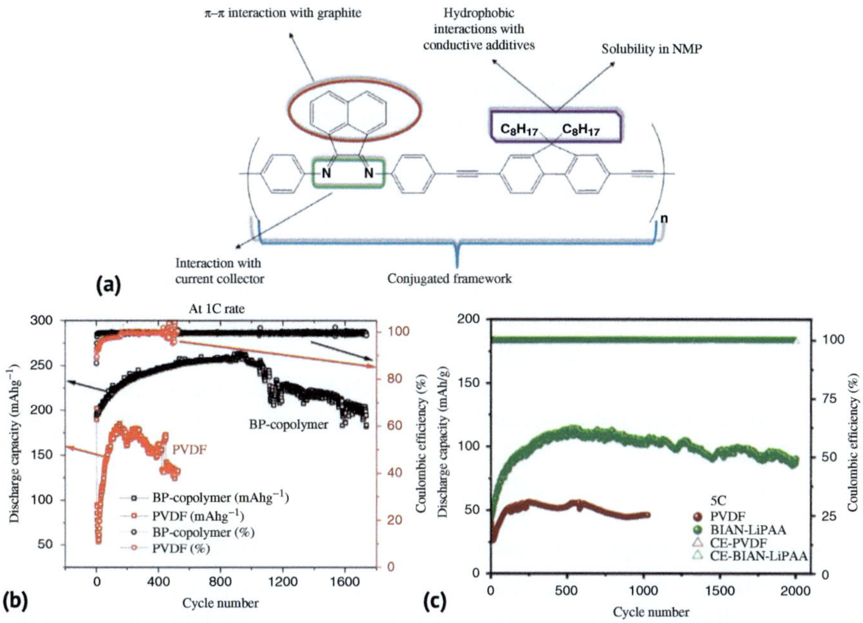

Figure 9.3 (a) Structural features of BIAN–fluorene [18] / Royal Society of Chemistry. (b) Charge–discharge studies of graphite-based anodic half-cell with BIAN–paraphenylene [16] / American Chemical Society. (c) Charge–Discharge studies of graphite-based anodic half-cell with BP-LiPAA binder [14] / American Chemical Society.

While the BIAN–fluorene binder shows excellent performance, its synthesis and purification are complex and time-consuming, making it economically challenging for large-scale production. In response to this issue, Gupta et al. at our group [16] demonstrated a simpler, acid-catalyzed Schiff base condensation method to synthesize a BIAN-based polymer, simplifying synthetic and purification steps. This simplified synthesis route retained the structural and electrochemical advantages of the BIAN–fluorene polymer while reducing the cost and complexity of its production. The BIAN-based polymer's low LUMO level continues to restrict SEI formation, minimize irreversible electrolyte loss, and improve the cycling stability of the battery, thus advancing the development of high-performance binders for LIBs. Further, density functional theory (DFT) calculations showed p-BIAN's low LUMO as −3.17 eV, this enables n-doping of p-BIAN prior to electrolyte reduction, facilitating the formation of a thin, conductive SEI that minimizes electrolyte decomposition during cycling.

Electrochemical testing showed that p-BIAN copolymer binders outperform PVDF in various critical parameters, including tensile strength, Young's modulus, and lithium-ion diffusion. The p-BIAN binder demonstrated a high reversible capacity of 260 mAh/g, a remarkable 95% capacity retention over 1735 cycles, and reduced interfacial resistance (as shown in Figure 9.3b). These impressive results highlight p-BIAN as a promising anodic binder for long-lasting, high-energy-density LIBs, offering a solution to the limitations of traditional binder systems.

Building upon these advancements, Sameer et al. at our group (2023) introduced a composite binder, BIAN–LiPAA, which combines the advantages of BIAN-based conjugated polymers with lithium poly(acrylate) (LiPAA) [14]. The addition of LiPAA provides an inherent lithium-ion reservoir, which significantly improves initial Coulombic efficiency and enhances the binder's ability to support fast lithium-ion diffusion, easy desolvation, and Li-ion uptake in SEI. This composite binder also facilitates the formation of a more stable and conductive SEI, further reducing impedance and enhancing charge transfer efficiency. The BIAN–LiPAA binder exhibited outstanding electrochemical performance, with discharge capacities of 276, 114.5, and 62.1 mAh/g at 1C, 5C, and 10C, respectively, while maintaining 94.2% capacity retention after 2000 cycles at 10C (as shown in Figure 9.3c). These results demonstrate the binder's potential to support fast-charging, high-performance LIBs, making it a promising alternative to conventional binders [14].

Thus, through the continuous evolution of BIAN-based conjugated polymers, we have addressed critical challenges such as mechanical robustness, adhesion, cycle life, and SEI stability.

In comparison, poly(acrylic acid) (PAA) is a non-conjugated polymeric binder that has also been widely investigated. The abundant carboxyl groups along the PAA chain can interact strongly with graphite particles via hydrogen bonding, to uniformly cover the graphite. This helps avoid direct contact between graphite and the electrolyte, leading to a higher reversible capacity and slower delithiation kinetics at elevated temperatures (55 °C). In addition to this, the efficient coverage of PAA on the graphite electrode surface, particularly at the edges, was found to reduce the electrode surface area, enhancing its reversibility. Further analysis of the SEI components revealed that the electrode with PAA exhibits a lower ratio of inorganic components in its SEI film compared to the electrode with PVDF, leading to unstable SEI and poor capacity retention [19].

However, as a downside, these non-conjugated polymers tend to have higher LUMOs and are also electrically insulating in nature. The electrically insulating nature of these non-conjugated polymer binders restricts the electrical conductivity in the anode, thus encouraging the undesirable internal resistance within the anode. The conjugated polymer electrodes, in contrast, offer less internal resistance by virtue of their electrically conducting nature and typically have lower LUMOs, thereby restricting electrolyte reduction. Further, through n-doping of the conjugated binder, interfacial resistance further decreases.

By modifying the structure and understanding the structure–activity relationship, these conjugated binders have advanced toward more multifunctional and high-performance solutions for LIBs.

9.2.2 Si Anodes

Si has emerged as one of the most promising anode materials for LIBs due to its high theoretical capacity, natural abundance, and low toxicity. However, its commercial viability has been hindered by several challenges, the most significant of which is the large volume expansion (>300%) that occurs during repeated cycling. This expansion leads to pulverization of Si particles, electrical contact loss, delamination from the current collector, and the formation of a thick and resistive SEI, all of which degrade battery performance over time [20].

As a binder, PVDF relies on weak van der Waals forces to bind electrode components, which is insufficient to maintain Si electrode structure during large volume changes [20]. Earlier, CMC [21, 22] and PAA [20, 23] were also considered as possible replacements for PVDF, and relatively improved performances were achieved. Later, alginate, a natural polysaccharide, was reported to effectively stabilize Si electrodes with superior stiffness as compared to PVDF. It delivered a high reversible capacity of 2000 mAh g^{-1} at 4200 mA g^{-1}, showing minimal capacity fade after 100 cycles. This stability is attributed to the high concentration of carboxylic groups along its polymer chain. Thus, all the linear polymers stated have exhibited strong mechanical strength via covalent bonding, which may theoretically extend the life of Si anode. However, in the due course of repeated volume changes, once the covalent bonding breaks, the bonds cannot recover. In addition, since there is no interaction between the adjacent polymer chains, the polymers tend to dissociate, leading to eventual electrode degradation [20].

Crosslinked polymer binders have accordingly been explored, as they offer great flexibility to the binder design. We can simultaneously obtain the advantages of two polymers, offering a stabilizing effect on the Si electrodes. Recently, Wang et al. designed a water-soluble binder, poly(acrylic acid)-poly(2-hydroxyethyl acrylate-co-dopamine methacrylate), which is referred to as PAA-P(HEA-co-DMA). Here the PAA segment ensures rigidity and adhesion, the HEA segment provides elasticity, and the DMA segment imparts self-healing capability. A network can be built via covalent bonding between the binder and the Si anode with localized hydrogen bonds formed in situ. When two broken polymer pieces are placed together, the modulus can be partially recovered; this can be improved in the wet condition [24].

Choi and coworkers investigated and designed a hyperbranched β-cyclodextrin polymer (β-CDP), for Si anodes. The polymer was functionalized with plenty of hydroxyl groups using epichlorohydrin (EPI). The hydroxyl groups interact with Si particles through multiple noncovalent bonds, leading to superior mechanical strength. Similarly, the massive hydrogen bonding formed between the binder and Si gives the electrode a self-healing effect [25]. Overall, test results report superior performance over alginate. As stated, functional polymers have been explored as binders for Si anodes to address these challenges by providing mechanical robustness and stabilizing the SEI via interesting strategies. Another such binder is the p-BIAN/PAA composite binder, developed by Gupta et al. at our group. This binder combines p-BIAN (poly(bisiminoacenaphthenequinone)), the n-type conducting polymer and PAA, a carboxylate-containing polymer. The p-BIAN/PAA binder provides several advantages, including mechanical flexibility to accommodate the large volume expansion of Si, electrical conductivity to facilitate electron transport, and a thin SEI that minimizes electrolyte decomposition. The electrostatic hydrogen bonding between p-BIAN and PAA further enhances the self-healing properties of the binder, allowing it to maintain integrity throughout cycling (as shown in Figure 9.4a)

In comparison to other binders, such as conducting polymers and self-healing polymers (SHPs), the p-BIAN/PAA composite offers superior performance. CPs like polyaniline (PANI) and poly(3,4-ethylenedioxythiophene) (PEDOT) improve conductivity but lack the mechanical robustness needed for Si anodes. On the

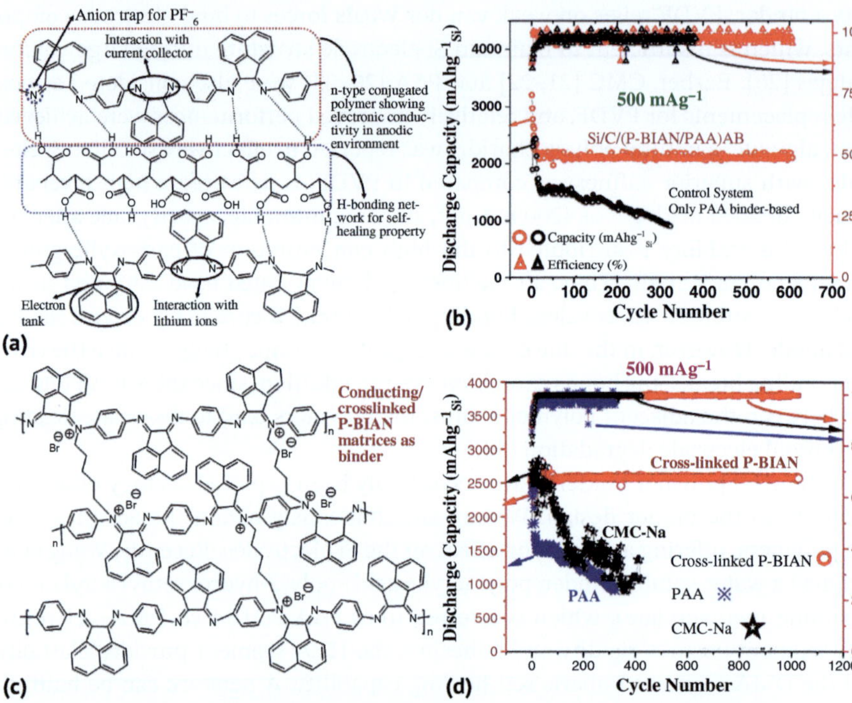

Figure 9.4 (a) Structural features of BP-PAA [26] / American Chemical Society. (b) Charge–discharge studies of silicon-based anodic half-cell with BP-LiPAA [26] / American Chemical Society. (c) Structural features of crosslinked p-BIAN [15] / John Wiley & Sons. (d) Charge–discharge studies of silicon-based anodic half-cell with crosslinked P-BIAN [15] / John Wiley & Sons.

other hand, SHPs do not adequately address the issue of SEI growth. The p-BIAN/PAA binder not only improves the mechanical properties of the electrode but also enhances lithium-ion diffusion and reduces interfacial impedance, leading to better cycling performance. The low-lying LUMO of p-BIAN allows for n-doping in the reducing environment, enabling the formation of a thin and conductive SEI that suppresses electrolyte decomposition [26]. This results in Si anodes with a high reversible capacity of around 2100 mAh/g, excellent capacity retention (~95%), and coulombic efficiency greater than 99% after over 600 cycles (shown in Figure 9.4b).

Further improvements to Si anode binders were made by Gupta et al., who developed a crosslinked p-BIAN binder that eliminates the need for PAA by using a simple crosslinking agent [15]. This approach results in a 3D network structure that offers superior mechanical properties and ionic conductivity (as shown in Figure 9.4c). The crosslinked p-BIAN binder effectively stabilizes Si anodes, preventing pulverization and maintaining electrical conductivity during cycling. The low LUMO of the crosslinked p-BIAN binder facilitates the formation of a stable and conductive SEI, which significantly improves cycling stability and capacity retention (Figure 9.4d).

Electrochemical testing revealed that Si anodes with the crosslinked p-BIAN binder achieved a high reversible capacity of around 2500 mAh/g over 1000 cycles at 500 mA/g, with a remarkable 99.1% capacity retention. The binder's unique combination of conductivity, mechanical stability, and SEI control makes it an ideal solution for enhancing the performance of Si-based anodes in LIBs.

In summary, the BIAN-based binder systems represent a significant advancement in the development of high-performance binders for graphite and Si-based anodes, effectively addressing the challenges of mechanical integrity, electrical conductivity, and SEI formation. This work paves the way for the commercialization of Si anodes in high-performance LIBs and offers promising directions for future conjugated polymer binder design and optimization.

9.3 Conjugated Polymers as Anodic Active Materials

9.3.1 Organic Anodes

In current LIBs, both capacity and fast-charging capabilities are limited by the available active materials. Organic conjugated polymers, with their structural tunability and high conductivity, have gained significant attention within the research community as potential solutions to these limitations [27–30]. Consequently, such organic polymers are expected to significantly improve fast-charging applications and offer high capacity, making them an attractive choice for use as anode materials. Recently, Mantripragada et al. at our group reported the synthesis and lithium storage application of BIAN–melamine-based porous organic polymer [31] (shown in Figure 9.5a). The polymer consists of 18 at% nitrogen in the framework. Further, the surface area was found to be 32 m^2/g with an average pore diameter of 1.7 nm. The effect of this structure and functional group was evaluated using DFT studies and electrochemical studies. Through DFT calculations, the theoretical capacity of BIAN–melamine-based porous organic polymer (PBM) was evaluated to be 1068.5 mAh/g, each C=C and C=N moieties in the monomer unit were found to be storing one lithium ion per moiety. Also, the DFT calculations revealed a lower LUMO energy level of PBM than the electrolyte, which can enable n-doping and restrict the excessive reduction of electrolyte. PBM-based anodic half-cells exhibited a reversible capacity of 740 mAh/g at 750 mA/g and 300 mAh/g at 1000 mA/g for 2000 and 1000 cycles, respectively (Figure 9.5c). The charge storage mechanism was evaluated to be a combination of diffusive and capacitive behaviors. The b value in power law was evaluated to be 0.81, which is intermediate to diffusive and capacitive behaviors. Furthermore, postmortem XPS was conducted to the electrodes with PBM as active material, and a clear N 1s peak was observed in XPS. As the X-ray penetration depth in XPS is typically 10 nm, detection of N1s indicates that the thickness of SEI is less than 10 nm. This observation proves the effect of utilizing n-type polymer, which can restrict the excessive reduction of the electrolyte. This shows the beneficial properties of conjugated organic polymers as anodic active materials to enable high charge storage (Figure 9.5b).

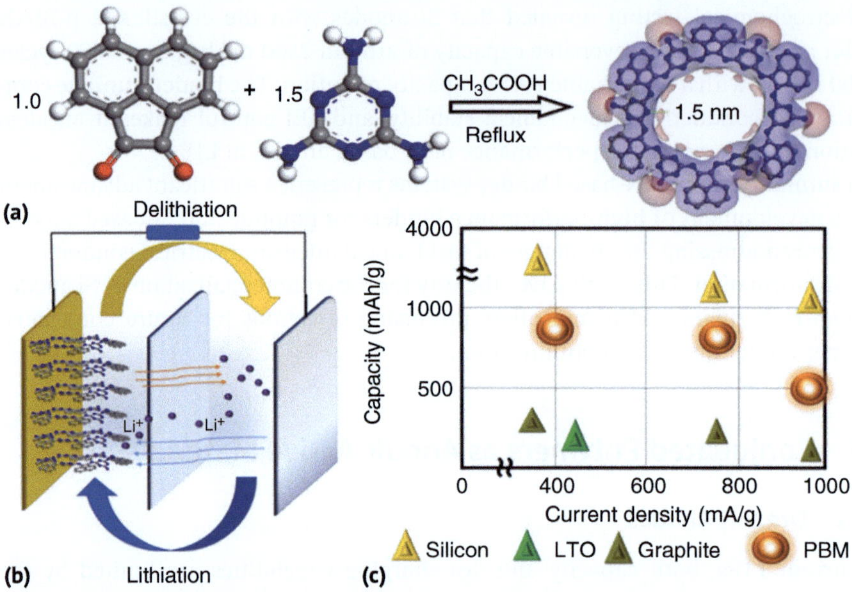

Figure 9.5 (a) Synthetic scheme of PBM (b) schematic representation of battery with PBM anode (c) comparison of reversible capacity delivered by commercial anodic active materials with PBM at different current densities.

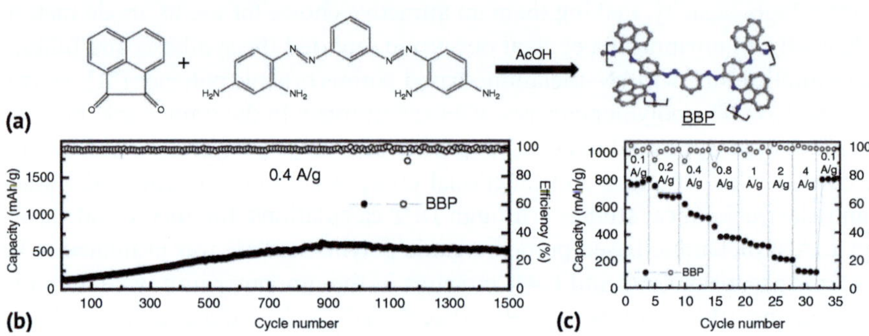

Figure 9.6 (a) Synthetic scheme of BBP (b) Charge–discharge studies of anodic half-cells with BBP active material (c) Rate studies of anodic half-cells with BBP active material [8] / Elsevier / CC BY 4.0.

Further, Mantripragada et al. also reported the application of BIAN-Bismarck brown-based imine and azo functional porous organic polymer (BBP) as anodic active material in LIB [8] (Figure 9.6a). Through DFT calculations, LUMO energy level of BBP was found to be much lower than that of electrolyte components. As discussed earlier, this can restrict the excessive reduction of the electrolyte. The morphology of the BBP was found to be amorphous in nature, and the structure contains 11 at% nitrogen consisting of both imine and azo functional nitrogen. BBP-based anodic half-cell delivers a reversible capacity of 500 mAh/g at current density of 400 mA/g till 1500 cycles (Figure 9.6b). Further, BBP-based anodic half-cells also exhibit excellent rate capability until 4000 mA/g current density (Figure 9.6c). Another interesting feature of BBP is its high specific energy and specific power.

BBP delivers a specific energy of 300 Wh/kg at a specific power of 7000 W/kg and a specific energy of 1300 Wh/kg at a specific power of 200 W/kg. Further, Dynamic Electrochemical Impedance Spectroscopy (DEIS) was conducted to the BBP-based anodic half-cells, and it was found that both charge-transfer and SEI resistances are much lower. Furthermore, postmortem XPS showed a significant N1s signal, indicating the thickness of SEI to be lesser than 10 nm. The effect of utilizing n-type polymer is clearly observed through these studies.

The introduction of a donor–acceptor structure into the polymer framework can achieve a very narrow band gap between the HOMO and LUMO levels, enhancing electronic interactions and thereby increasing power density. Recent research in the field of organic polymeric materials has predominantly focused on employing n-type polymers for energy storage. In contrast, relatively few studies have explored the design of donor–acceptor-type conjugated polymers with innate porosity. Moreover, the synthesis of such microporous polymers is not straightforward. While D–A-type conjugated porous polymers have been studied for their optical properties, their application in LIBs remains underexplored.

Hence, in the work reported by Saibrata et al. in our group, a donor–acceptor-based imine-linked conjugated network-type porous polymer, POL 202, was designed via a simple polycondensation reaction of 2,2′-bithiophene-5,5′-dicarboxaldehyde and 1,2,4,5-benzene tetraamine tetrahydrochloride [32]. A particularly novel aspect of this research is the investigation of D–A polymers with a rhombus-shaped pore structure for lithium storage and mass transport properties, which could significantly benefit fast-charging applications. These angles facilitate different diffusion pathways, allowing ions to move more efficiently through both narrow and wide channels. Additionally, the sharp corners of the acute angles enhance capillary action, drawing lithium ions into the pores more effectively.

In POL 202, it was reported that the charge storage mechanism was predominantly pseudocapacitive, a crucial factor for fast-charging applications in batteries. The polymer delivered a high capacity of 850 mAh/g at a current density of 100 mA/g with 100% capacity retention. Additionally, a reversible capacity of 205 mAh/g was observed at a current rate of 5000 mA/g, and a capacity of 270 mAh/g at 2000 mA/g. Moreover, the anode degradability studies conducted after cycling the cell at 2 A/g current density for 1000 cycles suggested that the volume expansion in such a material was observed to be 47.6%, which is significantly lower compared to conventional materials such as graphite and Si.

Such materials can be considered a very prospective anode material, as they show high capacity, support fast charging, and have stable electrochemical activity at the same time. Continuous rigorous research in such a field will not only result in the potential development of fundamental science but also lead to the development of industrially and commercially viable materials.

9.3.2 Hard Carbons

Amongst the various anode materials used in LIBs, graphite is the most widely known and commonly employed [33]. Its hexagonal layered structure, where carbon atoms are arranged in planes and held together by weak van der Waals forces, makes it highly suitable for lithium-ion intercalation. With a theoretical capacity of

372 mAh/g, excellent electronic conductivity, long cycle life, and, most importantly, its abundance and low cost, graphite has become the leading choice for commercialization in LIBs [34]. Although several other anode materials have been investigated, such as lithium titanium oxide (LTO), known for its exceptional fast-charging performance, its relatively low discharge capacity remains a significant drawback [35]. Additionally, alloy-based anodes, such as Si and tin (Sn), are considered promising alternatives for negative electrodes in LIBs. However, challenges such as volume expansion and contraction, which lead to particle pulverization and eventual electrode delamination, pose substantial concerns. As the demand for fast-charging capabilities increases, particularly for electric vehicles (EVs), graphite's performance at high charge–discharge rates proves to be inadequate. During fast charging, the influx of Li-ions increases significantly, and the distance they must travel before intercalating into graphite anode becomes large. This is because only edge-plane intercalation is available, while intercalation into the basal planes is not permitted in graphite (Figure 9.7). As a result, Li-ions may not intercalate quickly enough, leading to the formation of lithium dendrites on the surface of the anode, especially during fast charging [34]. In contrast, hard carbons (HCs) – materials characterized by a high degree of structural disorder, with numerous disordered carbon planes – demonstrate promising fast-charging performance (Figure 9.7). HCs generally

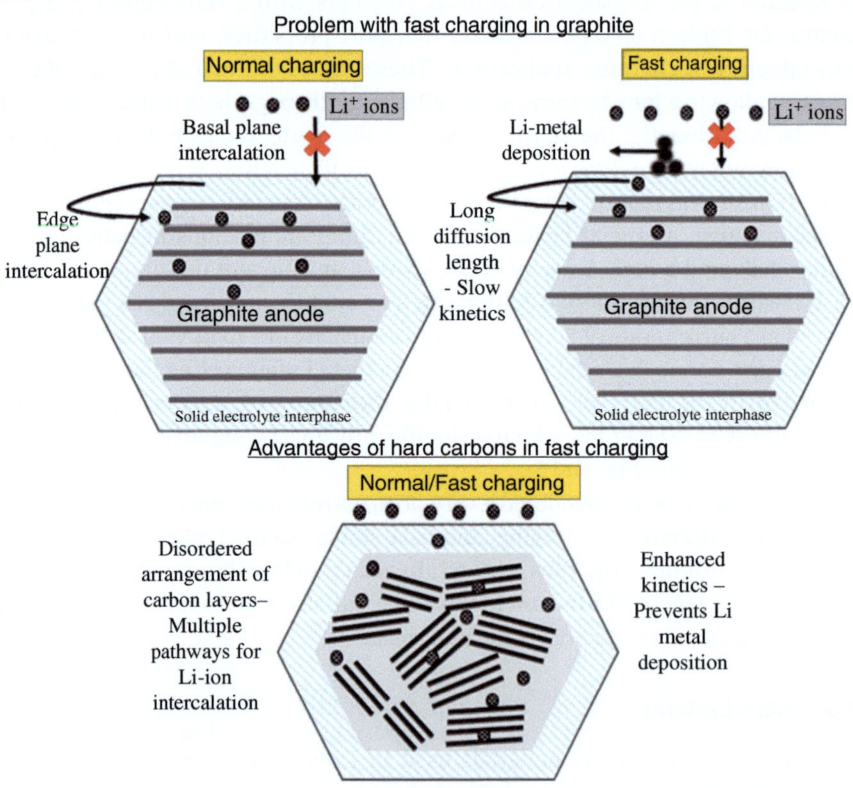

Figure 9.7 Illustration of problems with fast charging in graphite and advantages of hard carbons as anodes in LIBs.

contain a large number of randomly oriented disordered graphitic domains, which were classified as "house of cards" model by Dahn et al. The structure of HC is generally composed of non-graphitizable and disordered regions, with numerous structural defects, amorphous areas, and nanopores. Structural parameters, such as the d-spacing (d002), which represents the distance between graphitic planes, along with the thickness (Lc) and width (La) of the graphitic microcrystallites, can be determined through X-ray diffraction analysis (XRD). The porosity of HCs is typically divided into open porosity, where pores are accessible, and closed porosity, where pores are not accessible. While N_2 adsorption and desorption is commonly used to evaluate the open pore structure of HCs, CO_2 has been shown to be more effective in probing smaller pores due to its superior diffusion kinetics compared to N_2. Additionally, other gases like O_2 and H_2 have also been found to be efficient for this purpose. Notably, H_2, due to its smaller size and lower quadrupole moment, has been reported to be more suitable than CO_2 for detecting ultramicropores and supermicropores. However, closed porosity cannot be assessed through gas adsorption techniques [36]. Instead, small-angle X-ray scattering (SAXS) is a valuable tool for investigating closed porosity [37]. Along with this, HCs can also contain various types of defect sites, including Stone–Wales defects, multiple vacancies, line defects, etc. Generally, Raman spectroscopy is utilized for detecting different types of defects in HCs [36]. Overall, these structural irregularities lead to the formation of numerous nanovoids, defect sites, and edge sites, which enhance the ability of HCs to store and interact with ions. Heteroatom doping is also considered an effective strategy for enhancing charge storage in HCs. The most commonly doped heteroatoms include nitrogen (N), phosphorus (P), sulfur (S), and boron (B). Nitrogen and boron can be regarded as substitutional dopants, while phosphorus and sulfur act as interlayer dopants, as the C—S or C—P bond lengths are generally longer [38]. Given the drawbacks of graphite, HCs owing to their structural disorderliness appear to be a better option, particularly for fast-charging applications. The charge storage mechanism in HCs involves several processes: Li-ion adsorption on the surface of graphene planes, Li-ion intercalation between carbon layers, and Li-ion filling of both closed and open pores [39]. Because these diverse processes occur simultaneously during Li-ion storage in HCs, the charge–discharge curve for HCs typically exhibits a sloping profile, unlike the distinct plateau seen at 0.2 V in graphite-based anodes, which corresponds to the Li-ion intercalation between graphite layers.

The preparation processes, such as degree of graphitization, the pyrolysis temperature, the rate at which pyrolysis is performed, etc., also have a significant impact on the HC microstructure. Until now, various methods for obtaining HCs have been utilized, some of which involve pyrolysis of various precursors such as pitch, biomass, polymers, and resins [40]. Pitch-based HCs are typically derived from sources such as coal-tar pitch, petroleum pitch, or natural pitch [41] (summarized in Figure 9.8). These materials are cost-effective and readily available from a variety of sources. In a study, Wang et al. employed a solid-phase oxidation method combined with carbonization to synthesize coal tar pitch-derived HCs, achieving a high discharge capacity of 576 mAh/g. The enhanced performance was attributed to the microporous structure of the material. However, despite their many advantages, the

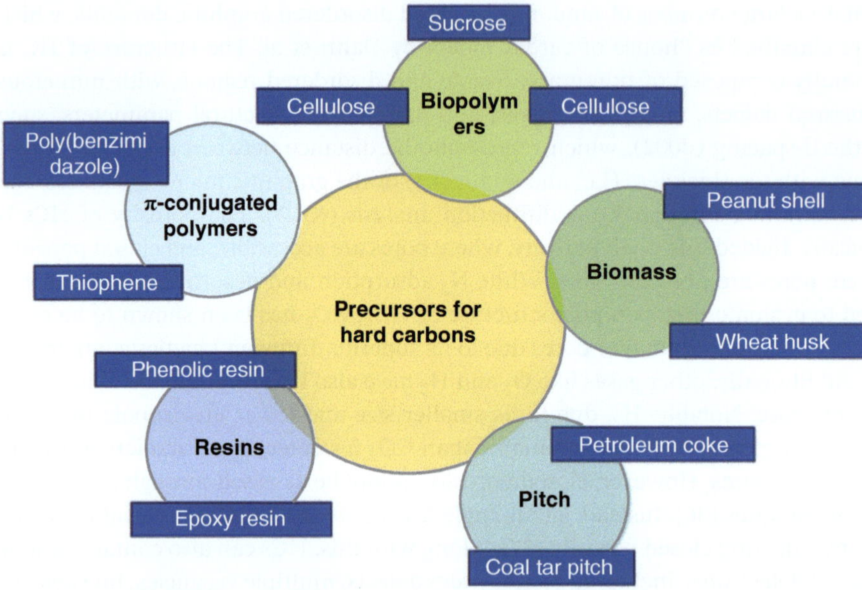

Figure 9.8 Indicates various sources of HCs with a few examples.

complex composition and the unclear reaction mechanisms in the preparation process pose challenges for the development of pitch-derived HCs [42]. Resin-based HCs are typically produced by pyrolyzing resins that are formed through the crosslinking and polymerization of organic monomers. For example, Gao et al. synthesized and pyrolyzed phenolic resin at 1100 °C, which was then used as an anode material for LIBs and supercapacitors, demonstrating a high capacity of 526 mAh/g. Biomass is another promising precursor for HCs, owing to its natural hierarchical porosity and high content of heteroatoms [43]. These characteristics make it an effective material for producing HCs with desirable properties. Recently, Li et al. studied the impact of pyrolysis temperature on the lithium-ion storage performance of PVDF-coated HCs derived from cellulose. The HCs exhibited excellent overall performance in terms of charge–discharge cycling, coulombic efficiency, rate capability, and cycling stability. Due to the diverse composition, morphology, and internal structure of biomass, it is not considered an ideal precursor for HCs. In addition to biomass, biopolymers such as sucrose [37], cellulose [44], glucose, and lignin [45] have also been widely used as precursors for HCs. While biopolymers have a more uniform composition and lower impurity content than biomass, they can be a good choice for HC production, as their char yields are improved to at least above 50%. Polymers offer an attractive alternative as precursors for HCs. One of their primary advantages is the ability to produce HCs with high purity and excellent char yields. The vast array of available polymers, combined with the flexibility to introduce diverse functional groups, provides exciting opportunities for doping heteroatoms and creating various internal microstructures in HCs during pyrolysis [46]. The resulting HC structure is largely determined by the polymer's original structure. Polymers with inherent porosity and heteroatom doping can be specifically designed and synthesized to optimize HC properties. A range of polymers, including poly(vinyl chloride) and poly(acrylonitrile) [47], have been investigated for this purpose. Moreover, incorporating π-conjugation into these polymers encourages a more ordered structure during pyrolysis, which is crucial for achieving high carbon content and stability in the resulting HC material. The polymer backbone, with its conjugated double bonds, can form a well-organized, graphitic framework when subjected to high-temperature treatment, leading to a higher degree of graphitization and minimizing the formation of amorphous regions. Moreover, the functional groups present in conjugated polymers (such as —OH, —COOH, $-NH_2$) can offer routes to further functionalize the resulting HC. The degree of aromaticity and the structure of the conjugated polymer can influence the final morphology of the HC material as well. In light of this, several groups utilized conjugated polymers to synthesize HCs by pyrolysis. For instance, Patnaik et al. at our group used π-conjugated bio-based poly(2,5-benzimidazole) (AB-PBI), as a single source of C and N to prepare a heavily N-doped carbon (~17 wt%) (PYPBI800) pyrolyzed at 800 °C with a high char yield (95%) [48]. The extremely high N-doping, enhanced d-spacing, and defects aided in high capacity and extremely fast Li-ion kinetics with ultralong cyclability for 3000 cycles with over 80% capacity retention at charge–discharge rates $\geq 3.7\ Ag^{-1}$. Expanding on this strategy, we also reported N- and O-dual-doped carbons derived from bio-based copolymers poly(benzimidazole-co-amide) (PYPBIPA) as single sources of C, N, and O [49]. This study emphasized the significance of the optimal balance of N and O doping. Several copolymers with different N and O content were utilized to synthesize N and O co-doped HCs. PYPBIPA 8.5–1.5 with 85% and 15 at% unit ratio of benzimidazole:amide outperformed others, delivering a delithiation capacity of 135 $mAhg^{-1}$ over 3000 cycles with a capacity retention of 90%. Though oxygen co-doping was found to be beneficial, excessive oxygen doping resulted in inferior electrochemical performance due to steric hindrance caused by the bulky oxygen groups leading to sluggish Li-ion diffusion. In another such work, Mantripragada et al. at our group reported a novel BIAN–melamine-based porous organic polymer (POP) as a single source of C and N to synthesize N-doped carbon. The POP was pyrolyzed at 600 °C (PyPBM600) and 800 °C (PyPBM800) to obtain HC with different morphologies and nitrogen content. PyPBM600, with a higher N-content of 14.35 at%, exhibited an increased amorphous nature and a predominant capacitive behavior resulting in a higher diffusion coefficient, low charge transfer resistance, and better rate performance [50]. However, the highly amorphous morphology of PyPBM600 resulted in excessive electrolyte reduction, greater irreversible capacity loss, and poor capacity retention. On the other hand, PyPBM800 demonstrated higher crystallinity and a balanced capacitive and diffusive behavior. It outperformed PyPBM600 in terms of current density, reversible capacity, and cyclability. Full cell fabricated with PyPBM800 anode delivered an excellent rate capability and fast charging in 15 minutes with good capacity retention and energy efficiency. This work emphasized the significance of N-content and the effect of synthesis temperature conditions on the HC morphology and the subsequent electrochemical performance. Further, our group synthesized a BIAN-Bismarck brown-based azo and imine functional POP (BBP) to obtain a high-capacity anode, delivering 500 mAh/g at 0.4 A/g after 1500 cycles. BBP was pyrolyzed at 600 °C (PyBBP600) and 800 °C (PyBBP800), where PyBBP800 delivered excellent rate capability and long cyclability at 4 A/g, delivering 125 mAh/g after 5000 cycles. Pyrolyzed POP (PyBBP800) showed superior capacity and enhanced cyclability, making them suitable for high-energy and high-power applications [8]. For instance, Nandan et al. at our group utilized polydopamine (PDA) for synthesizing carbon-coated silicon microparticles (SiMP). Initially, SiMP were coated with PDA and upon pyrolysis, they achieved a carbon-coated SiMP structure. Further, they also coated the amorphous carbon-coated SiMP with Si oxycarbide black glasses (BGs) embedded with acetylene black [51]. This meticulous design of SiMPs helped suppress the volume expansion and reduce overall resistance, demonstrating a high-capacity retention of 99.4% after 775th cycle at 750 mAg^{-1}. Our group extended the approach to synthesize β-silicon carbide (β-SiC), a zinc blende variant of SiC, using PDA and melamine, where melamine enhanced electronic conductivity of entire material through N-doping. Meanwhile, PDA's π-conjugated system combined with its extraordinary wet-adhesion characteristic made it ideal for surface coating and hence as a precursor of choice for β-SiC preparation [52]. Based on these research works, it is evident that heteroatom-doped π-conjugated polymers seem to be interesting options for utilization in anodes. π-conjugation in polymers is beneficial for the synthesis of HCs as well as carbon coatings on Si, as it helps control the structure, conductivity, stability, and porosity of the resulting material. The use of conjugated polymers can lead to more efficient and cost-effective synthesis of high-performance carbon materials for a range of advanced technological applications (summarized in Figure 9.9).

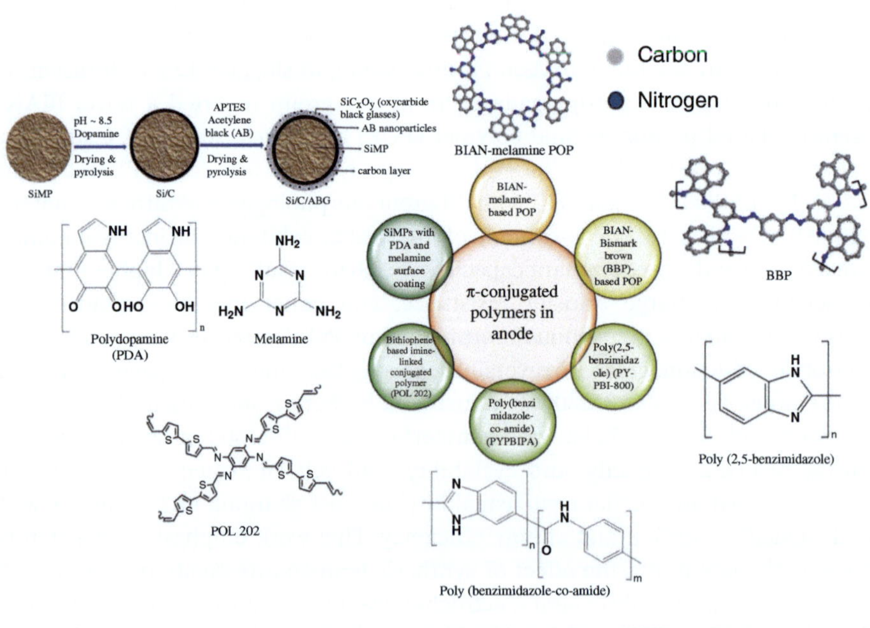

Figure 9.9 Various π-conjugated polymers utilized in anodes of lithium-ion batteries (LIBs).

9.4 Polymerizable Conjugated Molecules as Electrolyte Additives

Cathode materials in LIB have shown high power density, specific capacity, and high operating potential. However, they suffer from major issues during the reversible cycling, such as transition metal dissolution, catalytic electrolyte oxidation at transition metal centers on cathode and structural instability [53–55]. Trace amount of water initiates HF formation, which in turn dissolves CEI layer, increases cell impedance, and is also responsible for corrosion of Al current collector [56]. To address all these issues, strategies such as surface coatings are used. In this section, an overview on the electropolymerizable conjugated molecules that can lead to in situ formation of a passivation layer on electrodes leading to high voltage performance [57].

Compounds containing unsaturated bonds, free amine groups at the terminal positions, or heterocyclic five-membered rings with unoccupied 3 and 4 positions are highly prone to oxidative polymerization at the cathode during discharge. Polymer formation reduces the electrolyte degradation and improves cyclability.

Thiophene [58], 3,4-ethylenedioxythiophene (EDT), and *N*-methyl pyrole (MPL) [59] are known to undergo oxidative polymerization without any loss of Li-ions during the process and passivate the transition metal ions from dissolving into the electrolyte. These additives also reduce polarization by decreasing interfacial resistance [60].

Benzene derivatives such as biphenyl, o-terphenyl, dibenzyl, diphenyl ether, diphenyl disulfide, and *N*-(triphenylphosphoranylidene)aniline are also some of the known polymerizable additives [61]. These additives have lower oxidation potential than the electrolyte solvents, and they are oxidized on uneven high-voltage sites of cathode to form radicals, which in turn react with other monomer units to form a polymer [62]. Fumaronitrile (FN), a novel nitrile-containing additive reported by Wang et al., enhances cathode performance at high voltages by interacting through its nitrile group with transition metals [63]. Presence of two —CN groups in FN coordinates strongly with Co^{3+} compared to other solvent. This aids in forming a protective film that minimizes electrolyte decomposition of electrolyte and transition metal dissolution. The C=C gives FN an electrochemical oxidation preference over the carbonate-based solvents. Under oxidative conditions at high voltages, FN undergoes radical formation via cleavage of its double bond. These radicals initiate polymerization by reacting with monomers (structure of the additive: CN—CH=CH—CN).

In similar lines, Gupta et al. at our group demonstrated the application of microbially synthesized bio-based 2,5-dimethyl-3,6-bis(4-aminobenzyl) pyrazine (DMBAP) (Figure 9.10) as an electrolyte additive to stabilize $LiNi_{1/3}Mn_{1/3}Co_{1/3}O_2$ cathodes [64]. The distinguishing features of this additive include its involvement in the CEI formation (sacrificial in situ degradation) prior to the oxidation of electrolyte due to its high-lying HOMO energy level. The diamine group on the additive helps in interacting with the metal active centers on cathode. This additive undergoes oxidative polymerization on the surface of cathode to form a protective passivation layer. The same diamine group also acts as a Lewis base and helps in scavenging HF from the solution, keeping the cathode intact.

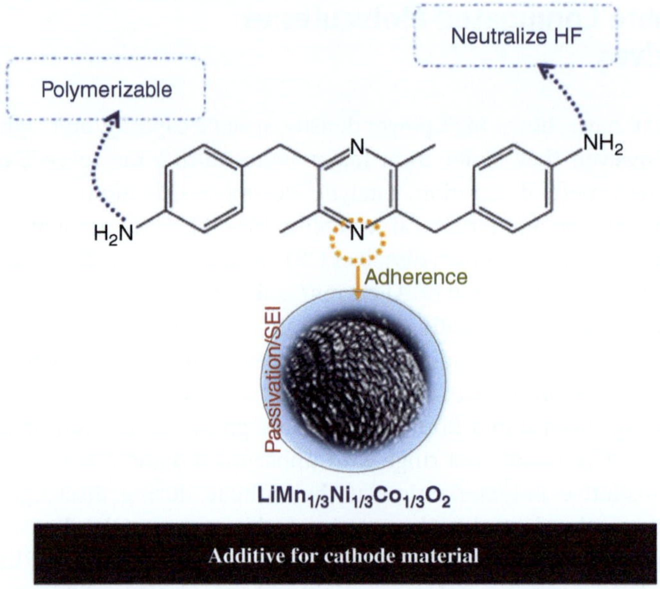

Figure 9.10 Structure activity relation of DMBAP [64] / Springer Nature/CC BY 4.0.

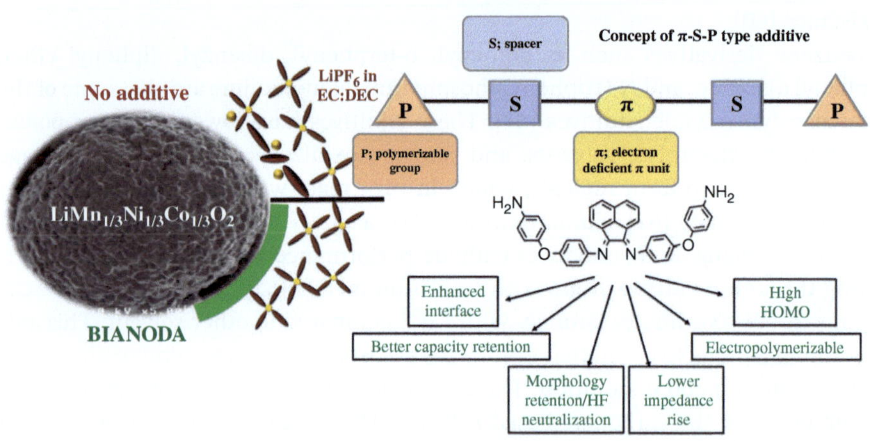

Figure 9.11 Structure–property relation of BIANODA [65] / Royal Society of Chemistry.

Similarly, another typical example of π-conjugated polymerizable additive is (acenaphthylene-1,2-diylidene-bis(4-(4-aminophenoxy)aniline) BIANODA reported by Gourang et al. at our group; it has a framework of π-s-p (π-spacer-polymerizable group) [65]. As shown in Figure 9.11, BIANODA has a basic amine group that polymerizes and also neutralizes the HF generated during the cycling. Presence of diimine moiety aids in coordinative with transition metal ions on cathode (LiNMC111). BIANODA (Figure 9.11) is highly effective in improving cell performance by enhancing capacity, interfacial properties, and morphology retention of the cathode. Higher HOMO level of BIANODA with respect to carbonate-based electrolyte makes oxidation of BIANODA inevitable. Electropolymerization on cathode happens during the 1st cycle of charge–discharge to form a CEI that exacerbates side reactions and electrolyte degradation. The interfacial CEI formation was studied by DEIS, indicating a better interface formation in presence of BIANODA. The structural stability of the electrode in presence of electrolyte was studied, and the electrode coated with BIANODA additive seems to keep the electrode intact even after seven days. The oxidative polymerization has been shown in ex situ studies.

These results indicate the beneficial properties of conjugated electropolymerizable additives to enable high-voltage performance of cathodes.

9.5 π-Conjugated Polymers as Electrocatalysts

In the previous sections, a detailed explanation on the application of conjugated polymers in various components of LIBs is presented. The structural flexibility of π-conjugated polymers can also be explored in the field of catalysis. Zhang et al. explored the mechanism of electrocatalysis in N-doped graphene materials using DFT calculations [67]. It was found that upon the substitution of nitrogen in the graphene framework, the electron distribution of the graphene is changed, and the carbon atoms adjacent to nitrogen tend to be relatively more electrophilic in nature. The change in the electron distribution leads to the adsorption of oxygen to the carbon atoms close to nitrogen atom, while on pure graphene sheets oxygen cannot be adsorbed. Furthermore, there are many reports related to the utilization of heteroatom-doped carbon materials for catalysis [68, 69]. In similar lines, Sai Gourang et al. at our group employed BIAN-based conjugated polymer with pyridinic nitrogen present in the structure of the polymer [17] (Figure 9.12). The polymer backbone consists of carbon atoms in two different environments, i.e. C=C and C=N. Similar to N-doped graphene, carbon atoms adjacent to pyridinic nitrogen exhibit electrophilicity and thereby act as active centers for ORR. As expected, BP-based catalyst showed two distinct peaks in cyclic voltammetry corresponding to the ORR activity. Further, to enhance the activity of the polymer-based catalyst, a composite of BP was prepared with graphene oxide (GO). The BP-GO composite showed significant enhancement of the current response from the two peaks

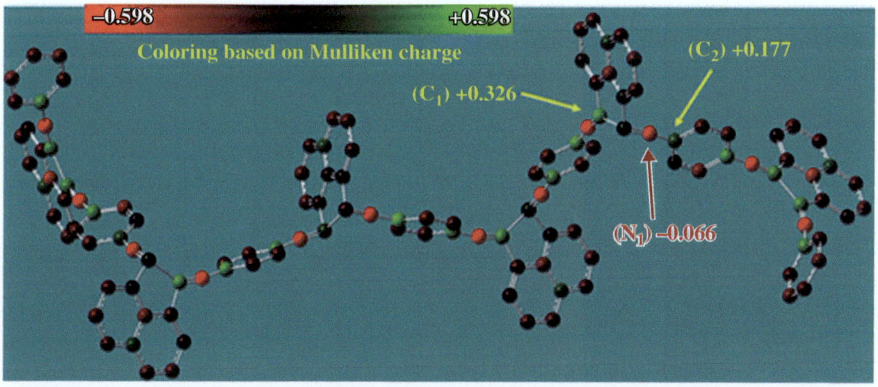

Figure 9.12 Mulliken charges of BP [17] / American Chemical Society.

in the cyclic voltammetry after low-temperature (400 °C) annealing. Two distinct peaks in cyclic voltammetry indicate the possibility of the presence of two different active sites. DFT calculations were performed to elucidate the origin of two peaks in cyclic voltammetry. Mülliken charge analysis showed that the charges on carbon atoms a and b are different (carbon a = +0.326 and carbon b = +0.177) (showed in Figure 9.13). Different electronegativities of two adjacent carbon atoms to nitrogen can be the plausible reason for the appearance of two peaks in cyclic voltammetry. Furthermore, rotating disk electrode-based experiments were also performed to elucidate the electron transfer process. KL plots indicated a $2e^-$ transfer process. ORR studies were conducted in oxygen-saturated solution of 0.1 M LiTFSI in TEGDME. BP-GO-based composite catalyst exhibited excellent stability for over 100 cycles. Furthermore, the imine moiety in the BP can also function as ligand for complex formation with transition metals. Rajashekhar et al. at our group, demonstrated the synthesis, catalytic properties, and application of lithium-air batteries of BP-Fe complex [66] (Figure 9.14). BP-Fe has atomically dispersed Fe(II) coordinated to the diamine moiety of the polymer chain. To understand the effect of this novel polymer–metal complex, charge–discharge analysis was conducted in Li-O_2 batteries. At a current density of 250 mA/g BP-Fe-based catalyst exhibited 100% columbic efficiency for 160 cycles when the capacity was limited to 500 mAh/g. The Fe^{2+} ions

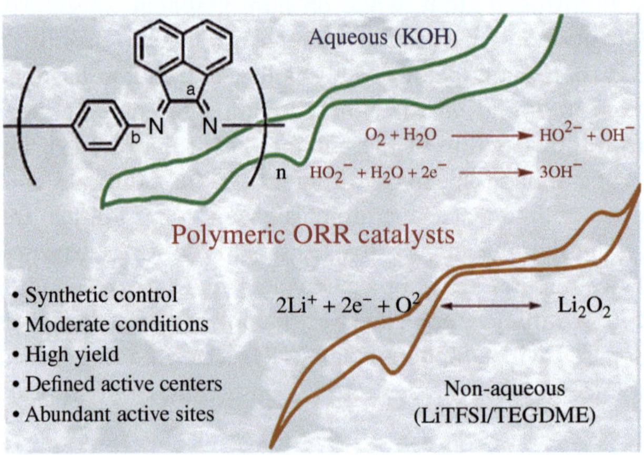

Figure 9.13 BP as catalyst for ORR/OER process [17] / American Chemical Society.

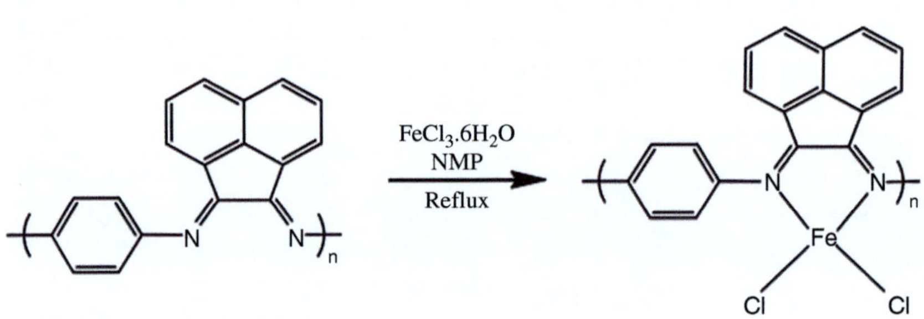

Figure 9.14 Reaction scheme for BIAN–Fe complex [66] / Springer Nature.

Figure 9.15 Mechanism of ORR/OER catalysis in BP-Fe complex [66] / Springer Nature.

$$Fe^{3+} + O_2 + e^- \leftrightarrow Fe^{2+}\text{–}O_{2(ads)}$$

$$Fe^{2+}\text{–}O_{2(ads)} + Li^+ + e^- \leftrightarrow Fe^{2+}\text{–O–O–}Li_{(ads)}$$

$$Fe^{2+}\text{–O–O–}Li_{(ads)} + Li^+ + e^- \leftrightarrow Fe^{2+}\text{–}Li_2O_{2(ads)}$$

$$Fe^{2+}\text{–}Li_2O_{2(ads)} \leftrightarrow Fe^{3+}\text{–}Li_2O_{2(ads)} + e^-$$

$$Fe^{3+}\text{–}Li_2O_{2(ads)} \leftrightarrow Fe^{3+} + 2Li^+ + O_2 + 2e^-$$

present in the complex catalyzing both ORR and OER, and the imine moiety helps in catalyzing ORR. Due to this synergistic effect, BP-Fe complex enabled reversible formation and decomposition of Li_2O_2 during charge–discharge process, which is reflected as 100% columbic efficiency. It was also observed that BP-Fe exhibits a constant discharge potential of 2.78 V at 50 mAh/g capacity for 160 cycles. However, the charge potential increases from 3.5 to 4.0 V. Cyclic voltammetry studies in three-electrode setup indicated good ORR activity for 200 cycles; however, the peak intensity of OER after 200 cycles was found to be decreased slightly. This indicates the incomplete dissociation of Li_2O_2. Ex situ XPS and XRD studies showed the electrical insulation due to Li_2O_2 and Li_2CO_3 decomposition, as mentioned earlier, this could be the reason for cell failure. The mechanism of catalysis was elucidated as shown in Figure 9.15.

9.6 Conclusion

In this chapter, the applications of conjugated polymers in various components of LIBs were presented with specific case studies. The examples demonstrate the interesting features of CPs and the effects on enhancing the battery performance in terms of cycle life, high capacity, high voltage, etc. Different aspects discussed in the chapter can be summarized as follows:

9.6.1 Binders

- Owing to the conjugated structure of these polymers, they can offer robust mechanical stability.
- The structural flexibility of n-type CPs can help in designing polymers with lower LUMO energy level than the electrolyte components, thereby leading to a thinner and robust SEI.
- Overall, a meticulous design of CPs can lead to enhancement of cycle life and reversible capacity of the battery.

9.6.2 Active Materials

- The geometry of the reactants can be selected in a manner that the resultant CP structure can be porous in nature.
- Porous polymers with significant heteroatom content enhances the lithium-ion storage.

- Furthermore, CPs can also be used as single sources of carbon and heteroatoms to synthesize highly heteroatom-doped carbons.
- Heteroatom-doped carbons can support extreme fast charging properties of the battery, thereby improving the battery's power density.

9.6.3 Electrolyte Additives

- CPs can be synthesized in situ by adding electropolymerizable additives in the electrolyte.
- These in situ synthesized CPs can form a protective layer between electrode and electrolyte and therefore act as an artificial SEI.
- This CP-based artificial SEI can help in increasing the voltage limit of the battery and therefore increase the energy density of the battery.

9.6.4 Catalysts in Li-O_2 Batteries

- The polarity induced in CPs through functional group inclusion can lead to oxygen adsorption. Thus, CPs can catalyze ORR/OER process.
- Furthermore, the functional group arrangement can be tuned in a manner that the polymer can behave as a ligand to complex with non-precious metals.
- These polymer–metal complexes can enhance the ORR/OER catalysis.

References

1 Hideki Shirakawa, B., Louis, E.J., Macdiarmid, A.G. et al. (1977). Synthesis of electrically conducting organic polymers: halogen derivatives of polyacetylene, (CH). *Journal of the Chemical Society, Chemical Communications* 16: 578–580.

2 Deng, D. (2015). Li-ion batteries: basics, progress, and challenges. *Energy Science & Engineering* 3 (5): 385–418.

3 Yang, Y., Yuan, W., Kang, W. et al. (2020). Silicon-nanoparticle-based composites for advanced lithium-ion battery anodes. *Nanoscale* 12 (14): 7461–7484.

4 Kwon, T.W., Choi, J.W., and Coskun, A. (2018). The emerging era of supramolecular polymeric binders in silicon anodes. *Chemical Society Reviews* 47 (6): 2145–2164.

5 Jie, J., Liu, Y., Cong, L. et al. (2020). High-performance PVDF-HFP based gel polymer electrolyte with a safe solvent in Li metal polymer battery. *Journal of Energy Chemistry* 49: 80–88.

6 Pan, Y., Ge, S., Rashid, Z. et al. (2020). Adhesive polymers as efficient binders for high-capacity silicon electrodes. *ACS Applied Energy Materials* 3 (4): 3387–3396.

7 Mantripragada, B.S., Badam, R., and Matsumi, N. (2021). Bis-imino-acenaphthenequinone based covalent organic framework for lithium-ion battery applications. *ECS Meeting Abstracts* MA2021-01 (2): 104.

8 Srimitra Mantripragada, B., Sumala Patnaik, K., Higashimine, K. et al. (2023). POP based nitrogen containing anodic materials for high capacity and fast charge-discharge applications in LIB. *Electrochemistry Communications* 157: 107616.

9 Elizalde-Segovia, R., Das, P., Zayat, B. et al. (2021). Understanding the role of π-conjugated polymers as binders in enabling designs for high-energy/high-rate lithium metal batteries. *Journal of the Electrochemical Society* 168 (11): 110541.

10 Jiang, F., Wang, Y., Qiu, T. et al. (2021). Superlithiation performance of covalent triazine frameworks as anodes in lithium-ion batteries. *ACS Applied Materials & Interfaces* 13 (41): 48818–48827.

11 Feng, X., Ding, X., and Jiang, D. (2012). Covalent organic frameworks. *Chemical Society Reviews* 41 (18): 6010–6022.

12 Shi, R., Liu, L., Lu, Y. et al. (2020). Nitrogen-rich covalent organic frameworks with multiple carbonyls for high-performance sodium batteries. *Nature Communications* 11: 178.

13 Khrizanforova, V.V., Morozov, V.I., Khrizanforov, M.N. et al. (2018). Iron complexes of BIANs: redox trends and electrocatalysis of hydrogen evolution. *Polyhedron* 154: 77–82.

14 Mishra, S.N., Punyasloka, S., Mantripragada, B.S. et al. (2023). Enabling ultrafast charging in graphite anodes using BIAN-based conjugated polymer/lithium polyacrylate as a binder. *ACS Applied Energy Materials* 6 (23): 11954–11962.

15 Gupta, A., Badam, R., Mantripragada, B.S. et al. (2024). Ultra-durability and reversible capacity of silicon anodes with crosslinked poly-BIAN binder in lithium-ion secondary batteries for sturdy performance. *Advanced Sustainable Systems* 9 (1): 2400263.

16 Gupta, A., Badam, R., Nag, A. et al. (2021). Bis-imino-acenaphthenequinone-paraphenylene-type condensation copolymer binder for ultralong cyclable lithium-ion rechargeable batteries. *ACS Applied Energy Materials* 4 (3): 2231–2240.

17 Patnaik, S.G., Vedarajan, R., and Matsumi, N. (2018). BIAN based electroactive polymer with defined active centers as metal-free electrocatalysts for oxygen reduction reaction (ORR) in aqueous and nonaqueous media. *ACS Applied Energy Materials* 1 (3): 1183–1190.

18 Patnaik, S.G., Vedarajan, R., and Matsumi, N. (2017). BIAN based functional polymer binder for high performance Li ion batteries. *Journal of Materials Chemistry A* 5 (34), 17909–17919.

19 Ui, K., Fujii, D., Niwata, Y. et al. (2014). Analysis of solid electrolyte interface formation reaction and surface deposit of natural graphite negative electrode employing polyacrylic acid as a binder. *Journal of Power Sources* 247: 981–990.

20 Zou, F. and Manthiram, A. (2020). A review of the design of advanced binders for high-performance batteries. *Advanced Energy Materials* 10 (45): 2002508.

21 Bridel, J.S., Azaïs, T., Morcrette, M. et al. (2010). Key parameters governing the reversibility of Si/carbon/CMC electrodes for Li-ion batteries. *Chemistry of Materials* 22 (3): 1229–1241.

22 Komaba, S., Shimomura, K., Yabuuchi, N. et al. (2011). Study on polymer binders for high-capacity SiO negative electrode of Li-Ion batteries. *Journal of Physical Chemistry C* 115 (27): 13487–13495.

23 Magasinski, A., Zdyrko, B., Kovalenko, I. et al. (2010). Toward efficient binders for Li-ion battery Si-based anodes: polyacrylic acid. *ACS Applied Materials & Interfaces* 2 (11): 3004–3010.

24 Xu, Z., Yang, J., Zhang, T. et al. (2018). Silicon microparticle anodes with self-Healing multiple network binder. *Joule* 2 (5): 950–961.

25 Jeong, Y.K., Kwon, T.W., Lee, I. et al. (2015). Millipede-inspired structural design principle for high performance polysaccharide binders in silicon anodes. *Energy & Environmental Science* 8 (4): 1224–1230.

26 Gupta, A., Badam, R., and Matsumi, N. (2022). Heavy-duty performance from silicon anodes using poly(BIAN)/poly(acrylic acid)-based self-healing composite binder in lithium-ion secondary batteries. *ACS Applied Energy Materials* 5 (7): 7977–7987.

27 Li, J., Jing, X., Li, Q. et al. (2020) Bulk COFs and COF nanosheets for electrochemical energy storage and conversion. *Chemical Society Reviews* 49 (11): 3565–3604.

28 Zhao, H., Luo, D., Xu, H. et al. (2022). A novel covalent organic framework with high-density imine groups for lithium storage as anode material in lithium-ion batteries. *Journal of Materials Science* 57 (22): 9980–9991.

29 Sharma, S.U., Chang, Y.L., Chaganti, S.V. et al. (2022). Cross-linked naphthalene diimide-based polymer as a cathode material for high-performance organic batteries. *ACS Applied Energy Materials* 5 (6): 7550–7558.

30 Chen, H., Zhang, Y., Xu, C. et al. (2019). Two π-conjugated covalent organic frameworks with long-term cyclability at high current density for lithium ion battery. *Chemistry–A European Journal* 25 (68): 15472–15476.

31 Mantripragada, B.S., Badam, R., and Matsumi, N. (2022) BIAN-based porous organic polymer as a high-performance anode for lithium-ion batteries. *ACS Applied Energy Materials* 5 (6): 6903–6912.

32 Punyasloka, S., Higashimine, K., and Matsumi, N. (2024) Donor-acceptor-based conjugated polymeric active material with enhanced pseudocapacitive contribution for ultrafast charging of lithium-ion batteries. *ACS Applied Energy Materials* 7 (13): 5379–5389.

33 Xu, J., Wang, X., Yuan, N. et al. (2019). Graphite-based lithium ion battery with ultrafast charging and discharging and excellent low temperature performance. *Journal of Power Sources* 430: 74–79.

34 Asenbauer, J., Eisenmann, T., Kuenzel, M. et al. (2020). The success story of graphite as a lithium-ion anode material-fundamentals, remaining challenges, and recent developments including silicon (oxide) composites. *Sustainable Energy & Fuels* 4 (11): 5387–5416.

35 Jung, H.-G., Jang, M.W., Hassoun, J. et al. (2011). A high-rate long-life Li 4 Ti 5 O 12 / Li[Ni 0.45 Co 0.1 Mn 1.45]O 4 lithium-ion battery. *Nature Communications* 2: 516.

36 Dou, X., Hasa, I., Saurel, D. et al. (2019). Hard carbons for sodium-ion batteries: structure, analysis, sustainability, and electrochemistry. *Materials Today* 23: 87–104.

37 Kubota, K., Shimadzu, S., Yabuuchi, N. et al. (2020). Structural analysis of sucrose-derived hard carbon and correlation with the electrochemical properties for lithium, sodium, and potassium insertion. *Chemistry of Materials* 32: 2961–2977.

38 Yuan, Y., Chen, Z., Yu, H. et al. (2020). Heteroatom-doped carbon-based materials for lithium and sodium ion batteries. *Energy Storage Materials* 32: 65–90.

39 Wahid, M., Puthusseri, D., Gawli, Y. et al. (2018). Hard carbons for sodium-ion battery anodes: synthetic strategies, material properties, and storage mechanisms. *ChemSusChem*, 11 (3): 506–526.

40 Alptekin, H., Au, H., Jensen, A.C. et al. (2020). Sodium storage mechanism investigations through structural changes in hard carbons. *ACS Applied Energy Materials* 3 (10): 9918–9927.

41 Wang, C., Cai, H., Zheng, S. et al. (2024). Pitch-based carbon anode for high-energy and long-life sodium-ion battery. *Electrochimica Acta* 504: 144956.

42 Xie, L., Tang, C., Bi, Z., et al. (2021). Hard carbon anodes for next-generation Li-ion batteries: review and perspective. *Advanced Energy Materials* 11 (38): 2101650.

43 Murugadoss, G., Rajaboopathi, M., Kumar, M.R. et al. (2022). Biomass-derived carbon-based materials for supercapacitor applications. *Biomass-Derived Carbon Materials* 253–274.

44 Simone, V., Boulineau, A., de Geyer, A. et al. (2016). Hard carbon derived from cellulose as anode for sodium ion batteries: dependence of electrochemical properties on structure. *Journal of Energy Chemistry* 25 (5): 761–768.

45 Lin, X., Liu, Y., Tan, H. et al. (2020). Advanced lignin-derived hard carbon for Na-ion batteries and a comparison with Li and K ion storage. *Carbon* 157: 316–323.

46 Wang, H., Shao, Y., Mei, S. et al. (2020). Polymer-derived heteroatom-doped porous carbon materials. *Chemical Reviews* 120 (17): 9363–9419.

47 Rao, X., Lou, Y., Chen, J. et al. (2020). Polyacrylonitrile hard carbon as anode of high rate capability for lithium ion batteries. *Frontiers in Energy Research* 8 (3): 515176.

48 Patnaik, K., Peng, Y., Badam, R. et al. (2021). Bio-derived lithium-ion battery anode material for fast charging and long-cycle life. *Electrochemical Society Meeting Abstracts* MA2021-01 (2): 167.

49 Patnaik, K.S., Mantripragada, B.S., Badam, R. et al. (2024). Bio-based poly(benzimidazole-co-amide)-derived N, O co-doped carbons as fast-charging anodes for lithium-ion batteries. *Nanoscale Advances* 6 (20): 5181–5192.

50 Mantripragada, B.S., Patnaik, M.K.S., Higashimine, K. et al. (2024). Energy efficient and fast charging nitrogen doped carbon anodes derived from BIAN-melamine based porous organic polymer for lithium-ion batteries. *Batteries & Supercaps* 7 (4): e202300515.

51 Nandan, R., Takamori, N., Higashimine, K. et al. (2022). Black glasses grafted micron silicon: a resilient anode material for high-performance lithium-ion batteries. *Journal of Materials Chemistry A* 10 (30): 15960–15974.

52 Nandan, R., Takamori, N., Higashimine, K. et al. (2022). Zinc blende inspired rational design of a β-SiC based resilient anode material for lithium-ion batteries. *Journal of Materials Chemistry A* 10 (10): 5230–5243.

53 Aurbach, D., Gamolsky, K., Markovsky, B. et al. (2000). The study of surface phenomena related to electrochemical lithium intercalation into Li x MO y host materials (M Ni, Mn). *Journal of The Electrochemical Society* 147 (4): 1322.

54 Shkrob, I.A., Wishart, J.F., and Abraham, D.P. (2015). What makes fluoroethylene carbonate different? *Journal of Physical Chemistry C* 119 (27): 14954–14964.

55 Wang, X., Zheng, X., Liao, Y. et al. (2017). Maintaining structural integrity of 4.5 V lithium cobalt oxide cathode with fumaronitrile as a novel electrolyte additive. *Journal of Power Sources* 338: 108–116.

56 Gauthier, M., Carney, T.J., Grimaud, A. et al. (2015). Electrode-electrolyte interface in li-ion batteries: current understanding and new insights. *Journal of Physical Chemistry Letters* 6 (22): 4653–4672.

57 Stich, M., Göttlinger, M., Kurniawan, M. et al. (2018). Hydrolysis of LiPF6 in carbonate-based electrolytes for lithium-ion batteries and in aqueous media. *Journal of Physical Chemistry C* 122 (16): 8836–8842.

58 Xing, L.Y., Hu, M., Tang, Q. et al. (2012). Improved cyclic performances of LiCoPO 4/C cathode materials for high-cell-potential lithium-ion batteries with thiophene as an electrolyte additive. *Electrochimica Acta* 59: 172–178.

59 Takeuchi, T., Kyuna, T., Morimoto, H. et al. (2011). Influence of surface modification of LiCoO2 by organic compounds on electrochemical and thermal properties of Li/LiCoO2 rechargeable cells. *Journal of Power Sources* 196 (5): 2790–2801.

60 Tong, B., Song, Z., Wan, H. et al. (2021). Sulfur-containing compounds as electrolyte additives for lithium-ion batteries. *InfoMat* 3 (12): 1364–1392.

61 Abe, K., Ushigoe, Y., Yoshitake, H. et al. (2006). Functional electrolytes: novel type additives for cathode materials, providing high cyclability performance. *Journal of power sources* 153 (2): 328–335.

62 Lee, J.N., Han, G.B., Ryou, M.H. et al. (2011). N-(triphenylphosphoranylidene) aniline as a novel electrolyte additive for high voltage LiCoO2 operations in lithium ion batteries. *Electrochimica Acta* 56 (14): 5195–5200.

63 Michan, A.L., Parimalam, B.S., Leskes, M. et al. (2016). Fluoroethylene carbonate and vinylene carbonate reduction: understanding lithium-ion battery electrolyte additives and solid electrolyte interphase formation. *Chemistry of Materials* 28 (22): 8149–8159.

64 Gupta, A., Badam, R., Takamori, N. et al. (2022). Microbial pyrazine diamine is a novel electrolyte additive that shields high-voltage $LiNi_{0.33}Co_{0.33}Mn_{0.33}O_2$ cathodes. *Scientific Reports* 12 (1): 19888.

65 Patnaik, S.G., Vedarajan, R., and Matsumi, N. (2019). Rational design of a BIAN-based multi-functional additive for higher durability and performance of high-voltage $LiNi_{0.33}Co_{0.33}Mn_{0.33}O_2$ cathodes. *Molecular Systems Design & Engineering* 4 (4): 939–950.

66 Badam, R., Shibuya, M., Mantripragada, B.S. et al. (2022). BIAN-based durable polymer metal complex as a cathode material for Li–O2 battery applications. *Polymer Journal* 54 (11): 1355–1366.

67 Zhang, J., Zhao, Z., Xia, Z. et al. (2015). A metal-free bifunctional electrocatalyst for oxygen reduction and oxygen evolution reactions. *Nature Nanotechnology* 10 (5): 444–452.

68 Ma, R., Lin, G., Zhou, Y., et al. (2019). A review of oxygen reduction mechanisms for metal-free carbon-based electrocatalysts. *npj Computational Materials* 5 (1): 78.

69 Xu, P., Chen, W., Wang, Q. et al. (2015). Effects of transition metal precursors (Co, Fe, Cu, Mn, or Ni) on pyrolyzed carbon supported metal-aminopyrine electrocatalysts for oxygen reduction reaction. *RSC Advances* 5 (8): 6195–6206.

10

Conjugated Polymers and Their Applications in Organic Field Transistors, Organic Electrochemical Transistors, and Organic Thermoelectrics

Yifei He[1,†], Isha Sanskriti[2,†], Nivedha Velmurugan[2,†], Tom Wilfling[2,†], and Christine K. Luscombe[2]

[1]*Materials Science and Engineering Department, University of Washington, Seattle, WA, USA*
[2]*pi-Conjugated Polymers Unit, Okinawa Institute of Science and Technology, Onna-son, Okinawa, Japan*

[†]Joint first authors

10.1 Introduction

Conjugated polymers (CPs), with backbones of alternating single and multiple bonds, exhibit π-delocalization along the carbon chain. This property enables electric and optical performance, driving innovations in organic and optoelectronic devices such as light-emitting diodes (LEDs), switches, solar cells, and sensors. Manipulating the basic molecular structure, i.e. the π-conjugated backbone and the side chains, tunes its optical, electrical, mechanical, and thermal properties. This tunability distinguishes them from traditional inorganic semiconductors, offering advantages such as solution processability, mechanical flexibility, and the potential for low-cost, large-area fabrication. In this chapter, we focus on the fundamentals of what makes a polymer semiconducting and then introduce applications of CPs that have not been covered in other chapters.

10.1.1 What Makes the CP Semiconducting?

The energy density gap, or bandgap (BG), is the difference between the conduction and valence bands (CB and VB, respectively). These bands are defined by the molecular orbitals that arise from the bonding and anti-bonding states of the carbon backbone and are often correlated to the highest occupied molecular orbital (HOMO) for the VB edge and the lowest occupied molecular orbital (LUMO) for the

Conjugated Polymer Synthesis and Materials Applications, First Edition. Edited by Yoshiki Chujo.
© 2026 Wiley-VCH GmbH. All rights reserved, including rights for text and data mining and training of artificial intelligence technologies or similar technologies. Published 2026 by Wiley-VCH GmbH.

CB edge. The BG provides indication on the semiconducting behavior by controlling how easily electrons can be excited above the Fermi level. This excitation generates charge carriers, enabling electrical conductivity and making the material responsive to thermal or optical stimuli.

In a simple *trans*-polyacetylene (*trans*-PA) molecule, the electronic structure renders complete delocalization (dispersed state), which expects a metallic character, in other words, a conductor with zero BG. However, this state is thermodynamically unstable, and Peierls' distortion occurs, rendering a BG of approximately 1.8 eV [1], making the polymer semiconducting (Figure 10.1). Typically, materials with BGs of 0.5–3.6 eV exhibit the properties of semiconductors. Peierls' transition is the conversion of the ideal into a real one-dimensional (1D) crystal lattice [2]. Peierls' transition or the inherent localization, along with other defects such as insufficient inter-chain transfer and charge carrier recombination, leads to low conductivity.

To overcome the low electrical conductivity, CP films are often

- oxidized (p-type) or reduced (n-type) to introduce equilibrium charge carriers on the π -conjugated backbone,
- injected with counterions to maintain the overall charge neutrality.

This process of incorporating foreign atoms, molecules, and ions into the lattice/matrix of the conjugated semiconductor is called doping.

10.1.2 What Are the Charge Carriers?

Charge carriers are the particles/quasi-particles responsible for charge transport [3]. The free movement of electrons and holes – the primary charge carriers – influences the conductivities of traditional inorganic semiconductors. Likewise, doped CPs also generate charge carriers, but their movement differs significantly from that in doped inorganic semiconductors, exhibiting non-linear and non-resonant dynamics. Charge transport in polymers is linked to excitations such as solitons, polarons, and bipolarons, which are generated as a result of chain deformations when charges are added or removed. Based on the reduction of energy by lattice distortion, the energy levels of the charge carriers from high to low can be written as soliton > polaron > ½ bipolaron [4].

10.1.3 Formation of Charge Carriers

A soliton is a localized charge defect, and its formation is well-studied in the case of systems with degenerate ground states, such as *trans*-PA. A neutral soliton naturally occurs in *trans*-PA when there is an odd number of conjugated carbons in a polymer chain, warranting an unpaired π-electron or a radical [5]. It is neutral and possesses

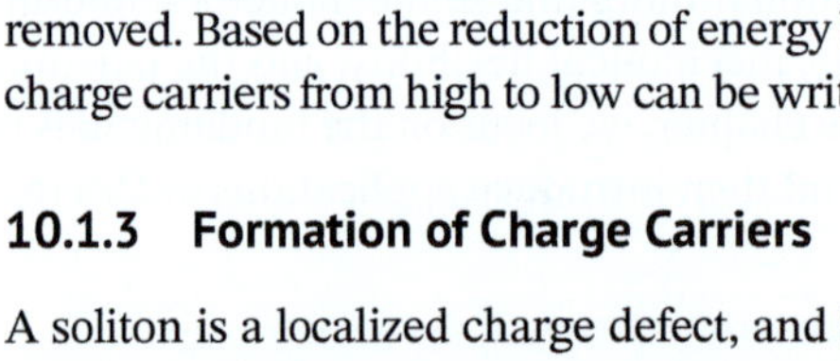

Figure 10.1 Schematic representation of Peirels' distortion on polyacetylene molecule.

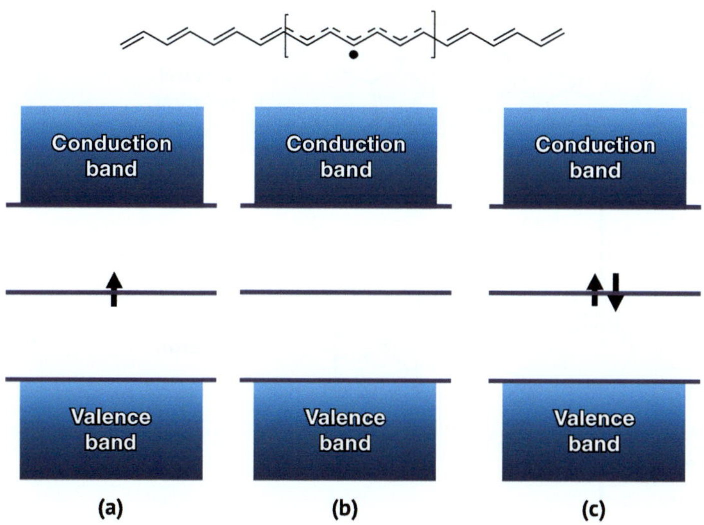

Figure 10.2 Top: Schematic representation of a neutral soliton on a *trans*-polyacetylene chain. Bottom: Band structure of *trans*-polyacetylene containing (a) neutral soliton, (b) a positively charged soliton, and (c) negatively charged soliton.

a half spin, and the nonbonding orbital associated with this soliton occupies a level in the middle of the BG (Figure 10.2). The soliton can move along the polymer backbone without causing deformation or dissipation, and the chain remains electrically neutral. However, this movement does not lead to macroscopic-scale charge transport.

Solitons represent low-energy excitations that can propagate through the polymer chain. According to the Su–Schrieffer–Heeger (SSH) model, the energy required to create a soliton in a polymer chain is less than half of the chemical potential, which is equal to the single particle gap [6]. This means that solitons can exist in the polymer chains after doping, as they require less energy to form compared to other charge carriers like electrons or holes. A soliton's spin density is not localized on a single carbon atom but rather dispersed on several carbons in the chain, giving it a width of seven times that of the in-chain lattice constant and an effective mass of six times that of an electron ($6m_e$) [6]. This intriguing spin–charge-separated species in PA, which could be generated by photoexcitation, is yet to find practical application in terms of optoelectronic devices.

While solitons are studied for degenerate ground state systems, polarons and bipolarons are the charge carriers for non-degenerate systems such as polythiophene (PT) and poly(*p*-phenylene vinylene) (PPV). In a classical p-type CP, such as poly(3-hexylthiophene) (P3HT), a dopant oxidizes the polymer backbone, removing one electron and locally changing the structure from aromatic to quinoidal to stabilize the polymer. Dopants with high electron affinity, such as iodine, antimony pentafluoride ($SbCl_5$), or 2,3,5,6-tetrafluoro-7,7,8,8-tetracyanoquinodi methane (F4-TCNQ), are commonly used in this process. The oxidation leaves behind an unpaired electron and a hole, generating a radical cation, as shown in Figure 10.3. This quasi-particle, along with the associated lattice deformation, is called "polaron." The polaron, now behaving like a single localized species, along

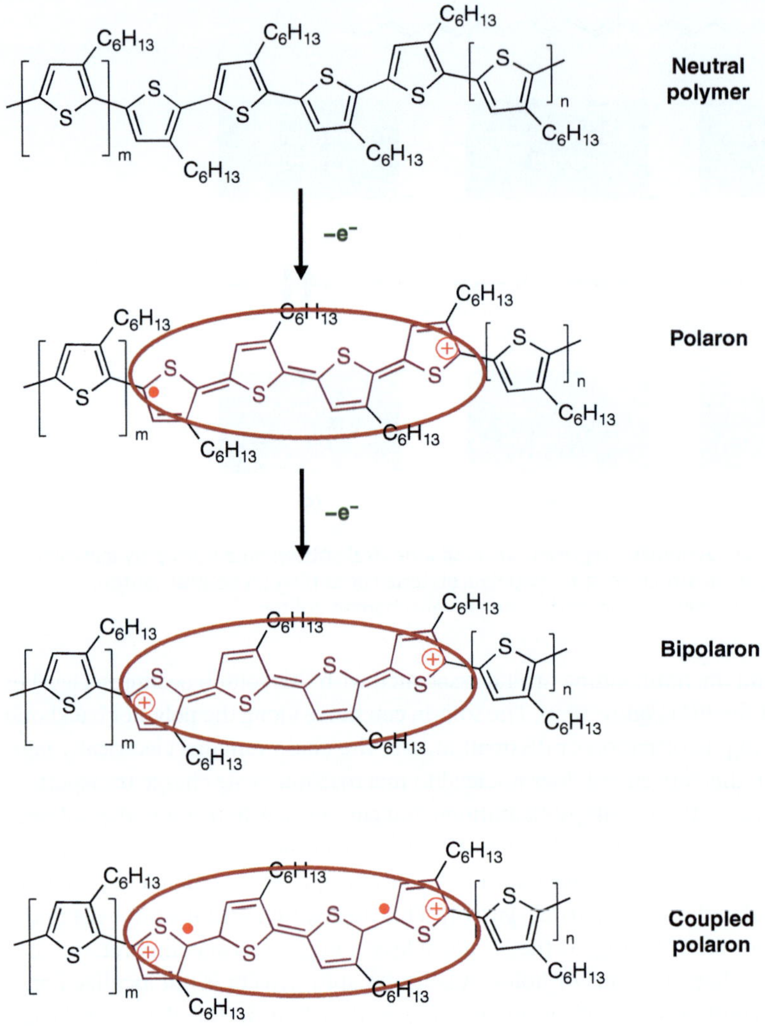

Figure 10.3 Schematic representation of the formation of charge carriers on doping P3HT.

with the structural changes, moves along the carbon chain, enabling the conduction of electric current.

Theoretically, a positive polaron has a spin of ½ with a charge e and uses an ionization potential IE^0. The hole creates a singly occupied energy level just above the VB edge, along with a similarly localized empty energy level that shifts downward from the CB edge into the polymer's fundamental energy gap. Conversely, a negative polaron is formed when an electron is added to the CB, gaining an energy of electron affinity EA^0. Similar to the positive polaron, the excess electron now locally rearranges its structure to stabilize the polymer chain. This stabilization leads to the formation of a singly occupied energy level slightly below the CB edge, along with a localized but now doubly occupied energy level that shifts upward from the VB edge because of the lattice deformation.

When the dopant has a strong oxidative potential, or, like in most cases, has a high concentration, it has the ability to remove another electron from the formed polaron; in other words, a second oxidation occurs to remove the radical. The resultant species called "bipolaron" is a spinless species with a double positive charge (2e), entirely devoid of radical behavior on the polymer backbone (Figure 10.3).

It is important to note that in the case of polarons, the conduction involves the transport of both charge and spin, whereas the bipolaron conduction carries only the charge. A bipolaron can also be formed when two polarons are paired up. In such a case, it is called a coupled polaron or a triplet bipolaron and has an open-shell structure.

10.1.4 Band Structure

An undoped CP shows a single absorption corresponding to the π–π* transition in the visible region (Figure 10.4a). When an electron is removed from the VB of the polymer, a positive polaron is formed, also creating two localized electronic states, $+\varepsilon$ and $-\varepsilon$, within the BG. The $+\varepsilon$ level is vacant, and the $-\varepsilon$ level is occupied with one electron. This also creates two newly allowed optical transitions, P1 and P2, where P1 is a low-energy transition. In most doped CPs, the low-energy P1 is absorbed in the mid-infrared (mid-IR) region, near 0.5 eV, while the higher energy P2 occurs an eV below the BG in the edge between the near-infrared (NIR) and visible region (Figure 10.4b).

A polaron is typically localized. However, due to the extended conjugation and the overlap of π-orbitals between adjacent polymer chains, the polaron can spread out or delocalize over neighboring chains, especially under certain conditions like increased temperature or external fields. The delocalization facilitates charge transport, and it also can hop in between the chains and conduct. In such cases, additional peaks may appear. During the singlet bipolaron formation, the two intra-gap states move further into the BG and away from the band edges, forming an optical transition BP1. BP1 is bluer than the P1 transition of the single polaron and typically

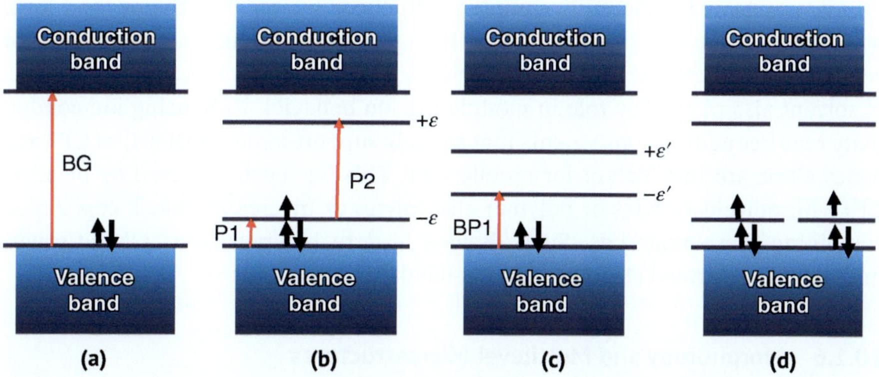

Figure 10.4 Schematic representation of energy level diagrams of (a) neutral polymer, (b) positive polaron, (c) singlet bipolaron, and (d) coupled polaron (triplet bipolaron) (BG = bandgap).

occurs between 0.8 and 1.1 eV (Figure 10.4c). At high doping densities, two polarons may not spin-pair; instead of behaving as a singlet bipolaron, they strongly interact, forming a coupled polaron (triplet bipolaron) (Figure 10.4d).

In steady-state spectroscopy, the peaks of various species often resemble each other and may overlap when multiple species are present at the same time, complicating the identification of each species. Time-resolved spectroscopy methods have recently been used to achieve high resolution of these overlapping peaks. Ultrafast transient absorption spectroscopy is also utilized to investigate the charge carriers.

10.1.5 Ion Transport

For many years, only electronic transport was considered for organic electronics. However, in the late 1990s and early 2000s, there was a growing interest in ionic transport, where the main charge carriers were ions. Organic materials, mainly CPs that conduct both ions and electrons, are very attractive for various bioelectronic applications due to their biodegradability and are called organic mixed ionic–electronic conductors (OMIECs). The most studied OMIEC poly (3,4-ethylenedioxythiophene): poly (styrenesulfonate) PEDOT:PSS has an electronically conductive PEDOT and an insulating PSS. PEDOT:PSS facilitates ionic transport through its polyelectrolyte component, PSS, while also enabling charge polarization between PEDOT (positively charged) and PSS (negatively charged).

Understanding the electrically insulating polyelectrolytes and polymer electrolytes is important to study the ionic transport of OMIECs. To conduct ions, a polymer must be able to dissolve salts. Poor solvation in mixed ionic–electronic conductors (MIECs) leads to ion injection and transport barriers, weakening ionic–electronic coupling. In protic electrolytes, hydrogen bonding aids ion solvation, while aprotic systems rely on dipole-functional groups like ether oxygens. Solid polymer electrolytes often use repeating ether groups to solvate cations, leaving anions unsolvated, or they function as ionomers or polyelectrolytes, incorporating ionic charges within the polymer backbone.

Ionic transport is inherently complex, influenced by ion size, charge, and form, with larger ions migrating more slowly through polymers. Multivalent ions and solvent choice further affect transport efficiency. The solvation process and the choice of solvent also play a key role in modulating ion behavior. Enhancing ion conductivity requires adding components that actively support ionic transport, as CP backbones alone are insufficient for mobile ions. This has been achieved by blending CPs with polyelectrolytes or polymer electrolytes or by creating block copolymers combining these materials. Such designs lead to heterogeneous OMIECs with microphase-separated electron- and ion-conducting domains.

10.1.6 Morphology and Multilevel Microstructures

In polymer systems, morphology and microstructure extend across multiple length scales. Morphology, observed at mm to nm scales, can be studied using techniques such as optical microscopy (OM), atomic force microscopy (AFM), scanning electron

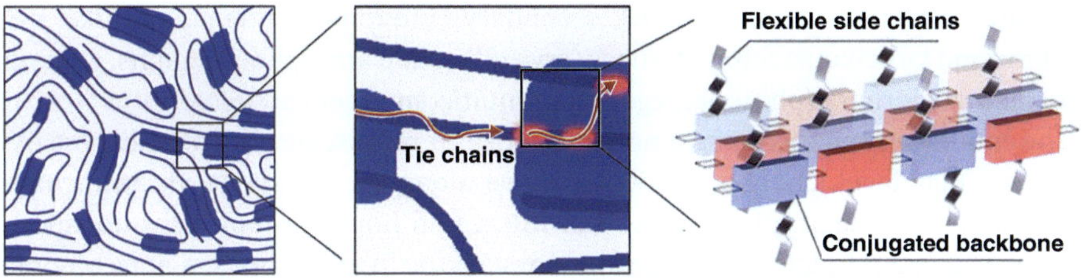

Figure 10.5 Schematic representation of microstructure and charge transport. *Source*: Adapted from [7].

microscopy (SEM), and transition electron microscopy (TEM). Microscopically, CPs exhibit both crystalline and amorphous regions, with domain sizes ranging from μm to nm, and are semicrystalline in nature. The ordered crystalline regions facilitate robust charge transport and structural strength, while the amorphous regions enhance flexibility and strain-hardening properties. This interplay and the synergy between order and disorder are key to their performance in solid-state functional devices (Figure 10.5). The adjacent ordered crystalline regions are connected by long polymer chains called tie chains to enable enhanced charge transport, as implicated by many studies.

The morphology and microstructure in solid-state thin films are complex due to factors such as:

- Uncontrolled molecular weight distribution
- Interaction of alkyl side chains with the conjugated main chain
- π–π interactions
- Side chain interactions
- Chain folding
- Entanglement
- Complex inter- and intramolecular reactions

CPs are usually solution-processed and subjected to growth methods to form thin films. In solution, the polymers are prone to form interchain aggregates that reflect on the solid-state film, affecting the microstructure, charge transport, and device performance. Poorly packed aggregates in solution can transfer to films, impeding charge transport, while controlled aggregation can enhance thin-film microstructures and improve transport properties. Common characterization techniques include ultraviolet-visible spectroscopy (absorption and fluorescence), light scattering methods (dynamic and static), small-angle scattering techniques (SAXS and SANS), and freeze-drying microscopy.

The concept of multilevel microstructure was introduced to understand the complex interactions of CPs. The complexities result in diverse microstructures, ranging from amorphous to highly crystalline, often spanning multiple scales – from molecular (Å–nm) to device-level (nm–mm). The aggregation structures are influenced by the backbone conformation, microscopic packing, and crystallization, affecting optoelectronic properties and device performance. Pei conceptualized the multilevel microstructures of CPs very similar to protein structures: primary (chemical structure and chain conformation), secondary (interchain interactions like π–π stacking and lamellar packing), tertiary (crystalline and amorphous regions, domain size), and quaternary (phase segregation in multicomponent systems).

The multilevel microstructure now gives rise to multiscale charge transport. At the molecular level, charges efficiently move along the backbone via intrachain coupling. On the aggregate scale, slower interchain hopping dominates transport. However, at macroscopic scales, disrupted π-stacking in amorphous regions hinders transport unless ordered domains are well-connected. These multiscale processes critically influence device performance.

10.2 Polymers for Organic Field Effect Transistors

10.2.1 Introduction

Organic field effect transistors (OFETs) are a class of thin film transistors (TFTs) based on organic semiconductors as channel material. Since the first report of such a device by Tsumura et al. almost four decades ago, this transistor type has gained significant attention in recent years due to its potential in flexible, printable, and biocompatible electronics and is the most widely used transistor type in the context of organic electrons [8, 9]. This section will discuss the working principle and design of OFETs and review relevant developments in OFET CP design.

10.2.2 Example p-type Polymers

10.2.2.1 Thiophene-based Polymers

Early CPs were mainly based on conjugated homopolymers. The first example of a p-type OFET using an organic semiconducting layer as a channel material was reported in 1987 by Tsumura et al. This first device utilized an electrochemically polymerized PT layer and reached a charge carrier mobility of $2'10^{-5}\ cm^2\ V^{-1}\ s^{-1}$ [8].

In a later study by Bao et al. in 1996, a regioregular P3HT film was used as channel material (Figure 10.6) [10]. Modifying the previously established PT with an alkyl chain showed improved charge carrier mobilities of up to $4.5'10^{-2}\ cm^2\ V^{-1}\ s^{-1}$.

10.2.2.2 Indacenodithiophene-based Polymers

The first report of an indacenodithiophene-based (IDT) polymer as a semiconducting layer in an OFET came from Zang et al. in 2010 [11]. IDT copolymers with benzothiadiazole (BT) and thieno[3,2-b]thiophene (TT) were synthesized, and their charge carrier mobilities were evaluated in a top gate, bottom contact (TGBC) device (Figure 10.7). Interestingly, PIDT-BT displayed high mobilities (1.0 $cm^2\ V^{-1}\ s^{-1}$)

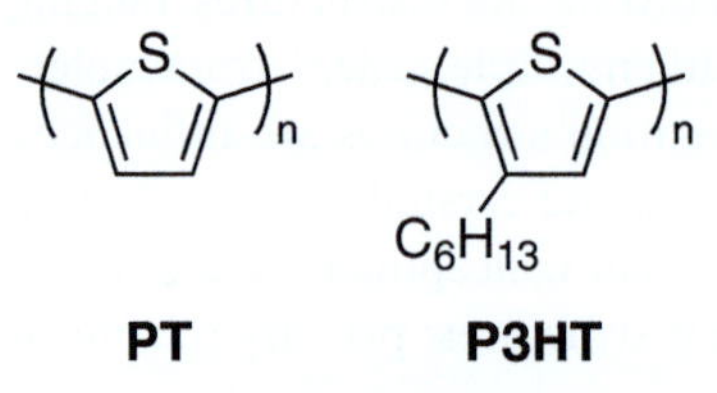

Figure 10.6 Chemical structure of thiophene-based p-type polymers.

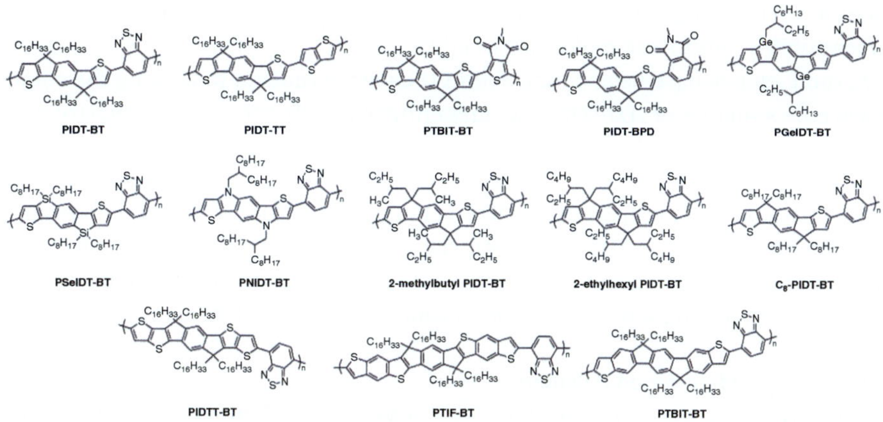

Figure 10.7 Chemical structures of IDT-based p-type polymers.

while lacking the crystallinity commonly associated with high charge mobilities. Later, Zhang et al. increased the mobility of the PIDT-BT polymer to an exceptional 3.6 $cm^2 V^{-1} s^{-1}$ in a TGBC device by increasing the polymer's molecular weight [12]. In the same study, the group aimed to elucidate the origin of this unusual high-charge mobility. The IDT-BT backbone was found to possess a low degree of backbone torsion, which led to quasi-1D-charge transport along the polymer backbone, only requiring occasional intermolecular π-stacking to maintain high charge carrier mobility. These findings were confirmed in follow-up studies using IDT-BT and other semicrystalline polymers [13, 14].

Following these findings, IDT research has gained significant interest, focusing on using different electron-deficient acceptor units and modifying the IDT scaffold itself. In a study in 2018, Li et al. designed three D–A polymers using BT, thienopyrroledione (TPD), and benzopyrrolodione (BPD) as acceptor units to make the copolymers PIDT-BT, PIDT-TPD, and PIDT-BPD (Figure 10.7) [15]. It was found that exchanging the BT unit for TPD helped increase the polymer backbone's planarity further due to interactions of the keto-oxygen of the TPD unit with the sulfur of the IDT thiophene. This leads to higher charge carrier mobilities when compared to PIDT-BT at similar molecular weights (0.20 $cm^2 V^{-1} s^{-1}$ vs. 0.17 $cm^2 V^{-1} s^{-1}$, respectively).

Modifications to the IDT scaffold are commonly approached in three ways: (1) by replacing the bridging carbon, (2) by changing the length and structure of the side chains, or (3) by extending the conjugated structure. Fei et al. and McCulloch et al. in 2012 synthesized IDT variants in which the carbon was replaced with germanium (PGeIDT-BT), silicon (PSeIDT-BT), and nitrogen (PNIDT-BT) (Figure 10.7) [16–18]. While this strategy proved effective in modifying the HOMO levels of the resulting copolymers, charge carrier mobilities were below what was achieved with the standard IDT monomers. Bronstein et al. conducted a study in 2011 comparing the influence of different branched (2-methylbutyl PIDT-BT and 2-ethylhexyl PIDT-BT) and straight (PIDT-BT and C_8-IDT-BT) alkyl side chains on the bridging atom of the conjugated IDT scaffold on polymer performance (Figure 10.7) [19]. Long alkyl side chains led to higher hole mobilities thanks to improved solubility and solution processability, improving molecular weights.

Several studies have been conducted in the past decade on modifying the conjugated core structure of IDT. This includes either extending the conjugated structure by, e.g. adding two additional thiophenes (PIDTT-BT) or thienobenzyl (PTIF-BT) groups to each side of the IDT scaffold, as well as replacing the flanking thiophenes with thienobenzyl groups (PTBIT-BT) (Figure 10.7) [20–22]. These copolymers based on these modified IDT monomers showed promising results with high charge mobilities of up to 2.8 $cm^2\ V^{-1}\ s^{-1}$ (PTIF-BT) and higher on currents compared to PIDT-BT, making these modifications a promising motif for future research.

10.2.2.3 Isoindigo-based Polymers

Its first report as an active layer in OFETs was in 2011 by Lei et al. [23]. Two isoindigo-based (IID) polymers were synthesized through Stille coupling with either thiophene (PIID-T) or bithiophene (PIID-2T) as comonomers (Figure 10.8). After modification with two branched 2-octyldecyl side chains, bottom gate, top contact (BGTC) devices were fabricated and measured hole mobilities of up to 0.02 and 0.79 $cm^2\ V^{-1}\ s^{-1}$, respectively. Following this study, more research was conducted to further increase the charge carrier mobility of IID polymers up to 8.06 $cm^2\ V^{-1}\ s^{-1}$ with high on/off ratios of up to 10^6 [24]. In 2016, Wu et al. achieved these mobilities by replacing the alkyl side chains with carbosilane side chains (PIID2T-C8) (Figure 10.8). The resulting IID monomer was polymerized using microwave-assisted Stille coupling with 2T, and the resulting polymer was tested in a BGTC device. The high charge carrier mobility was attributed to the high number-average molecular weight of the polymer ($M_n = 434$ kg mol^{-1}) and the strong ordering effect of the carbosilane side chains in the solid state.

Besides side chain modification, several studies have investigated the effect of other structural modifications such as halogenation of the IID core, substituting the benzene with other heterocycles, extension of the IID core structure, and exchange

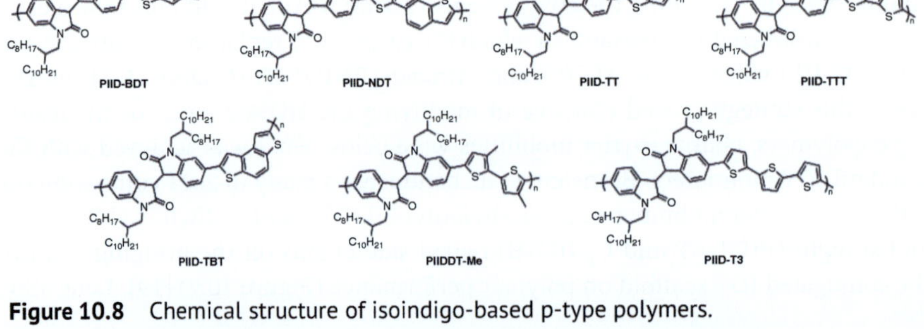

Figure 10.8 Chemical structure of isoindigo-based p-type polymers.

of donor units. However, many structurally modified IID polymers display ambipolar charge transport and will be discussed in detail in Section 10.2.4.

In 2012, Lei et al. performed a comprehensive study using 10 different selenophene and thiophene donor units to synthesize the semiconducting copolymers PIID-T, PIID-2T, PIID-Se, PIID-BDT, PIID-NDT, PIID-TT, PIID-TTT, PIID-TBT, PIIDDT-Me, and PIID-3T (Figure 10.8) [25]. They utilized a "molecular docking" strategy, where small conjugated units (donor units in this case) can fit in the cavity between larger aromatic units and their side chains to increase the interchain π–π stacking and improve performance. It was shown that by using this strategy, they could increase charge carrier mobility to up to 1.06 $cm^2\ V^{-1}\ s^{-1}$ in IIDDT.

10.2.2.4 Cyclopentadithiophene-based Polymers

The first report on the use of cyclopentadithiophene-based (CDT) polymers in OFET devices dates back to 2007 [26]. Zhang et al. synthesized a PCDT-BT copolymer and tested its performance in TGBC OFET, where they found hole mobilities of up to 0.17 $cm^2\ V^{-1}\ s^{-1}$ (Figure 10.9). The performance of CDT-containing polymers has steadily increased throughout the next decade, reaching up to 36.3 $cm^2\ V^{-1}\ s^{-1}$ for PCDT-PT (Figure 10.9) [27]. These impressive values were achieved using capillary action to control the polymer nanostructure and ensure unidirectional alignment.

10.2.3 Example n-type Polymers

While organic p-type semiconductors were historically more prevalent, advances in material design in the last decade have led to a vast number of n-type semiconductors readily available for use in OFETs. N-type organic CPs have been more challenging to realize due to the higher susceptibility of organic anions to atmospheric oxidants and the incompatibility in the work function of most metals used to make the electrodes (Au, Ag, other precious metals) [28, 29].

One of the earliest examples of a n-type CP used in OFET was poly(benzobis imidazobenzophenanthroline) (BBL) in 2002 (Figure 10.10) [30]. The fabricated OFET device showed an electron mobility of up to $5'10^{-4}\ cm^2\ V^{-1}\ s^{-1}$. Lei et al. in 2013 reported a modified PPV polymer, BDPPV, as the first case of an n-type-based OFET reaching electron mobilities of above 1 $cm^2\ V^{-1}\ s^{-1}$ under ambient conditions (Figure 10.10) [31]. These mobilities were achieved by introducing intramolecular hydrogen bonding to reduce conformational disorder, improve interchain interactions, and lower the LUMO level. When tested in a TGBC device, BDPPV

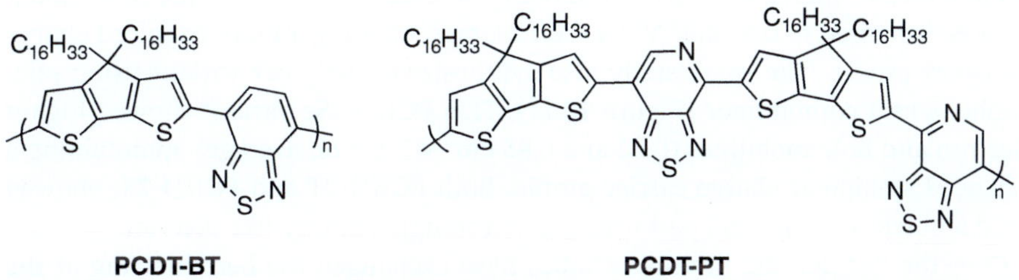

Figure 10.9 Chemical structures of cyclopentadithiophene-based p-type polymers.

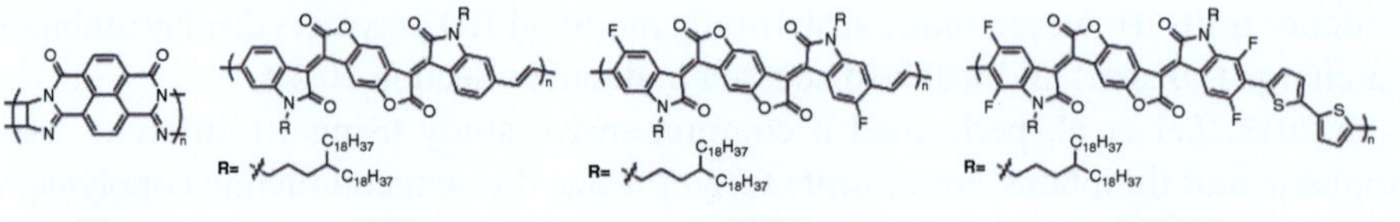

Figure 10.10 Chemical structures of n-type polymers.

could reach electron mobilities of up to 1.1 $cm^2 V^{-1} s^{-1}$. In 2014, Lei et al. further increased the performance of BDPPV by introducing fluorine to the core structure (FBDPPV) (Figure 10.10) [32]. This led to electron mobilities of up to 1.7 $cm^2 V^{-1} s^{-1}$ when tested in a TGBC device due to the formation of intramolecular hydrogen bonds, further planarizing the backbone.

The current highest-performing n-type is a tetrafluorinated benzodifurandione-based oligo(p-phenylene vinylene) polymer with 2T as a comonomer (F_4BDPPV-2T) (Figure 10.10) [33]. Zheng et al. measured electron mobilities of up to 14.9 $cm^2 V^{-1} s^{-1}$ in a TGBC device. The high charge carrier mobility was attributed to the improved planarizing effect of the interactions between the fluorine of F_4BDOPV and the sulfur of the 2T comonomer.

10.2.4 Example Ambipolar Polymers

Ambipolar materials bring several advantages over unipolar materials for use in integrated circuits by removing the need for advanced patterning techniques, simplifying circuit fabrication, lowering power consumption, increasing noise margins, and higher robustness [34].

10.2.4.1 Isoindigo-based Polymers

As mentioned in Section 10.2.2.3, many structurally modified IID polymers display effective ambipolar charge transfer. The first report of ambipolar charge transfer in IID-containing CPs was a study by Lei et al. in 2012 [35]. Here, an IID monomer was fluorinated in the 7 and 7′ positions and polymerized with 2T using Stille coupling. The resulting copolymer PFIID-2T was tested using a TGBC device (Figure 10.11). It was shown that adding fluorine led electron mobility of 0.38 $cm^2 V^{-1} s^{-1}$, while hole mobilities were found to be 1.41 $cm^2 V^{-1} s^{-1}$ on average, leading to an unbalanced charge carrier profile. In a follow-up study, the same copolymer was chlorinated in the 5 and 5′ positions to form PCIID-2T (Figure 10.11) [36]. This led to a further increase in electron mobility to 0.42 $cm^2 V^{-1} s^{-1}$ while hole mobility was decreased to 0.51 $cm^2 V^{-1} s^{-1}$ on average, creating a more balanced charge transport profile. The same study also synthesized a polymer variant using selenophene as a comonomer (Figure 10.11). This PCIID-2Se variant showed higher electron and hole mobilities (0.62 and 0.85 $cm^2 V^{-1} s^{-1}$ on average), maintaining a balanced ambipolar charge carrier profile. Both PCIID-2T and PCIID-2Se showed good ambient stability and performance in a complementary-like inverter.

Over the last decade, multiple studies have exchanged the benzene ring of the IID unit with heterocyclic aromatic rings. One of the first was reported by Ashraf

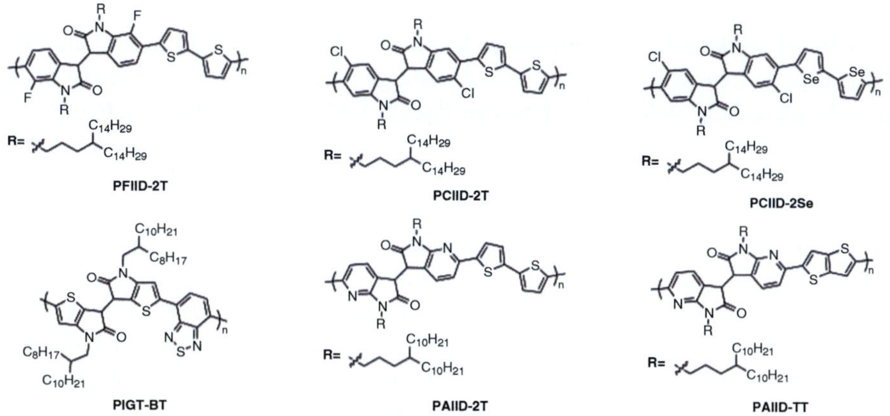

Figure 10.11 Chemical structure of ambipolar polymers.

et al. in 2012, where benzene was substituted with thiophene to synthesize a dithienoisoindigo (IGT) [37]. The monomer was polymerized with BT using Stille coupling and tested in a TGBC device. PIGT-BT showed highly balanced ambipolar charge carrier transport with hole mobilities of 0.1 and electron mobilities of 0.14 $cm^2\ V^{-1}\ s^{-1}$ (Figure 10.11). Their high mobilities can be explained through the increased planarity of the IGT monomer due to the intramolecular interaction of the thiophene sulfur with the carbonyl oxygen of the lactam ring, increasing the planarity. In 2016, Huang et al. reported ambipolar charge transfer in pyridine-substituted IID copolymers [38]. They exchanged the benzene ring of the IID scaffold with pyridine to create diazaindigo (AIID) monomers with nitrogens in the 7,7′ positions. After polymerized with 2T through Stille coupling, the ambipolar performance of PAIID-2T was tested using a TGBC device (Figure 10.11). This version showed high but unbalanced hole and electron mobilities (1.89 $cm^2\ V^{-1}\ s^{-1}$ and 0.48 $cm^2\ V^{-1}\ s^{-1}$, respectively). A 2017 follow-up study used TT as comonomer. When tested in a TGBC device, this PAIID-TT copolymer showed high, nearly balanced ambipolar charge carrier transport with hole mobilities of 0.78 $cm^2\ V^{-1}\ s^{-1}$ and electron mobilities of 1.12 $cm^2\ V^{-1}\ s^{-1}$ (Figure 10.11).

10.2.5 Key Approaches to Improve OFET Performance

Throughout the last decades, multiple strategies have been adapted to improve the performance of organic CPs. These approaches can be generally classified as either modifications to the polymer backbone, addition or modification of side chains, or changes in the device manufacturing process. One generally effective way to improve charge carrier mobility in organic CPs is to increase the molecular weight of the polymer itself. Higher molecular weights correspond to longer conjugated backbone structures, reducing the need for interchain charge transfer and higher charge carrier mobility.

10.2.5.1 Backbone Modifications

A coplanar backbone is an important feature for charge mobility in organic CPs. A rigid planar backbone structure favors charge carrier transport along the backbone by increasing molecular orbital overlap. Planar backbones also tend to increase coplanarity within the polymer bulk, leading to a more efficient intermolecular overlap. Multiple strategies have been implemented over the years to increase the planarity of the backbone structure.

Ladder-style polymers consist of a continuous sequence of adjacent aromatic ring structures fused by two or more atoms. While this strategy creates a fully π-conjugated, highly planar, and rigid backbone structure, this usually leads to poor solubility, reducing molecular weights and limiting their charge carrier transport. Another strategy to increase planarity is the introduction of double bond linkers. Double bonds reduce the rotational angles between (co)monomers and are often realized through aldol condensation polymerization. Conformationally locked CPs use non-covalent interactions between the comonomers to lock the backbone conformation. They are similar to ladder-style polymers; however, the lack of covalent bonds allows for some flexibility. This usually improves solubility, leading to better solution processibility, and higher molecular weights compared to the more rigid ladder-style polymers.

10.2.5.2 Side Chain Modifications

Implanting various side chains can help the CP's solubility and influence the polymer film's microstructure [39, 40]. In general, long, straight alkyl side chains, e.g. hexadecyl, are preferred, but some polymers can benefit from the bulkiness provided by branched side chains. Many modern high-performance CPs implement a combination of these design principles. PIDT-BT, for example, uses an extended ladder-style monomer (IDT), which is conformationally locked to its comonomer (BT) and is modified with long alkyl side chains.

10.2.5.3 Device Architecture

Besides modifications to the polymer itself, adjusting the device architecture impacts the device performance as well. This includes changes to the device architecture as well as changes to the dielectric and electrode materials.

10.2.6 Application and Future Outlook

With the rising interest in flexible and wearable electronics, OFETs present promising devices thanks to their mechanical properties, solution processability, and low-temperature fabrication. However, while a lot of research has been done to improve the performance and properties of organic polymers for OFET applications, current materials present some major weaknesses, which must be overcome to make them viable alternatives to classic inorganic semiconductors. Many current high-performing polymers for OFET applications are oxygen- and/or light-sensitive and can degrade during day-to-day use due to environmental factors or after many on/off cycles. Additionally, they often possess complex chemical structures, making their synthesis procedures complex. This limits their up-scalability and can lead to non-uniformity between batches. These elaborate synthesis paths also require large amounts of organic solvents and other environmentally impactful chemicals. Lastly, more research into ambipolar OFETs is needed to make more complex circuitry such as inverters.

While the mobilities of organic semiconducting materials have steadily improved over the years to a point where they can compete with inorganic transistors, their adaptability for large-scale applications is currently still limited. However, their feasibility has been proven in multiple studies over the last decades, showing great potential for future applications for flexible devices, like screens and skin-like electronics, printable electronics, and other biocompatible devices.

10.3 Organic Electrochemical Transistors

10.3.1 Design Principle

The development of organic electrochemical transistor (OECT) channel materials has advanced significantly for the past two decades, due to the improved understanding of the structure–property relationship and the synthesis methods that help achieve the material design. Similar to the requirement of other organic TFTs, the performance of OECT partially relies on the high conductivity of the channel material. In other words, previous design strategies and CP library for other organic electronics could serve as an adequate starting point for OECT. Hence, approaches including improved backbone planarity and energetic engineering are correspondingly applied and have promoted better OECT material design. On the other hand, channel materials of OECT are responsible for the volumetric doping of ions from an aqueous medium, which subsequently requires the improvement of ionic conduction while maintaining the charge mobility and device stability. Thus, the design of novel OECT materials demands additional techniques to finely couple the mixed conduction.

10.3.2 Example Polymers

10.3.2.1 P-type

10.3.2.1.1 PEDOT:PSS

Being the most applied p-type material in OECT, PEDOT:PSS has multiple advantages, including commercial availability, high performance in mixed conductance, and exceptional transconductance in OECT. In OECTs, PEDOT:PSS exhibits remarkable μC^* around 47 F cm^{-1} V^{-1} s^{-1} in its pristine form [41] and up to 1500 F cm^{-1} V^{-1} s^{-1} with proper processing condition and morphology control [42]. The outstanding device performance of PEDOT:PSS in OECT originates from the combined impact of efficient electronic transport of PEDOT domain and ion migration through PSS domain. However, due to its intrinsically doped state, PEDOT:PSS-based OECT operates in depletion mode in which the device is constantly on its ON state when no bias is applied. Such operation behavior results in higher power consumption compared to its accumulation counterpart and restricts its application in multiple scenarios of complementary circuit and biological sensing. Furthermore, the heterogeneous nature of PEDOT:PSS blend limits systematic chemical modifications for tuning OECT characteristics. This has motivated exploration of synthetic non-doped OECT materials, enabling accumulation mode operation with satisfactory performance and stability.

10.3.2.1.2 Polythiophene-based Polyelectrolyte

Given the high conductivity, high thermal stability [43], and the variety of synthesis routes [44–47] that allow both backbone and side chain modification, PT and its derivatives have thus been the most widely studied model p-type materials for organic electronics. Indeed, the study of polythiophene-based polyelectrolytes is one of the early attempts of synthetic OECT material to assemble the mixed conducting feature of PEDOT:PSS. In 2014, Thelakkat et al. first reported a polythiophene-based polyelectrolyte, poly[6-(thiophen-3-yl)hexane-1-sulfonate tetraalkylammonium]($PTHS^-M^+$), synthesized via Kumada catalyst transfer polymerization (KCTP) [48]. The resulting controlled backbone achieves benchmark hole mobility compared to previously reported polythiophene-based polyelectrolytes [49]. More importantly, OECT device fabricated based on $PTHS^-M^+$ exhibits accumulation mode with normalized transconductance around 0.4 ± 0.02 mS at relatively high gate voltage [50]. When copolymerized with hydrophobic P3HT, $PTHS^-TMA^+$-co-P3HT shows more ordered domains in the thin film, which results in similar volumetric capacitance but significantly improved hole mobility of $1.7 \pm 0.36 \times 10^{-2}\ cm^2\,V^{-1}\,s^{-1}$ and higher transconductance of up to 70.5 S/cm of the fabricated OECT [51]. The introduction of the hydrophobic component also lowers the threshold voltage to −0.15 V, which better improves the stability of the material when in contact with water. The systematic study developed based on PTHS series has shown that the efficient ion transport via the ionic side chain and the potential of the well-controlled backbone to form ordered structures leads to high performance of the device.

10.3.2.1.3 Glycolated Polythiophenes

Following the idea of utilizing the hydration ability of hydrophilic side chains, ethylene glycol (EG) side chain has been integrated onto a variety of PT derivatives and later become one of the most successful strategies for OECT channel material development. Both reported in 2016, glycolated PT p(g2T-TT) and p(g2T-T) demonstrated the first synthetic mixed conductive polymers outperforming PEDOT:PSS [52, 53]. The imparted EG side chains significantly enhance the ionic conduction of the two p-type materials, with volumetric capacitance both exceeding 200 F cm^{-3} [54]. Diverse side chain parameters, including the total chain length [55, 56], the linearity or branching [57], the percentage in the bulk material [58, 59], and finally the engineered position along the backbone [60], have since been optimized on EG side chains on p-type materials and comprehensively reviewed elsewhere [61, 62]. With the improved fundamental understanding of EG side chain impact, the OECT figure-of-merit of glycolated p-type material has drastically progressed for over threefold for the past decade. The benchmark p-type material born during the systematic research era of EG side chain is p(g2T2-g4T2), reported by Moser et al., with the superior μC^* of 522 F $cm^{-1}\,V^{-1}\,s^{-1}$ and high stability [56]. Redistributed EG side chains also impact the electronic transport along the backbone by extending S–O interaction and hence improving the backbone planarity. Over fivefold increment of μC* has been observed for the isomeric PT pgBTTT (502 ± 18 F cm^{-1} V^{-1} s^{-1}) of p(g2T-TT) due to such modification [60].

Two limitations of EG side chain engineering are noteworthy from the previous investigations. First, increasing the EG chain length and amount percentage do not lead to direct scaling of the volumetric capacitance, since excessive water uptake will interfere with the ion stabilization and dilute the stored charge content in the bulk material. Second, there is a trade-off relationship between hole mobility and volumetric capacitance upon EG side chain addition. Hydration-induced swelling tends to result in morphology distortion, which in turn decreases the effective charge transport. Moreover, water could be retained in the bulk material during cyclic doping–doping process, further interrupting charge transport pathways and deteriorating device stability [63].

With the demanding careful adjustment of the hydrophilicity of the p-type material, hybrid side chains have been employed. In a series of systematic study, PT homopolymers have been engineered with hybrid side chains covalently combining different lengths of alkyl "spacers" in between the EG component and the thiophene backbone [64–66]. In the study reported by Schmode et al., extending the alkyl linkage length to at least ethyl linkage significantly boosts both of the hole mobility and the volumetric capacitance at the same time, with the resulting μC* achieving two orders of magnitude higher [66]. The existence of a sufficiently long alkyl spacer provides support from two aspects to circumvent the previously mentioned challenges. Firstly, the alkyl spacer controls water penetration near the backbone region, significantly improves and retains the ordered morphology upon ion injection, which results in the enhanced hole mobility. On the other hand, EG side chains far away from the backbone provide more accessibility to the ions in the aqueous environment. While more considerations are required, including the relative oxygen content, the relative alkyl spacer length, and the position of the alkyl spacers [64], the incorporation of a linear alkyl spacer between the EG side chain and the thiophene backbone is another promising technique to develop intrinsically high-performance OMIEC.

The performance of the material is also benefited from the synthesis technique and workup process. In a recent study, Griggs et al. reported that p(g4T2-TT) synthesized from Stille polymerization could achieve another breakthrough of μC* up to 2008 F cm^{-1} V^{-1} s^{-1} with sufficient purification of the Pd catalyst [67]. Same improvement has been seen on n-type materials as well. Notably, the above observation and insights related to hydrophilic side chain engineering for OECT are equivalently essential to the design of n-type channel, with a couple of analog studies emerging at the same time, and will be discussed in the n-type material section later.

10.3.2.1.4 P-type Donor–Acceptor Copolymer

While less discussed as the all-donor structures in p-type material design, D–A configuration combining strong donor units with electron acceptors has been applied to the conjugated backbone primarily for the reason of enhanced hole mobility and reduced faradaic side reactions in p-type material design. Specifically, D–A p-type

CPs have advantages including tunable energy level offset and controllable polaron distribution, therefore improving the electronic conductance and device durability [68–74].

A commonly studied moiety is thiophene-flanked diketopyrrolopyrrole (TDPP) due to researchers' interest in its high planarity and electron-accepting ability in the development of other TFTs [75]. For example, Moser et al. designed a series of p-type D–A copolymers by copolymerizing glycolated TDPP moiety with different thiophene donors [72]. In this study, the values of volumetric capacitance of the three materials are close to each other, while μ_{OECT} has been varied over fivefold among this family. p(gDPP-T2) in which TDPP unit is copolymerized with unsubstituted bithiophene moiety possesses the highest charge mobility of 1.55 $cm^2\,V^{-1}s^{-1}$, which leads to high μC^* of 342 $F\,cm^{-1}\,V^{-1}\,s^{-1}$ comparable with all-donor polythiophenes OECT material. The high hole mobility results from the closer energy level between T2 unit and TDPP, which provides better positive polaron delocalization in the material.

Following the preceding work of p(gDPP-T2), another breakthrough of TDPP-based D–A copolymer in OECT has been achieved by Jo et al., where they combine the strategy of D–A configuration with hybrid side chains [76]. Among the synthesized polymer family, PDPP-4EG has shown the best OECT performance with a μC^* value of 684 $F\,cm^{-1}\,V^{-1}\,s^{-1}$ with medium chain length and balanced hydrophobic and hydrophilic portions. The figure-of-merit is comparable to other best-performing all-donor material in OECT [56, 60]. Notably, PDPP-3EG and PDPP-4EG retain 99 and 88% initial current, respectively, after 1.5-hour stability test. This study is an example demonstrating the importance of comprehensive consideration of multiple design strategies.

10.3.3 N-type

While p-type materials have been extensively studied and optimized, the research progress of efficient n-type counterparts in OECT remains relatively behind. However, the development of high-performing n-type materials faces significant challenges, particularly their potential to undergo electrochemical degradation in aqueous environment and the substantially restricted electron mobility in water. In addition to the hydrophilic side chain engineering as discussed in p-type section, the primary strategy employed to overcome the difficulty of n-type material is the adaptation of D–A configuration for energetic engineering. Given the requirement of operation stability, LUMO of a n-type CP is expected to be deep-lying below −4.0 eV to efficiently avoid electron loss to oxygen and water [77].

In this section, progress of n-type material research will be categorized based on the major electron-deficient unit on the polymer backbone, which covers naphthalene diimide (NDI), bithiophene imide (BTI), IID, and DPP. Additionally, side chain free ladder type n-type OECT will be discussed as an emerging design trend.

10.3.3.1 Imide-based

10.3.3.1.1 Naphthalene Diimide

As one of the most extensively studied electron-deficient unit in OPVs and OFETs, NDI polymers engineered with glycolated side chains were also the first reported n-type CP applied in OECT with satisfactory operation stability [68]. In this study, D–A polymer p(gNDI-gT2) copolymerized from NDI and bithiophene units is fabricated into OECT, which supports ambipolar charge transport and stability in the aqueous environment. The p(gNDI-gT2)-based OECT is able to maintain negligible degradation upon operation in the aqueous environment for long term. Despite the encouraging discovery in the stability, the conductivity of p(gNDI-gT2) as OECT is very restricted. The results of this founding study both provide a promising design strategy for future stable n-type OECT as well as lessons of careful tuning of the chemical structure rather than simply adding hydrophilic side chains to the electron-deficient backbone.

Following the discovery of the potential application with glycolated NDI, a systematic study of the impact of hydrophilic side chain content has been performed to further understand the structure–property relationship for n-type CP as mixed conductance material [58]. By controlling the amount of glycolated NDI units above 50% in the polymer, the operation mode of the material switches from OFET to OECT, which corresponded to the early study regarding operation mode switch on polythiophenes [53]. Among the controlled glycolated NDI polymers, P-90 showed the best OECT performance with a normalized transconductance of 0.210 mS, volumetric capacitance of 198.2 F cm^{-3}, and electron mobility of 2.38×10^{-4} $cm^2\,V^{-1}\,s^{-1}$, setting the benchmark performance metrics for n-type OECT by the time of report.

However, due to the nature of D–A copolymer, polarons are more likely to be trapped in the acceptor units, which result in restricted electron transport [78]. Researchers have thus attempted to achieve energetic engineering of the NDI copolymer to minimize electron traps and extend the delocalization of charge carrier via donor selection and backbone functionalization. Copolymerizing with fluorinated selenophene instead of the regular bithiophene gives the novel PNDI polymer increased electron mobility of up to 0.32 $cm^2\,V^{-1}\,s^{-1}$, which is two orders of magnitude higher than P-90, and remarkably lowered V_{th} of −4.6 eV [79]. Thus, the novel P(NDIOD-FSVS) polymer provides better figure-of-merit (> 0.2 F $cm^{-1}\,V^{-1}\,s^{-1}$) compared to the previous PNDI benchmark P-90.

Another strategy is to increase the coplanarity of the glycolated NDI backbone to facilitate the charge transport in the bulk material, a strategy that has been successfully employed for p-type pgBTTT as well [60]. Glycolated NDI (gNDI) copolymerized with bithiophene (gNDI-T) and vinylene (gNDI-V) provides copolymers both having deep-lying LUMO below −4.2 eV and higher electron mobility compared to P-90 [80]. The leap of the electron mobility in gNDI-V gives rise to a figure-of-merit exceeding 2.0 F $cm^{-1}\,V^{-1}\,s^{-1}$, highest of the reported value for NDI family in OECT application. Further examination of the copolymers reveals that coupling with rigid vinylene unit reduces backbone distortion between NDI and the donor portion, which is the major limiting factor for the electron mobility along the polymer backbone in gNDI-T. The enhanced intramolecular hydrogen bonding in gNDI-V hence provides more compact π–π stacking and crystallinity, allowing further effective charge transport pathway in the bulk material.

10.3.3.1.2 Bithiophene Imide

Similar donor functionalization strategy has been successfully applied to BTI, another strong electron-deficient unit widely studied for other TFTs with remarkably high electron mobility attributed to its high backbone planarity [81, 82]. By copolymerizing glycolated BTI units with cyano-functionalized thienylene–vinylene–thienylene (TVTCN), the resulting polymer f-BTI2g-TVTCN achieved a breakthrough of n-type OECT performance by the time of report, with electron mobility exceeding 0.2 cm^2 V^{-1} s^{-1} and μC^* of 41.3 F cm^{-1} V^{-1} s^{-1} [83]. By lowering the LUMO level and hence promoting polaron delocalization, cyano-functionalized TVT unit has become the key to the huge progress achieved in imide-based n-type OECT material.

10.3.3.2 Isoindigo

One of the most famous lactam-based n-type moieties is isoindigo-benzodifurandione (IID-BDF), since such structure provides both strong electron-deficient features and rigid backbones. Another unique advantage of IID-BDF-based CPs is its potential for green synthesis, primarily via aldo condensation [84]. The representative example of this class that leads to huge progress on n-type OECT other than ladder type CPs is p(C2F-V) [85]. Further fluorination of the glycolate IID-BDF moiety with a planar vinylene as the donor group, p(C2F-V), achieves record-high electron mobility of 0.912 cm^2 V^{-1} s^{-1} and a μC^* of 107.56 F cm^{-1} V^{-1} s^{-1}, first time comparable to p-type OECT materials. In the case of p(C2F-V), fluorination of acceptor units enhances the electron deficiency feature, which consequently lowers the LUMO level and enhances charge mobility. Moreover, fluorine atom provides sites for the formation of intermolecular hydrogen bonding within the material, increasing the coplanarity of the backbone and promoting more compact chain packing.

10.3.3.3 Diketopyrrolopyrrole

As discussed in Section 10.2.1.4, DPP moieties, more commonly researched for p-type D–A copolymers, could be switched to n-type transport with proper doped state engineering with high OECT performance [86]. By fluorination of the thiophene unit in the TDPP moiety, P(gTDPP2FT) in the doped state adopts n-type transport instead of p-type. Such transition renders equivalently high OECT performance as its unfluorinated p-type analogy P(gTDPPT) without further compromise. The effectiveness of the doped state engineering rises from the control of charge delocalization and stabilization in the doped state of the channel material, which leads to new discussion of efficient energetic tuning strategy for n-type CP in OECT.

10.3.3.4 Ladder Type

As D–A configuration is predominant in current n-type material investigation, a couple of limitations stemmed from such structure are worth mentioning. First, donor and acceptor units along the backbone are usually connected by single bonds, which tend to have greater distortion that weakens the backbone planarity. Second, the nature of D–A copolymer easily results in localized LUMO interfering electron hopping in between chains. Adapting side chain free ladder type polymer is an efficient approach to achieve high-performance and high-stability n-type CPs for OECT while circumventing the challenges with D–A configuration. First reported by Sun et al., BBL, having exceptionally planar backbone and no pendant side chains, was fabricated into OECT device [78]. Surprisingly, the side chain free configuration in turn leads to outstanding volumetric capacitance of the device while maintaining comparable charge mobility as it is in OFET devices. Since then, the BBL-based ladder type polymer has been further optimized by controlling the molecular weight [87]. It is shown from the systematic study that increasing BBL molecular weight provides enhanced electron mobility, lower threshold voltage, higher normalized transconductance, faster switch response, and, more importantly, higher figure-of-merit μC^*, for over an order of magnitude. The fundamental mechanism behind the comprehensive enhancement is related to the stronger π–π interaction and higher crystallinity in higher molecular weight BBL that facilitates charge transport. Although still lagging behind average p-type material performance, side chain free ladder type material marks a new design strategy for n-type OECT as it omits complex tuning of side chain property and requires simpler and greener synthesis procedure.

Another remarkable ladder type n-type moiety is poly(benzodifurandione) (PBFDO) with currently the benchmark μC^* of 180 F cm^{-1} V^{-1} s^{-1}, setting the record high OECT performance of n-type material due to the unprecedently improved conductivity of 2000 S cm^{-1} [88]. The high conductivity of PBFDO results from the high efficiency of in situ n-doing during the oxidative polymerization and the existence of metallic state due to the redox-reversible quinones along the backbone structure. Moreover, the synthesis preparation process of PBFDO is significantly simpler than traditional n-type material, making PBFDO a promising candidate for a wide range of organic TFTs, including OECT.

10.3.3.5 Acceptor–Acceptor

Another efficient strategy to promote more delocalized LUMO as well as enhanced coplanarity along the backbone is employing acceptor–acceptor (A–A) configuration. Chen et al. in 2021 raised the concept of A–A n-type OECT material and synthesized PgNaN and PgNgN, two polymers based on fused lactam rings bearing different ratios of glycolated side chains with double bond connection in between the electron-deficient units [89]. Both polymers are prepared from a benign, metal-free aldo condensation, which avoids complicated procedure and negative effect of residual catalyst in metal-assisted synthesis of D–A polymer [67]. Between the two polymers, PgNaN, having balanced hydrophilic and hydrophobic side chains, shows higher μC^* value of 0.662 ± 0.113 F cm^{-1} V^{-1} s^{-1}. Further study investigating the balance between EG and alkyl side chains on PgNaN series has addressed the significance of fine-tuning hydrophobicity in the polymer, with μC^* optimized 1.83 F cm^{-1} V^{-1} s^{-1} when reducing the alkyl side chain on PgNaN from 12 to 10 units [90].

10.3.4 Summary Tables of Reviewed p- and n-type Materials

10.3.4.1 P-type Materials

Name	$g_{m, norm}$ [mS/cm]	$g_{m, peak}$ [mS]	V_{th} [V]	μ_{OECT} [$cm^2 V^{-1} s^{-1}$]	C* [$F cm^{-3}$]	$[\mu C^*]_{OECT}$ [$F cm^{-1} V^{-1} s^{-1}$]	$[\mu_{OECT}][C^*]$ [$F cm^{-1} V^{-1} s^{-1}$]	HOMO [eV]	LUMO [eV]	Ref.
PTHS$^-$M$^+$	—	0.4 ± 0.02	—	—	—	—	—	—	—	[50]
PTHS$^-$TMA$^+$-co-P3HT	—	70.5	−0.15	0.017 ± 0.0036	100 ± 7	1.7	—	—	—	[51]
p(g2T-TT)	—	—	—	0.94 ± 0.25	241 ± 94	261 ± 29	227 ± 107	−4.16	−2.11	[53, 54]
p(g2T-T)	—	—	—	0.28 ± 0.1	220 ± 30	167 ± 65	62 ± 24	−4.38	−2.65	[52, 54]
p(g2T2-g4T2)	—	—	+0.02	1.72 ± 0.31	187 ± 8	522	—	—	—	[56]
pgBTTT	—	—	−0.243 ± 0.0049	3.44 ± 0.13	164 ± 7	502 ± 18	563 ± 33	—	—	[60]
P3MEET	0.02	—	−0.18	—	80 ± 9	0.04 ± 0.01	—	—	—	[66]
P3MEEET	20.4	—	−0.57	—	242 ± 17	11.5 ± 1.4	—	—	—	[66]
P(g4T2-TT)-F3	1457 ± 6	—	—	6.53 ± 0.065	308 ± 20	2008 ± 130	—	—	—	[67]

10.3.4.2 N-type Materials

Name	$g_{m, norm}$ [S/cm]	$g_{m, peak}$ [S]	V_{th} [V]	μ_{OECT} [$cm^2 V^{-1} s^{-1}$]	C* [$F cm^{-3}$]	$[\mu C^*]_{OECT}$ [$F cm^{-1} V^{-1} s^{-1}$]	$[\mu_{OECT}][C^*]$ [$F cm^{-1} V^{-1} s^{-1}$]	HOMO [eV]	LUMO [eV]	Ref
p(gNDI-gT2)	—	2.17×10^{-5}	—	1.0×10^{-5}	397	—	0.00397	−4.83	−4.12	[68]
P-90	0.210	—	—	2.38×10^{-4}	198.2	—	0.0472	−5.57	−4.23	[58]
P(NDIEG7-FSVS)	≈ 0.04	—	0.16	—	—	≈0.21–0.25	—	—	−4.62	[79]
gNDI-T	0.11 ± 0.01	—	0.20	$1.5 \times 10^{-3} \pm 2.0 \times 10^{-4}$	237 ± 8	0.42	—	−5.94	−4.21	[80]
f-BTI2g-TVTCN	11.9 ± 1.2	—	0.68	0.23 ± 0.02	170 ± 22	38.4 ± 3.9	—	−5.57	−3.81	[83]
P(C2F-V)	25.67 ± 21.52	—	0.02	0.912	117.96 ± 2.93	107.56 ± 3.96	—	−6.12	−4.45	[85]
P(gTDPP2FT)	—	—	0.64 ± 0.01	0.27 ± 0.04	156 ± 24	42.2 ± 6.5	—	−5.20	−3.86	[86]
BBL	—	9.7	—	7×10^{-4}	930 ± 40	—	—	—	—	[78]
BBL_{152}	11.1 ± 0.2	—	0.151 ± 0.005	$(4.40 \pm 0.34) \times 10^{-3}$	589 ± 26	25.9	—	—	—	[87]
PBFDO	1970	—	—	—	—	180	—	−5.18	−4.67	[88]
PgNaN	0.212	—	0.37 ± 0.01	$(6.50 \pm 1.01) \times 10^{-3}$	100 ± 6	0.662 ± 0.113	0.652 ± 0.107	−5.14	−4.28	[89]
PgNgN	0.007	—	0.21 ± 0.02	$(1.89 \pm 0.23) \times 10^{-4}$	239 ± 97	0.037 ± 0.016	0.046 ± 0.024	−5.22	−4.35	[89]
$P(C_6g_3NDI\text{-}T)\text{-}F3$	0.095 ± 0.024	—	—	$(60.6 \pm 1.58) \times 10^{-4}$	521 ± 160	0.316 ± 0.127	—	—	—	[67]
$P(g_7NC_{10}N)$	0.370 ± 0.023	—	0.300 ± 0.0030	$(1.20 \pm 0.07) \times 10^{-3}$	153 ± 34	1.83 ± 0.101	—	−5.12	−4.23	[90]

10.4 Organic Thermoelectrics

10.4.1 Introduction

The depletion of fossil fuels, combined with a growing population and industrialization, necessitates innovative technologies to enhance energy efficiency and reduce waste heat. Organic thermoelectric (OTE) materials are gaining attention for their ability to convert low-grade heat into electricity, providing a sustainable solution to rising energy demands. In the past two decades, inorganic semiconductors have been developed as efficient TE materials [91, 92], but they face drawbacks such as low natural abundance, toxicity, rigidity, high-temperature processability, and production cost [93–95]. Conversely, CPs offer several advantages, such as abundant availability of their atomic elements, low toxicity, durability, flexibility, lower cost, solution processibility, and tunable structure–property, making them suitable for various applications in OTEs [95, 96]. Moreover, their inherent low thermal conductivity provides an additional advantage [7]. Most high-performing inorganic TE materials are typically suited for high temperatures, while the low-temperature energy conversion applications are effectively managed by organic materials.

10.4.2 TE Figure-of-merit

The efficiency of a TE material is determined by the dimensionless figure-of-merit (ZT), defined as $ZT = \sigma S^2 T/k$, where S is the Seebeck coefficient, σ is the electrical conductivity, k is the thermal conductivity, and T is the working temperature [95, 97]. A good TE material should have a high ZT, but optimizing σ, S, and k independently is challenging as these three are interlinked with carrier concentration. The evaluation of the TE performance of polymers often relies on the power factor (PF), defined as $PF = \sigma S^2$, which is often used in place of ZT because of the low inherent k of conducting polymers. For a considerable number of OTE materials, particularly the CPs, the relationship between S and σ is empirically expressed as $S \propto \sigma^{-1/4}$ [97, 98] leading to $PF \propto \sigma^{1/2}$ [95]. Consequently, improving σ is crucial to achieving higher PFs, which is largely the focus of current research.

10.4.3 Progress in TE Properties of CPs

The development of OTE materials is still in its early phases, and some long-lasting challenges remain to overcome. Achieving a higher σ is essential for enhanced PF. However, the inverse relationship between S and σ as a function of charge density (n) creates a trade-off between σ and S as increasing one often leads to a decrease in the other [95]. The expression, $\sigma = n\mu q$ shows the dependency of σ on both charge density (n) and charge carrier mobility (μ). Although, increasing n by adding dopants can enhance μ, it simultaneously lowers S. Additionally, a higher dopant concentration may alter the material's morphology, which can reduce μ and, result in an unintended decrease in σ. Therefore, dissociating S and σ to resolve this conflict is a crucial step toward achieving enhanced PFs [7, 95]. The challenge for researchers is to develop strategies that optimize these interdependent TE parameters without compromising one for the other. To design an efficient TE generator module, complementary pairs of p- and n-type TE materials are needed [99, 100]. While, the TE performance of p-type CPs has been extensively explored, with impressive σ exceeding 4000 S cm^{-1} [101–109], n-type CPs have lagged behind with only a few reports of σ exceeding 20 S cm^{-1} [110 –113]. Some improvements have been noted with σ reaching over 90 S cm^{-1} [114] and even 100 S cm^{-1} [99, 115]. We will explore in detail the gradual advancements in the following sections.

10.4.3.1 P-type TE CPs

Until now, several p-type CPs have been explored as TE materials. In this section, we will first discuss the most widely studied polymers based on PEDOT, then explore significant advancements in polythiophenes and their derivatives, such as P3HT and poly(2,5-bis(thiophen-2-yl)thieno-[3,2-b]thiophene) (PBTTT), as well as cover donor–acceptor (D–A) copolymers, particularly represented by DPP-based copolymers in the subsequent paragraphs (Figure 10.12).

PEDOT-based materials have been widely explored for TE applications [116–119]. PEDOT, synthesized from EDOT monomer, was insoluble, so oxidative polymerization of EDOT was performed with aqueous PSS, resulting in a stable dispersion of PEDOT:PSS [120]. This approach improved the processibility of PEDOT, but the surplus amount of PSS disturbed the molecular arrangement of PEDOT,

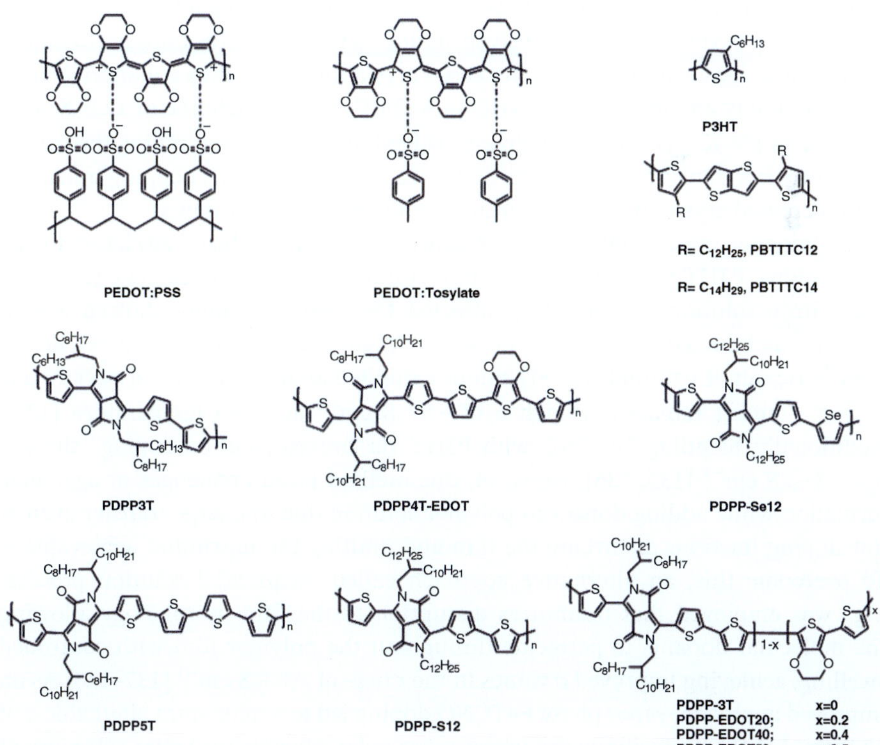

Figure 10.12 Chemical structure of p-type TE polymers.

causing inefficient TE performances with $\sigma = 10^{-1} - 1$ S cm^{-1} , S < 20 μV K^{-1}, and PF = ~10^{-2} μW m^{-1} K^{-2} [121, 122]. Chemical dedoping has been effectively exploited for adjusting the high doping levels (>30 %) of pristine PEDOT:PSS, which otherwise reduces its suitability for achieving a high PF. This tunes the material's band structure and density of states (DOS), which directly affects S and thus enhances TE performance [123, 124]. Various reducing agents like hydrazine [125, 126], $NaHCO_3$, Na_2SO_3, $NaBH_4$ [127], and tetrakis(dimethylamino)ethylene (TDAE) [128] have been used. For instance, PEDOT:PSS, when doped with p-toluenesulfonic acid monohydrate (TSA), achieved a high σ of 1260 S cm^{-1}, and PF of 70.7 μW mK^{-2} and S of 24.1 μV K^{-1} [126]. Then, the dedoping with hydrazine/DMSO led to an increase in PF to 318.4 μW mK^{-2} with S of 49.3 μV K^{-1}.

The literature reported replacing PSS with tosylate (Tos) to diminish the insulating phase. PEDOT:Tos, obtained by mixing an oxidative solution of iron p-toluenesulfonate ($Fe(Tos)_3$) to EDOT monomers, displayed a PF of 38 μW m^{-1} K^{-2}, which increased to 324 μW m^{-1} K^{-2} after TDAE reduction [117]. The altered polymer's electronic structure allowed the attainment of desired optimum oxidation state and thus enhanced PF as σ decreases whereas S increases at lower oxidation state. Similar to PEDOT:PSS, both secondary dopant and dedoping have been used to improve σ of PEDOT:Tos [117, 129–132]. For example, adding high-boiling solvents resulted in higher crystallinity and a threefold increase in σ with a maximum value of 640 S cm^{-1} with DMF, compared to 230 S cm^{-1} for the pristine film and leaving S unaltered, and thus enhanced PF is achieved with increasing σ [130]. In another study, adding pyridine along with solvent additives to PEDOT:Tos films led to higher molecular weight PEDOT chains, increased crystallinity, and thus enhanced charge carrier mobility and improved S. Pyridine + DMF showed highest σ of 1225 S cm^{-1} and S of 44 μV K^{-1} [132]. These chains bridged ordered regions, aiding the carrier transport via a transition from quasi-1D to two dimensional (2D) hopping transport, with increased crystallinity contributing to efficient TE performance.

Polythiophenes and their derivatives have also gained significant attraction for TE applications. P3HT when doped with nitrosonium hexafluorophosphate ($NOPF_6$)/acetonitrile solution wherein PF_6^- affected the chain assembly showed a σ of 1 S cm^{-1} and a maximum PF of ~1.4×10^{-7} W m^{-1} K^{-2} [133]. Later, P3HT doped with ferric salt of triflimide (TFSI) anions exhibited an improved PF of over 20 μW m^{-1} K^{-2} and a remarkably high σ of ~87 S cm^{-1} at room temperature [134]. Additionally, blending F4TCNQ with P3HT via "mixed solution doping" showed σ = ~1–2 S cm^{-1} [135, 136]. However, this method faced challenges of aggregate formation while adding dopant to polymer solution due to charge transfer even at low doping fractions disturbing the film and limiting the maximum achievable σ. To overcome this, an alternative approach called "sequential solution processing" was employed that minimizes disruption to the film morphology allowing the molecular dopants to permeate throughout the polymer film with controlled swelling, achieving improved σ values in the range of ~3–9 S cm^{-1} [137–140] As an improved approach, vapor phase F4TCNQ doping led to a maximum attainable σ of 48 S cm^{-1} [141]. To facilitate the use of more polar solvents for better solubility of dopants, tetraethylene glycol side chains substituted polythiophenes were synthesized and doped with F4TCNQ, resulting into very high σ of up to 100 S cm^{-1} [142].

Other intriguing parameters, such as molecular weight, regioregularity, and the influence of heteroatoms on the TE performance of doped P3HT, are also studied [143–146].

Simultaneously, the TE performance of PBTTT has also been investigated, similar to P3HT. Thermal annealing of the $NOPF_6$-doped film of PBTTT-C14 induced structural ordering leading to an enhanced σ value from ~33 S cm^{-1} to 55 S cm^{-1} and an improvement in PF from 0.31 μW m^{-1} K^{-2} to 0.98 μW m^{-1} K^{-2} [147]. Furthermore, PBTTT-C12 doped with ferric salt of TFSI anions achieved σ of 220 S cm^{-1} and PF of 14 μW m^{-1} K^{-2} [148]. Both "mixed solution doping" [149, 150] and "solid-state diffusion" [151] methods have been explored for doping PBTTT with F4TCNQ, with the latter method yielding σ of 250 S cm^{-1} [151]. A comparison between F4TCNQ vapor-doped PBTTT films exhibited σ greater than two orders of magnitude (670 S cm^{-1}) compared to solution-doped films and a large PF of 120 μW m^{-1} K^{-2} [152]. Here, the alignment of ordered domains in the polymer film measured through orientation correlation length was found to be larger in the vapor-doped film and thus have better charge transport and enhanced σ and PF. PBTTT-C14 when vapor doped with a different dopant (tridecafluoro-1,1,2,2-tetrahydrooctyl) trichlorosilane (FTS) displayed σ of ~1000 S cm^{-1} and PF of 110 μW m^{-1} K^{-2} [153].

Donor–acceptor CPs, particularly DPP-based copolymers, are gaining attention as efficient TE materials. Poly(diketopyrrolopyrrole-terthiophene), PDPP3T, doped with $FeCl_3$ exhibited high mobility and a low BG, with a PF of 276 μW m^{-1} K^{-2} and σ of 55 S cm^{-1} [154]. $FeCl_3$-doped PDPP-5T and PDPP-4T-EDOT (one thiophene unit substituted by EDOT) showed a significant enhancement in σ from 119.6 S cm^{-1} to 271.5 S cm^{-1} [155]. Substitution of selenium for sulfur displayed enhanced performance attributed to the larger atomic radius of selenium leading to better intermolecular interaction and packing [156]. For instance, $FeCl_3$-doped PDPPSe-12 showed σ of 997 S cm^{-1} and a PF of 300 μW m^{-1} K^{-2}, whereas PDPPS-12 has σ of 318 S cm^{-1} and a PF of 126 μW m^{-1} K^{-2}. In another study, $FeCl_3$-doped PDPP-3T exhibited a lower σ of ~0.28 S cm^{-1}, PF of 1.6 μW m^{-1} K^{-2}, and S of 242 μV K^{-1}, whereas the synthesized random copolymers with EDOT (PDPP-EDOT-20, PDPP-EDOT-40, PDPP-EDOT-50) when doped displayed higher σ of 0.76 S cm^{-1}, 2.8 S cm^{-1}, and 4.12 S cm^{-1}, respectively, albeit with decreasing S of 177 μV K^{-1}, 152 μV K^{-1}, and 105 μV K^{-1} [157]. The random copolymer incorporating 40% EDOT achieved a PF of 6.4 μW m^{-1} K^{-2}, about four times higher than the reference material. This incorporation facilitated an orientation change from face-on dominated to edge-on dominated and improved the copolymer chains packing, resulting in enhanced charge transport.

10.4.3.2 N-type TE CPs

As in the other applications, the development of n-type TE materials lags behind p-type due to poor conductivity and stability in air [99, 100, 94, 158]. Chemical structures of n-type TE polymers discussed here are shown in Figure 10.13. A 2014 report highlighted an n-type polymer, poly{[N,N′-bis(2-octyldodecyl)-naphthalene-1,4,5,8 bis(dicarboximide)-2,6-diyl]-alt-5,5′-(2,2′-bithiophene)} (P(NDI2OD-T2) doped with (4-(1,3-dimethyl-2,3-dihydro-1H-benzoimidazol-2-yl)phenyl) dimethylamine, N-DMBI exhibiting a PF of 0.6 μW m^{-1} K^{-2} and a σ of ~10^{-3} S cm^{-1} [159].

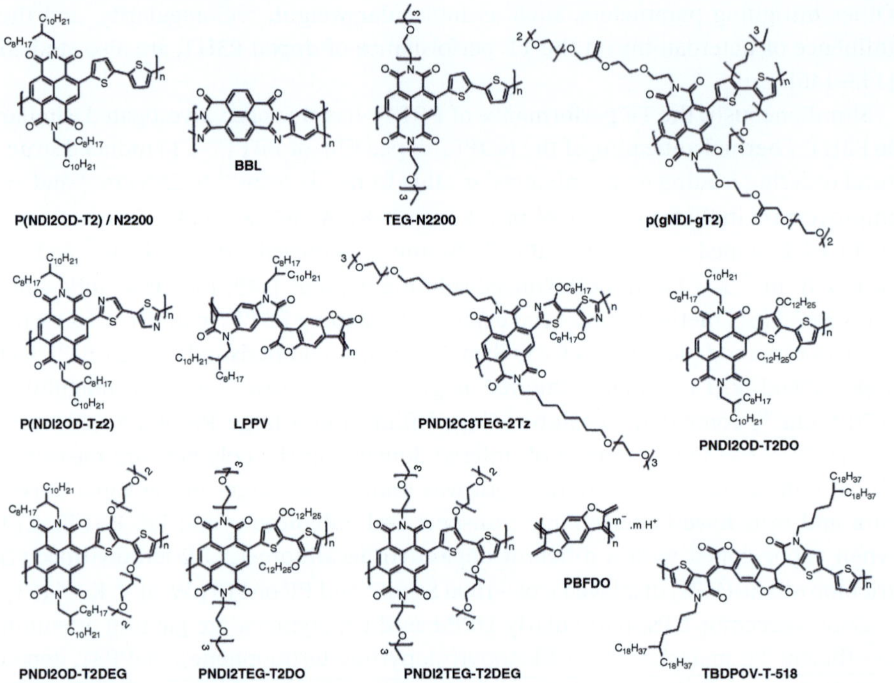

Figure 10.13 Chemical structure of n-type TE polymers.

However, the dopants aggregated on the surface, leading to poor TE performance. Another study explored TDAE-doped (BBL) – a ladder type solution-processable n-type polymer – and found that it has a σ of 2.4 S cm^{-1} while doped P(NDI2OD-T2) had ~0.003 S cm^{-1} due to the larger computed polarization length in linear-torsion-free homopolymer BBL, suggesting higher intrachain polaron mobility [160]. To improve dopant and polymer miscibility, researchers tailored PNDI-based polymers [161]. For instance, modifying alkyl side chains to polar triethylene glycol-based ones in poly{[N,N′-bis(2-octyldodecyl)-naphthalene-1,4,5,8-bis(dicarboximide)-2,6-diyl](NDI)-alt-5,5′ (2,2′-bithiophene)} N2200 resulted in a high σ of 0.17 S cm^{-1} and a PF of 0.4 μW m^{-1} K^{-2} in TEG-N2200 after doping with N-DMBI. Further enhancing miscibility, naphthalenediimide–bithiophene copolymer p(gNDI-gT2) (a structural analogue of p(NDI2OD-T2)) with polar oligoethylene glycol-containing side chains on both the NDI acceptor and the bithiophene donor unit after doping with N-DMBI yielded a higher σ of ~0.3 S cm^{-1} [162].

A NDI-bithiazole (Tz2)-based polymer, P(NDI2OD-Tz2), was synthesized and doped with TDAE, achieving a σ of ~0.1 S cm^{-1} higher than P(NDI2OD-T2) [163]. The 5,5′-di(1,3-thiazolyl) (Tz2) moiety reduces internal steric repulsion compared to NDI-T2, facilitating a more planar backbone, which improved electron affinity, and decreases the polymer's D–A feature due to its electron deficiency. A rigid coplanar PPV derivative, LPPV-1, featuring an almost torsion-free backbone and low-lying LUMO, when doped with N-DMBI, showed a high conductivity (up to 1.1 S cm^{-1}) and PF of ~1.96 μW m^{-1} K^{-2} [164]. Additionally, amphipathic side chains were used to enhance S [165]. These side chains with an alkyl chain segment as a spacer between the backbone and an EG-type chain segment in a D–A copolymer (PNDI2C8TEG-2Tz) minimize energetic disorder and maintain the dopants away from the polymer backbone, addressing both the issues of polymer-dopant miscibility and polaron delocalization, leading to a large S of about $-326\ \mu V\ K^{-1}$ an enhanced PF of ~16.5 $\mu W\ m^{-1}\ K^{-2}$ in the N-DMBI-doped polymer film. The impact of strategic modification of chemical identity and regiochemistry of non-polar(izable) and polar(izable) pendant groups of D–A copolymer backbone NDI-alt-T2 on molecular doping in thin films was also investigated in another report [166]. Four different D–A copolymers synthesized with various combinations of alkyl and glycol ether pendant groups and their doping behavior analyzed at diverse concentrations of N-DMBI reported a maximum σ of 0.08 S cm^{-1} and PF of 0.24 $\mu W\ m^{-1}\ K^{-2}$ for the copolymer with both monomers carrying glycol ether side chains. The doping efficiency was significantly impacted by the relative arrangement of side chains. In 2022, n-type synthesized polymer PBFDO achieved a remarkable σ over 2000 S cm^{-1} and a PF of 90 $\mu W\ m^{-1}\ K^{-2}$ demonstrating high air stability and solution processibility without extra side chains or surfactants [88]. Most recently in 2023, good electrical conductivities (both n- and p-type) and PFs were reported in polymers derived from thiophene-fused benzodifurandione oligo(p-phenylenevinylene) (TBDOPV-T-518) by reducing the crystallization rate of the side chains and thus enabling the precise control over the degree of interchain p–p interactions, achieving n-type conductivities of 114 S cm^{-1} and PFs exceeding 200 $\mu W\ m^{-1}\ K^{-2}$ with N-DMBI, and p-type conductivities of 452 S cm^{-1} and PFs exceeding 100 $\mu W\ m^{-1}\ K^{-2}$ with $FeCl_3$ within a single polymer [115].

Studies on N-DMBI aim to strategically design next-generation dopants and organic materials and improve the doping efficiency [167, 168]. A study comparing n-dopants TAM and N-DMBI-H found that TAM-doped UFBDPPV achieved σ of ~22 S cm^{-1} and PF of over 80 $\mu W\ m^{-1}\ K^{-2}$ through reduced engagement with the polymer backbone [110]. These examples highlight the ongoing efforts and advancements in developing high-performance n-type CPs.

10.4.4 Applications and Outlook

OTE materials offer advantages such as flexibility, stretchability, and the ability to convert low-grade heat sources – like body heat or infrared radiation – into electricity [95]. Although the typical power output is low, it is often sufficient for use in wearable and portable electronics. Several studies have investigated the applications of OTE materials in flexible self-powered/multiparameter sensors (temperature, pressure, humidity) [169–172] and stretchable TE generators [88, 111, 173, 174]. Additionally, there is growing interest in exploring OTE materials for use in solar organic TE generators (SOTEGs) [175]. Despite notable advancements in this field, the development of OTE materials and devices remains in its early stages. A deeper, systematic understanding of the correlation of structure–property is crucial for addressing the challenges of designing market-ready commercial devices, requiring collaborative efforts from diverse scientific backgrounds.

References

1 Vogl, P. and Leising, G. (1989). The intrinsic energy gap of crystalline trans-polyacetylene. *Synthetic Metals* 28 (3): D209–D214.

2 Vohlídal, J., Graeff, C.F.O., Hiorns, R.C. et al. (2022). Glossary of terms relating to electronic, photonic and magnetic properties of polymers (IUPAC Recommendations 2021). *Pure and Applied Chemistry* 94 (1): 15–69.

3 Liu, Y., Gao, S., Zhang, X. et al. (2022). Probing the nature of charge carriers in one-dimensional conjugated polymers: a review of the theoretical models, experimental trends, and thermoelectric applications. *Journal of Materials Chemistry C* 11 (1): 12–47.

4 Bredas, J.L. and Street, G.B. (1985). Polarons, bipolarons, and solitons in conducting polymers. *Accounts of Chemical Research* 18: 309–315.

5 Meier, E.J., An, F.A., and Gadway, B. (2016). Observation of the topological soliton state in the Su–Schrieffer–Heeger model. *Nature Communications* 7 (1): 1–6.

6 Kivelson, S. et al. (1981). Electron hopping conduction in the soliton model of polyacetylene. *Physical Review Letters* 46 (20):1344–1348.

7 Ding, L., Yu, Z.-D., Wang, X.-Y. et al. (2023). Polymer semiconductors: synthesis, processing, and applications. *Chemical Reviews* 123(12): 7421–7497.

8 Tsumura, A., Koezuka, H., and Ando, T. (1986). Macromolecular electronic device: field-effect transistor with a polythiophene thin film. *Applied Physics Letters* 49 (18): 1210–1212.

9 Wang, S., Xu, J., Wang, W. et al. (2018). Skin electronics from scalable fabrication of an intrinsically stretchable transistor array. *Nature* 555 (7694): 83–88.

10 Bao, Z., Dodabalapur, A., and Lovinger, A.J. (1996). Soluble and processable regioregular poly(3-hexylthiophene) for thin film field-effect transistor applications with high mobility. *Applied Physics Letters* 69 (26): 4108–4110.

11 Zhang, W., Smith, J., Watkins, S.E. et al. (2010). Indacenodithiophene semiconducting polymers for high-performance, air-stable transistors. *Journal of the American Chemical Society* 132 (33): 11437–11439.

12 Zhang, X., Bronstein, H., Kronemeijer, A.J. et al. (2013). Molecular origin of high field-effect mobility in an indacenodithiophene–benzothiadiazole copolymer. *Nature Communications* 4 (1): 2238.

13 Noriega, R., Rivnay, J., Vandewal, K. et al. (2013). A general relationship between disorder, aggregation and charge transport in conjugated polymers. *Nature Materials* 12 (11): 1038–1044.

14 Venkateshvaran, D., Nikolka, M., Sadhanala, A. et al. (2014). Approaching disorder-free transport in high-mobility conjugated polymers. *Nature* 515: 384–388.

15 Li, Y., Tatum, W.K., Onorato, J.W. et al. (2018). Low elastic modulus and high charge mobility of low-crystallinity indacenodithiophene-based semiconducting polymers for potential applications in stretchable electronics. *Macromolecules* 51 (16): 6352–6358.

16 McCulloch, I., Ashraf, R.S., Biniek, L. et al. (2012). Design of semiconducting indacenodithiophene polymers for high performance transistors and solar cells. *Accounts of Chemical Research* 45 (5): 714–722.

17 Fei, Z., Ashraf, R.S., Huang, Z. et al. (2012). Germaindacenodithiophene based low band gap polymers for organic solar cells. *Chemical Communications* 48 (24): 2955–2957.

18 Wadsworth, A., Chen, H., Thorley, K.J. et al. (2020). Modification of indacenodithiophene-based polymers and its impact on charge carrier mobility in organic thin-film transistors. *Journal of the American Chemical Society* 142 (2): 652–664.

19 Bronstein, H., Leem, D.S., Hamilton, R. et al. (2011). Indacenodithiophene-co-benzothiadiazole copolymers for high performance solar cells or transistors via alkyl chain optimization. *Macromolecules* 44 (17): 6649–6652.

20 Zhang, W., Han, Y., Zhu, X. et al. (2016). A novel alkylated indacenodithieno[3,2-b]thiophene-based polymer for high-performance field-effect transistors. *Advanced Materials* 28 (20): 3922–3927.

21 Chen, H., Hurhangee, M., Nikolka, M. et al. (2017). Dithiopheneindenofluorene (TIF) semiconducting polymers with very high mobility in field-effect transistors. *Advanced Materials* 29 (36): 1702523.

22 Chen, H., Wadsworth, A., Ma, C. et al. (2019). The effect of ring expansion in thienobenzo[b]indacenodithiophene polymers for organic field-effect transistors. *Journal of the American Chemical Society* 141 (47): 18806–18813.

23 Lei, T., Cao, Y., Fan, Y. et al. (2011). High-performance air-stable organic field-effect transistors: isoindigo-based conjugated polymers. *Journal of the American Chemical Society* 133: 6099–6101.

24 Wu, H.-C., Hung, C.-C., Hong, C.-W. et al. (2016). Isoindigo-based semiconducting polymers using carbosilane side chains for high performance stretchable field-effect transistors. *Macromolecules* 49 (22): 8540–8548.

25 Lei, T., Cao, Y., Zhou, X. et al. (2012). Systematic investigation of isoindigo-based polymeric field-effect transistors: design strategy and impact of polymer symmetry and backbone curvature. *Chemistry of Materials* 24: 1762–1770.

26 Zhang, M., Tsao, H.N., Pisula, W. et al. (2007). Field-Effect transistors based on a benzothiadiazole–cyclopentadithiophene copolymer. *Journal of the American Chemical Society* 129 (12): 3472–3473.

27 Luo, C., Kyaw, A.K.K., Perez, L.A. et al. (2014). General strategy for self-assembly of highly oriented nanocrystalline semiconducting polymers with high mobility. *Nano Letters* 14 (5): 2764–2771.

28 de Leeuw, D.M., Simenon, M.M.J., Brown, A.R. et al. (1997). Stability of n-type doped conducting polymers and consequences for polymeric microelectronic devices. *Synthetic Metals* 87 (1): 53–59.

29 Newman, C.R., Frisbie, C.D., da Silva Filho, D.A. et al. (2004). Introduction to organic thin film transistors and design of n-channel organic semiconductors. *Chemistry of Materials* 16 (23): 4436–4451.

30 Babel, A. and Jenekhe, S.A. (2002). Electron transport in thin-film transistors from an n-type conjugated polymer. *Advanced Materials* 14 (5): 371–374.

31 Lei, T., Dou, J.-H., Cao, X.-Y. et al. (2013). Electron-deficient poly(p-phenylene vinylene) provides electron mobility over 1 $cm^2\ v^{-1}\ s^{-1}$ under ambient conditions. *Journal of the American Chemical Society* 135 (33): 12168–12171.

32 Lei, T., Xia, X., Wang, J.Y. et al. (2014). "Conformation Locked" strong electron-deficient poly(p -phenylene vinylene) derivatives for ambient-stable n-type field-effect transistors: synthesis, properties, and effects of fluorine substitution position. *Journal of the American Chemical Society* 136 (5): 2135–2141.

33 Zheng, Y.Q., Lei, T., Dou, J.H. et al. (2016). Strong electron-deficient polymers lead to high electron mobility in air and their morphology-dependent transport behaviors. *Advanced Materials* 28: 7213–7219.

34 Crone, B., Dodabalapur, A., Lin, Y.-Y. et al. (2000). Large-scale complementary integrated circuits based on organic transistors. *Nature* 403 (6769): 521–523.

35 Lei, T., Dou, J.H., Ma, Z.J. et al. (2012). Ambipolar polymer field-effect transistors based on fluorinated isoindigo: high performance and improved ambient stability. *Journal of the American Chemical Society* 134 (49): 20025–20028.

36 Lei, T., Dou, J.H., Ma, Z.J. et al. (2013). Chlorination as a useful method to modulate conjugated polymers: balanced and ambient-stable ambipolar high-performance field-effect transistors and inverters based on chlorinated isoindigo polymers. *Chemical Science* 4 (6): 2447–2452.

37 Ashraf, R.S., Kronemeijer, A.J., James, D.I. et al. (2012). A new thiophene substituted isoindigo based copolymer for high performance ambipolar transistors. *Chemical Communications* 48: 3939–3941.

38 Huang, J., Mao, Z., Chen, Z. et al. (2016). Diazaisoindigo-based polymers with high-performance charge-transport properties: from computational screening to experimental characterization. *Chemistry of Materials* 28: 2209–2218.

39 Yang, Y., Liu, Z., Zhang, G. et al. (2019). The effects of side chains on the charge mobilities and functionalities of semiconducting conjugated polymers beyond solubilities. *Advanced Materials* 31 (46): 1903104.

40 Lei, T., Wang, J.-Y., and Pei, J. (2014). Roles of flexible chains in organic semiconducting materials. *Chemistry of Materials* 26 (1): 594–603.

41 Rivnay, J., Inal, S., Collins, B.A. et al. (2016). Structural control of mixed ionic and electronic transport in conducting polymers. *Nature Communications* 7 (1): 11287.

42 Kim, Y., Noh, H., Paulsen, B.D. et al. (2021). Strain-engineering induced anisotropic crystallite orientation and maximized carrier mobility for high-performance microfiber-based organic bioelectronic devices. *Advanced Materials* 33 (10): 2007550.

43 Nurazzi, N.M., Harussani, M.M., Demon, S.Z.N. et al. (2022). Research progress on polythiophene and its application as chemical sensor. *Journal of Defence Science Engineering & Technology* 5(1): 48–68.

44 Liang, Z., Neshchadin, A., Zhang, Z. et al. (2023). Stille polycondensation: a multifaceted approach towards the synthesis of polymers with semiconducting properties. *Polymer Chemistry* 14 (40): 4611–4625.

45 Nakabayashi, K. (2018). Direct arylation polycondensation as conjugated polymer synthesis methodology. *Polymer Journal* 50 (7): 475–483.

46 Cheng, S., Zhao, R., and Seferos, D.S. (2021). Precision synthesis of conjugated polymers using the kumada methodology. *Accounts of Chemical Research* 54 (22): 4203–4214.

47 Kesornsit, S., Direksilp, C., Phasuksom, K. et al. (2022). Synthesis of highly conductive poly(3-hexylthiophene) by chemical oxidative polymerization using surfactant templates. *Polymers (Basel)* 14 (18): 3860.

48 Brendel, J.C., Schmidt, M.M., Hagen, G. et al. (2014). Controlled synthesis of water-soluble conjugated polyelectrolytes leading to excellent hole transport mobility. *Chemistry of Materials* 26 (6): 1992–1998.

49 Thalluri, G.K.V.V., Bolsée, J.C., Gadisa, A. et al. (2011). Opto-electrical and morphological characterization of water soluble conjugated polymers for eco-friendly hybrid solar cells. *Solar Energy Materials and Solar Cells* 95 (12): 3262–3268.

50 Inal, S., Rivnay, J., Leleux, P. et al. (2014). A high transconductance accumulation mode electrochemical transistor. *Advanced Materials* 26 (44): 7450–7455.

51 Schmode, P., Ohayon, D., Reichstein, P.M. et al. (2019). High-performance organic electrochemical transistors based on conjugated polyelectrolyte copolymers. *Chemistry of Materials* 31 (14): 5286–5295.

52 Nielsen, C.B., Giovannitti, A., Sbircea, D.-T. et al. (2016). Molecular design of semiconducting polymers for high-performance organic electrochemical transistors. *Journal of the American Chemical Society* 138 (32): 10252–10259.

53 Giovannitti, A., Sbircea, D.-T., Inal, S. et al. (2016). Controlling the mode of operation of organic transistors through side-chain engineering. *Proceedings of the National Academy of Sciences United States of America* 113 (43): 12017–12022.

54 Inal, S., Malliaras, G.G., and Rivnay, J. (2017). Benchmarking organic mixed conductors for transistors. *Nature Communications* 8 (1): 1–7.

55 Moser, M., Savagian, L.R., Savva, A. et al. (2020). Ethylene glycol-based side chain length engineering in polythiophenes and its impact on organic electrochemical transistor performance. *Chemistry of Materials* 32 (15): 6618–6628.

56 Moser, M., Hidalgo, T.C., Surgailis, J. et al. (2020). Side chain redistribution as a strategy to boost organic electrochemical transistor performance and stability. *Advanced Materials* 32 (37): 2002748.

57 Jones, A.L., De Keersmaecker, M., Savagian, L.R. et al. (2021). Branched oligo(ether) side chains: a path to enhanced processability and elevated conductivity for polymeric semiconductors. *Advanced Functional Materials* 31 (35): 1–11.

58 Giovannitti, A., Maria, I.P., Hanifi, D. et al. (2018). The role of the side chain on the performance of n-type conjugated polymers in aqueous electrolytes. *Chemistry of Materials* 30 (9): 2945–2953.

59 Savva, A., Hallani, R., Cendra, C. et al. (2020). Balancing ionic and electronic conduction for high-performance organic electrochemical transistors. *Advanced Functional Materials* 30 (11): 1–9.

60 Hallani, R.K., Paulsen, B.D., Petty, A.J. et al. (2021). Regiochemistry-driven organic electrochemical transistor performance enhancement in ethylene glycol-functionalized polythiophenes. *Journal of the American Chemical Society* 143 (29): 11007–11018.

61 He, Y., Kukhta, N.A., Marks, A. et al. (2022). The effect of side chain engineering on conjugated polymers in organic electrochemical transistors for bioelectronic applications. *Journal of Materials Chemistry C* 10 (7): 2314–2332.

62 Kukhta, N.A., Marks, A., and Luscombe, C.K. (2022). Molecular design strategies toward improvement of charge injection and ionic conduction in organic mixed ionic-electronic conductors for organic electrochemical transistors. *Chemical Reviews* 122 (4): 4325–4355.

63 Flagg, L.Q., Bischak, C.G., Onorato, J.W. et al. (2019). Polymer crystallinity controls water uptake in glycol side-chain polymer organic electrochemical transistors. *Journal of the American Chemical Society* 141 (10): 4345–4354.

64 Chen, S.E., Flagg, L.Q., Onorato, J.W. et al. (2022). Impact of varying side chain structure on organic electrochemical transistor performance: a series of oligoethylene glycol-substituted polythiophenes. *Journal of Materials Chemistry A* 10 (19): 10738–10749.

65 Dong, B.X., Nowak, C., Onorato, J.W. et al. (2019). Influence of side-chain Chemistry on structure and ionic conduction characteristics of polythiophene derivatives: a computational and experimental study. *Chemistry of Materials* 31 (4): 1418–1429.

66 Schmode, P., Savva, A., Kahl, R. et al. (2020). The key role of side chain linkage in structure formation and mixed conduction of ethylene glycol substituted polythiophenes. *ACS Applied Materials & Interfaces* 12 (11): 13029–13039.

67 Griggs, S., Marks, A., Meli, D. et al. (2022). The effect of residual palladium on the performance of organic electrochemical transistors. *Nature Communications* 13 (1): 7964.

68 Giovannitti, A., Nielsen, C.B., Sbircea, D.-T. et al. (2016). N-type organic electrochemical transistors with stability in water. *Nature Communications* 7 (1): 13066.

69 Wang, Y., Zeglio, E., Liao, H. et al. (2019). Hybrid alkyl-ethylene glycol side chains enhance substrate adhesion and operational stability in accumulation mode organic electrochemical transistors. *Chemistry of Materials* 31 (23): 9797–9806.

70 Giovannitti, A., Rashid, R.B., Thiburce, Q. et al. (2020). Energetic control of redox-active polymers toward safe organic bioelectronic materials. *Advanced Materials* 32 (16): 1908047.

71 Wang, Y., Wustoni, S., Surgailis, J. et al. (2024). Designing organic mixed conductors for electrochemical transistor applications. *Nature Reviews Materials* 9 (4): 249–265.

72 Moser, M., Savva, A., Thorley, K. et al. (2021). Polaron delocalization in Donor–acceptor polymers and its impact on organic electrochemical transistor performance. *Angewandte Chemie International Edition* 60 (14): 7777–7785.

73 Wu, X., Liu, Q., Surendran, A. et al. (2021). Enhancing the electrochemical doping efficiency in diketopyrrolopyrrole-based polymer for organic electrochemical transistors. *Advanced Electronic Materials* 7 (1): 2000701.

74 Jia, H., Huang, Z., Li, P. et al. (2021). Engineering donor–acceptor conjugated polymers for high-performance and fast-response organic electrochemical transistors. *Journal of Materials Chemistry C* 9 (14): 4927–4934.

75 Liu, Q., Bottle, S.E., and Sonar, P. et al. (2020). Developments of diketopyrrolopyrrole-dye-based organic semiconductors for a wide range of applications in electronics. *Advanced Materials* 32 (4): 1903882.

76 Jo, I.-Y., Jeong, D., Moon, Y. et al. (2024). High-performance organic electrochemical transistors achieved by optimizing structural and energetic ordering of diketopyrrolopyrrole-based polymers. *Advanced Materials* 36 (4): 2307402.

77 Zhan, X., Facchetti, A., Barlow, S. et al. (2011). Rylene and related diimides for organic electronics. *Advanced Materials* 23 (2): 268–284.

78 Sun, H., Vagin, M., Wang, S. et al. (2018). Complementary logic circuits based on high-performance n-type organic electrochemical transistors. *Advanced Materials* 30 (9): 1704916.

79 Kim, J., Ren, X., Zhang, Y. et al. (2023). Efficient n-type organic electrochemical transistors and field-effect transistors based on PNDI-copolymers bearing fluorinated selenophene-vinylene-selenophenes. *Advanced Science* 10 (29): 2303837.

80 Chen, J., Cong, S., Wang, L. et al. (2023). Backbone coplanarity manipulation via hydrogen bonding to boost the n-type performance of polymeric mixed conductors operating in aqueous electrolyte. *Materials Horizons* 10 (2): 607–618.

81 Feng, K., Guo, H., Sun, H. et al. (2021). N-type organic and polymeric semiconductors based on bithiophene imide derivatives. *Accounts of Chemical Research* 54 (20): 3804–3817.

82 Guo, H., Yang, C.Y., Zhang, X. et al. (2021). Transition metal-catalysed molecular n-doping of organic semiconductors. *Nature* 599 (7883): 67–73.

83 Feng, K., Shan, W., Wang, J. et al. (2022). Cyano-functionalized n-type polymer with high electron mobility for high-performance organic electrochemical transistors. *Advanced Materials* 34 (24): 2201340.

84 Wang, Y., Zeglio, E., Wang, L. et al. (2022). Green synthesis of lactone-based conjugated polymers for n-type organic electrochemical transistors. *Advanced Functional Materials* 32 (16): 2111439.

85 Wang, Y., Koklu, A., Zhong, Y. et al. (2024). Acceptor functionalization via green chemistry enables high-performance n-type organic electrochemical transistors for biosensing, memory applications. *Advanced Functional Materials* 34 (15): 2304103.

86 Li, P., Shi, J., Lei, Y. et al. (2022). Switching p-type to high-performance n-type organic electrochemical transistors via doped state engineering. *Nature Communications* 13 (1): 5970.

87 Wu, H.-Y., Yang, C.-Y., Li, Q. et al. (2022). Influence of molecular weight on the organic electrochemical transistor performance of ladder-type conjugated polymers. *Advanced Materials* 34 (4): 2106235.

88 Tang, H., Liang, Y., Liu, C. et al. (2022). A solution-processed n-type conducting polymer with ultrahigh conductivity. *Nature* 611 (7935): 271–277.

89 Chen, X., Marks, A., Paulsen, B.D. et al. (2021). n-type rigid semiconducting polymers bearing oligo(ethylene glycol) side chains for high-performance organic electrochemical transistors. *Angewandte Chemie International Edition* 60 (17): 9368–9373.

90 Marks, A., Chen, X., Wu, R. et al. (2022). Synthetic nuances to maximize n-type organic electrochemical transistor and thermoelectric performance in fused lactam polymers. *Journal of the American Chemical Society* 144 (10): 4642–4656.

91 Tan, G., Zhao, L.D., and Kanatzidis, M.G. (2016). Rationally designing high-performance bulk thermoelectric materials. *Chemical Reviews* 116 (19): 12123–12149.

92 Yang, L., Chen, Z.G., Dargusch, M.S. et al. (2018). High performance thermoelectric materials: progress and their applications. *Advanced Energy Materials* 8 (6): 1701797.

93 Beretta, D., Neophytou, N., Hodges, J.M. et al. (2019). Thermoelectrics: from history, a window to the future. *Materials Science and Engineering: R: Reports* 138: 100501.

94 Yao, C.J., Zhang, H.L., and Zhang, Q. (2019). Recent progress in thermoelectric materials based on conjugated polymers. *Polymers* 11 (1): 107.

95 Wang, S., Zuo, G., Kim, J. et al. (2022). Progress of conjugated polymers as emerging thermoelectric materials. *Progress in Polymer Science* 129: 101548.

96 Wu, W.N., Zheng, Q.B., and Liu, C.L. (2024). Recent progress in p-type doped conjugated polymer-based thermoelectric thin films. *Synthetic Metals* 307: 117682.

97 Russ, B., Glaudell, A., Urban, J.J. et al. (2016). Organic thermoelectric materials for energy harvesting and temperature control. *Nature Reviews Materials* 1 (10): 1–14.

98 Thomas, E.M., Popere, B.C., Fang, H. et al. (2018). Role of disorder induced by doping on the thermoelectric properties of semiconducting polymers. *Chemistry of Materials* 30 (9): 2965–2972.

99 Sun, Y., Di, C.A., Xu, W. et al. (2019). Advances in n-type organic thermoelectric materials and devices. *Advanced Electronic Materials* 5 (11): 1800825.

100 Tang, J., Chen, Y., McCuskey, S.R. et al. (2019). Recent advances in n-type thermoelectric nanocomposites. *Advanced Electronic Materials* 5 (11): 1800943.

101 Kim, N., Kee, S., Lee, S.H. et al. (2014). Highly conductive PEDOT:PSS nanofibrils induced by solution-processed crystallization. *Advanced Materials* 26 (14): 2268–2272.

102 Worfolk, B.J., Andrews, S.C., Park, S. (2015). Ultrahigh electrical conductivity in solution-sheared polymeric transparent films. *Proceedings of the National Academy of Sciences United States of America* 112 (46): 14138–14143.

103 Cho, B., Park, K.S., Baek, J. et al. (2014). Single-crystal poly(3,4-ethylenedioxythiophene) nanowires with ultrahigh conductivity. *Nano Letters* 14 (6): 3321–3327.

104 Jin Bae, E., Hun Kang, Y., Jang, KS., Yun Cho, S. (2016). Enhancement of thermoelectric properties of PEDOT:PSS and tellurium-PEDOT:PSS hybrid composites by simple chemical treatment. *Scientific Reports* 6: 18805.

105 Gueye, M.N., Carella, A., Massonnet, N. et al. (2016). Structure and dopant engineering in PEDOT thin films: practical tools for a dramatic conductivity enhancement. *Chemistry of Materials* 28 (10): 3462–3468.

106 Wang, X., Zhang, X., Sun, L. et al. (2018). High electrical conductivity and carrier mobility in oCVD PEDOT thin films by engineered crystallization and acid treatment. *Science Advances* 4 (9): 1–9.

107 Schultheiss, A., Carella, A., Pouget, S. et al. (2020). Water content control during solution-based polymerization: a key to reach extremely high conductivity in PEDOT thin films. *Journal of Materials Chemistry C* 8 (48): 17254–17260.

108 Hinckley, A.C., Andrews, S.C., Dunham, M.T. et al. (2021). Achieving high thermoelectric performance and metallic transport in solvent-sheared PEDOT:PSS. *Advanced Electronic Materials* 7 (3): 2001190.

109 Huang, Y., Tjhe, D.H.L., Jacobs, I.E. et al. (2021). Design of experiment optimization of aligned polymer thermoelectrics doped by ion-exchange. *Applied Physics Letters* 119 (11): 111903.

110 Lu, Y., Yu, Z.D., Liu, Y. et al. (2020). The critical role of dopant cations in electrical conductivity and thermoelectric performance of n-doped polymers. *Journal of the American Chemical Society* 142 (36): 15340–15348.

111 Yang, C.-Y., Ding, Y.-F., Huang, D. et al. (2020). A thermally activated and highly miscible dopant for n-type organic thermoelectrics. *Nature Communications* 11 (1): 3292.

112 Feng, K., Guo, H., Wang, J. et al. (2021). Cyano-functionalized bithiophene imide-based n-type polymer semiconductors: synthesis, structure-property correlations, and thermoelectric performance. *Journal of the American Chemical Society* 143 (3): 1539–1552.

113 Xiong, M., Yan, X., Li, J.T. et al. (2021). Efficient n-doping of polymeric semiconductors through controlling the dynamics of solution-state polymer aggregates. *Angewandte Chemie International Edition* 60 (15): 8189–8197.

114 Lu, Y., Yu, Z.-D., Un, H.-I. et al. (2021). Persistent conjugated backbone and disordered lamellar packing impart polymers with efficient n-doping and high conductivities. *Advanced Materials* 33 (2): 2005946.

115 Yu, Z.D., Lu, Y., Wang, Z.Y. et al. (2023). High n-type and p-type conductivities and power factors achieved in a single conjugated polymer. *Science Advances* 9 (8): eadf3495.

116 Chang, K.C., Jeng, M.S., Yang, C.C. et al. (2009). The thermoelectric performance of poly(3,4-ethylenedi oxythiophene)/poly(4-styrenesulfonate) thin films. *Journal of Electronic Materials* 38 (7): 1182–1188.

117 Bubnova, O., Khan, Z.U., Malti, A. et al. (2011). Optimization of the thermoelectric figure of merit in the conducting polymer poly(3,4-ethylenedioxythiophene). *Nature Materials* 10 (6): 429–433.

118 Park, T., Park, C., Kim, B. et al. (2013). Flexible PEDOT electrodes with large thermoelectric power factors to generate electricity by the touch of fingertips. *Energy Environmental Science* 6 (3): 788–792.

119 Culebras, M., Gómez, C.M., and Cantarero, A. (2014). Enhanced thermoelectric performance of PEDOT with different counter-ions optimized by chemical reduction. *Journal of Materials Chemistry A* 2 (26): 10109–10115.

120 Jona, F. and Morrisonb, J.T. (1997). 3,4-polyethylenedioxythiophene (PEDT): conductive coatings technical applications and properties. *Synthetic Metals* 85: 1397–1398.

121 Kirchmeyer, S. and Reuter, K. (2005). Scientific importance, properties and growing applications of poly(3,4-ethylenedioxythiophene). *Journal of Materials Chemistry A* 15 (21): 2077–2088.

122 Xia, Y. and Ouyang, J. (2012). Significant different conductivities of the two grades of poly(3,4-ethylenedioxythiophene): poly(styrenesulfonate), Clevios P and Clevios PH1000, arising from different molecular weights. *ACS Applied Materials & Interfaces* 4 (8): 4131–4140.

123 Poehler, T.O. and Katz, H.E. (2012). Prospects for polymer-based thermoelectrics: state of the art and theoretical analysis. *Energy & Environmental Science* 5 (8): 8110–8115.

124 Upadhyaya, M., Boyle, C.J., Venkataraman, D. et al. (2019). Effects of disorder on thermoelectric properties of semiconducting polymers. *Scientific Reports* 9 (1): 5820.

125 Park, H., Lee, S.H., Kim, F.S. et al. (2014). Enhanced thermoelectric properties of PEDOT:PSS nanofilms by a chemical dedoping process. *Journal of Materials Chemistry A* 2 (18): 6532–6539.

126 Lee, S.H., Park, H., Kim, S. et al. (2014). Transparent and flexible organic semiconductor nanofilms with enhanced thermoelectric efficiency. *Journal of Materials Chemistry A* 2 (20): 7288–7294.

127 Saxena, N., Keilhofer, J., Maurya, A.K. et al. (2018). Facile optimization of thermoelectric properties in PEDOT:PSS thin films through acido-base and redox dedoping using readily available salts. *ACS Applied Energy Materials* 1 (2): 336–342.

128 Massonnet, N., Carella, A., Jaudouin, O. et al. (2014). Improvement of the Seebeck coefficient of PEDOT:PSS by chemical reduction combined with a novel method for its transfer using free-standing thin films. *Journal of Materials Chemistry C* 2 (7): 1278–1283.

129 Fabretto, M.V., Evans, D.R., Mueller, M. et al. (2012) Polymeric material with metal-like conductivity for next generation organic electronic devices. *Chemistry of Materials* 24 (20): 3998–4003.

130 Petsagkourakis, I., Pavlopoulou, E., Portale, G. et al. (2016). Structurally-driven enhancement of thermoelectric properties within poly(3,4-ethylenedioxythiophene) thin films. *Scientific Reports* 6: 30501.

131 Lee, Y.H., Oh, J., Lee, S.S. et al. (2017). Highly ordered nanoconfinement effect from evaporation-induced self-assembly of block copolymers on in situ polymerized PEDOT:Tos. *ACS Macro Letters* 6 (4): 386–392.

132 Petsagkourakis, I., Pavlopoulou, E., Cloutet, E. et al. (2018). Correlating the Seebeck coefficient of thermoelectric polymer thin films to their charge transport mechanism. *Organic Electronics* 52: 335–341.

133 Xuan, Y., Liu, X., Desbief, S. et al. (2010). Thermoelectric properties of conducting polymers: the case of poly(3-hexylthiophene). *Physical Review B Condensed Matter* 82 (11): 115454.

134 Zhang, Q., Sun, Y., Xu, W. et al. (2012). Thermoelectric energy from flexible P3HT films doped with a ferric salt of triflimide anions. *Energy & Environmental Science* 5 (11): 9639–9644.

135 Aziz, E.F., Vollmer, A., Eisebitt, S. et al. (2007). Localized charge transfer in a molecularly doped conducting polymer. *Advanced Materials* 19 (20): 3257–3260.

136 Duong, D.T., Wang, C., Antono, E. et al. (2013). The chemical and structural origin of efficient p-type doping in P3HT. *Organic Electronics* 14 (5): 1330–1336.

137 Scholes, D.T., Hawks, S.A., Yee, P.Y. et al. (2015). Overcoming film quality issues for conjugated polymers doped with F4TCNQ by solution sequential processing: hall effect, structural, and optical measurements. *Journal of Physical Chemistry Letters* 6 (23): 4786–4793.

138 Jacobs, I.E., Aasen, E.W., Oliveira, J.L. et al. (2016). Comparison of solution-mixed and sequentially processed P3HT:F4TCNQ films: effect of doping-induced aggregation on film morphology. *Journal of Materials Chemistry C* 4 (16): 3454–3466.

139 Scholes, D.T., Yee, P.Y., Lindemuth, J.R. et al. (2017). The effects of crystallinity on charge transport and the structure of sequentially processed F4TCNQ-doped conjugated polymer films. *Advanced Functional Materials* 27 (44): 1702654.

140 Fontana, M.T., Stanfield, D.A., Scholes, D.T. et al. (2019). Evaporation vs solution sequential doping of conjugated polymers: F4TCNQ doping of micrometer-thick P3HT films for thermoelectrics. *Journal of Physical Chemistry C* 123 (37): 22711–22724.

141 Lim, E., Peterson, K.A., Su, G.M. et al. (2018). Thermoelectric properties of poly(3-hexylthiophene) (P3HT) doped with 2,3,5,6-tetrafluoro-7,7,8,8-tetracyanoquinodimethane (F4TCNQ) by vapor-phase infiltration. *Chemistry of Materials* 30 (3): 998–1010.

142 Kroon, R., Kiefer, D., Stegerer, D. et al. (2017). Polar side chains enhance processability, electrical conductivity, and thermal stability of a molecularly p-doped polythiophene. *Advanced Materials* 29 (24): 1700930.

143 Hynynen, J., Kiefer, D., Yu, L. et al. (2017). Enhanced electrical conductivity of molecularly p-doped poly(3-hexylthiophene) through understanding the correlation with solid-state order. *Macromolecules* 50 (20): 8140–8148.

144 Mardi, S., Pea, M., Notargiacomo, A. et al. (2020). The molecular weight dependence of thermoelectric properties of poly (3-hexylthiophene). *Materials* 13 (6): 1404.

145 Lim, E., Glaudell, A.M., Miller, R. et al. (2019). The role of ordering on the thermoelectric properties of blends of regioregular and regiorandom poly (3-hexylthiophene). *Advanced Electronic Materials* 5 (11): 1800915.

146 Gregory, S.A., Menon, A.K., Ye, S. et al. (2018). Effect of heteroatom and doping on the thermoelectric properties of poly(3-alkylchalcogenophenes). *Advanced Energy Materials* 8 (34): 1802419.

147 Zhang, Q., Sun, Y., Jiao, F. et al. (2012). Effects of structural order in the pristine state on the thermoelectric power-factor of doped PBTTT films. *Synthetic Metals* 162 (9–10): 788–793.

148 Zhang, Q., Sun, Y., Xu, W. et al. (2014). What to expect from conducting polymers on the playground of thermoelectricity: lessons learned from four high-mobility polymeric semiconductors. *Macromolecules* 47 (2): 609–615.

149 Cochran, J.E., Junk, M.J.N., Glaudell, A.M. et al. (2014). Molecular interactions and ordering in electrically doped polymers: blends of PBTTT and F4TCNQ. *Macromolecules* 47 (19): 6836–6846.

150 Glaudell, A.M., Cochran, J.E., Patel, S.N. et al. (2015). Impact of the doping method on conductivity and thermopower in semiconducting polythiophenes. *Advanced Energy Materials* 5 (4): 1401072.

151 Kang, K., Watanabe, S., Broch, K. et al. (2016). 2D coherent charge transport in highly ordered conducting polymers doped by solid state diffusion. *Nature Materials* 15 (8): 896–902.

152 Patel, S.N., Glaudell, A.M., Peterson, K.A. et al. (2017). Morphology controls the thermoelectric power factor of a doped semiconducting polymer. *Science Advances* 3 (6): e1700434.

153 Patel, S.N., Glaudell, A.M., Kiefer, D. et al. (2016). Increasing the thermoelectric power factor of a semiconducting polymer by doping from the vapor phase. *ACS Macro Letters* 5 (3): 268–272.

154 Jung, I.H., Hong, C.T., Lee, U.H. et al. (2017). High thermoelectric power factor of a diketopyrrolopyrrole-based low bandgap polymer via finely tuned doping engineering. *Scientific Reports* 7: 44704.

155 Liu, Z., Hu, Y., Li, P. et al. (2020). Enhancement of the thermoelectric performance of DPP based polymers by introducing one 3,4-ethylenedioxythiophene electron-rich building block. *Journal of Materials Chemistry C* 8 (31): 10859–10867.

156 Ding, J., Liu, Z., Zhao, W. et al. (2019). Selenium-substituted diketopyrrolopyrrole polymer for high-performance p-type organic thermoelectric materials. *Angewandte Chemie International Edition* 58 (52): 18994–18999.

157 Hu, Y., Gao, Y., Li, P. et al. (2022). Enhancement of thermoelectric properties of D-A conjugated polymer through constructing random copolymers with more electronic donors. *Journal of Polymer Science* 60 (6): 1002–1012.

158 Jia, H. and Lei, T. (2019). Emerging research directions for n-type conjugated polymers. *Journal of Materials Chemistry C* 7 (41): 12809–12821.

159 Schlitz, R.A., Brunetti, F.G., Glaudell, A.M. et al. (2014). Solubility-limited extrinsic n-type doping of a high electron mobility polymer for thermoelectric applications. *Advanced Materials* 26 (18): 2825–2830.

160 Wang, S., Sun, H., Ail, U. et al. (2016). Thermoelectric properties of solution-processed n-doped ladder-type conducting polymers. *Advanced Materials* 28 (48): 10764–10771.

161 Liu, J., Qiu, L., Alessandri, R. et al. (2018). Enhancing molecular n-type doping of donor–acceptor copolymers by tailoring side chains. *Advanced Materials* 30 (7): 1704630.

162 Kiefer, D., Giovannitti, A., Sun, H. et al. (2018). Enhanced n-doping efficiency of a naphthalenediimide-based copolymer through polar side chains for organic thermoelectrics. *ACS Energy Letters* 3 (2): 278–285.

163 Wang, S., Sun, H., Erdmann, T. et al. (2018). A chemically doped naphthalenediimide-bithiazole polymer for n-type organic thermoelectrics. *Advanced Materials* 30 (31): 1801898.

164 Lu, Y., Yu, Z., Zhang, R. et al. (2019). Rigid coplanar polymers for stable n-type polymer thermoelectrics. *Angewandte Chemie International Edition* 58 (33): 11390–11394.

165 Liu, J., Ye, G., Potgieser, H.G.O. et al. (2021). Amphipathic side chain of a conjugated polymer optimizes dopant location toward efficient n-type organic thermoelectrics. *Advanced Materials* 33 (4): 2006694.

166 Ye, G., Liu, J., Qiu, X. et al. (2021). Controlling n-type molecular doping via regiochemistry and polarity of pendant groups on low band gap donor-acceptor copolymers. *Macromolecules* 54 (8): 3886–3896.

167 Naab, B.D., Guo, S., Olthof, S. et al. (2013). Mechanistic study on the solution-phase n-doping of 1,3-dimethyl-2-aryl-2, 3-dihydro-1H-benzoimidazole derivatives. *Journal of the American Chemical Society* 135 (40): 15018–15025.

168 Bardagot, O., Aumaître, C., Monmagnon, A. et al. (2021). Revisiting doping mechanisms of n-type organic materials with N-DMBI for thermoelectric applications: photo-activation, thermal activation, and air stability. *Applied Physics Letters* 118 (20): 203904.

169 Zhang, F., Zang, Y., Huang, D. et al. (2015). Flexible and self-powered temperature-pressure dual-parameter sensors using microstructure-frame-supported organic thermoelectric materials. *Nature Communications* 6 (1): 8356.

170 Khan, Z.U., Edberg, J., Hamedi, M.M. et al. (2016). Thermoelectric polymers and their elastic aerogels. *Advanced Materials* 28 (22): 4556–4562.

171 Han, S., Jiao, F., Khan, Z.U. et al. (2017). Thermoelectric polymer aerogels for pressure–temperature sensing applications. *Advanced Functional Materials* 27 (44): 1703549.

172 Han, S., Alvi, N.U.H., Granlöf, L. et al. (2019). A multiparameter pressure–temperature–humidity sensor based on mixed ionic–electronic cellulose aerogels. *Advanced Science* 6 (8): 1802128.

173 Kim, N., Lienemann, S., Petsagkourakis, I. et al. (2020). Elastic conducting polymer composites in thermoelectric modules. *Nature Communications* 11 (1): 1424.

174 Yang, C.Y., Stoeckel, M.A., Ruoko, T.P. et al. (2021). A high-conductivity n-type polymeric ink for printed electronics. *Nature Communications* 12 (1): 2354.

175 Jurado, J.P., Dörling, B., Zapata-Arteaga, O. et al. (2019). Solar harvesting: a unique opportunity for organic thermoelectrics? *Advanced Energy Materials* 9 (45) 1902385.

11

π-Conjugated Polymers with Controlled Higher-order Structures: Poly(acetylene)s and Poly(aryleneethynylene)s

Fumio Sanda

Department of Chemistry and Materials Engineering, Faculty of Chemistry, Materials and Bioengineering, Kansai University, Suita, Osaka, Japan

11.1 Introduction

π-Conjugated polymers have attracted much attention due to their potentials for a broad range of photoelectrical applications [1, 2], including light-emitting diodes [3], photovoltaic materials [4], polyelectrolytes [3], and bio/chemo sensors [5]. Most naturally occurring biomacromolecules, including proteins and DNA, form higher-order structures, representatively helices, which significantly contribute to elegant functions for maintaining living systems. Synthetic helical polymers possibly also exhibit features and functions based on the regulated structures in a fashion similar to biomacromolecules. Highly stereoregulated and moderately rigid polymers form helical structures. The first example is isotactic poly(propylene) developed by Natta et al. [6]. Advance in synthetic and organometallic chemistry allows us to design monomers and catalysts with sophisticated functions. As a result, various optically active synthetic helical polymers have been developed [7], which include poly(methacrylate)s [8], poly(isocyanate)s [9], poly(isocyanide)s [10], poly(quinoxaline)s [11], poly(silane)s [12], and poly(acetylene)s [13–18] obtained by chain-growth polymerizations; poly(phenylenevinylene)s [19], poly(aryleneethynylene)s [20], and poly(fluorene)s [21] obtained by step-growth polymerizations (Figure 11.1); metal-coordinating double helical polymers [22]; supramolecular helices transferring chirality [23]; and poly(thiophene)s detecting optical activity of proteins and DNA [24]. This chapter focuses on poly(acetylene)s and poly(aryleneethynylene)s and describes their controlled higher-order structures and functions.

Conjugated Polymer Synthesis and Materials Applications, First Edition. Edited by Yoshiki Chujo.
© 2026 Wiley-VCH GmbH. All rights reserved, including rights for text and data mining and training of artificial intelligence technologies or similar technologies. Published 2026 by Wiley-VCH GmbH.

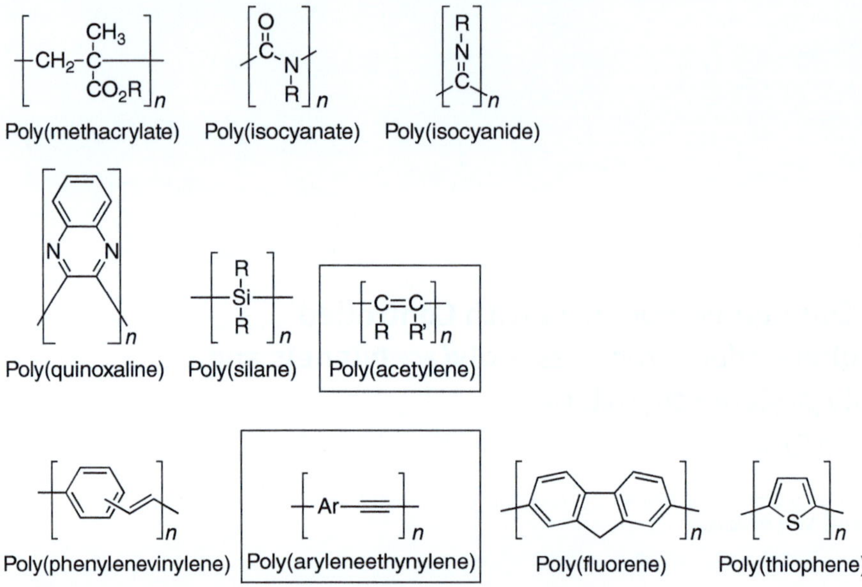

Figure 11.1 Representative polymers that possibly form helical structures, including poly(acetylene) and poly(aryleneethynylene).

11.2 Helical Substituted Poly(acetylene)s

Non-substituted poly(acetylene) is synthesized by polymerization of acetylene with Ziegler–Natta catalyst. Poly(acetylene) is called electronic conductive polymer, but the value is not so high compared with metals. Heeger, MacDiarmid, Shirakawa, and coworkers studied the electronic properties and gas-phase doping of poly(acetylene) films. By doping Cl_2, Br_2, I_2, and AsF_5, the electronic conductivity of poly(acetylene) films enhances as large as 10^7 to reach 500 S/cm, close to those of metals [25]. For practical usages as photoelectronic materials, it is necessary to introduce substituents because non-substituted poly(acetylene) is easily oxidized under air to undergo degradation. By introducing appropriate substituents in poly(acetylene), the main chain is twisted due to steric repulsion between the side chains. Substituted poly(acetylene)s are commonly synthesized by the polymerization of the corresponding substituted acetylene monomers using transition metal catalysts. The stereoregularity of poly(acetylene) main chain, i.e. control of *cis/trans*-geometry, is important for inducing regulated structures in substituted poly(acetylene)s. Rhodium (Rh) catalysts afford cis-stereoregulated poly(acetylene)s. Poly(acetylene) helix is dynamic, because the barrier for helix inversion is small. Helix inversion easily occurs, especially in the solution state. Poly(acetylene)s substituted with optically inactive groups contain equal amounts of right- and left-handed helices. On the other hand, poly(acetylene)s substituted with optically active groups commonly feature predominance of helical sense. To bias the torsion of poly(acetylene) main chain in one direction, it is necessary to introduce point chirality or axial chirality as an optically active site in the side chain substituents, or to add optically active compounds and/or solvents that interact with the side chain substituents. Amino acids and peptides are widely used as chiral sources due to their easy availability. Controlled twisting direction of substituted poly(acetylene)s is also achieved by topological chirality of interlocked molecules [26]. Twisting of poly(acetylene) backbone is brought about by steric repulsion between the side chains. By introducing amide [27], sulfamide [28], phosphonamidite [29], and carbamate [30] moieties in the side chain substituents (Figure 11.2), intramolecular hydrogen-bonding interaction becomes strong to stabilize the helical structures in addition to steric repulsion between the side chains, as shown in Figure 11.3.

The helical structures of substituted poly(acetylene)s are remarkably affected by conformation of the side chain substituents. The helix sense, radius, and pitch are

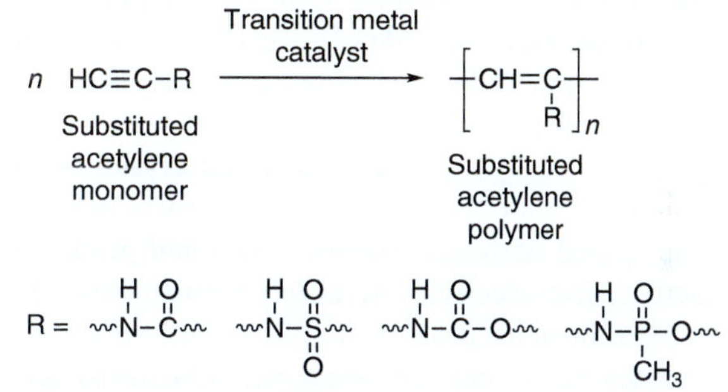

Figure 11.2 Synthesis of substituted acetylene polymers bearing hydrogen-bonding forming groups.

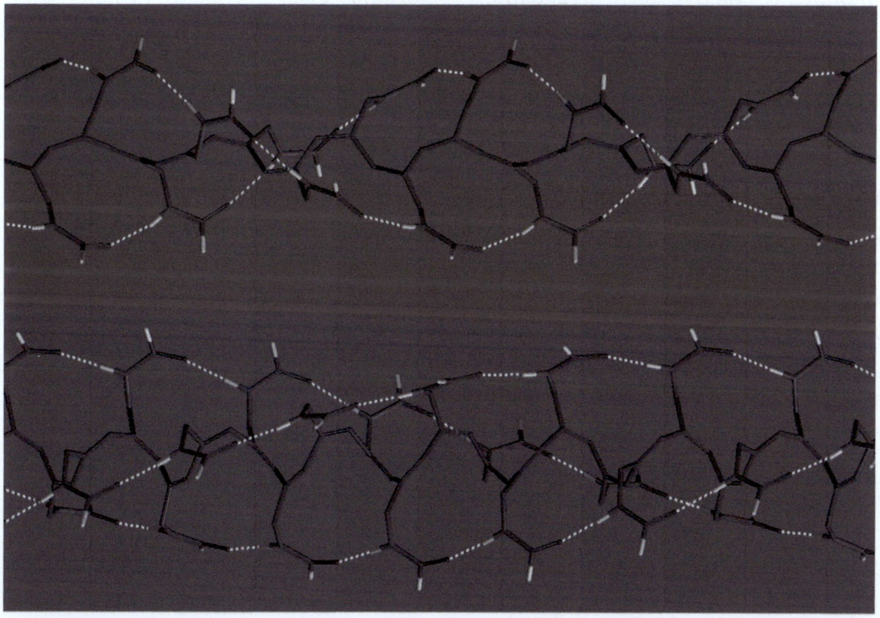

Figure 11.3 Images of helical poly(acetylene)s substituted with amide groups. Top: helix-forming intramolecular hydrogen-bonding strands between *i*th and *i* + 2nd units. Bottom: helix-forming hydrogen-bonding strands between *i*th and *i* + 3rd units.

changed in response to external stimuli such as solvent, temperature, pH, and pressure. These chirality changes are commonly examined by circular dichroism (CD) and ultraviolet (UV)-visible absorption spectroscopy. They are also examined by vibrational circular dichroism (VCD) [31] and Raman [32] spectroscopy. Further, the helical structures are directly observed using atomic force microscopy [33] and examined by quantum mechanical [34, 35] and molecular dynamics (MD) [36] simulations.

The UV-vis absorptions of helical poly(acetylene)s are red- and blueshifted when the degree of twisting of the main chain becomes small and large, respectively. Namely, π-conjugation becomes large when poly(acetylene) backbone becomes planar, and vice versa. The changes of twisting are tunable by temperature, solvent, and organic/inorganic additives, and the color change is detectable by the naked eye. Thus, helical poly(acetylene)s are applicable as sensing materials [37, 38]. The functions of substituted poly(acetylene)s with photoluminescent and redox properties are controllable and associated with changes of helical structures. Substituted poly(acetylene)s that exhibit circularly polarized luminescence (CPL) are expected as polarized light sources for liquid crystalline displays, next-generation optical communication technology, and medical materials [39–41]. Helical poly(acetylene)s substituted with catalytically functional groups are developed, and the catalytic activity and selectivity are well controlled [42–44].

Non-linear asymmetric amplification is often observed in the copolymers of substituted acetylenes. For example, substituted acetylene copolymers, obtained by copolymerizing a small amount of optically active phenylacetylene derivative with an optically inactive derivative, exhibit remarkably biased direction of helix sense, even if the ratio of optically active units is very small. In addition, copolymers of enantiomeric (*R*)- and (*S*)-acetylene monomers with a low enantiomer excess ratio also exhibit strongly biased helix direction [45–47]. These phenomena of asymmetric amplification were first discovered in poly(isocyanate)s and referred as "Sergeants-and-Soldiers Principle" [48] and "Majority Principle" [49], respectively. They have attracted attention due to the possible relation with the origin of homochirality. In addition to these asymmetric amplifications, "chiral memory effect" [50] is also reported. Namely, when optically active compounds are added to achiral substituted poly(acetylene)s, the direction of helix sense is biased. After that, even when the optically active compounds are removed, substituted poly(acetylene)s still exhibit chirality. Such substituted poly(acetylene)s that show this memory effect are applicable as stationary phase in high-performance liquid chromatography, wherein reversible control over elution order of enantiomers is possible [51].

Most of the research on control of higher-order structures of substituted poly(acetylene)s is carried out concerning monosubstituted acetylene polymers obtained using Rh catalysts, which afford cis-stereoregulated substituted poly(acetylene)s. Ruthenium (Ru) and palladium (Pd) catalysts are often used to obtain substituted poly(acetylene)s with controlled main chain stereoregularity [52, 53]. The research on disubstituted acetylene polymers is limited compared to monosubstituted ones, because it is difficult to control the main chain stereoregularity of disubstituted acetylene polymers. Recently, the synthesis of structure-controlled disubstituted acetylene polymers [54] and the formation of helical structures are achieved [55]. Since disubstituted acetylene polymers exhibit excellent fluorescent properties, further development is expected utilizing the helix nature to photoluminescent functions.

11.3 Poly(aryleneethynylene)s

Poly(aryleneethynylene)s are also widely studied π-conjugated polymers that exhibit useful photoelectric functions [20]. The conjugation length of poly(aryleneethynylene)s is controllable by the arylene units, leading to tuning of photoelectric properties, especially UV-vis absorption and photoluminescence emission. Poly(aryleneethynylene)s with hydrophilic side chains such as oligo(ethylene glycohol)s [56], ammonium salts [57], and carboxylates [58] fold into helices in hydrophilic solvents, wherein solvophobic interactions drive the folding transition. On the contrary, poly(aryleneethynylene)s with amide-linked hydrophobic alkyl groups fold into helices in hydrophobic solvents, wherein the helical structures are stabilized by van der Waals interaction between the alkyl groups and hydrogen bonds between the amide groups [59]. When hydrophilic solvent is added, the helical structures are converted into random structures due to collapse of hydrogen-bonding strands. Stability against environmental media is required to utilize polymers with controlled helical structures. Poly(phenyleneethynylene)s with long alkylamide groups also form stable folded helical structures in hydrophobic solvents [60]. The hydrogen bonds between the side chains of poly(phenyleneethynylene)s with side chains containing optically active amide and ethylene glycol units are weakened in hydrophilic solvents, but the hydrophilic side chains and solvent show solvophobic effect that increases the π–π interactions of the main chain, maintaining the helical structures. The secondary structures of peptides are stabilized using hydrocarbon staples, i.e. side chain cross-linking [61, 62]. Such peptides are expected as inhibitors of protein–protein interactions [63]. Methods of cross-linking of polymer side chains include metathesis reactions, lactonization, and the cycloaddition of azides and alkenes [64]. Poly(phenyleneethynylene)s with cinnamic acid moieties in the side chains form stable helical structures by cross-linking between the side chains through photoirradiation [65]. Optically active poly(phenyleneethynylene)s with diyne moieties in the side chains undergo selective intramolecular cross-linking via metathesis reaction under dilute conditions (Figure 11.4) [66].

Photoresponsive polymers are expected as optical memory and molecular machines. As shown in Figure 11.5, poly(phenyleneethynylene)s with azobenzene moieties in the side chains fold into helices, and the azobenzene moieties are reversibly isomerized by UV and visible light irradiation [67]. However, the change in higher-order structures associated with azobenzene isomerization is small. In contrast, the higher-order structure of poly(phenyleneethynylene)s with azobenzene units in the main chain reversibly switches between folded helical structures and random structures in response to the photoisomerization of the azobenzene units [68]. Figure 11.6 shows the stable conformation of an azobenzene-linked polymer, whose

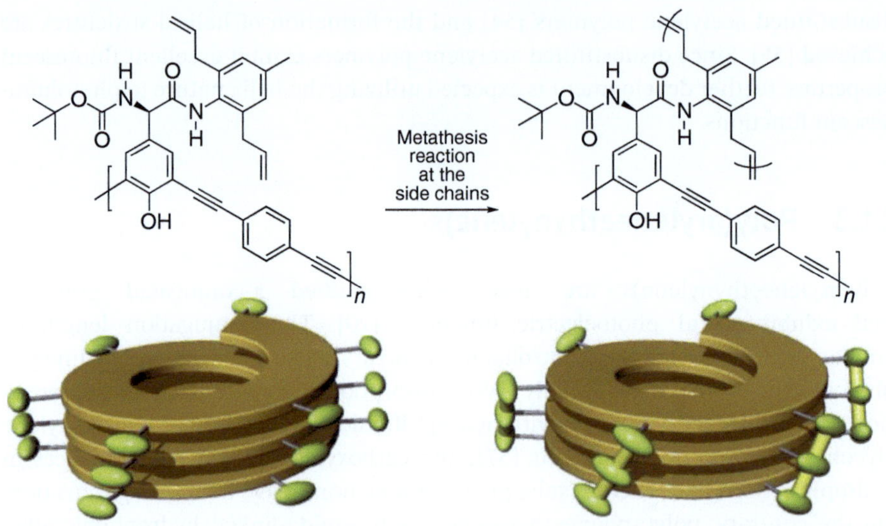

Figure 11.4 Stabilization of helically folded conformation by intramolecular cross-linking at the side chains of a poly(phenyleneethynylene) utilizing metathesis reaction.

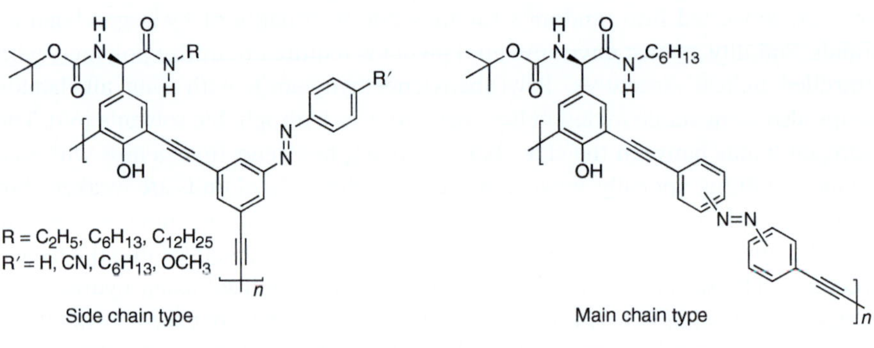

Figure 11.5 Poly(phenyleneethynylene)s bearing azobenzene moieties.

geometries were optimized by density functional theory (DFT) calculations. The CD simulated using the semi-empirical quantum mechanical method is in good agreement with the measured spectrum, indicating the properness of the conformation.

The conjugation lengths of the poly(aryleneethynylene)s are controllable by changing the aromatic rings [69, 70]. Poly(aryleneethynylene)s with 1,4-phenylene and 2,7-naphthylene units emit purple to blue light, a poly(aryleneethynylene) with 1,4-naphthylene units emits green light, and a poly(aryleneethynylene) with 1,6-pyrenylene units emits yellow light (Figure 11.7). The poly(aryleneethynylene) with 1,4-phenylene units forms a supramolecular structure in the solid state that is arranged in the axial direction of the helix. Poly(aryleneethynylene)s with perylene side chains transfer energy from the main chain to side chains [71].

Terphenylenes are rigid conjugated building blocks that are frequently used to construct pillar- and cage-like molecules. *Para*-linked polymers obtained by oxidative coupling polymerization of diethynylterphenylene monomers show CD signals

Figure 11.6 Possible conformer of an azobenzene-containing poly(phenyleneethynylene) after geometry optimization by DFT method (M06-2X/6-31G*). *Source*: [68]/with permission of American Chemical Society.

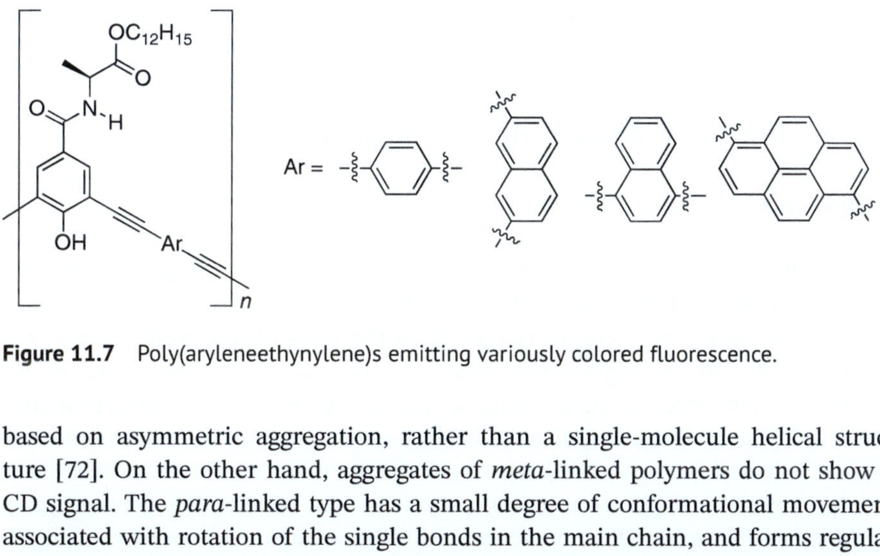

Figure 11.7 Poly(aryleneethynylene)s emitting variously colored fluorescence.

based on asymmetric aggregation, rather than a single-molecule helical structure [72]. On the other hand, aggregates of *meta*-linked polymers do not show a CD signal. The *para*-linked type has a small degree of conformational movement associated with rotation of the single bonds in the main chain, and forms regular aggregates of chiral molecules, whereas the *meta*-linked type has a large degree of conformational freedom, and it is thought that it cannot form regular aggregates.

2,2′-Bipyridine and its derivatives efficiently coordinate to metals to give a wide variety of organometallic compounds utilized as catalysts and photo- and electroluminescent materials. Organometallic compounds with chiral 2,2′-bipyridine derivatives catalyze asymmetric reactions, and compounds bearing chiral 2,2′-bipyridine moieties coordinate to metals to give chiral supramolecular assemblies. 2,2′-Bipyridine derivatives are also widely employed as building blocks in

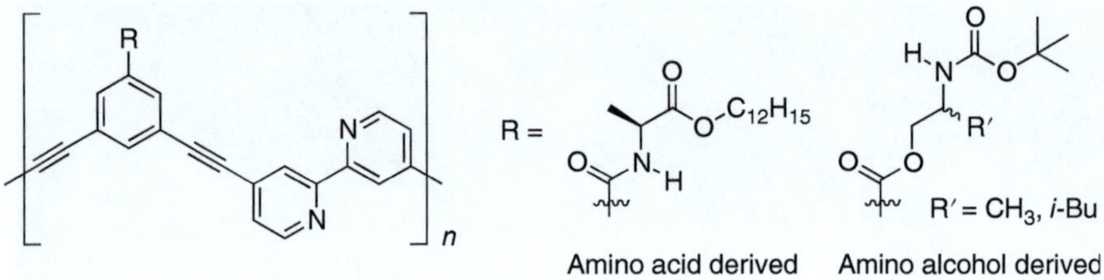

Figure 11.8 Bipyridine-containing poly(phenyleneethynylene)s derived from amino acid and amino alcohol.

polymer chemistry. Most bipyridine-containing π-conjugated polymers exhibit photoluminescence, and the luminescence color changes upon coordination of metals. 2,2′-Bipyridine-containing poly(phenyleneethynylene)s (Figure 11.8, left) adopt folded helical structures, and simultaneously form aggregates depending on solvents as supported by DFT calculations and MD simulations [73]. Bipyridine-containing π-conjugated polymers with optically active amino-alcohol-derived side chains (Figure 11.8, right) also adopt a folded helical conformation and form aggregates [74]. The polymer films show differences in contact angles with aqueous solutions of optically active alcohols. Addition of some metal chlorides and perchlorates changes the intensity and color of photoluminescence of the polymers. These features are well explained by quantum mechanical and MD simulations, and suggest the applicability of the polymers to chiral recognition and metal ion sensing materials.

11.4 Platinum-containing Poly(aryleneethynylene)s

Various platinum (Pt) acetylide π-conjugated polymers and oligomers have been developed since the first report by Fujikura, Sonogashira, and Hagihara in 1975 [75]. Pt acetylide π-conjugated systems transport energy through the conjugated main chain [76] and find applications in nonlinear optical materials and photovoltaic devices. Although there are many reports on the properties of π-conjugated polymers related with the higher-order structures as mentioned above, there are few reports on the effect of higher-order structures on the photophysical properties of Pt acetylide conjugated polymers [77]. A series of *meta*- and *para*-phenylene-linked poly(aryleneethynylene)s bearing Pt in the main chain are synthesized by the Sonogashira–Hagihara coupling polymerization of bis[(ethynylphenyl)ethynyl]bis(butylphosphine)Pt monomers with dibromoarylene monomers using Pd and copper (Cu) catalysts (Figure 11.9) [78], and the relationship between the optical and aggregation properties is investigated. The wavelengths at the maximum point in the spectrum (λ_{max}'s) of the *meta*-phenylene-linked polymers are shorter than those of the *para*-phenylene-linked counterparts. This result indicates that the conjugation length of the *meta*-phenylene-linked polymers is shorter than that of the corresponding *para*-phenylene-linked compounds. The order of wavelength maxima of absorption and emission of the *meta*- and *para*-phenylene linked polymers is as follows: Ar = 1,4-phenylene <1,4-naphthalene <9,10-anthracene. This agrees with the order of λ_{max} of the corresponding aromatics: benzene

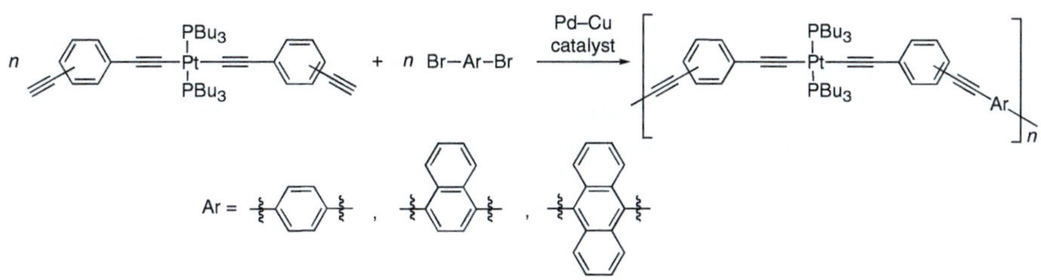

Figure 11.9 Synthesis of poly(aryleneethynylene)s bearing Pt in the main chain by Sonogashira–Hagihara coupling polymerization. *Source*: [78]/American Chemical Society.

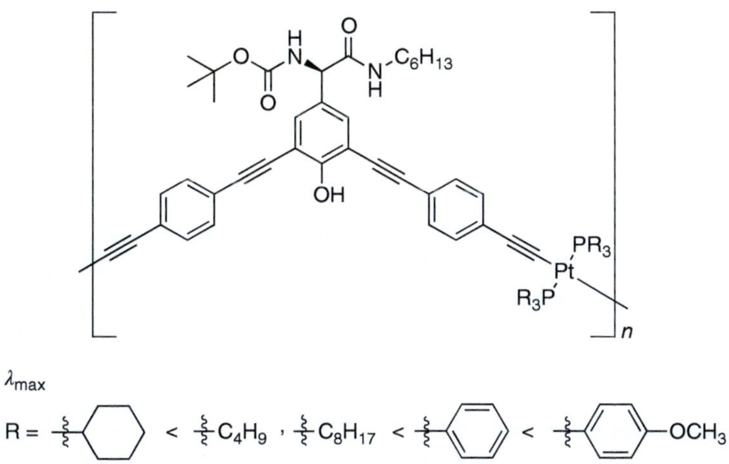

Figure 11.10 Pt-containing optically active poly(phenyleneethynylene)s with various phosphine ligands. *Source*: [81]/Royal Society of Chemistry.

(255 nm) <naphthalene (286 nm) <anthracene (375 nm). The number of fused aromatic rings apparently affects the absorption and emission properties of the polymers.

The σ-donating and π-accepting characters of the phosphine ligands in Pt complexes remarkably affect the HOMO → LUMO π–π* transition energy and the conjugation length of the backbone [79]. The bulkiness of trialkylphosphine ligands plays an important role in the supramolecular polymerization and macroscopic gelation behavior [80]. The UV-vis absorption spectra of Pt-containing D-hydroxyphenylglycine-derived poly(phenyleneethynylene)s bearing P(butyl)$_3$, P(octyl)$_3$, P(cyclohexyl)$_3$, P(phenyl)$_3$, and P(4-methoxyphenyl)$_3$ are measured to get information about the effect of the ligand on the optical properties (Figure 11.10) [81]. The order of λ_{max} of the monomers is as follows: P(cyclohexyl) <P(butyl)$_3$, P(octyl)$_3$ <P(phenyl)$_3$ <P(4-methoxyphenyl)$_3$. The λ_{max}'s of monomers with triarylphosphine ligands are positioned at longer wavelengths than those of monomers with trialkylphosphine ligands. The phosphine aryl groups seem to extend the conjugation more than the phosphine alkyl groups. The HOMO–LUMO gaps of the monomers calculated by the DFT method support this trend. These results indicate that the

conjugation length and HOMO–LUMO gaps of the monomers are affected by the substituents on the phosphine ligands.

Poly(phenyleneethynylene)s substituted with optically active groups tend to form chiral higher-order structures [82], i.e. predominantly one-handed helices and/or chiral aggregates, due to the small mobility of the conjugated main chains and solvophobic effects. The higher-order structures are further stabilized by intra- and/or intermolecular interactions, such as hydrogen bonding and π-stacking [83]. The chiral higher-order structures and photoelectric properties of the above-mentioned D-hydroxyphenylglycine-derived polymers containing Pt in the main chain are investigated [84]. The polymers exhibit λ_{max} around 370 nm, indicating that the substituents hardly affect the conjugation length, as supported by the very small difference of HOMO–LUMO band gaps between the model compounds calculated by the DFT method. Negligibly small CD signals are observed in THF, while intense CD signals are observed at the absorption region of the main chain chromophore around 320–420 nm in THF/MeOH = 1/9. The UV-vis absorbance is small in THF/MeOH = 1/9 compared with that in THF. In THF/toluene, the CD intensity gradually increases with increasing toluene content, accompanied by a small decrease of the UV-vis absorbance. Poly(*meta*-phenyleneethynylene)s form aggregates or foldamers under certain conditions, as mentioned above [20]. The λ_{max} of UV-vis absorbance of poly(phenyleneethynylene)s commonly depends on the conformation (random coil and helix) [58], while the λ_{max} values of the present polymers in helically folded and aggregated forms are not so different. One possible reason is the presence of amide groups, which suppress the molecular mobility of the polymers by forming inter/intramolecular hydrogen bonding, resulting in almost the same λ_{max}'s irrespective of helically folded and aggregated forms. Particles with hydrodynamic diameters around 100 nm are detected in a polymer solution of THF/MeOH = 1/9, while no particles are detected in a polymer solution of THF/toluene = 1/9. These results indicate that the polymer forms chiral aggregates in THF/MeOH = 1/9 and a predominantly one-handed helix in THF/toluene = 1/9.

The polymers emit blue-colored fluorescence in THF/MeOH mixtures. The shape and intensity of the fluorescence spectra are not affected by the solvent composition. On the other hand, in THF/toluene mixtures, the fluorescence intensity decreases, and another signal appears at a wavelength longer by 55 nm as the toluene content is increased. As mentioned above, toluene induces formation of a folded helical structure. The emission peak at the longer wavelength is attributed to intramolecular π-stacking. In THF/MeOH mixtures, fluorescence seems to be negligibly small, presumably because of the decreased efficiency of intermolecular (compared to intramolecular) π-stacking. Thus, the fluorescence properties of the polymers are affected by the change of secondary structures driven by solvent composition.

Pt-containing poly(aryleneethynylene)s can be also synthesized by the dehydrochlorination polymerization of $PtCl_2$ complex monomers with diethynylarylene monomers [75]. Pt ethynyl complexes with pyridine, bipyridine, and phenanthroline ligands are synthesized by the dehydrochlorination coupling polymerization of a *meta*-diethynylbenzene derivative with $PtCl_2$ complexes (Figure 11.11) [85]. The polymers exhibit regularly twisted ribbon-like structures in the solid states by AFM measurement. Molecular modeling study suggests that it is difficult for the polymer

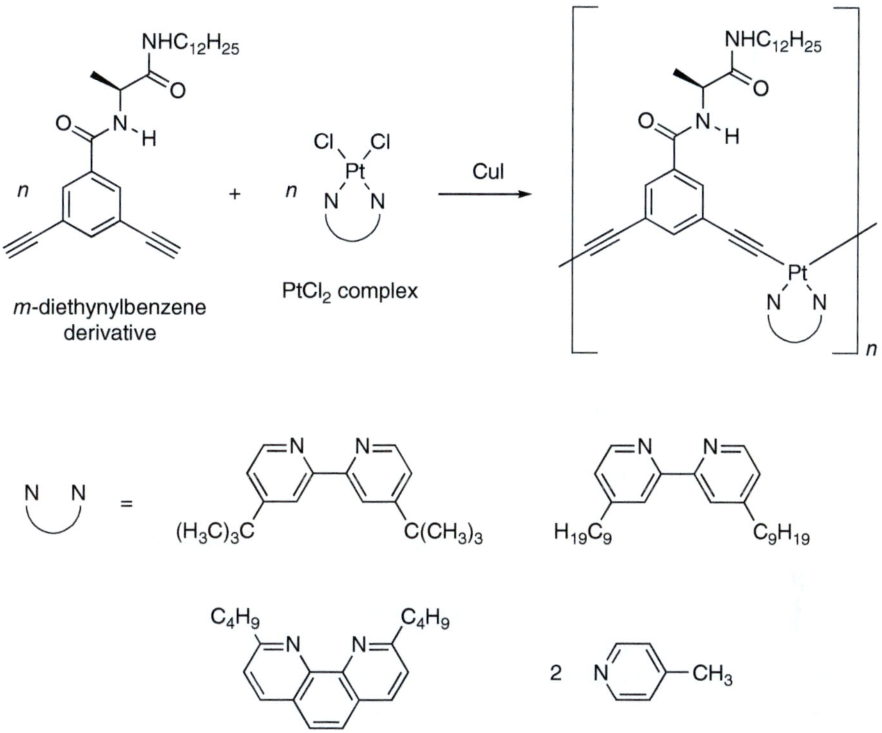

Figure 11.11 Synthesis of poly(aryleneethynylene)s bearing Pt in the main chain by dehydrochlorination polymerization. *Source*: [85]/Royal Society of Chemistry.

molecules to be folded into a helix due to the combination of *meta*-phenylene and bipyridine-ligated Pt linkages. It seems that the polymer molecules possibly adopt extended

Both *cis*- and *trans*-configurations are possible at the Pt centers of Pt acetylide complexes. Employment of a diphosphine ligand, $Ph_2P{-}(CH_2)_2{-}PPh_2$ enables a *cis*-Pt-configuration due to geometrical restrictions [86], but Pt-polyynes with monophosphine ligands are likely to adopt *trans*-configuration due to the thermodynamic favorableness. Although the configuration of the precursor (PtL_2Cl_2 L = ligand) is *cis*, Pt-polyynes transform into the *trans*-configuration, presumably by isomerization during the dehydrochlorination process. There are various reports of the exchange reactions of PPh_3 ligands for PR_3 (R=alkyl), because PPh_3 has a low coordination ability compared to PR_3. Other examples include replacement of ligands with diphosphines to form stable chelated structures [87]. Ligand exchange reactions are also applicable to the control of double helical structures of Pt-containing dimers [88]. Since *cis*/*trans*-geometries and ligand structure remarkably affect the photoelectric properties and shapes of Pt-polyynes, one can expect to gain control of the conformation and properties utilizing the ligand exchange reaction. Pt-containing optically active polymers are synthesized by ligand exchange of a $Pt(PPh_3)_2$-containing polymer (Figure 11.12) [89], and the conformation is investigated. The geometry of the Pt center is *trans*, which is confirmed by ^{31}P NMR

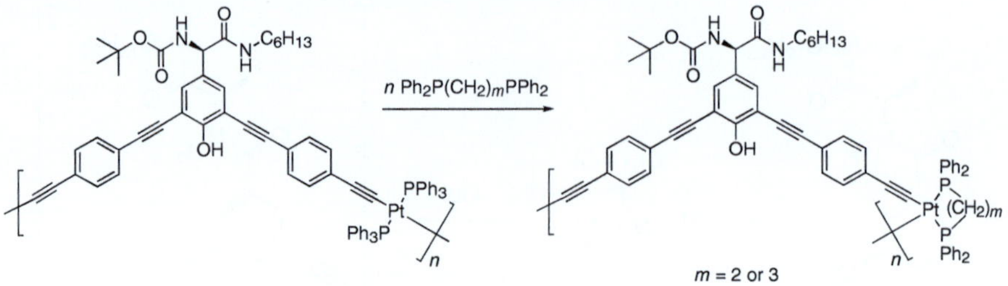

Figure 11.12 Ligand exchange reaction of a $Pt(PPh_3)_2$-containing polymer with $Ph_2P(CH_2)_mPPh_2$ (m = 2 or 3). *Source*: [89]/American Chemical Society.

spectroscopy. After ligand exchange, the geometry of the Pt center turns into *cis*. On the other hand, the geometry at the Pt center of $Pt[Ph_2P(CH_2)_3PPh_2]$-containing polymer keeps *trans*.

Figure 11.13 shows the possible structures of Pt-containing polymers having *trans* Pt centers bridged with $Ph_2P(CH_2)_2PPh_2$ and $Ph_2P(CH_2)_3PPh_2$ ligands. One $Ph_2P(CH_2)_2PPh_2$ molecule coordinates to one Pt atom with a *cis*-configuration. On the other hand, one $Ph_2P(CH_2)_3PPh_2$ molecule coordinates to two Pt atoms at the ith and $i + 6$th units with a *trans* configuration, which is supported by molecular modeling as shown in Figure 11.14.

As described above, the ligand exchange reaction of optically active polymers with optically inactive phosphine ligands successfully gives polymers with controlled secondary structures. The ligand exchange reaction of optically inactive polymers with optically active phosphine ligands also gives structure-controlled polymers (Figure 11.15) [90]. Intramolecular bridging reaction takes place to give a predominantly one-handed folded helical polymer exhibiting intense CD signals based on exciton coupling. The conformation is supported by the TD-DFT-simulated CD spectrum. The methodology leads to a new concept of control of inter/intramolecular cross-linking/bridging at the side chains of polymers, as well as induction of chirality in optically inactive metal-containing polymers by utilizing optically active ligands.

There is another type of optically active phosphine ligands that have P-chirality, including 1,2-bis(*tert*-butylmethylphosphino)benzene (BenzP*) and 2,3-bis (*tert*-butylmethylphosphino)quinoxaline (QuinoxP*) [91]. BenzP* and QuinoxP* are air-stable, especially QuinoxP*, and applicable to ligands of transition metal complexes that catalyze enantioselective hydrogenation [92, 93], aldol reaction [94], and borylation [95]. QuinoxP* is also usable as a building block of Pt-containing polymers, in which QuinoxP* plays a significant role for chirality induction to the polymer main chain [96]. BenzP* and QuinoxP* are employed as chiral phosphines for ligand exchange reactions to induce chirality to the main chain of Pt-containing poly(aryleneethynylene)s (Figure 11.16) [97]. The chirality of BenzP* and QuinoxP* is transferred to the main chains of naphthalene- and carbazole-containing polymers after ligand exchange reactions, as confirmed by CD and UV-vis absorption spectroscopies. The chiral twist of the main chain is supported by the thermodynamic stability of the model compound, as well as the CD and UV-vis absorption spectra simulated by the DFT method. The QuinoxP*-based polymers emit

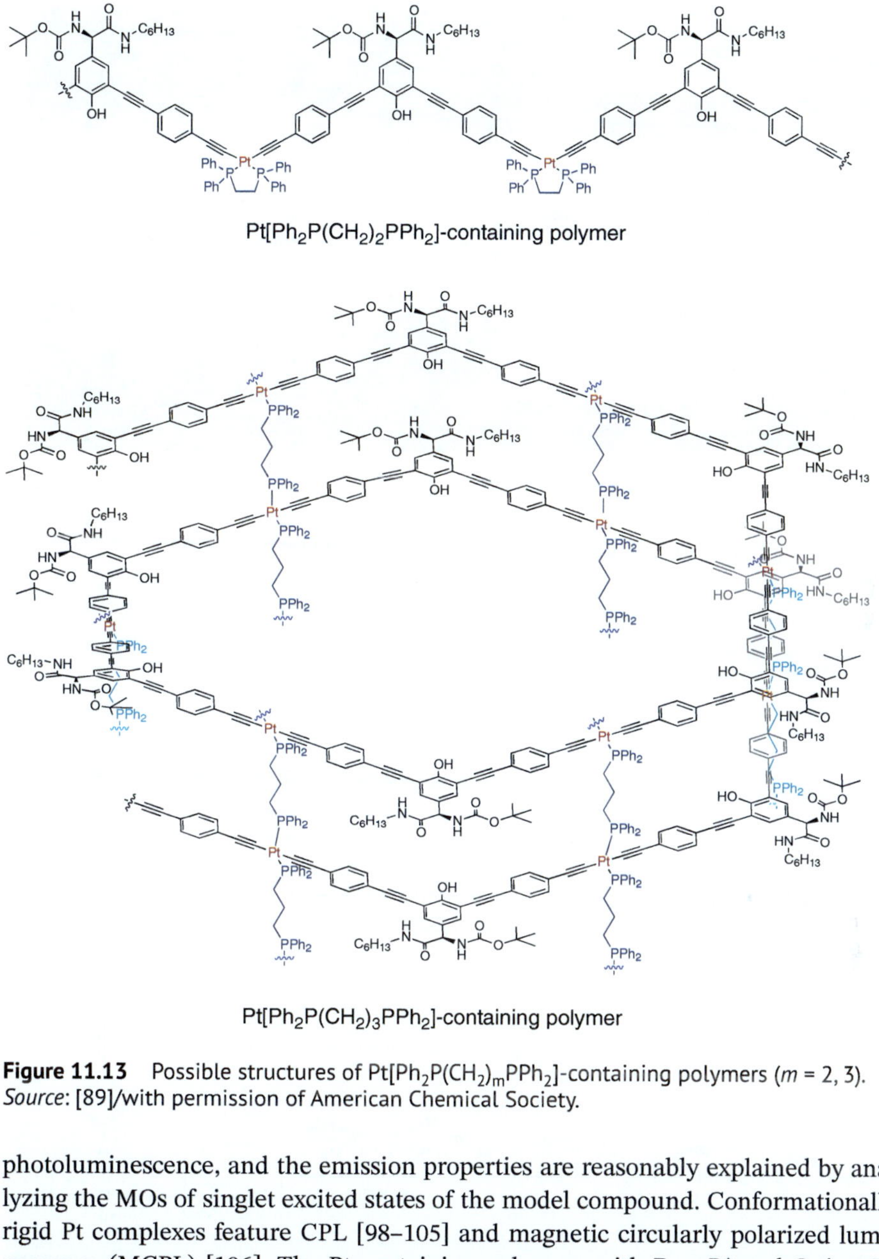

Figure 11.13 Possible structures of Pt[$Ph_2P(CH_2)_mPPh_2$]-containing polymers (m = 2, 3). *Source*: [89]/with permission of American Chemical Society.

photoluminescence, and the emission properties are reasonably explained by analyzing the MOs of singlet excited states of the model compound. Conformationally rigid Pt complexes feature CPL [98–105] and magnetic circularly polarized luminescence (MCPL) [106]. The Pt-containing polymers with BenzP* and QuinoxP* exhibit CPL and MCPL. Thus, the ligand exchange reaction is an efficient method to synthesize geometrically controlled Pt-functionalized optically active polymers from the optically inactive precursor polymer.

Poly(phenyleneethynylene)s bearing Pt-complex moieties ligated with phosphines substituted with L-valine-derived groups are synthesized (Figure 11.17) [107]. Geometrically *trans*-controlled polymer displayed a negative CD signal in the main chain chromophore region in (CH_2 Cl_2)/MeOH mixed solvent. The dynamic light

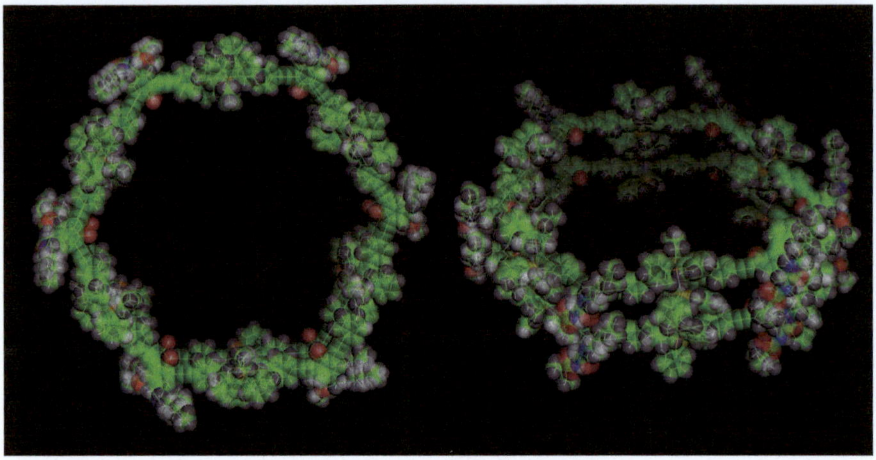

Figure 11.14 Top and tilt views of a 12-mer model for a $Pt[Ph_2P(CH_2)_3PPh_2]$-containing polymer. The geometries are optimized by the DFT method [B3LYP/6-31G* (H, C, N, O, P)-LANL2DZ (Pt)]. *Source*: [89]/with permission of American Chemical Society.

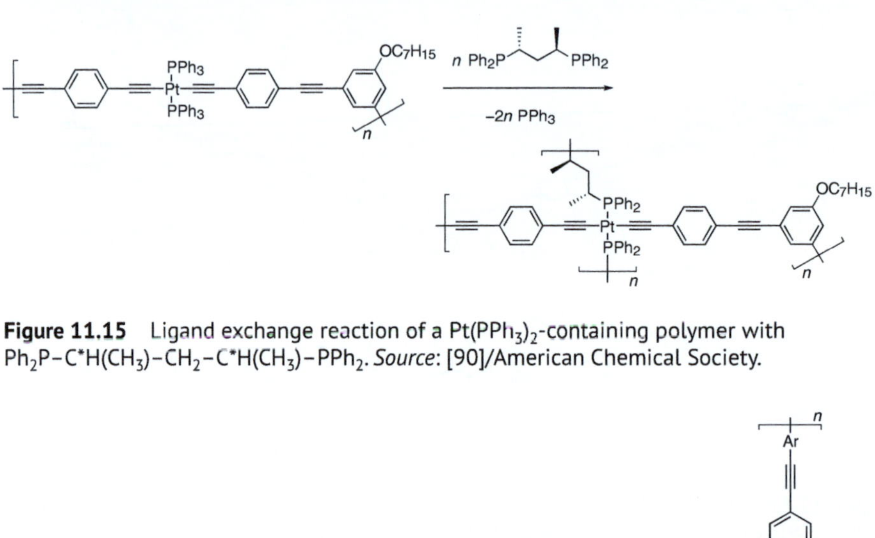

Figure 11.15 Ligand exchange reaction of a $Pt(PPh_3)_2$-containing polymer with $Ph_2P-C^*H(CH_3)-CH_2-C^*H(CH_3)-PPh_2$. *Source*: [90]/American Chemical Society.

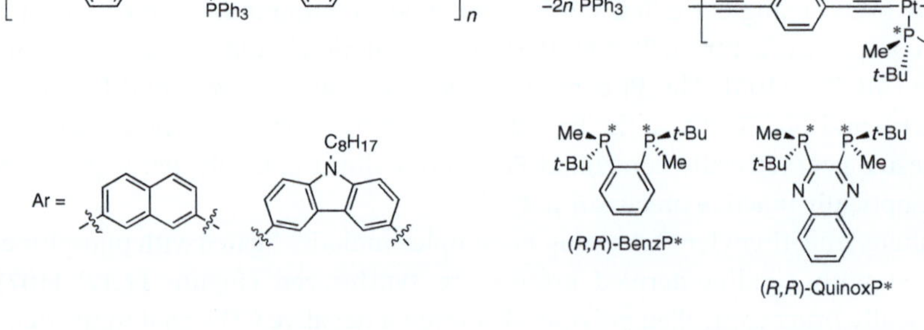

Figure 11.16 Ligand exchange reaction of a $Pt(PPh_3)_2$-containing polymers with (*R*,*R*)-BenzP* and (*R*,*R*)-QuinoxP*. *Source*: [97]/with permission of Elsevier.

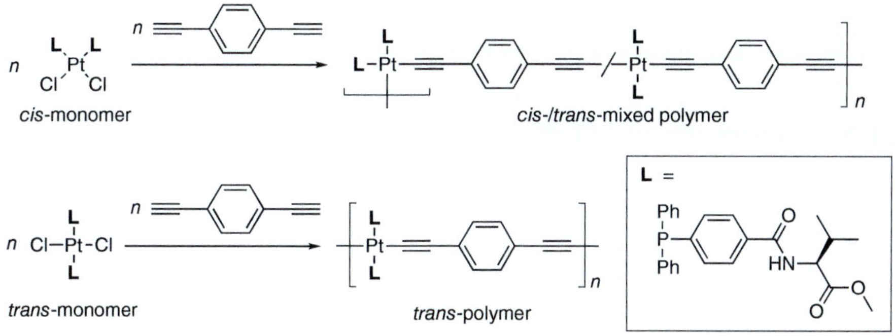

Figure 11.17 Synthesis of Pt-containing polymers bearing amino acid-functionalized phosphine ligands. *Source*: [107]/American Chemical Society.

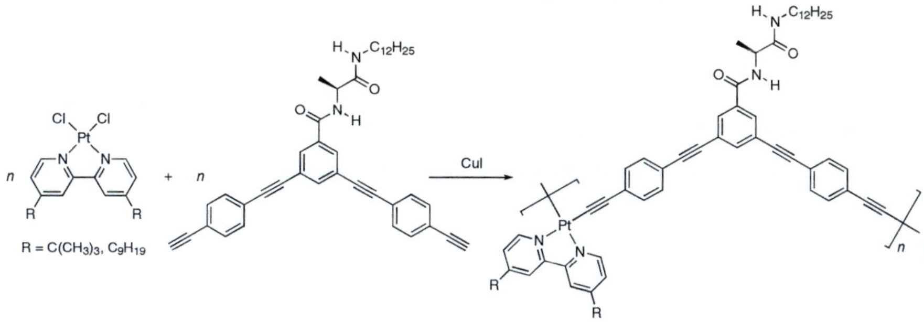

Figure 11.18 Synthesis of Pt(bipyridine)-containing polymers bearing amino acid-functionalized side chains. *Source*: [108]/American Chemical Society.

scattering (DLS) measurements, single crystal X-ray structure analysis, and DFT calculations reveal that the CD signal is induced by chirality transfer from the phosphine ligands to the polymer main chain-based conformational fixation by aggregation. Thus, the employment of an amino acid as a chiral source of a phosphine ligand is effective for induction of chiral higher-order structures of Pt-containing conjugated polymers. In the future, by the design of phosphine ligands carrying di-, tri-, and oligopeptide moieties that form multiple intramolecular hydrogen bonds, it is expected to effectively suppress the *cis*-to-*trans*-isomerization of the Pt centers, leading to the synthesis of completely *cis*-geometry controlled $[-C \equiv C-Pt(PR_3)_2-C \equiv C-Ar-]_n$ from *cis*-$Pt(PR_3)_2$ and $Ar(C \equiv C-H)_2$.

The other types of Pt-containing conjugated polymers bearing amino acid-derived optically active amide groups are synthesized by the dehydrochlorination coupling polymerization of the corresponding monomers with 1,3-bis[2-(4-ethynylphenyl) ethynyl]benzene derivative using CuI as a catalyst (Figure 11.18) [108]. The polymers show bisignate CD signals assignable to the main chain chromophore (300–450 nm) upon aggregation in DMF/MeOH = 80/20–50/50 (v/v), which is confirmed by DLS measurement and filtering experiments. The single-crystal X-ray analysis in conjunction with the DFT calculations of the model compounds reveals that the polymer molecules aggregate mainly by the interaction between the Pt complex and aromatic moieties, wherein the Pt complex moieties align slightly clockwise twisted, leading to the metal-to-ligand charge transfer- and ligand-to-ligand charge transfer-based CD signal in a positive exciton coupling manner.

11.5 Summary and Outlook

This chapter overviewed the control of higher-order structures and properties of π-conjugated polymers, focusing on substituted poly(acetylene)s and poly(aryleneethynylene)s. These polymers adopt helical conformation driven by various intramolecular interactions, including steric repulsion, hydrogen bonding, van der Waals interaction between the side chains, and π-stacking between the main chains. Introduction of chiral substituents and addition of chiral compounds induce bias of helical sense, which is detectable by CD, CPL, and VCD spectroscopy. Some of helical polymers exhibit chirality amplification and memory phenomena. π-Conjugated helical polymers feature sophisticated optically active functions as well as photoelectric properties, and expected as sensing, information transmission, and medical materials. The author wishes the chapter contributes to the molecular design and methodology of development of next-generation π-conjugated polymers with well-defined structures that exhibit useful functions.

References

1 Chujo, Y. (ed.) (2011). *Conjugated Polymer Synthesis: Methods and Reactions*. Weinheim, Germany: Wiley-VCH.

2 Müllen, K., Reynolds, J.R., and Masuda, T. (eds.) (2013). *Conjugated Polymers: A Practical Guide to Synthesis*. Cambridge, UK: RSC Publishing.

3 Jiang, H., Taranekar, P., Reynolds, J.R. et al. (2009). Conjugated polyelectrolytes: synthesis, photophysics, and applications. *Angewandte Chemie International Edition* 48: 4300–4316. https://doi.org/10.1002/anie.200805456.

4 Wong, W.-Y. and Ho, C.-L. (2010). Organometallic photovoltaics: a new and versatile approach for harvesting solar energy using conjugated polymetallaynes. *Accounts of Chemical Research* 43: 1246–1256. https://doi.org/10.1021/ar1000378.

5 Rochat, S. and Swager, T.M. (2013). Conjugated amplifying polymers for optical sensing applications. *ACS Applied Materials & Interfaces* 5: 4488–4502. https://doi.org/10.1021/am400939w.

6 Natta, G., Pino, P., Corradini, P. et al. (1955). Crystalline high polymers of α-olefins. *Journal of the American Chemical Society* 77: 1708–1710. https://doi.org/10.1021/ja01611a109.

7 Yashima, E., Ousaka, N., Taura, D. et al. (2016). Supramolecular helical systems: helical assemblies of small molecules, foldamers, and polymers with chiral amplification and their functions. *Chemical Reviews* 116: 13752–13990. https://doi.org/10.1021/acs.chemrev.6b00354.

8 Nakano, T. and Okamoto, Y. (2000). Asymmetric polymerization of methacrylates. *Macromolecular Rapid Communications* 21: 603–612. https://doi.org/10.1002/1521-3927(20000601)21:10<603::aid-marc603>3.0.co;2-h.

9 Green, M.M., Peterson, N.C., Sato, T. et al. (1995). A helical polymer with a cooperative response to chiral information. *Science* 268: 1860–1866. https://doi.org/10.1126/science.268.5219.1860.

10 Nolte, R.J.M. (1994). Helical poly(isocyanides). *Chemical Society Reviews* 23: 11–19. https://doi.org/10.1039/cs9942300011.

11 Suginome, M., Yamamoto, T., and Nagata, Y. (2015). Poly(quinoxaline-2,3-diyl)s: a fascinating helical macromolecular scaffold for new chiral functions. *Journal of Synthetic Organic Chemistry, Japan* 73: 1141–1155. https://doi.org/10.5059/yukigoseikyokaishi.73.1141.

12 Fujiki, M. (2001). Optically active polysilylenes: state-of-the-art chiroptical polymers. *Macromolecular Rapid Communications* 22: 539–563. https://doi.org/10.1002/1521-3927(20010501)22:8<539::aid-marc539>3.0.co;2-k.

13 Lam, J.W.Y. and Tang, B.Z. (2005). Functional polyacetylenes. *Accounts of Chemical Research* 38: 745–754. https://doi.org/10.1021/ar040012f.

14 Aoki, T., Kaneko, T., and Teraguchi, M. (2006). Synthesis of functional π-conjugated polymers from aromatic acetylenes. *Polymer* 47: 4867–4892. https://doi.org/10.1016/j.polymer.2006.04.047.

15 Masuda, T. (2007). Substituted polyacetylenes. *Journal of Polymer Science Part A: Polymer Chemistry* 45: 165–180. https://doi.org/10.1002/pola.21782.

16 Liu, K.-P., Yu, Z.-Q., Liu, J.-H. et al. (2009). Molecular shapes of monosubstituted polyacetylenes in their liquid crystalline phases. *Macromolecular Chemistry and Physics* 210: 707–716. https://doi.org/10.1002/macp.200900011.

17 Xu, A., Masuda, T., and Zhang, A. (2017). Stimuli-responsive polyacetylenes and dendronized poly(phenylacetylene)s. *Polymer Reviews* 57: 139–159. https://doi.org/10.1080/15583724.2016.1169547.

18 Akagi, K. (2019). Interdisciplinary chemistry based on integration of liquid crystals and conjugated polymers: development and progress. *Bulletin of the Chemical Society of Japan* 92: 1509–1655. https://doi.org/10.1246/bcsj.20190092.

19 Praveen, V.K., Babu, S.S., Vijayakumar, C. et al. (2008). Helical supramolecular architectures of self-assembled linear π-systems. *Bulletin of the Chemical Society of Japan* 81: 1196–1211. https://doi.org/10.1246/bcsj.81.1196.

20 Hill, D.J., Mio, M.J., Prince, R.B. et al. (2001). A field guide to foldamers. *Chemical Reviews* 101: 3893–4011. https://doi.org/10.1021/cr990120t.

21 Knaapila, M., Stepanyan, R., Lyons, B.P. et al. (2006). Towards general guidelines for aligned, nanoscale assemblies of hairy-rod polyfluorene. *Advanced Functional Materials* 16: 599–609. https://doi.org/10.1002/adfm.200500419.

22 Greenfield, J.L. and Nitschke, J.R. (2022). Self-assembly of double-helical metallopolymers. *Accounts of Chemical Research* 55: 391–401. https://doi.org/10.1021/acs.accounts.1c00657.

23 Yang, Y., Zhang, Y., and Wei, Z. (2013). Supramolecular helices: chirality transfer from conjugated molecules to structures. *Advanced Materials* 25: 6039–6049. https://doi.org/10.1002/adma.201302448.

24 Ho, H.-A., Najari, A., and Leclerc, M. (2008). Optical detection of DNA and proteins with cationic polythiophenes. *Accounts of Chemical Research* 41: 168–178. https://doi.org/10.1021/ar700115t.

25 Chiang, C.K., Fincher, C.R. Jr., Park, Y.W. et al. (1977). Electrical conductivity in doped polyacetylene. *Physical Review Letters* 39: 1098–1101. https://doi.org/10.1103/PhysRevLett.39.1098.

26 Ishiwari, F., Nakazono, K., Koyama, Y. et al. (2017). Induction of single-handed helicity of polyacetylenes using mechanically chiral rotaxanes as chiral sources. *Angewandte Chemie International Edition* 56: 14858. https://doi.org/10.1002/anie.201707926.

27 Nomura, R., Tabei, J., and Masuda, T. (2001). Biomimetic stabilization of helical structure in a synthetic polymer by means of intramolecular hydrogen bonds. *Journal of the American Chemical Society* 123: 8430–8431. https://doi.org/10.1021/ja015688+.

28 Deng, J., Tabei, J., Shiotsuki, M. et al. (2004). Synthesis and characterization of poly(*N*-propargylsulfamides). *Macromolecules* 37: 5538–5543. https://doi.org/10.1021/ma049739+.

29 Yue, D., Fujii, T., Terada, K. et al. (2006). Synthesis and polymerization of optically active *N*-propargylphosphonamidates: a novel helical polymer carrying a P-chiral center. *Macromolecular Rapid Communications* 27: 1460–1464. https://doi.org/10.1002/marc.200600347.

30 Nomura, R., Nishiura, S., Tabei, J. et al. (2003). Stereoregular poly(*N*-propargylcarbamates. having helical conformation stabilized by the intramolecular hydrogen bonds. *Macromolecules* 36: 5076–5080. https://doi.org/10.1021/ma0209085.

31 Maeda, K., Wakasone, S., Shimomura, K. et al. (2014). Chiral amplification in polymer brushes consisting of dynamic helical polymer chains through the long-range communication of stereochemical information. *Macromolecules* 47: 6540–6546. https://doi.org/10.1021/ma501612e.

32 Palomo, L., Rodríguez, R., Medina, S. et al. (2020). Optical activity (ROA) as a new tool to elucidate the helical structure of poly(phenylacetylene)s. *Angewandte Chemie International Edition* 59: 9080–9087. https://doi.org/10.1002/anie.202000651.

33 Sakurai, S., Okoshi, K., Kumaki, J. et al. (2006). Two-dimensional surface chirality control by solvent-induced helicity inversion of a helical polyacetylene on graphite. *Journal of the American Chemical Society* 128: 5650–5651. https://doi.org/10.1021/ja061238b.

34 Tsuji, Y. and Hoffmann, R. (2016). Helical oligoenes: conformations, bond alternation, and competing through-bond and through-space transmission. *Chemistry: A European Journal* 22: 4878–4888. https://doi.org/10.1002/chem.201600042.

35 Porsev, V.V. and Evarestov, R.A. (2022). Ab initio modeling of helical polyacetylenes: Peierls and Mott-Hubbard metal–insulator transitions. *Computational Materials Science* 213: 111642. https://doi.org/10.1016/j.commatsci.2022.111642.

36 Goto, M., Mino, S., Sogawa, H. et al. (2023). Synthesis of homo polymers and block copolymers of chiral/achiral phenylacetylene derivatives. Spectroscopic and molecular modeling study on solvent-dependent predominance of helical sense. *Journal of Polymer Science* 61: 1270–1278. https://doi.org/10.1002/pol.20220765.

37 Kakuchi, R., Tago, Y., Sakai, R. et al. (2009). Effect of the pendant structure on anion signaling property of poly(phenylacetylene)s conjugated to α-amino acids through urea groups. *Macromolecules* 42: 4430–4435. https://doi.org/10.1021/ma900648h.

38 Sogawa, H., Shiotsuki, M., and Sanda, F. (2012). α-Propargyl amino acid-derived optically active novel substituted polyacetylenes: synthesis, secondary structures, and responsiveness to ions. *Journal of Polymer Science Part A: Polymer Chemistry* 50: 2008–2018. https://doi.org/10.1002/pola.25975.

39 Maeda, K., Nozaki, M., Hashimoto, K. et al. (2020). Helix-sense-selective synthesis of right- and left-handed helical luminescent poly(diphenylacetylene)s with memory of the macromolecular helicity and their helical structures. *Journal of the American Chemical Society* 142: 7668–7682. https://doi.org/10.1021/jacs.0c02542.

40 Zhong, H., Zhao, B., and Deng, J. (2020). Polymer-based circularly polarized luminescent materials. *Advanced Optical Materials* 8: 2000858. https://doi.org/10.1002/adom.202202787.

41 Wang, S., Hu, D., Guan, X. et al. (2021). Brightening up circularly polarized luminescence of monosubstituted polyacetylene by conformation control: mechanism, switching, and sensing. *Angewandte Chemie International Edition* 60: 21918. https://doi.org/10.1002/anie.202108010.

42 Terada, K., Masuda, T., and Sanda, F. (2009). Asymmetric reduction of aromatic ketimines in the presence of helical polymer as catalyst. *Journal of Polymer Science Part A: Polymer Chemistry* 47: 4971–4981. https://doi.org/10.1002/pola.23549.

43 Tang, Z., Iida, H., Hu, H. et al. (2012). Remarkable enhancement of the enantioselectivity of an organocatalyzed asymmetric henry reaction assisted by helical poly(phenylacetylene)s bearing cinchona alkaloid pendants via an amide linkage. *ACS Macro Letters* 1: 261–265. https://doi.org/10.1021/mz200161s.

44 Zhou, L., He, K., Liu N. et al. (2022). Recent advances in asymmetric organocatalysis based on helical polymers. *Polymer Chemistry* 13: 3967–3974. https://doi.org/10.1039/D2PY00483F.

45 Gao, G., Sanda, F., and Masuda, T. (2003). Copolymerization of chiral amino acid-based acetylenes and helical conformation of the copolymers. *Macromolecules* 36: 3938–3943. https://doi.org/10.1021/ma021432s.

46 Ohsawa, S., Sakurai, S., Nagai, K. et al. (2011). Hierarchical amplification of macromolecular helicity of dynamic helical poly(phenylacetylene)s composed of chiral and achiral phenylacetylenes in dilute solution, liquid crystal, and two-dimensional crystal. *Journal of the American Chemical Society* 133: 108–114. https://doi.org/10.1021/ja1087453.

47 Ishida, R., Markvoort, A.J., Maeda, K. et al. (2019). Unexpectedly strong chiral amplification of chiral/achiral and chiral/chiral copolymers of biphenylylacetylenes and further enhancement/inversion and memory of the macromolecular helicity. *Journal of the American Chemical Society* 141: 7605–7614. https://doi.org/10.1021/jacs.9b02904.

48 Green, M.M., Reidy, M.P., Johnson, R.D. et al. (1989). Macromolecular stereochemistry: the out-of-proportion influence of optically active comonomers on the conformational characteristics of polyisocyanates. The sergeants and soldiers experiment. *Journal of the American Chemical Society* 111: 6452–6454. https://doi.org/10.1021/ja00198a084.

49 Green, M.M., Garetz, B.A., Munoz, B. et al. (1995). Majority rules in the copolymerization of mirror image isomers. *Journal of the American Chemical Society* 117: 4181–4182. https://doi.org/10.1021/ja00119a039.

50 Yashima, E. and Maeda, K. (2021). Helical polymers with dynamic and static macromolecular helicity memory: the power of helicity memory for helical polymer synthesis and applications. *Bulletin of the Chemical Society of Japan* 94: 2637–2661. https://doi.org/10.1246/bcsj.20210282.

51 Shimomura, K., Ikai, T., Kanoh, S. et al. (2014). Switchable enantioseparation based on macromolecular memory of a helical polyacetylene in the solid state. *Nature Chemistry* 6: 429–434. https://doi.org/10.1038/nchem.1916.

52 Katsumata, T., Shiotsuki, M., Sanda, F. et al. (2009). Polymerization of ortho-substituted phenylacetylenes with well-defined ruthenium-alkylidene catalysts and related metathesis initiators. *Macromolecular Chemistry and Physics* 210: 1891–1902. https://doi.org/10.1002/macp.200900245.

53 Castanon, J.R., Kuwata, K., Shiotsuki, M. et al. (2012). Synthesis of end-functionalized polyacetylenes that contain polar groups by employing well-defined palladium catalysts. *Chemistry: A European Journal* 18: 14085–14093. https://doi.org/10.1002/chem.201202256.

54 Miyairi, M., Taniguchi, T., Nishimura, T. et al. (2020). Revisiting the polymerization of diphenylacetylenes with tungsten(VI) chloride and tetraphenyltin: an alternative mechanism by a metathesis catalytic system. *Angewandte Chemie International Edition* 59: 14772. https://doi.org/10.1002/anie.202005964.

55 Tarrío, J.J., Rodríguez, R., Fernández, B. et al. (2022). Dissymmetric chiral poly(diphenylacetylene)s: secondary structure elucidation and dynamic luminescence. *Angewandte Chemie International Edition* 61: e202115070. https://doi.org/10.1002/anie.202115070.

56 Nelson, J.C., Saven, J.G., Moore, J.S. et al. (1997). Solvophobically driven folding of nonbiological oligomers. *Science* 277: 1793–1796. https://doi.org/10.1126/science.277.5333.1793.

57 Arnt, L. and Tew, G.N. (2004). Conformational changes of facially amphiphilic meta-poly(phenylene ethynylene)s in aqueous solution. *Macromolecules* 37: 1283–1288. https://doi.org/10.1021/ma034436l.

58 Zhao, X. and Schanze, K.S. (2006). Meta-linked poly(phenylene ethynylene) conjugated polyelectrolyte featuring a chiral side group: helical folding and guest binding. *Langmuir* 22: 4856–4862. https://doi.org/10.1021/la060031t.

59 Liu, R., Shiotsuki, M., Masuda, T. et al. (2009). Synthesis and chiroptical properties of hydroxyphenylglycine-based poly(*m*-phenyleneethynylene-*p*-phenyleneethynylene). *Macromolecules* 42: 6115–6122. https://doi.org/10.1021/ma900910k.

60 Banno, M., Yamaguchi, T., Nagai, K. et al. (2012). Optically active, amphiphilic poly(*meta*-phenylene ethynylene)s: synthesis, hydrogen-bonding enforced helix stability, and direct AFM observation of their helical structures. *Journal of the American Chemical Society* 134: 8718–8728. https://doi.org/10.1021/ja303204m.

61 Schafmeister, C.E., Po, J., and Verdine, G.L. (2000). An all-hydrocarbon cross-linking system for enhancing the helicity and metabolic stability of peptides. *Journal of the American Chemical Society* 122: 5891–5892. https://doi.org/10.1021/ja000563a.

62 Walensky, L.D., Kung, A.L., Escher, I. et al. (2004). Activation of apoptosis in vivo by a hydrocarbon-stapled BH3 helix. *Science* 305: 1466. https://doi.org/10.1126/science.1099191.

63 Walensky, L.D. and Bird, G.H. (2004). Hydrocarbon-stapled peptides: principles, practice, and progress. *Journal of Medicinal Chemistry* 57: 6275. https://doi.org/10.1021/jm4011675.

64 Lau, Y.H., de Andrade, P., Wu Y. et al. (2015). Peptide stapling techniques based on different macrocyclisation chemistries. *Chemical Society Reviews* 44: 91–102. https://doi.org/10.1039/C4CS00246F.

65 Hecht, S. and Khan, A. (2003). Intramolecular cross-linking of helical folds: an approach to organic nanotubes. *Angewandte Chemie International Edition* 42: 6021–6024. https://doi.org/10.1002/anie.200352983.

66 Hashimoto, A., Sogawa, H., Shiotsuki, M. et al. (2012). Stabilization of higher-order structure of poly(phenyleneethynylene)s by metathesis polymerization at the side chains. *Polymer* 53: 2559–2566. https://doi.org/10.1016/j.polymer.2012.04.028.

67 Sogawa, H., Shiotsuki, M., Matsuoka, H. et al. (2011). Synthesis, chiroptical properties, and photoresponsiveness of optically active poly(*m*-phenyleneethynylene)s containing azobenzene moieties. *Macromolecules* 44: 3338–3345. https://doi.org/10.1021/ma200281e.

68 Sogawa, H., Shiotsuki, M., and Sanda, F. (2013). Synthesis and photoresponse of helically folded poly(phenyleneethynylene)s bearing azobenzene moieties in the main chains. *Macromolecules* 46: 4378–4387. https://doi.org/10.1021/ma4002947.

69 Sogawa, H., Shiotsuki, M., Hirao, T. et al. (2013). Synthesis of optically active poly(*m*-phenyleneethynylene–aryleneethynylene)s bearing hydroxy groups and examination of the higher order structures. *Macromolecules* 46: 8161–8170. https://doi.org/10.1021/ma4017295.

70 Sogawa, H., Miyagi, Y., Shiotsuki, M. et al. (2013). Synthesis of novel optically active poly(phenyleneethynylene–aryleneethynylene)s bearing hydroxy groups. Examination of the chiroptical properties and conjugation length. *Macromolecules* 46: 8896–8904. https://doi.org/10.1021/ma401730e.

71 Hashimoto, A., Miyagi, Y., Sogawa, H. et al. (2014). Synthesis and properties of poly(phenyleneethynylene)s bearing perylene moieties at the side chains. *Chemistry Letters* 43: 1622–1624. https://doi.org/10.1246/cl.140601.

72 Miyagi, Y., Sogawa, H., Shiotsuki, M. et al. (2014). Synthesis of optically active conjugated polymers bearing *m*-terphenylene moieties by acetylenic coupling polymerization: chiral aggregation and optical properties of the product polymers. *Macromolecules* 47: 1594–1603. https://doi.org/10.1021/ma402123w.

73 Sotani, T., Otoba, M., Hosotani, M. et al. (2024). π-Conjugated polymers with optically active substituents possibly form chiral higher-order structures. *Polymer Chemistry* 15: 915–923. https://doi.org/10.1039/d3py01374j.

74 Hosotani, M., Sotani, T., Sogawa, H. et al. (2024). Bipyridine-containing optically active π-conjugated polymers derived from amino alcohols: examination of the higher order structures, chiral recognition and metal coordination based on quantum mechanical and molecular dynamics calculations. *The Journal of Physical Chemistry A*. 128: 11020–11029. https://doi.org/10.1021/acs.jpca.4c06980.

75 Fujikura, Y., Sonogashira, K., and Hagihara, (1975). Preparation and UV spectra of some oligomer-complexes composed of platinum group metals and conjugated poly-yne systems. *Chemistry Letters* 10: 1067–1070. https://doi.org/10.1246/cl.1975.1067.

76 Ho, C.-L. and Wong, W.-Y. (2013). Charge and energy transfers in functional metallophosphors and metallopolyynes. *Coordination Chemistry Reviews* 257: 1614–1649. https://doi.org/10.1016/j.ccr.2012.08.023.

77 Wong, W.-Y. (2007). Mercury alkynyls as versatile templates for new organometallic materials and polymers. *Coordination Chemistry Reviews* 251: 2400–2427. https://doi.org/10.1016/j.ccr.2007.01.006.

78 Miyagi, Y., Shibutani, Y., Otaki, Y. et al. (2016). Synthesis of platinum-containing poly(phenyleneethynylene)s having various chromophores: aggregation and optical properties. *Polymer Chemistry* 7: 1070–1078. https://doi.org/10.1039/c5py01584g.

79 Schull, T.L., Kushmerick, J.G., Patterson, C.H. et al. (2003). Ligand effects on charge transport in platinum(II) acetylides. *Journal of the American Chemical Society* 125: 3202–3203. https://doi.org/10.1021/ja029564o.

80 Gao, Z., Zhu, J., Han, Y. et al. (2016). Ligand effects on cooperative supramolecular polymerization of platinum(II) acetylide complexes. *Polymer Chemistry* 7: 5763–5767. https://doi.org/10.1039/C6PY01440B.

81 Miyagi, Y., Sotani, T., Yajima, T. et al. (2018). Effect of phosphine ligand on the optical absorption/emission properties of platinum-containing conjugated polymers. *Polymer Chemistry* 9: 1772–1799. http://doi.org/10.1039/C7PY01886J.

82 Prince, R.B., Brunsveld, L., Meijer, E.W. et al. (2000). Twist sense bias induced by chiral side chains in helically folded oligomers. *Angewandte Chemie International Edition* 39: 228–230. https://doi.org/10.1002/(sici)1521-3773(20000103)39:1<228::aid-anie228>3.0.co;2-b.

83 Wilson, J.N., Steffen, W., McKenzie, T.G. et al. (2002). Chiroptical properties of poly(*p*-phenyleneethynylene) copolymers in thin films: large g-values. *Journal of the American Chemical Society* 124: 6830–6831. https://doi.org/10.1021/ja026532s.

84 Miyagi, Y., Hirao, T., Haino, T. et al. (2015). Synthesis of optically active conjugated polymers containing platinum in the main chain: control of the higher-order structures by substituents and solvents. *Journal of Polymer Science Part A: Polymer Chemistry* 53: 2452–2461. https://doi.org/10.1002/pola.27759.

85 Otaki, Y., Marumoto, M., Miyagi, Y. et al. (2016). Synthesis and properties of novel optically active platinum-containing poly(phenyleneethynylene)s. *Chemistry Letters* 45: 937–939. https://doi.org/10.1246/cl.160385.

86 Zhou, G.-J., Wong, W.-Y., Lin, Z. et al. (2006). White metallopolyynes for optical limiting/transparency trade-off optimization. *Angewandte Chemie International Edition* 45: 6189–6193. https://doi.org/10.1002/anie.200601651.

87 Campbell, K., McDonald, R., Ferguson, M.J. et al. (2003). Using ligand exchange reactions to control the coordination environment of Pt(II) acetylide complexes: applications to conjugated metallacyclynes. *Journal of Organometallic Chemistry* 683: 379–387. https://doi.org/10.1016/S0022-328X(03)00792-7.

88 Furusho, Y., Tanaka, Y., and Yashima, E. (2006). Double helix-to-double helix transformation, using Pt(II) acetylide complexes as surrogate linkers. *Organic Letters* 8: 2583–2586. https://doi.org/10.1021/ol0608294.

89 Miyagi, Y., Ishida, T., Marumoto, M. et al. (2018). Ligand exchange reaction for controlling the conformation of platinum-containing polymers. *Macromolecules* 51: 815–824. https://doi.org/10.1021/acs.macromol.7b02249.

90 Ishida, T., Sotani, T., Sano, N. et al. (2022). Control of higher-order structures of platinum-containing conjugated polymers by ligand exchange reactions: chirality transfer from optically active ligands to optically inactive polymers. *Macromolecules* 55: 309—321. https://doi.org/10.1021/acs.macromol.1c01773.

91 Imamoto, T., Sugita, K. and Yoshida, K. (2005). An air-stable P-chiral phosphine ligand for highly enantioselective transition-metal-catalyzed reactions. *Journal of the American Chemical Society* 127: 11934—11935. https://doi.org/10.1021/ja053458f.

92 Imamoto, T., Tamura, K., Zhang, Z. et al. (2012). Rigid P-chiral phosphine ligands with *tert*-butylmethylphosphino groups for rhodium-catalyzed asymmetric hydrogenation of functionalized alkenes. *Journal of the American Chemical Society* 134: 1754—1769. https://doi.org/10.1021/ja209700j.

93 Llopis, Q., Guillamot, G., Phansavath, P. et al. (2017). Enantioselective synthesis of α-acetal-β'-amino ketone derivatives by rhodium-catalyzed asymmetric hydrogenation. *Organic Letters* 19: 6428–6431. https://doi.org/10.1021/acs.orglett.7b03332.

94 Yanagisawa, A., Lin, Y., Takeishi, A. et al. (2016). Enantioselective nitroso aldol reaction catalyzed by a chiral phosphine–silver complex. *European Journal of Organic Chemistry* 32: 5355–5359. https://doi.org/10.1002/ejoc.201601143.

95 Iwamoto, H., Endo, K., Ozawa, Y. et al. (2019). Copper(I)-catalyzed enantioconvergent borylation of racemic benzyl chlorides enabled by quadrant-by-quadrant structure modification of chiral bisphosphine ligands. *Angewandte Chemie International Edition* 58: 11112–11117. https://doi.org/10.1002/anie.201906011.

96 Marumoto, M., Sotani, T., Miyagi, Y. et al. (2020). Synthesis of Pt-containing conjugated polymers having QuinoxP* and bipyridine ligands. Chirality transfer from the phosphine ligand to the polymer backbone. *Macromolecules* 53: 2031—2038. https://doi.org/10.1021/acs.macromol.0c00096.

97 Aoki, R., Horiuchi, T., Makino, S. et al. (2023). Chirality induction in Pt-containing polyaryleneethynylenes by exchange from achiral phosphine ligands to P-chiral phosphine ligands. *Polymer* 265: 125576. https://doi.org/10.1016/j.polymer.2022.125576.

98 Fu, H.L-K., Po, C., He, H. et al. (2016). Tuning of supramolecular architectures of L-valine-containing dicyanoPt(II) 2,2'-bipyridine complexes by metal–metal, π–π stacking, and hydrogen-bonding interactions. *Chemistry: A European Journal* 22: 11826–11836. https://doi.org/10.1002/chem.201601983.

99 Biet, T., Cauchy, T., Sun, Q. et al. (2017). Triplet state CPL active helicene–dithiolene Pt bipyridine complexes. *Chemical Communications* 53: 9210–9213. https://doi.org/10.1039/C7CC05198K.

100 Schulte, T.R., Holstein, J.J., Krause, L. et al. (2017). Chiral-at-metal phosphorescent square-planar Pt(II)-complexes from an achiral organometallic ligand. *Journal of the American Chemical Society* 139: 6863–6866. https://doi.org/10.1021/jacs.7b03963.

101 Zhang, X-P., Wang, L-L., Qi, X-W. et al. (2018). Pt⋯Pt interaction triggered tuning of circularly polarized luminescence activity in chiral dinuclear Pt(II) complexes. *Dalton Transactions* 47: 10179–10186. https://doi.org/10.1039/C8DT02277A.

102 Han, J., Lu, H., Xu, Y. et al. (2020). Concentration-dependent circularly polarized luminescence of chiral cyclometalated Pt(II) complexes for electroluminescence. *Journal of Organometallic Chemistry* 915: 121240. https://doi.org/10.1016/j.jorganchem.2020.121240.

103 Li, B., Li, Y., Chan, MH-Y. et al. (2021). Phosphorescent cyclometalated Pt(II) enantiomers with circularly polarized luminescence properties and their assembly behaviors. *Journal of the American Chemical Society* 143: 21676–21684. https://doi.org/10.1021/jacs.1c10943.

104 Ikeshita, M., Furukawa, S., Ishikawa, T. et al. (2022). Enhancement of chiroptical responses of *trans*-bis[(β-iminomethyl)naphthoxy]Pt(II) complexes with distorted square planar coordination geometry. *ChemistryOpen* 11: e202100277. https://doi.org/10.1002/open.202100277.

105 Morikubo, J. and Tsubomura, T. (2022) Circularly polarized luminescence of cyclometalated Pt(II) complex excimers: large difference between isomers. *Inorganic Chemistry* 61: 17154–17165. https://doi.org/10.1021/acs.inorgchem.2c02662.

106 Matsudaira, K., Mimura, Y., Hotei, J. et al. (2021). Magnetic circularly polarized luminescence from Pt^{II}OEP and F_2-ppyPt^{II}(acac) under north-up and south-up faraday geometries. *Chemistry: An Asian Journal* 16: 926–930. https://doi.org/10.1002/asia.202100172.

107 Makino, S., Horiuchi, T., Ishida, T. et al. (2022). Synthesis of Pt-containing conjugated polymers bearing chiral phosphine ligands. Study of geometries and intermolecular interactions leading to aggregation. *Organometallics* 41: 1699–1709. https://doi.org/10.1021/acs.organomet.2c00173.

108 Sotani, T., Yajima, T., Sogawa, H. et al. (2020). Synthesis of Pt-containing conjugated polymers bearing optically active amide groups: a mechanistic study of chiral aggregation. *Macromolecules* 53: 11077–11088. https://doi.org/10.1021/acs.macromol.0c00911.

12

Advances in Organic Solar Cell Performance Involving Anthracene and Anthanthrone Core Polycyclic Aromatic Materials

Vivian S. John-Denk, Morongwa E. Ramoroka, and Emmanuel I. Iwuoha

SensorLab (UWC Sensor Laboratories), University of the Western Cape, Bellville, Cape Town, South Africa

12.1 Introduction

The declining availability of fossil fuels, along with their detrimental environmental impacts, necessitates immediate action. A critical approach to addressing the increasing global demand for energy and environmental issues is the direct conversion of solar energy through photovoltaic technology [1, 2]. The successful and efficient utilization of the principles underlying the photovoltaic effect for electricity generation from sunlight presents a viable solution to the escalating requirement for clean, sustainable, and renewable energy sources [3]. However, the substantial expense associated with silicon-based solar cells has resulted in a mere fraction of the world's energy consumption being derived from solar power [4]. An emerging alternative to silicon-based solar cells is organic solar technology. Organic solar cells (OSCs), particularly those utilizing polymers or small molecules, have gained recognition as a viable photovoltaic solution. This is attributed to their potential for flexible devices through solution-processing methods, the adaptability in the design of organic materials, minimal environmental impact, and their lightweight nature, which facilitates the efficient conversion of sunlight into electricity. The high extinction coefficient of organic semiconductor materials allows for the effective absorption of photons with a relatively small quantity of material [5, 6].

Furthermore, through chemical synthesis, the molecular structure of these organic materials can be precisely modified, thereby enhancing their optical and electrical characteristics. Numerous studies have concentrated on the design and synthesis of conjugated semiconducting polymers featuring novel molecular architectures, with the objective of capturing solar radiation within the visible range of the electromagnetic spectrum [7–9]. These efforts, combined with advanced device design strategies, have led to enhanced efficiency in solar cells, which has recently

Conjugated Polymer Synthesis and Materials Applications, First Edition. Edited by Yoshiki Chujo.
© 2026 Wiley-VCH GmbH. All rights reserved, including rights for text and data mining and training of artificial intelligence technologies or similar technologies. Published 2026 by Wiley-VCH GmbH.

surpassed 20% [10, 11]. Nevertheless, to achieve higher efficiency and facilitate widespread commercialization, it is essential to investigate the development of novel materials and their integration into effective device architectures. Researchers have introduced several new polymers and small molecules designed to enhance characteristics such as solubility, light absorption, and charge transport in organic semiconductors [12, 13]. This review will concentrate on polycyclic aromatic compounds (PACs) as foundational elements for organic materials (small molecules or polymers) in OSCs and perovskite solar cells (PSCs), examining how structural variations influence the spectral, electronic, charge transport, and photovoltaic properties of these organic materials.

PACs serve as highly effective building blocks for the synthesis of organic semiconducting materials utilized in various organic electronic applications, including organic field-effect transistors, organic light-emitting diodes, PSCs, and OSCs. This distinctive class of organic compounds is characterized by planar geometries, rendering them particularly promising for optoelectronic applications [14, 15]. PACs are composed of fused aromatic rings, which allow for well-organized stacking and demonstrate favorable charge transport properties in thin films, while also exhibiting light absorption capabilities across the visible spectrum [16, 17]. The optical and electronic properties of these materials are influenced by several structural and electronic features, such as effective conjugation length, overlap of π-orbitals, a rigid π-conjugated backbone, and the specific way the rings are fused [18]. Economically, PACs are typically cost-effective and possess properties that facilitate the introduction of new functional groups, thereby enabling fine-tuning of their electronic and optical characteristics [19]. Furthermore, they can be readily functionalized through established reactions involving C—C cross-couplings. Modifications to their structural properties allow for access to a broad range of the visible spectrum. The combination of low cost and tunable electronic and optical properties across the entire visible range presents numerous advantages for applications in organic electronics [20, 21]. Notable examples of PACs emerging as foundational building blocks for donor materials in photovoltaic design include anthanthrone, anthracene (as depicted in Figure 12.1), and their respective derivatives [22, 23].

Herein, we present an overview of the applications of PACs in solar cells, highlighting recent trends and advancements. While PACs were defined as π-conjugated molecules containing three or more condensed aromatic rings, we will shed light on

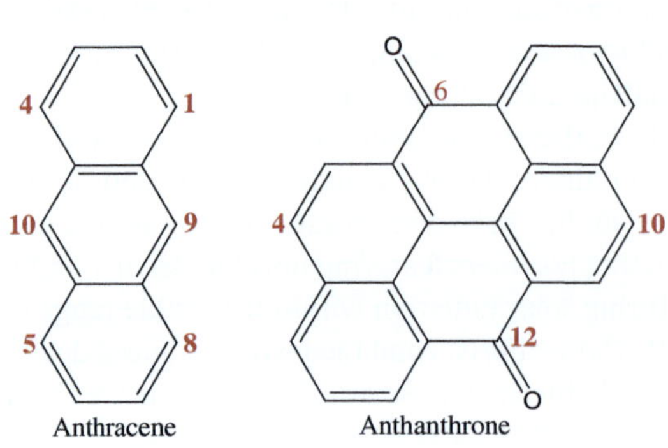

Figure 12.1 Chemical structures of PACs with labeled carbon positions.

the application of recently developed PACs containing materials that have proven useful or bear unique properties for solar cells usage. This review is focused on application of anthracene- and anthanthrone-based materials in the active layer of OSCs or hole transport layer (HTL) in PSCs.

12.2 Anthracene-containing Materials for OSCs Applications

Anthracenes are highly attractive as structural components due to their excellent thermal and device stability. The inflexible framework of anthracenes can be readily modified at its 9,10 sites. They exhibit a high fluorescence quantum yield, and their emission can be adjusted during polymerization through the addition of other arylene-building blocks [24]. The excellent photoconductive properties, elevated fluorescence quantum yields in thin films, and high absorption coefficients near 100,000 $M^{-1}\,cm^{-1}$ of anthracene-derived polymers render them perfect for the development of organic electronic devices. Anthracene and its derivatives have been studied and utilized in light-emitting diodes as well as other thin-film organic electronic devices such as transistors and solar cells [25]. Nonetheless, quite limited research has been conducted using these materials in the field of OSC device fabrication. The primary constraint is the low solubility of anthracene-based polymers, resulting in challenges during the thin-layer preparation via the spin-coating method, as well as the high crystallinity of anthracene components, which hinders the creation of stable, flexible thin films. Jiang et al. [26a] designed four 2D non-fullerene acceptors with an anthracene core (TEA-SIC (**1a**), TEA-SIC-8F (**1b**), TEA-SIC-OH-8F (**1d**), TEA-SIC-OH (**1c**)), as shown in Figure 12.2a. The designed 2D non-fullerenes exhibited enhanced sunlight absorption, more efficient exciton dissociation, and superior electron-accepting abilities compared to Y6 previously reported by Yuan et al. [26b]. TEA-SIC-8F and TEA-SIC demonstrated impressive electron mobilities of $1.18 \times 10^{-3}\,cm^2\,V^{-1}\,s^{-1}$ and $1.76 \times 10^{-3}\,cm^2\,V^{-1}\,s^{-1}$, respectively, suggesting their suitability for developed high-performing OSCs. Lower electron mobility obtained for TEA-SIC-OH-8F ($2.77 \times 10^{-5}\,cm^2\,V^{-1}\,s^{-1}$) and TEA-SIC-OH ($1.10 \times 10^{-4}\,cm^2\,V^{-1}\,s^{-1}$), efficient exciton separation, and their high light absorption indicate their capability as third components in the active layer of ternary OSCs. The solubility of four 2D non-fullerene acceptors with an anthracene core was predicted using machine learning model (see Figure 12.2a). These designed 2D non-fullerene acceptors demonstrate notable solubility in various alcohol-based solvents, suggesting their suitability for unique uses in the production of stacked OSCs. These results offer important perspectives for the future creation of synthesizable high-efficiency 2D non-fullerene acceptors. Liu et al. [27] synthesized the anthracene-terminated polymer as a donor material and studied their photovoltaic performance in different solvents (3-methylthiophene, o-xylene, and chloroform) using BO-4Cl (**3**) as an acceptor material. The anthracene-terminated polymer showed a good solubility in 3-methylthiophene, leading to a champion power conversion efficiency (PCE) of 16.87% (current density (J_{SC}) = 26.98 mA cm^{-2}, fill factor (FF) = 72.91%, and open-circuit voltage (V_{OC}) = 0.86 V), which is higher than the performance of devices fabricated using o-xylene (PCE = 15.69%, J_{SC} = 26.28 mA cm^{-2}, FF = 70.83%, and V_{OC} = 0.84 V) and chloroform (PCE = 16.18%, J_{SC} = 26.52 mA cm^{-2},

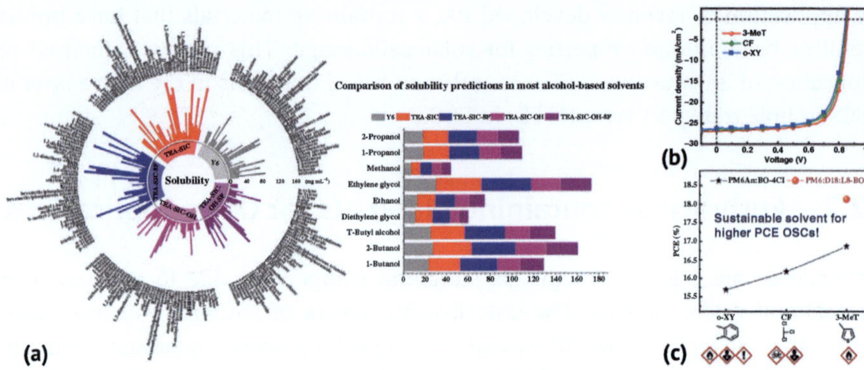

Figure 12.2 (a) The solubility predictions of TEA-SIC-OH-8F, TEA-SIC-OH, TEA-SIC-8F, TEA-SIC, and Y6 in common solvents along with solubility comparison in most alcohol-based solvents. *Source*: [26a]/American Chemical Society. (b) Current density–voltage and (c) PCE curves of PM6An:BO-4Cl in different solvents. *Source*: [27]/American Chemical Society.

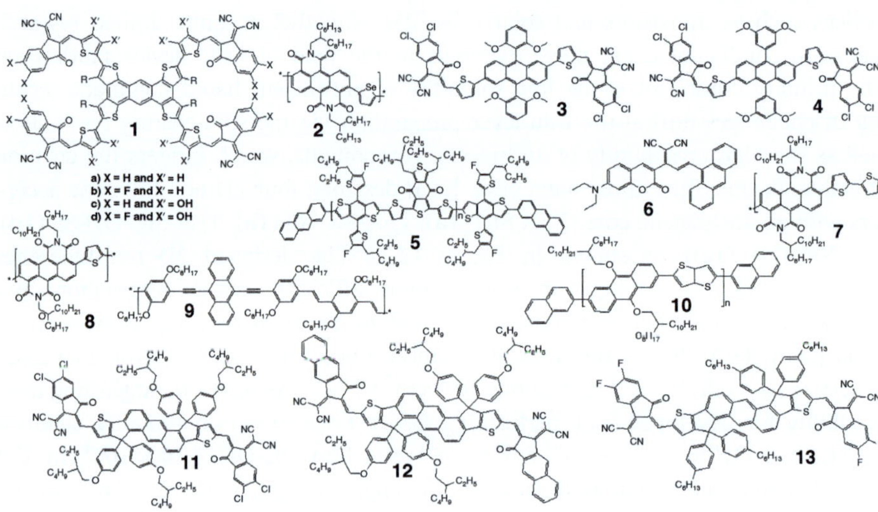

Scheme 12.1 The chemical structures of polycyclic aromatic active layer materials based on anthracenes.

FF = 71.99%, and V_{OC} = 0.85 V) as solvents (see Figure 12.2b and c). Since harmful halogenated solvents limit the industrial-scale manufacturing of OSCs, 3-methylthiophene offers a low-toxicity and eco-friendly solvent for the green fabrication of the OSCs active layer. The effect of substituents on the anthracene core toward the development of new active layer materials is presented in Table 12.1. The investigations recorded in Table 12.1 revealed that different substituents on anthracene core help in fine-tuning the photovoltaic parameters and properties of a material required for photovoltaic application, such as optical band gap and energy levels (Scheme 12.1).

Table 12.1 Application of anthracenes-based materials in an active layer.

Active layer (Donor:acceptor)	Optical bandgap (eV)	E_{LUMO} (eV)	E_{HOMO} (eV)	Active layer (Donor:acceptor)	FF (%)	J_{SC} (mA cm^{-2})	V_{OC} (V)	PCE (%)	Ref.
o-AT-2Cl (**4**)	1.53	−4.04	−5.68	D18:o-AT-2Cl	69.60	20.10	0.91	12.80	[28]
m-AT-2Cl (**5**)	1.61	−4.05	−5.72	D18:m-AT-2Cl	65.10	13.00	0.90	7.66	
PADIHD-Se (**2**)	1.59	−3.94	−5.53	J61:PADIHD-Se	64.50	16.10	0.84	8.72	[29]
CA-CN (**6**)	1.95	−3.95	−5.76	PTB7-Th:CA-CN	64.00	18.91	0.92	11.13	[23]
				PTB7-Th:CA-CN:DICTF	72.00	21.78	0.95	14.91	
PADIOD-2T (**7**)	1.40	−3.93	−5.33	PTB7-Th:PADIOD-2T	49.00	3.80	0.75	1.40	[30]
				J61:PADIOD-2T	48.00	4.68	0.83	1.86	
PADIOD-T (**8**)	1.61	−4.00	−5.61	PTB7-Th:PADIOD-T	53.00	7.08	0.73	2.71	
				J61:PADIOD-T	60.00	11.42	0.78	5.36	
AnE-PVstat (**9**)	–	−3.50	−5.30	AnE-PVstat:PCBM	51.00	9.05	0.83	3.83	[31]
ODATT (**10**)	2.01	−3.86	−5.55	ODATT:PCBM	50.00	5.70	0.76	2.2	[32]
AT-NC (**12**)	1.62	−3.80	−5.44	PBDB-T:AT-NC	69.40	17.11	0.92	10.91	[33]
AT-4Cl (**11**)	1.60	−3.89	−5.71	PBDB-TF:AT-4Cl	75.50	19.52	0.90	13.27	
ANT-4F (**13**)	1.68	−4.01	−5.69	PBDB-TF:ANT-4F	73.90	19.00	0.93	13.10	[34]

12.3 Anthanthrone-containing Materials for OSCs Applications

Anthanthrones are a class of PACs characterized by their extensive conjugation and symmetrical configurations. These compounds are recognized for their applications as pigments and vat dyes. The substantial conjugated planar architecture of anthanthrones enhances intermolecular interactions, and their prolonged π-conjugation is thought to confer benefits in photovoltaic applications, as electronic coupling tends to improve with increasing molecular size [35]. Despite their potential, anthanthrones have not been extensively investigated as components in organic semiconductors. Anthanthrone derivative, 4,10-dibromanthranthrone (ANT), has been widely used as a starting point to develop new materials. ANT presents several notable advantages: first, the presence of bromine atoms at the 4- and 10-positions facilitates the potential for coupling with additional conjugated units, thereby enhancing the effective conjugation length; second, the ketone groups located at the 6- and 12-positions permit the incorporation of alkyl chains to improve solubility; third, the symmetrical nature of the structure likely promotes ease of crystallization; and finally, the extensive π-conjugation, both in terms of length and area, is advantageous for charge transport. It is noteworthy that there are limited studies in which anthanthrone has been utilized as a precursor for the development of organic photovoltaic materials [36, 37]. Liu et al. [25] developed four different anthanthrone-based polymers for use in OSC as donor materials. PANT (**17**) denotes the self-coupled homopolymer, whereas the remaining three (PANT-TffBT (**20**), PANT-TBT (**19**), and PANT-TBO (**18**)) are classified as donor–acceptor type copolymers. From the UV-Vis spectroscopy studies, PANT-TBT displayed a wider bathochromic effect from solution in chloroform to thin films compared with PANT-TffBT, PANT-TBO, and PANT, suggesting a more efficient backbone stacking (see Figure 12.3a). The absorption wavelength edges of PANT, PANT-TffBT, PANT-TBT, and PANT-TBO were determined to be 484, 663, 653, and 664 nm, which correspond to the optical bandgaps of 2.56, 1.87, 1.90, and 1.87 eV, respectively. Because there are no acceptor groups along the backbone, PANT has a wider optical bandgap. The surface morphology of PANT, PANT-TffBT, PANT-TBT, and PANT-TBO was investigated using atomic force

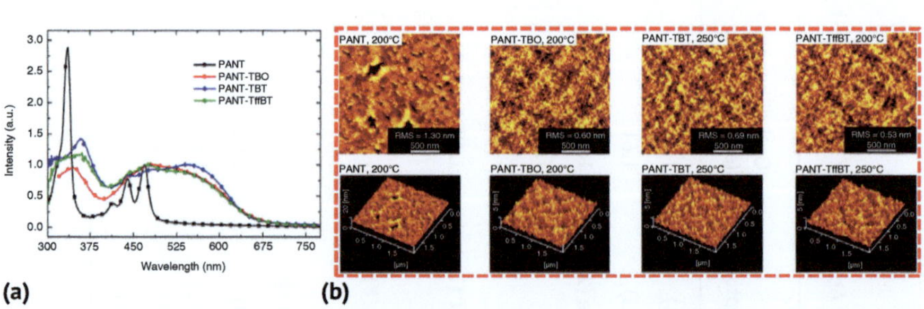

Figure 12.3 (a) The UV-Vis absorption spectra of synthesized polymers as thin films. (b) Tapping-mode AFM topography images of PANT, PANT-TffBT, PANT-TBT, and PANT-TBO with their corresponding 3D profiles. *Source*: [25]/Royal Society of Chemistry.

microscopy as depicted in Figure 12.3b. The root-mean-square values were determined to be 1.30, 0.53, 0.69, and 0.60 nm for PANT, PANT-TffBT, PANT-TBT, and PANT-TBO, respectively. The higher root-mean-square of PANT suggests its higher crystallinity, while the lower root-mean-square of PANT-TffBT indicates its amorphous morphology. The PANT, PANT-TffBT, PANT-TBT, and PANT-TBO exhibit p-type properties with hole mobility values of 2.00×10^{-3}, 4.50×10^{-4}, 3.60×10^{-3}, and $2.30 \times 10^{-4}\ cm^2\,V^{-1}\,s^{-1}$ leading to PCE of 0.28, 0.56, 4.57, and 1.18%, respectively. This suggests that the efficiency can reach a higher value through further structural modifications by selecting suitable comonomers to produce new polymers. The potential of ANT-based polymers in OSCs is demonstrated by the successful operation of devices with these four polymers, despite their generally poor performance, which was attributed to their large dihedral angles and low crystallinity.

Li et al. [38] designed three novel donor materials with anthanthrene as a core, namely AANT(TDPP)$_2$ (**16**), AAN(T-DPP)$_2$ (**15**), and AAN-DPP$_2$ (**14**), for OSC applications. Differential scanning calorimetry was used to study the thermal responses of the three designed novel donor materials, as shown in Figure 12.4a. During heating process, the endothermic peaks were observed at 203, 222, and 218 °C for AANT(TDPP)$_2$, AAN(T-DPP)$_2$, and AAN-DPP$_2$, which correspond to melting temperatures, respectively. However, AAN(T-DPP)$_2$ and AAN-DPP$_2$ showed an exothermal peak at 209 and 201 °C, respectively, during cooling process. The higher melting temperatures obtained for AAN(T-DPP)$_2$ with sharpest exothermal peak indicate its better crystallinity than AANT(TDPP)$_2$ and AAN-DPP$_2$, which in turn facilitates π–π stacking and improves charge transport. These findings were supported by higher hole mobility value obtained for AAN(T-DPP)$_2$ ($7.61 \times 10^{-5}\ cm^2\ V^{-1}\,s^{-1}$) than AANT(TDPP)$_2$ ($5.07 \times 10^{-5}\ cm^2\,V^{-1}\,s^{-1}$) and AAN-DPP$_2$ ($1.23 \times 10^{-5}\ cm^2\,V^{-1}\,s^{-1}$) (see Figure 12.4b). As illustrated in Figure 12.4c, the OSC device based on AAN(T-DPP)$_2$ (PCE = 2.33%, J_{SC} = 6.82 mA cm^{-2}, FF = 39.80%, and V_{OC} = 0.86 V) showed best photovoltaic performance compared to AANT(TDPP)$_2$ (PCE = 2.03%, J_{SC} = 6.25 mA cm^{-2}, FF = 36.12%, and V_{OC} = 0.90 V) and AAN-DPP$_2$ (PCE = 1.92%, J_{SC} = 6.12 mA cm^{-2}, FF = 34.25%, and V_{OC} = 0.92 V) based devices due to better molecular backbone planarity and broader absorption. The findings suggest π-bridges and side chain substitution strategy as effective method to improve photovoltaic performance (Scheme 12.2).

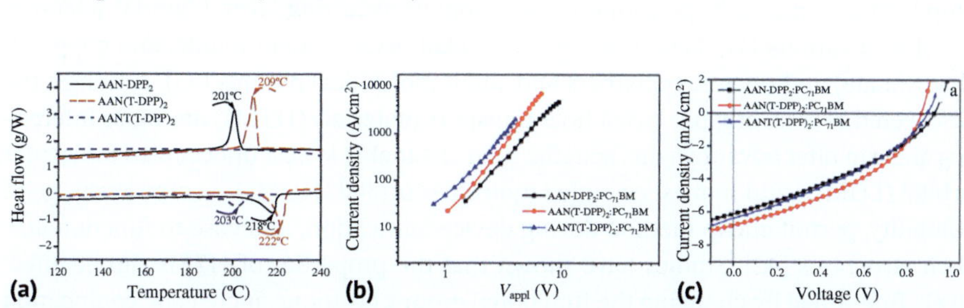

Figure 12.4 (a) Differential scanning calorimetry thermograms of AANT(TDPP)$_2$, AAN(T-DPP)$_2$, and AAN-DPP$_2$ under nitrogen environment at a heating rate of 10 °C min^{-1}. (b) Current density–voltage curves of the hole-only devices. (c) Current density–voltage curves of the synthesized donors/PC$_{71}$BM devices. *Source*: [38]/Royal Society of Chemistry.

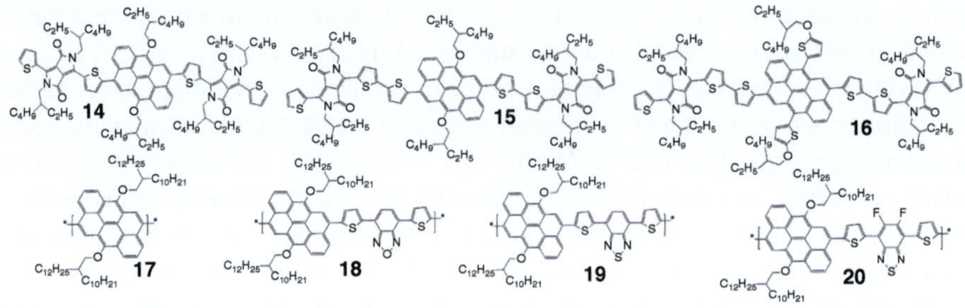

Scheme 12.2 The chemical structures of polycyclic aromatic active layer materials based on anthanthrone.

12.4 Hole-transporting Materials

PSCs have received a lot of attention in recent years from industries and academics because of their possible cost saving, ease of processing, and capacity to perform similarly to solar cell technologies that are now utilized around the world. The development of PSCs technology is almost comparable to traditional solar energy technologies such as silicon, cadmium telluride, and copper indium gallium selenide. The PCE of PSCs has increased from the first 3.8% reported in 2009 to over 22.0% [39, 40]. The inclusion of a hole-transporting layer in the devices is a major factor contributing to their quick success. Hole-transporting layer not only assists in hole transportation but also prevents the absorber layer from electrode penetration and moisture invasion. In addition, it plays an important role in maintaining absorber layer stability and inhibiting charge recombination [41]. Studies revealed that the presence of hole-transporting layer within the device improves the PCE by over 30%. The PCE is associated with the physical interaction between the absorber layer and the hole-transporting layer. The hole-transporting layers should have good solubility and cheap synthetic methods for large-scale production. Hole-transporting layers favor high photochemical and thermal stability, as well as high hydrophobicity to prevent degradation and enhance the overall long-term stability of the device [42]. Many researchers carried out studies on hole-transporting layers to understand the relationship between solar cells performance and hole-transporting layer. Material scientists developed various HTMs with inorganic (nickel oxide, copper iodide, and copper(I) thiocyanate) and organic (polymers and small molecules) natures [43]. Amongst the discovered HTMs, organic-based hole transport materials (HTMs) are more interesting and can offer advantageous benefits such as tunable lowest unoccupied molecular orbital (LUMO) and highest occupied molecular orbital (HOMO) energy levels, good solubility, permit cheap method during device fabrication, and ease to functionalize their structures [42]. Studies have shown that the properties of HTMs can be effectively fine-tuned by changing the functional groups positions, increasing conjugation length, introducing alkyl chain (to adjust and control the solubility, intermolecular stacking, and morphology), and introducing atoms with varying electronegativity to the molecular structure [43, 44]. Among the reported HTMs, 2,2′,7,7′-tetrakis-(N,N-dip-methoxyphenylamine)-9,9′-spiro-bifluorene (spiro-OMeTAD) is the most studied HTM because of its definite molecular weight, good batch-to-batch reproducibility, promising tunable HOMO and LUMO energy levels, and high purity. Spiro-OMeTAD gives high-performing devices as HTM only in the presence of some dopants, which are moisture sensitive. In addition, multiple synthetic steps lead to high production cost of spiro-OMeTAD [45].

In this section, we look at the reported HTMs derived from anthracene and anthanthrone as cores. Singh et al. [46] reported new anthracene-based HTMs, denoted as 2-AF (**36**), 2,6-diAF (**37**), and 9-AF (**35**), with arylamine as a donor unit. The arylamine substituents were introduced to assist with hole-hopping, and the anthracene core provides intermolecular assembly, causing an increase in intermolecular electronic coupling. Singh and coworkers investigated the energy levels, photophysical properties, hole transport responses, and photovoltaic performance of the newly synthesized HTMs. The UV-Vis absorption spectra of 2-AF, 2,6-diAF, and 9-AF are depicted in Figure 12.5a. The 2-AF, 2,6-diAF, and 9-AF showed absorption peak at 383, 430, and 416 nm, respectively, associated with intramolecular charge transfer from arylamine units to the anthracene. The large twist angle between the arylamine and anthracene units of 9-AF leads to shortest maximum absorption wavelength associated with intramolecular charge transfer. The bi-substituted 2,6-diAF with arylamine leads to the longest maximum absorption wavelength related to intramolecular charge transfer. The energy level diagram of HTMs as shown in Figure 12.5b was obtained from reduction and oxidation potentials of

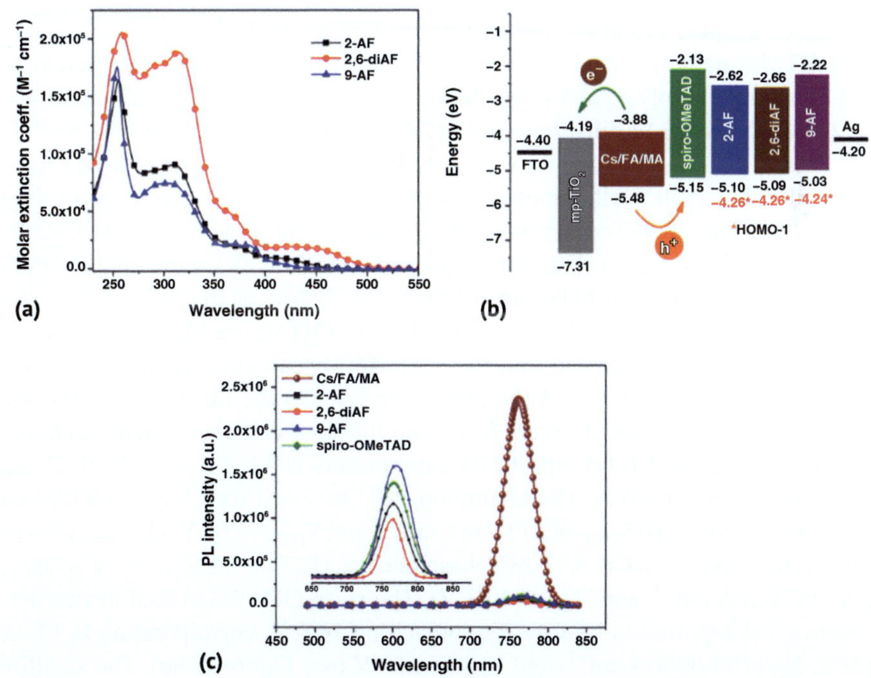

Figure 12.5 (a) UV-Vis absorption spectra, (b) the energy level diagram, and (c) the steady-state photoluminescence. *Source*: [46]/American Chemical Society.

cyclic voltammetry using ferrocene as an internal standard. The HOMO energy levels were estimated to be e −5.10, −5.09, and −5.03 eV for 2-AF, 2,6-diAF, and 9-AF, respectively (see Figure 12.5b). These HOMO energy levels are higher than that of perovskite (−5.48 eV), indicating that holes can be successfully injected from the perovskite into the HTL. The space-charge limited current method was employed to investigate the hole-transporting performance without any additives. Higher hole mobility of 9.05×10^{-4} and 8.31×10^{-4} $cm^2 V^{-1} s^{-1}$ were obtained for 2,6-diAF and 2-AF, respectively, than 9-AF (2.68×10^{-4} $cm^2 V^{-1} s^{-1}$) and spiro-OMeTAD (5.69×10^{-4} $cm^2 V^{-1} s^{-1}$). The lowest hole mobility (9-AF) is attributed to large twist angle leading to looser 9-AF molecular packing and less effective hole hopping. The steady-state photoluminescence spectra of perovskite layer on a glass substrate without and with HTMs are presented in Figure 12.5c. Quenching efficiency of over 95% perovskite emission was observed after capping perovskite absorber layer with either 2-AF, 2,6-diAF, 9-AF, or spiro-OMeTAD, indicating successful transfer of holes from HOMO energy levels of perovskites to HOMO energy levels of HTMs. The photovoltaic performance was investigated using 2-AF, 2,6-diAF, 9-AF, and spiro-OMeTAD, separately, as HTLs. The 2-AF-based device showed a PCE of 18.11% corresponding to FF of 73.78%, J_{SC} of 23.38 mA cm^{-2}, and V_{OC} of 1.05 V. 2.6-diAF-based device showed PCE of 18.64% corresponding to FF of 75.15%, J_{SC} of 23.63 mA cm^{-2}, and V_{OC} of 1.05 V. 9-AF-based device showed PCE of 13.86% corresponding to FF of 71.03%, J_{SC} of 19.90 mA cm^{-2}, and V_{OC} of 0.98 V, and finally, spiro-OMeTAD-based device showed PCE of 17.90% corresponding to FF of 74.47%, J_{SC} of 22.67 mA cm^{-2}, and V_{OC} of 1.06 V.

Pham et al. [47] designed and synthesized novel and cheap anthanthrone-based HTMs, namely, 4,10-bis(1,2-dihydroacenaphthylen-5-yl)-6,12-bis(octyloxy)-6,12-dihydronaphtho[7,8,1,2,3-nopqr]tetraphene (ACE-ANT-ACE, **23**) and 4,4′-(6,12-bis(octyloxy)-6,12-dihydronaphtho[7,8,1,2,3-nopqr]tetraphene-4,10-diyl)bis(N,N-bis(4-methoxyphenyl)aniline) (TPA-ANT-TPA, **24**). The energy levels were estimated using cyclic voltammetry and photoelectron spectroscopy in air (see Figure 12.6a). The HOMO energy levels of TPA-ANT-TPA (−5.41 eV) and ACE-ANT-ACE (−5.32 eV) are higher than that of perovskite (−5.46 eV), with small HOMO offset suggesting sufficient hole transfer from the HOMO of the perovskite to the HOMO of the HTL. The LUMO energy levels of the HTMs are higher than that of the perovskite layer, suggesting that the novel HTMs can successfully block electron transfer from the LUMO of the perovskite layer to the LUMO of HTMs (see Figure 12.6a). Pham and coworkers fabricated the solar cell devices with architecture comprising of FTO|compact-TiO_2|mesoporous-TiO_2| $CH_3NH_3PbI_3$|HTL|Ag using different HTM layers. The champion PCE achieved for TPA-ANT-TPA was 17.50% with FF of 79.60%, J_{SC} of 21.07 mA cm^{-2}, and V_{OC} of 1.03 V whereas a lower PCE was achieved for ACE-ANT-ACE-based device (PCE of 13.10%, FF of 67.00%, J_{SC} of 18.70 mA cm^{-2}, and V_{OC} of 1.03 V). The spiro-OMeTAD-based device performance was reported to have PCE approaching 16.80%, corresponding to FF of 76.80%, J_{SC} of 21.09 mA cm^{-2}, and V_{OC} of 1.04 V (see Figure 12.6b). The stability

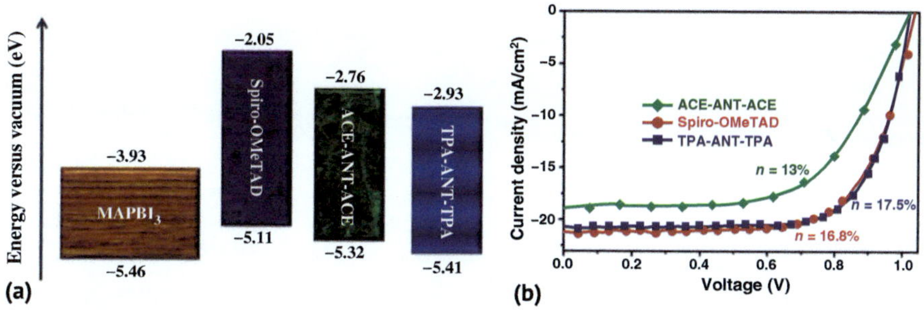

Figure 12.6 (a) The energy level diagrams and (b) current density versus voltage of ACE-ANT-ACE, TPA-ANT-TPA, spiro-OMeTAD. *Source*: [47]/John Wiley and Sons.

Table 12.2 Application of anthracenes anthracene- and anthanthrone-based materials as HTLs in PSCs.

HTL	E_{HOMO} (eV)	Hole mobility ($cm^2 V^{-1} s^{-1}$)	FF (%)	J_{SC} (mA cm^{-2})	V_{OC} (V)	PCE (%)	Ref.
C1 (**21**)	−5.35	9.77×10^{-4}	70.30	21.42	1.06	15.96	[48]
C2 (**22**)	−5.32	4.65×10^{-4}	67.00	20.42	0.94	12.86	
PCBZANT (**25**)	−5.15	1.31×10^{-4}	77.50	21.03	0.95	15.50	[49]
PTPAANT (**26**)	−5.24	5.89×10^{-5}	77.84	11.66	1.00	14.52	
DPA-ANT-DPA (**27**)	−5.20	2.60×10^{-4}	71.00	16.20	1.00	11.50	[50]
X2 (**28**)	−4.94	3.10×10^{-4}	66.17	16.57	0.94	10.25	[51]
X1 (**29**)	−5.03	2.80×10^{-4}	67.54	23.80	1.01	16.10	
DC1 (**30**)	−4.99	4.05×10^{-5}	78.00	21.89	1.06	18.09	[52]
DC2 (**31**)	−5.29	1.32×10^{-5}	74.00	17.51	1.02	13.21	
DC3 (**32**)	−5.09	2.70×10^{-5}	73.00	20.03	1.02	14.91	
HL31 (**33**)	−5.01	3.00×10^{-4}	68.00	21.20	1.12	16.60	[45]
HL22 (**34**)	−4.83	4.90×10^{-4}	75.00	23.30	1.11	19.20	

studies revealed that the PCE of TPA-ANT-TPA was ~14.00% after exposure to 58% room humidity conditions for more than 50 h, while spiro-OMeTAD-based device diminished to ~2.00% after 10 h exposure. Their study suggests that TPA-ANT-TPA provides a promising alternative to HTL materials for best-performing solar technology. Table 12.2 shows the investigations reported on developments of new HTMs based on anthracene and anthanthrone cores and their performance in photovoltaic applications. Studies revealed that different substituents on anthracene and anthanthrone cores alter the HOMO energy levels, hole mobility, and photovoltaic performance (Scheme 12.3).

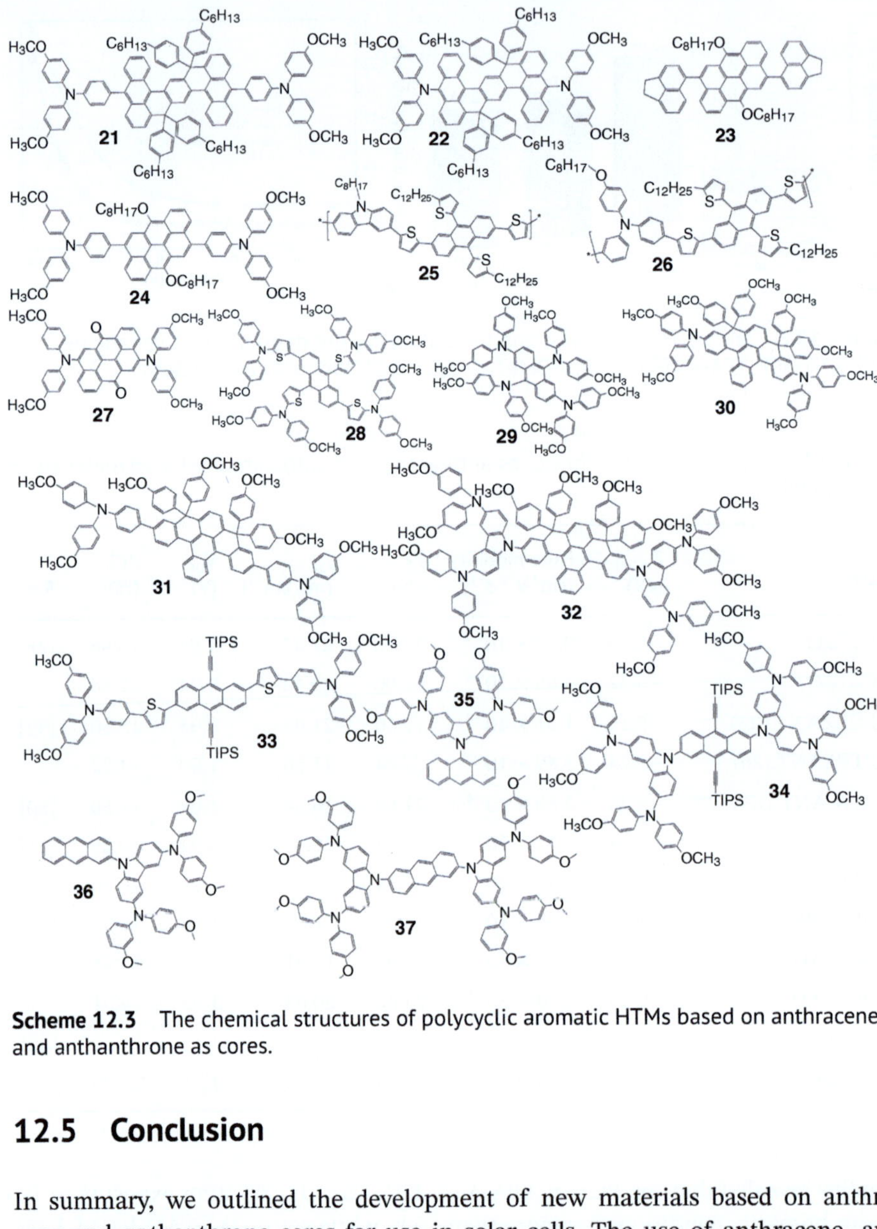

Scheme 12.3 The chemical structures of polycyclic aromatic HTMs based on anthracene and anthanthrone as cores.

12.5 Conclusion

In summary, we outlined the development of new materials based on anthracene and anthanthrone cores for use in solar cells. The use of anthracene- and anthanthrone-based cores as starting materials for development of new active layer materials in OSCs and HTLs in PSCs shows great potential for producing high-performing devices with improved stability. In OSCs, designed donor materials developed using anthracene and anthanthrone cores show better performance when blended with non-fullerene acceptor materials, while anthracene- and anthanthrone-based acceptor materials perform better with improved solubility in common solvents when blended with small molecule-based donors. In PSCs, different substituents on anthracene and anthanthrone cores for developed HTMs alter the HOMO energy levels and hole mobility, permitting optimization of overall device performance. In comparison to the most popular HTM (spiro-OMeTAD), the anthracene- and anthanthrone-based HTMs are economical to produce and exhibit higher PCE. The development of new PACs using low-cost anthracene and anthanthrone cores can help for large-scale production of solar cells for energy harvesting applications and commercialization.

Acknowledgment

The authors most gratefully acknowledge the financial support of the National Research Foundation of South Africa.

Conflict of Interest

The authors declare no conflict of interest.

References

1 Wang, J. and Azam, W. (2024). Natural resource scarcity, fossil fuel energy consumption, and total greenhouse gas emissions in top emitting countries. *Geoscience Frontiers* 15 (2): 101757.

2 van Asselt, H., Fragkos, P., Peterson, L. et al. (2024). The environmental and economic effects of international cooperation on restricting fossil fuel supply *International Environmental Agreements: Politics, Law and Economics* 24 (1): 141–166.

3 Liu, M., Sun, Y., Shao, K. et al. (2024). Synergistic solar-powered water-electricity generation: an integrated floating system on water. *Nano Energy* 119: 109074.

4 Di Sabatino, M., Hendawi, R., and Garcia, A.S. (2024). Silicon solar cells: trends, manufacturing challenges, and AI perspectives. *Crystals* 14 (2): 167.

5 Li, Q., Li, T., Kutlu, A. et al. (2024). Life cycle cost analysis and life cycle assessment of ETFE cushion integrated transparent organic/perovskite solar cells: comparison with PV glazing skylight. *Journal of Building Engineering* 87: 109140.

6 Nkinyam, C.M., Ujah, C.O., Nnakwo, K.C. et al. (2025). Insight into organic photovoltaic cell: Prospect and challenges. *Unconventional Resources* 5: 100121.

7 Hosseinnezhad, M., Nasiri, S., Nutalapati, V. et al. (2024). A review of the application of organic dyes based on naphthalimide in optical and electrical devices. *Progress in Color, Colorants, and Coatings* 17 (4): 417–433.

8 Hosny, M.M., Abdel-Aziz, M.H., Zoromba, M.S. et al. (2024). Design and synthesis of high-performance polymer blend solar cell. *Journal of Materials Research and Technology* 28: 4199–211.

9 Nozaki, N., Uva, A., Matsumoto, H. et al. (2024). Thienoisoindigo-based recyclable conjugated polymers for organic electronics. *RSC Applied Polymers* 2 (2): 163–171.

10 Wei, N., Chen, J., Cheng, Y. et al. (2024). Constructing multiscale fibrous morphology to achieve 20% efficiency organic solar cells by mixing high and low molecular weight D18. *Advanced Materials* 36 (41): 2408934.

11 Fu, J., Yang, Q., Huang, P. et al. (2024). Rational molecular and device design enables organic solar cells approaching 20% efficiency. *Nature Communications* 15 (1): 1830.

12 Li, X., Ke, H., Li, S. et al. (2024). Intrinsically stretchable organic photovoltaic cells with improved mechanical durability and stability via dual-donor polymer blending. *Advanced Functional Materials* 34 (29): 2400702.

13 Zhang, C., Zhong, X., Sun, X. et al. (2024). Designing a novel wide bandgap small molecule guest for enhanced stability and morphology mediation in ternary organic solar cells with over 19.3% efficiency. *Advanced Science* 11 (23): 2401313.

14 Elbert, S.M., Bolgert, E.A., Paine, O.T. et al. (2024). Z-shaped polycyclic aromatic hydrocarbons with embedded five-membered rings and their application in organic thin-film transistors. *Organic Chemistry Frontiers* 11 (19): 5340–5349.

15 Darvish, A., Lirette, F., Fernández, I. et al. (2024). Synthesis of contorted polycyclic aromatic compounds via alkyne benzannulation on zigzag-edged dibenzochrysene derivatives. *Chemistry: A European Journal* 17: e202403456.

16 Li, H., Zhao, G., Tang, Q. et al. (2024). Isomeric diazapyrene–thiophene conjugated systems: synthesis, characterization, and transport properties. Organic *Chemistry Frontiers* 11 (24): 7205–7213.

17 Xue, N., Chen, K., Liu, G. et al. (2024). Molecular engineering of rylene diimides via sila-annulation toward high-mobility organic semiconductors. *Small* 20 (17): 2307875.

18 Pourebrahimi, S. and Pirooz, M. (2024). Exploring the optoelectronic properties of flavylium cations as acceptors in organic solar cells: DFT/TD-DFT investigations. *Solar Energy* 275: 112617.

19 Huang, R., Liu, T., Peng, H. et al. (2024). Molecular design and architectonics towards film-based fluorescent sensing. *Chemical Society Reviews* 53 (13): 6960–6991.

20 Brucoli, J., Gariboldi, D., Puglisi, A. et al. (2024). In continuo Pd-catalysed cross coupling reactions of organolithium reagents with aryl bromides under aerobic conditions. *European Journal of Organic Chemistry* 27 (6): e202301289.

21 Jansen-van Vuuren, R.D., Liu, S., Miah, M.A.J. et al. (2024). The versatile and strategic *O*-carbamate directed metalation group in the synthesis of aromatic molecules: an update. *Chemical Reviews* 124 (12): 7731–7828.

22 Rashid, M.A.M., Min, S., Namgoong, S.K. et al. (2024). Effect of substituting donors on the hole mobility of hole transporting materials in perovskite solar cells: a DFT study. *Physical Chemistry Chemical Physics* 26 (2): 1352–1363.

23 Niharika Bhuyan, N., Shankar, S.S., Jyoti Panda, S. et al. (2024). An asymmetric coumarin-anthracene conjugate as efficient fullerene-free acceptor for organic solar cells. *Angewandte Chemie International Edition* 63 (40): e202406272.

24 Giguère, J., Verolet, Q., and Morin, J. (2013). 4,10-Dibromoanthanthrone as a new building block for p-type, n-type, and ambipolar π-conjugated materials. *Chemistry: A European Journal* 19 (1): 372–381.

25 Liu, Q., Wang, Y., Arunagiri, L. et al. (2020). Versatile nature of anthanthrone based polymers as active multifunctional semiconductors for various organic electronic devices. *Advanced Materials* 1 (9): 3428–3438.

26 (a) Jiang, Y., Yao, C., Wang, X. et al. (2024). A Perspective on 2D fused-ring quad-rotor-shaped non-fullerene acceptors with an anthracene core. *The Journal of Physical Chemistry A* 128 (39): 8446–8456; (b) Yuan, J., Zhang, Y., Zhou, L. et al. (2019). Single-junction organic solar cell with over 15% efficiency using fused-ring acceptor with electron-deficient core. *Joule* 3 (4): 1140–1151.

27 Liu, T., Liu, J., Li, Y. et al. (2024). 3-Methylthiophene: a sustainable solvent for high-performance organic solar cells. *ACS Applied Materials & Interfaces* 16 (38): 50916–50925.

28 Zheng, R., Zhang, C., Zhang, A. et al. (2023). Effect of steric hindrance at the anthracene core on the photovoltaic performance of simple nonfused ring electron acceptors. *ACS Applied Materials & Interfaces* 15 (3): 4275–4283.

29 Tu, D., Qiao, Y., Ni, Y. et al. (2022). Structural engineering of anthracene diimide polymers for molecular ordering manipulation. *Macromolecules* 55 (10): 4102–4110.

30 Tu, D., Feng, Z., Feng, Z. et al. (2021). Crystallinity and orientation manipulation of anthracene diimide polymers for all-polymer solar cells. *Advanced Functional Materials* 31 (22): 2011049.

31 Abdallah, R.B., Radaoui, M., Abdelaziz, B.B. et al. (2023). Effect of active layer thickness on the charge recombination and dissociation in bulk heterojunction polymer solar cells under open circuit conditions. *Synthetic Metals* 295: 117353.

32 Cha, H., Li, J., Li, Y. et al. (2020). Effects of bulk heterojunction morphology control via thermal annealing on the fill factor of anthracene-based polymer solar cells. *Macromolecular Research* 28 (9): 820–825.

33 Feng, H., Yi, Y., Ke, X. et al. (2019). New anthracene-fused non-fullerene acceptors for high-efficiency organic solar cells: energy level modulations enabling match of donor and acceptor. *Advanced Energy Materials* 9 (12): 1803541.

34 Yao, C., Liu, B., Zhu, Y. et al. (2019). Highly fluorescent anthracene derivative as a non-fullerene acceptor in OSCs with small non-radiative energy loss of 0.22 eV and high PCEs of over 13%. *Journal of Materials Chemistry A* 7 (17): 10212–10216.

35 Kim, Y.J., Lee, J.S., Hong, J. et al. (2016). Two dibenzo[Def, Mno]chrysene-based polymeric semiconductors: surprisingly opposite device performances in field-effect transistors and solar cells. *Journal of Polymer Science Part A: Polymer Chemistry* 54 (16): 2559–2570.

36 Stewart, D.J., Shi, J., Naranjo, T.R. et al. (2018). Manipulating triplet states: tuning energies, absorption, lifetimes, and annihilation rates in anthanthrene derivatives. *Physical Chemistry Chemical Physics* 20 (45): 28412–28418.

37 Gagnon, F., Tremblay, V., Soldera, A. et al. (2022). 2,9-Dibenzo [b, def] chrysene as a building block for organic electronics. *Advanced Materials* 3 (1): 599–603.

38 Li, M., Xiao, M., and Li, Z. (2021). Adjusting the photovoltaic performance of big fused ring-based small molecules by tailoring with different modifications. *RSC Advances* 11 (63): 39625–39635.

39 Velusamy, A., Afraj, S.N., Guo, Y.S. et al. (2024). Bicyclopentadithiophene-based organic semiconductor for stable and high-performance perovskite solar cells exceeding 22%. *ACS Applied Materials & Interfaces* 16 (5): 6162–6175.

40 Wang, F., Wang, T., Sun, Y. et al. (2024). Two-step perovskite solar cells with > 25% efficiency: unveiling the hidden bottom surface of perovskite layer. *Advanced Materials* 36 (31): 2401476.

41 Tiep, N.H., Ku, Z., and Fan, H.J. (2016). Recent advances in improving the stability of perovskite solar cells. *Advanced Energy Materials* 6 (3): 1501420.

42 Tumen-Ulzii, G., Matsushima, T., and Adachi, C. (2021). Mini-review on efficiency and stability of perovskite solar cells with spiro-OMeTAD hole transport layer: recent progress and perspectives. *Energy Fuels* 35 (23): 18915–18927.

43 Kung, P., Li, M., Lin, P. et al. (2018). A review of inorganic hole transport materials for perovskite solar cells. *Advanced Materials Interfaces* 5 (22): 1800882.

44 Wu, C., Chen, C., Tao, L. et al. (2020). Highly efficient perovskite solar cells based on symmetric hole transport material constructed with indaceno [1,2-b:5,6-b′] dithiophene core building block. *Journal of Energy Chemistry* 43: 98–103.

45 Aktas, E., Le, T.H., Frigoli, M. et al. (2023). Triisopropylsilylethynyl-functionalized anthracene-based hole transport materials for efficient hybrid lead halide perovskite solar cells. *Chemistry of Materials* 35 (21): 9378–9389.

46 Singh, A., Abate, S.Y., Pavan Kumar, C. et al. (2020). Bis(diphenylamine)-tethered carbazolyl anthracene derivatives as hole-transporting materials for stable and high-performance perovskite solar cells. *ACS Applied Energy Materials* 3 (11): 10752–10764.

47 Pham, H.D., Do, T.T., Kim, J. et al. (2018). Molecular engineering using an anthanthrone dye for low-cost hole transport materials: a strategy for dopant-free, high-efficiency, and stable perovskite solar cells. *Advanced Energy Materials* 8 (16): 1703007.

48 Chandrasekaran, D., Chiu, Y., Yu, C. et al. (2021). Polycyclic arenes dihydrodinaphthopentacene-based hole-transporting materials for perovskite solar cells application. *Chemistry: An Asian Journal* 16 (22): 3719–3728.

49 Tong, T., Tan, C., Keller, T. et al. (2018). Two anthracene-based copolymers as the hole-transporting materials for high-performance inverted (p-i-n) perovskite solar cells. *Macromolecules* 51 (18): 7407–7416.

50 Pham, H.D., Hayasake, K., Kim, J. et al. (2018). One step facile synthesis of a novel anthanthrone dye-based, dopant-free hole transporting material for efficient and stable perovskite solar cells. *Journal of Materials Chemistry C* 6 (14): 3699–3708.

51 Chandrasekaran, D., Chiu, W.H., Lee, K.M. et al. (2022). Effect of thiophene insertion on X-shaped anthracene-based hole-transporting materials in perovskite solar cells. *Polymers* 14 (8): 1580.

52 Chandrasekaran, D., Liou, S.J., Chiu, W.H. et al. (2023). Ladder-type dihydronaphtho [1,2,3,4-rst]pentaphene as building block to construct hole-transporting materials for perovskite solar cells. *Journal of Power Sources* 581: 233496.

13

Conjugated Polymers for Biomedical Applications

*Qi Shen and Shu Wang**

Beijing National Laboratory for Molecular Sciences, Key Laboratory of Organic Solids, Institute of Chemistry, Chinese Academy of Sciences, Beijing, P.R. China

*Corresponding author: wangshu@iccas.ac.cn

13.1 Introduction

Conjugated polymers (CPs) have achieved significant advancements over the past few decades, emerging as promising candidates for biomedical applications such as biological detection, bioimaging, and disease therapy. As a promising category of multifunctional materials for biomedical applications, CPs can be fabricated to display excellent properties through simple preparation methods and surface modification [1]. The backbone structure of CPs determines the absorption and fluorescence emission spectra, as well as the fluorescence quantum efficiency. To diversify the types of CPs and further modulate their energy band structures, many researchers have adopted the strategy of altering the backbone structure based on conventional CPs [2]. Generally, short-wavelength-emitting CPs possess higher fluorescence quantum efficiencies compared to long-wavelength-emitting CPs, which is more beneficial for biological detection. Furthermore, near-infrared (NIR) fluorescence-emitting CPs can be employed as photothermal agents for deep-tissue imaging and therapeutic applications [3]. More importantly, modifying side chains with various drug molecules or recognition groups (e.g. phospholipids, oligopeptides, small-molecule ligands, or antibodies) significantly enhances the functionality of CPs. To date, CPs have been utilized for DNA/RNA detection, protein or cell monitoring, fluorescence imaging, photoacoustic imaging (PAI), photodynamic/photothermal therapy (PDT/PTT), antimicrobial therapy, and antiviral therapy due to their outstanding photochemical stability, tunable spectral properties, and superior fluorescence efficiency [4].

Conjugated polymer nanoparticles (CPNs) have emerged with the aim of expanding the application of CPs in biomedical fields. Scientists have employed the assembly approach to integrate the superior photophysical properties of CPs with the advantages of nanoparticles (NPs), resulting in improved outcomes such as good biocompatibility,

Conjugated Polymer Synthesis and Materials Applications, First Edition. Edited by Yoshiki Chujo.
© 2026 Wiley-VCH GmbH. All rights reserved, including rights for text and data mining and training of artificial intelligence technologies or similar technologies. Published 2026 by Wiley-VCH GmbH.

excellent photostability, and small particle size effect [5]. Preparation methods to produce CPNs are primarily founded on two main strategies: mini-emulsion method and nano-precipitation method. Among them, mini-emulsion approach is the most traditional way for CPN preparation. In 2002, Landfester et al. demonstrated the feasibility of the mini-emulsion process for the preparation of CPNs [6]. The mini-emulsion method for the preparation of CPNs entails the transfer of hydrophobic polymers into an aqueous solution of surfactant [7]. Subsequently, the solutions are mixed with the assistance of ultrasound to form mini-emulsions composed of polymer droplets. The particle size of CPN prepared using this method can be as small as approximately 20 nm. In comparison to the mini-emulsion method, the nano-precipitation method is capable of obtaining smaller sized NPs. Nano-precipitation method involves dissolving the hydrophobic polymer in organic solvent that is miscible with water. Next, polymers are aggregated through hydrophobic interactions among the polymer molecules under ultrasound assistance to form CPNs. In addition to the above two methods, other material compositions can be utilized for the preparation of CPNs, including lipid [8] and protein [9]. CPNs derived from CPs exhibit augmented properties in aqueous media, facilitating improved visualization in living organisms and serving as effective carriers for drug delivery. CPNs are regarded as revolutionary materials in the field of therapeutic diagnostics [4b].

To date, CPs have been rapidly developed from the design, synthesis, and application of polymers. The chapter involves various applications of CPs in biomedical fields, focusing on biosensing mechanisms and detection methods that utilize the optical and electrochemical properties of CPs. It also explores fluorescence imaging, NIR imaging, PAI, and CP-based therapies for anticancer, antimicrobial, and antiviral treatments. Figure 13.1 summarizes the structures of relevant CPs for biomedical applications.

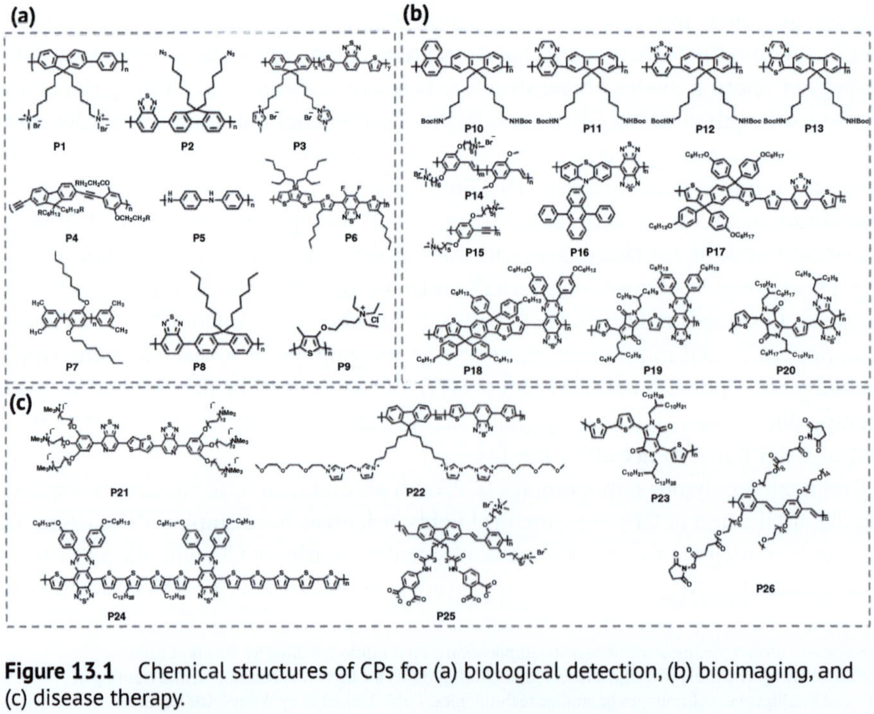

Figure 13.1 Chemical structures of CPs for (a) biological detection, (b) bioimaging, and (c) disease therapy.

13.2 Biosensing and Biological Detection

13.2.1 Fluorescence Signal Amplification Based on CPs

CPs have been employed as fluorescent probes for the detection of biomarkers due to their outstanding optoelectronic properties and stability [10]. The backbone structure of CPs facilitates exciton migration, promoting energy transfer to the lower electron/energy receptor, which quenches CPs' fluorescence or amplifies the receptor signal [11]. In comparison to small-molecule dyes, CPs dramatically enhance the detection sensitivity and lower the limit of detection (LOD) [12]. The most common detection strategy includes fluorescence resonance energy transfer (FRET). Since FRET was first proposed in the early twentieth century [13], the technique has emerged as a widely utilized technique in the fields of biological detection and live cell imaging [14]. Three conditions must be met for FRET to occur: (1) there needs to be overlap between the emission spectra of donor molecule and absorption spectra of receptor molecule; (2) the distance between donor and receptor molecules needs to be extremely close; (3) the dipole moment orientation of the donor and acceptor in CPs should be suitable for energy transfer [11, 15]. Several factors affecting FRET efficiency in CPs-based biosensors include fluorescent energy receptor, number of repeating units of CPs, and length of side chain. Wang et al. investigated the close relationship between the number of repeating units of CPs and FRET efficiency. As the number of repeating units increased, FRET efficiency also increased [16]. Regarding intramolecular energy transfer, the effective conjugated repeating units of CPs were 6–8 [17]. Building on the high FRET efficiency observed between CPs and small-molecule fluorescent dyes (fluorescein [FL], rhodamine, and Cy3), an increasing number of FRET-based biosensing methods have been developed [18].

13.2.1.1 DNA Detection

DNA plays a crucial role in storing the genetic information of species and in encoding various biological functions, which are primarily determined by the base sequences of DNA. Several major human diseases, including malignant tumors, cardiovascular diseases, and immunodeficiency disorders, are associated with genomic DNA. Acquisition of DNA information is anticipated to guide the prevention, early diagnosis, and drug screening of these diseases, providing powerful tools for individualized treatment of disease. Fluorescence analysis possesses high sensitivity and is widely used in DNA detection systems [19]. To achieve accurate fluorescence detection signals, it is essential to establish surface interactions between CPs and DNA. Both DNA/RNA and proteins carry negatively charged due to the presence of phosphate groups or carboxyl groups [20]. Therefore, positively charged side chains (quaternary ammonium, phosphonium, ammonium, and iminium) modified CPs interact electrostatically with DNA [21]. Bazan et al. used cationic water-soluble polyfluorene derivatives for DNA detection [4a]. In this design, FL-labeled peptide nucleic acid (DNA analogs of peptide nucleic acid (PNA) [22]) was used as probes for the detection of cationic CPs to the target DNA. Before hybridization, PNA was electrically neutral, resulting in no electrostatic interaction with CPs. However, upon the addition of target DNA, PNA hybridized with the negatively charged DNA, bringing CPs and FL closer together and generating FRET. Fluorescence intensity of added CPs was more than 25-fold higher compared to fluorescence intensity of directly excited FL.

In CPs-based FRET systems, CPs serve as donors while various small-molecule fluorescent dyes act as acceptors, which are combined in different configurations to yield distinct fluorescence detection signals. Based on this strategy, Wang et al. attempted to construct a detection strategy for DNA methylation using FRET effect between CPs and fluorescent dyes (Figure 13.2a) [23]. The methylation morphology of DNA has emerged as key markers of cancer pathogenesis, including the CpG island methylation phenotype [26]. This assay employed the restriction endonuclease *HpaII* to specifically recognize and cleave unmethylated CpG sites, distinguishing between methylated and unmethylated DNA. Then, cationic CPs poly{(1,4-phenylene)-2,7-[9,9-bis (6-N, N, N-trimethyl ammonia)-hexyl fluorene] dibromide} (PFP, P1) were utilized to electrostatically interact with negatively charged fluorescently labeled DNA and achieved efficient FRET. Compared with the sodium bisulfite-treated method, this method was able to quantitatively obtain the methylation level of samples. As shown in Figure 13.2b, fluorescence intensity of FL gradually increased with the methylation level. To detect DNA with low methylation levels, PCR cycling conditions were optimized and the standard curve was reconstructed using repeated measurements (Figure 13.2c). The improved protocol was capable of detecting DNA methylation as low as 1.25%, increasing the sensitivity of hypomethylated DNA detection.

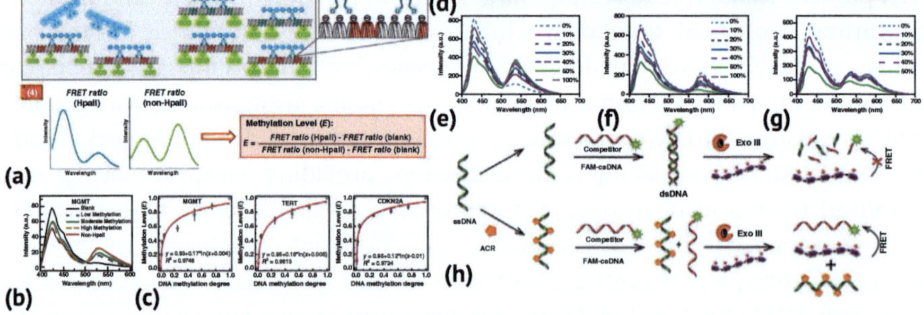

Figure 13.2 (a) Design strategies for CPs-based DNA methylation detection method. (b) Detection of fluorescence spectra with different methylation degrees based on CPs. (c) Standard curves of methylation of MGMT, CDKN2A, and TERT genes. *Source*: [23]/with permission of American Chemical Society. (d) Detection strategies for CPs-based single-base polymorphism. (e) FL-labeled samples of different mutation proportions. (f) Cy3-labeled samples of different mutation proportions. (g) double labeled (FL and Cy3) samples of different mutation proportions. *Source*: [24]/with permission of American Association for the Advancement of Science. (h) Schematic representation of fluorescence signal amplification strategy combined with Exo III was used to detect ACR. *Source*: [25]/with permission of Elsevier.

Single-base mutations in DNA can also be detected using CPs-FRET method. Single nucleotide polymorphisms (SNPs) are polymorphisms in DNA sequences due to mutations in a single nucleotide, numbering more than 1.26 trillion in the human genome [27]. For the analysis of SNPs, the detection system needs to be highly selective and capable of accurately and sensitively identifying target DNA bases. Shen et al. developed a sensitive and cost-effective method for SNP genotyping by combining poly{(1,4-phenylene)-2,7-[9,9-bis (6-N, N, N-trimethyl ammonia)-hexyl fluorene] dibromide}(PFP) with allele-specific PCR [24]. The detection principle is illustrated in Figure 13.2d, where allele-specific PCRs occur exclusively when the 3′ end of the specific primer perfectly matches target DNA. To detect multiple SNP loci simultaneously, cationic CPs, FL-labeled nucleotides, and Cy3-labeled nucleotides were used for multiplex FRET effects. In this detection system, both FL and Cy3 acted as energy acceptors for CPs, while FL served as energy donor for Cy3. In this study, SNPs associated with hypertension medication were selected. For these SNPs, there are four genotypes that could be detected: wild-type loci, two single-mutant loci, and double-mutant loci. For FL-labeled single-mutant loci, the fluorescence emission intensity of FL increased with the proportion of mutations, while the fluorescence emission intensity of CPs decreased (Figure 13.2e). The same trend was observed for Cy3-labeled single-mutant loci (Figure 13.2f). For the detection of double-mutation sites, both the fluorescence emission intensity of FL and Cy3 increased simultaneously (Figure 13.2g). After retrospective study of clinical samples and sequence alignment, the method was applied to medication guidance for hypertension disease. Using this detection system, the type and dosage of medication could be adjusted according to the results of gene detection, demonstrating the feasibility of this approach for individualized treatment. Furthermore, Song et al. established FRET fingerprinting based on CPs that detects more than four mutation sites simultaneously [28]. This method integrated the properties of multi-step FRET from CPs with multiple fluorescent dyes with double deoxy sequencing reaction. For nonadjacent mutations, allele-specific primers could be used for extension to achieve one-tube multiple detections of four mutations in *PIK3CA* gene. For adjacent mutations sites, single-primer extension was utilized to achieve multiple detections of 12 mutations in KRAS gene. The technique was highly sensitive and could detect mutations in 5% of the total DNA.

The exonuclease activity of exonuclease III (Exo III) assists in amplifying FRET signals from cationic CPs [29]. Liu et al. developed a fluorescence biosensor for the detection of ACR using cationic CPs PFP in combination with Exo III [25]. The detection system contained single-stranded DNA (ssDNA, S1), FAM-csDNA (S2, complementary to S1), Exo III, and ACR. S1 and S2 could hybridize to form double-stranded DNA (dsDNA), which served as a substrate for Exo III digestion into single nucleotides (Figure 13.2h), leading to inefficient FRET. Meanwhile, ACR strongly bound to the N-7 site of ssDNA and inhibited hybridization between S1 and S2, reducing the possibility of Exo III digestion. The substantial concentration of free S2 in solution interacted with the positively charged PFP through electrostatic interactions, resulting in the generation of FRET. The detection limit of the fluorescence sensor was 0.16 μM.

13.2.1.2 RNA Detection

Ribonucleic acid (RNA) is an important molecule involved in gene expression and has been identified as a potential biomarker for various diseases, including early cancer diagnosis and screening [30]. RNA can be divided into several forms, such as messenger RNA (mRNA) [31], microRNA (miRNA) [32], and transfer RNA (tRNA) [33]. MicroRNA is an endogenous non-protein-coding RNA molecule [34]. MicroRNA has received widespread attention as a biomarker for cancer, diabetes, cardiovascular disease, and neurological disorders [35]. Despite the development of various analytical methods for miRNA detection, these methods remain inadequate for high-throughput miRNA detection [36]. The substantial number of energy-absorbing CPs units facilitate the transfer of excitation energy to the chromophore, improving detection sensitivity [37]. Feng et al. proposed a multiplexed miRNA detection scheme that employed high-tension hybridization and dual signal amplification [38]. This approach utilized recoverable self-catalytic DNAzyme and light-harvesting CPs to achieve high-specificity and high-sensitivity detection of miRNA (Figure 13.3a). Initially, the fluorescent dye was in the "off" state. The target miRNA bound to the toehold region of the cyclooctyne modified ring probe (CRP), which resulted in the release of the enzyme sequence 17ES. Subsequently, 17ES was bound to the biotin-modified substrate sequence. With the assistance of metal ions, DNAzyme catalyzed the cleavage of the substrate chain for the removal of the biotin label. After several catalytic amplifications, a large amount of DNA sequence labeled with dye was generated. Finally, after separation via magnetic beads, Y-DNA and PFP were added to the supernatant and fluorescence of the dye was observed. This method holds a promising application prospect in miRNA detection within biological samples.

The advantage of biosensing with CPNs lies in their ability to maintain the signal amplification properties of CPs while also incorporating the beneficial characteristics

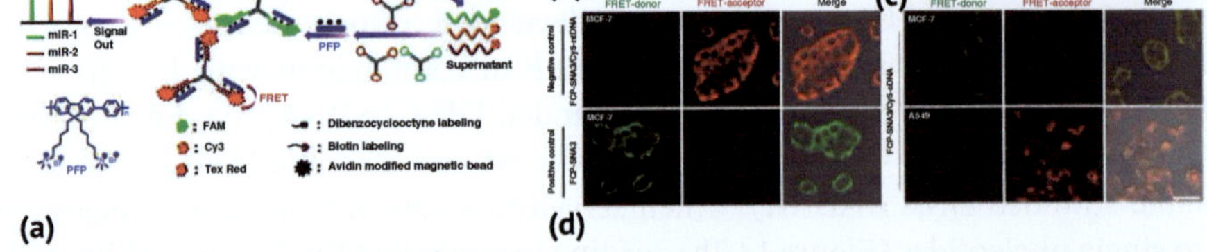

Figure 13.3 (a) Detection strategies of CPs-based FRET technique for multiple miRNA detection. *Source*: [38]/with permission of Royal Society of Chemistry. (b) Schematic diagram of CPN-based FRET technique for miRNA detection. (c) Schematic illustration of CPN-based FRET technique for single-cell level detection and imaging. (d) Cellular-level imaging using CPN-based FRET technique. *Source*: Reproduced with permission from [39]/John Wiley & Sons.

of nanoparticles. CPNs exhibit excellent photostability and low toxicity, which allows them to be employed for long-term monitoring in cellular-level imaging or detection. Furthermore, the surface of CNPs can be easily modified, allowing researchers to customize the surface functionality to meet the specific requirements of biological detection. Tian et al. coupled DNA to the side chain of azide-functionalized poly(fluorene-alt-benzothiadiazole) (PFBT-N_3, P2) via a "click reaction" to construct spherical nucleic acids based on CPs (Figure 13.3b) [39]. The size and light-harvesting ability of spherical nucleic acids (SNAs) were manipulated by adjusting the hydrophobicity of the grafted polymers, which subsequently amplified the fluorescent signals. By optimizing the structure of the fluorescent CPs, the signal amplification could be as high as 37-fold, and the detection limit of 1.7 pM in miRNA detection was achieved. CPNs fluorescent probes were then further applied to in situ miRNA detection at the single-cell level (Figure 13.3c). Compared with the negative control, CPNs fluorescent probe demonstrated weaker acceptor emission and stronger donor emission in MCF-7 cells (Figure 13.3d). Additionally, different FRET levels were detected for various cells, indicating a higher sensitivity of the nanoprobe. This nanomaterial provides an effective tool for the highly sensitive and specific detection of nucleic acids in biological samples.

DNA-mediated nanoparticle assembly has attracted extensive attention from researchers [40]. The ssDNA is modified onto nanoparticles, which are then spontaneously assembled into nanoparticle structures through DNA hybridization. Additionally, the DNA double helix structure can control the distance between nanoparticles [41]. Ohta et al. developed a DNA-mediated sequential assembly of CPNs on membranes for sensitive detection of cellular markers. The ssDNA-coupled antibody was first utilized to label the cell surface. Subsequently, DNA-modified CPNs were sequentially assembled via DNA hybridization. This method utilized DNA as a connector to regulate the assembly distance of CPNs, preventing fluorescence quenching. Detection of CD19 on the cell surface was accomplished through two rounds of DNA-mediated assembly of CPNs, resulting in fluorescence signal amplification of up to 31-fold without decreasing the binding capacity of the antibody. DNA-mediated CNPs demonstrated the advantages of the sequential assembly strategy in avoiding spatial site-blocking and proposed an effective strategy for sensitive detection of low-expression surface markers.

13.2.1.3 Protein Detection

Identifying and detecting proteins hold significant importance for both biology and medicine. Proteins constitute essential components of cells and serve as the material basis of life. The occurrence or development of certain major diseases is intimately associated with the activity of specific proteins [42]. Developing drugs targeting these protein molecules and conducting drug screening has emerged as a crucial strategy for disease treatment. Protein detection based on CPs primarily involves designing a sensing system using proteins or protein-recognizing target molecules as analytes. Currently functional CPs probes are mostly used as fluorescent reporter molecules in protein sensing. Hussain et al. proposed a strategy using self-assembly between cationic CPs (PF–DBT–Im) and anionic surfactants sodium dodecyl sulfate (SDS) to amplify and selectively monitor serum albumin [43].

This assembly triggered intermolecular FRET between the cationic CPs and anionic surfactants, resulting in fluorescence changes. When PF-DBT-Im and SDS formed a complex through electrostatic interaction, the solution color changed from purple to deep red. Then serum albumin was added to bind to the complex: the color of the solution shifted from dark red to magenta due to the intermolecular reverse FRET effect. This unique reverse FRET process was attributed to the synergistic contribution of charge interactions, hydrophobicity, and the relatively high intrinsic quantum yield of serum albumin. Moreover, the potential of this system for long-term protein monitoring in clinical applications was validated through the development of a smartphone-integrated point-of-care device. Huang et al. [44] developed a cationic CPs poly{[9,9-bis(6′-(N,N,N-diethylmethylammonium)hexyl)-2,7-fluorenylene ethynylene]-alt-co-[2,5-bis(3′-(N,N,N-diethylmethylammonium)-1′-oxapropyl)-1,4-phenylene]} tetraiodide (PFEP, P4) and FL amide-hyaluronan HA complexes were probed for detecting cell surface marker CD44. Electronic interactions between PFEP and HA induced the generation of complexes, facilitating FRET from PFEP to FA-HA. The interaction between CD44 and hyaluronic acid (HA) is essential for cell adhesion to the complex. The method only requires mixing the sample with a solution containing the complex, allowing the fluorescence to be observed with the naked eye.

13.2.2 Electrochemical Sensing Based on CPs

With advancements in science and technology, the use of analytical sensor devices for data monitoring has become increasingly straightforward. Biosensors generally consist of three components: (1) a sensor element that identifies the target analyte and converts it into a measurable signal; (2) an electronic circuit for signal processing; and (3) a presentation component that displays the output signal [45]. The functionality of a biosensor depends on the affinity of the biorecognition element for the target analyte, as well as signal conduction and amplification. As sensor materials, CPs can capture measurable signals generated by target analytes interacting with the bioreceptors immobilized on their surfaces, converting these signals into electrical outputs (Figure 13.4a) [46]. The presence of CPs contributes to the sensitivity of detection. Polyaniline (PAn, P5) demonstrates excellent electrical conductivity when doped with acid or nanomaterials [47]. Huang et al. investigated PAn as an electroactive probe for the detection of *Mycobacterium tuberculosis* MPT64 antigen [48]. Fullerene (C_{60})-doped PAn nanocomposites were utilized as probes and catalysts for electrochemical signal generation and amplification (Figure 13.4b). Tracer labels were fabricated through decorating the nanocomposites with gold nanoparticles and labeling them with signal aptamers. The introduction of the target MPT54 antigen captured the sandwich reaction between aptamer and the tracer tag, allowing the observation of detection signal. In addition, the detection signal was greatly enhanced through the efficient electrocatalytic oxidation of ascorbic acid using the nanocomposites. The detection limit of this sensor for MPT64 antigen was 20 fg/mL. An electrochemical enzyme biosensor based on CPs constitutes an ideal analytical instrument. Gu et al. focused on the design and application of CPs-enhanced electrochemical biosensors

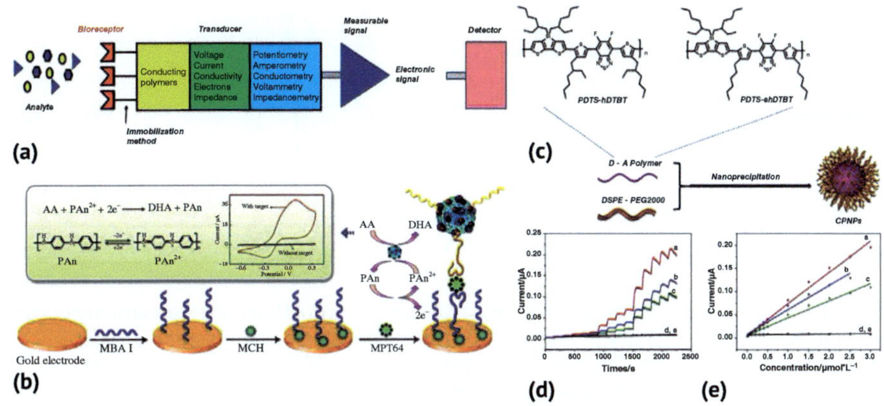

Figure 13.4 (a) Schematic representation of electrochemical biosensor. *Source*: [46]/MDPI/CC BY 4.0. Copyright 2020, MDPI. (b) Schematic representation of electrochemical aptamer sensor. *Source*: [48]/with permission of Elsevier . Copyright 2017, Elsevier. (c) CPs-based electrochemical sensing for bisphenol A detection in water environments. (d) Current–time curves of different concentrations of bisphenol A on various modified electrodes. (e) Calibration curve of BPA concentration. *Source*: [49]/with permission of Elsevier.

for tyrosinase, aiming for selective and rapid detection of bisphenol A (BPA) [49]. In the design of the sensors, CPNs were prepared using poly[(4,4′-bis(2-ethylhexyl)-dithieno[3,2-b:2′,3′-d]silole)-2,6-diyl-alt-4,7-di(4-(2-ethylhexyl)-2-thienyl)-5,6-difluoro-2,1,3-benzothiadiazole] (PDTS-ehDTBT, P6) with liposomal DSPE-mPEG2000 via nanoprecipitation method (Figure 13.4c). The larger specific surface area of BPA was able to form π–π interactions with CPNs, which promoted enzyme immobilization, material dispersion, electron transfer, and BPA enrichment. After validation with various modified electrodes, the incorporation of CPNs increased the specific surface area of the electrodes, thus enhancing the signal amplification of the electrochemical biosensors (Figure 13.4d). Figure 13.4e depicts the concentration-current correction curves for BPA detection with various modified electrodes. The results indicated that the strong π–π interaction between CPNs and BPA facilitated BPA accumulation, which improved the sensitivity of the biosensor. The LOD of the electrochemical biosensor based on PDTS-ehDTBT CPNs was 11.83 nM. The CPs-enhanced biosensor provides a reliable approach for rapid detection of BPA in real samples.

Photoelectrochemical (PEC) cytosensor is an evolutionary product of electrochemical methods. Compared to conventional electrochemical techniques, PEC significantly reduces background signals [50]. PEC process involves the electronic excitation and charge transfer of a photoexcited material induced by applied light, leading to the conversion of light into electricity [51]. Currently, various PEC biosensors have been developed for the detection of biomolecules, metal ions, and cells [52]. CPs with π-conjugated backbone and charged side chains have excellent photostability and good biocompatibility compared to inorganic photoelectrical responsive materials [53]. CPs have the ability to generate electron–hole pairs under light irradiation. When electrons migrate through the main chain of CPs and are captured by electron acceptor, holes are eliminated, resulting in a stable photocurrent [54].

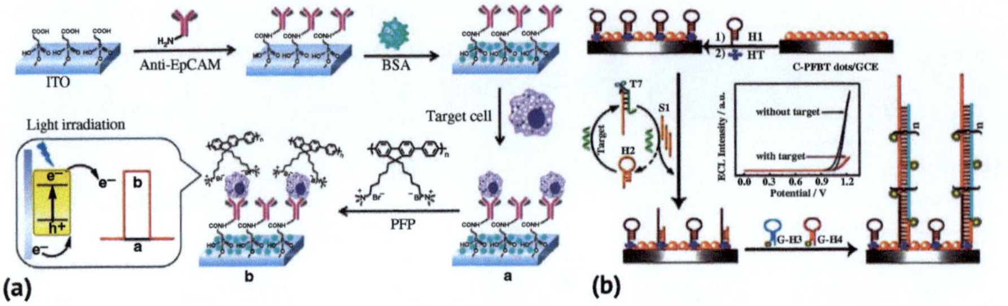

Figure 13.5 (a) Schematic diagram of PEC cytosensor using CPs as photoelectric response material. *Source*: [55]/with permission of American Chemical Society. Copyright 2018, American Chemical Society. (b) Schematic representation of ECL light-emitting biosensor construction. *Source*: [60]/with permission of Elsevier.

Wang et al. developed a PEC cytosensor based on CPs [55]. The sensor was accomplished by grafting 3-phosphonopropionic acid onto the electrode, which was then modified with anti-EpCAM via amide condensation reaction to capture SKBR-3 cells exhibiting high expression of EpCAM antigen (Figure 13.5a). The electrostatic interaction between the sensor and the negatively charged cell membrane was achieved by adding positively charged photoelectric response CPs PFP. The introduction of PFP generated photocurrents in the presence of light, providing turn-on signals for cell detection. CPs-based PEC cytosensor demonstrated high specificity and excellent sensitivity for detecting specific cells. The LOD was 24 cells/mL. The study shows the potential for constructing multifunctional PEC sensors and for cancer prognosis analysis.

Electrochemiluminescence (ECL) is a technique that combines electrochemical and luminescent phenomena. ECL phenomenon results from the electron transfer reactions of substances generated at the electrode surface, forming excited-state species that emit light upon returning to their ground state [56]. Most ECL systems introduce co-reactants to amplify ECL signal. Chen et al. focused on enhancing ECL of CPs for the detection of SARS-CoV-2 nucleocapsid protein using a hydrophobically localized enrichment strategy to enrich co-reactants [57]. Among them, carboxyl-functionalized CPs poly[2,5-dioctyl-1,4-phenylene] polymer nanoparticles (PDP PNPs, P7) were used as the luminescent group. Researchers utilized Pt nanoparticles to grow in situ at the metal ion coordination site of β-cyclodextrin for the preparation of the nanocomplexes. The hydrophobic cavity inside the nanocomplex facilitated efficient enrichment of co-reactant 3-(dibutylamino) propylamine. ECL emission of co-reactant concentrated through hydrophobic enrichment significantly enhanced ECL emission of CPs and improved the detection sensitivity. Additionally, the high charge-carrier mobility property of CPs is beneficial for ECL applications. Poly(9,9-dioctylfluorenyl-2,7-diyl) (PFO) can be used as a classical CPs for ECL-biosensing applications [58]. Xu et al. discovered that carboxyl-functionalized PFO could serve as a signal marker for ECL detection of Cu^{2+} without adding any co-reactants [59]. In addition to the aforementioned CPs for ECL, poly[(9,9-dioctylfluorenyl-2,7-diyl)-co-(1,4-benzo-{2,1′-3}-thiadiazole)] (PFBT, P8) has also been demonstrated to serve as an effective ECL sensor for the detection of miRNA. Yuan et al. demonstrated that functionalized PFBT exhibited strong ECL emission without the requirement for exogenous co-reactants or dissolved oxygen (Figure 13.5b) [60]. ECL detection of miRNAs was achieved by combining target cycling and hybridization chain reaction (HCR) to introduce large amounts of glucose oxidase (GOD) for in situ generation of the bursting agent H_2O_2. This ECL sensor could detect miRNA in specific cell lysates with LOD as low as 33 aM.

13.2.3 Colorimetric Sensing Based on CPs

Colorimetric detection is one of the most intuitive methods among various biological detection techniques, typically relying on the observation of color changes either with the naked eye or under UV irradiation. Several situations can contribute to colorimetric sensing of CPs, including electron transfer, energy transfer, conformational changes in polymer chains, or polymer–analyte complexation. For example, the binding of polythiophene derivatives to DNA leads to a conformational change in the molecular chain of CPs, which increases the π-electron overlap and conjugation length change [61]. This conformational change significantly affects the absorption and emission spectra of CPs. Zhu et al. utilized the conformational change of CPs with ssDNA to develop optical probes that can be used for DNA cleavage [62]. Poly[3-(3′-N, N, N-triethylamino-1′-propyloxy)-4-methyl-2,5-thiophene hydrochloride] (PMNT, P9) in CPs exhibited a change in complexation with ssDNA from an irregular coil conformation to planar conformation, resulting in a relative redshift of absorption wavelengths (Figure 13.6a). The color of the solution was pink upon

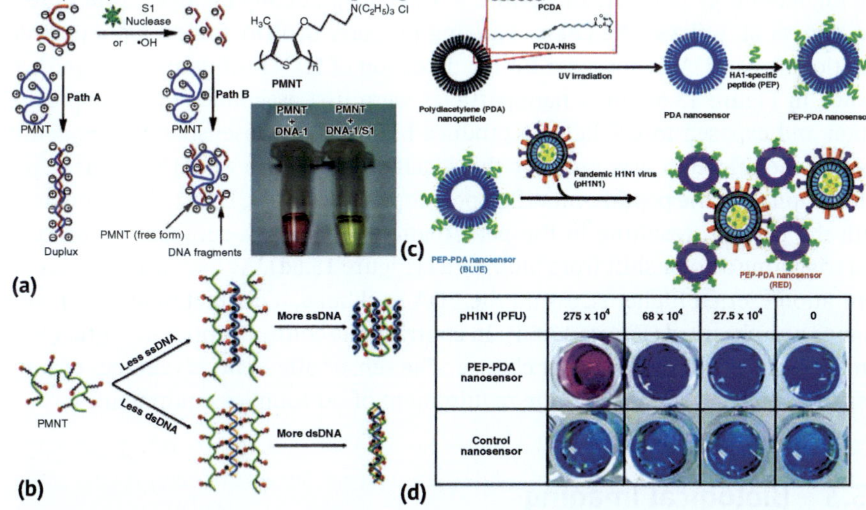

Figure 13.6 (a) Schematic illustration of PMNT for colorimetric sensing. *Source*: [62]/with permission of American Chemical Society. (b) Schematic illustration of PMNT for DNA detection. *Source*: [63]/with permission of American Chemical Society. (c) Schematic mechanism of CPs nanosensor colorimetric detection of virus. (d) Color transition images of CPs-based nanosensors after incubation with different concentrations of viruses. *Source*: [66]/with permission of Royal Society of Chemistry.

naked-eye observation. However, after ssDNA was cleaved into small fragments by S1 nuclease, PMNT maintained the irregular coil conformation and exhibited a relatively short absorption wavelength. After S1 nuclease hydrolysis for 30 minutes, the solution color remained yellow. In addition, ssDNA and dsDNA exhibit distinct binding mode to PMNT. Liu et al. developed label-free fluorescent biosensors using the different binding modes of PMNT to DNA (Figure 13.6b) [63]. Upon addition of low concentration of DNA, excess PMNT stretched into a highly conjugated and planar conformation. As the concentration of DNA increased, the electrostatic interactions between PMNT and DNA tended toward saturation. When high concentrations of ssDNA were complexed with PMNT, a conformational change was undergone that resulted in a redshift in emission. The structural alteration observed in the PMNT–ssDNA complex was attributed to interactions between the aromatic units of the PMNT backbone and the exposed bases of the ssDNA. When complexed with high-concentration dsDNA, PMNT could not interact with the unexposed bases of double-stranded structure, leading to the formation of dense aggregates. This approach is also applicable to aptamer binding assays [64]. The conformational change of the aptamer induced by target binding could be reflected in the fluorescence of PMNT, demonstrating the significant potential of CPs as label-free fluorescent biosensors.

In colorimetric detection, polydiacetylene (PDA) has been widely used as a colorimetric sensor material in biological detection due to its distinctive color-changing properties [65]. PDA usually appears blue due to the absorption of electrons through π—π^* transitions of electrons in the conjugated backbone. However, external stimuli (temperature, pH, and ligand–receptor interactions) induce strain in the backbone structure of PDA, consequently increasing the energy gap and absorbing higher-energy light. At this point, PDA color gradually changes from blue to red. Lim et al. utilized the color-changing property of PDA to develop a peptide-functionalized PDA nanosensor for the detection of H1N1 influenza virus [66]. As shown in Figure 13.6c, PDA nanoparticles were first assembled via nanoprecipitation and exposed to UV light to produce blue PDA nanosensors. Subsequently, peptide modification was used for the specific recognition of H1N1 virus. Upon interaction with the peptide-modified polydopamine (PDA), H1N1 virus interacted with the peptide, resulting in the perturbation of the PDA-conjugated backbone and triggering a color shift from blue to red (Figure 13.6d). As increasing virus concentration exerted higher stress on the PDA backbone, a distinct blue to red color change was observed (Figure 13.6d). In contrast, the control peptide-modified PDA nanosensor did not exhibit color change. The sensor offers the advantages of rapid response, high sensitivity, and the requirement of no complex equipment.

13.3 Biological Imaging

Biological imaging plays a crucial role not only in the study of cellular and tissue functions but also in the clinical diagnosis and treatment of diseases. There are many methods for cell imaging, among which traditional fluorescent imaging reagents primarily consist of organic small-molecule dyes. However, these organic small-molecule dyes are susceptible to photobleaching, which is detrimental to long-term observation [67]. Therefore, it is important to develop imaging materials with low toxicity and good stability. Compared with small organic dye molecules, CPs demonstrate superior photostability and fluorescence signal amplification effect, which render them excellent materials for biological imaging. In biological imaging applications, CPs can be modified to establish associations with biological targets. Under physiological circumstances, the surface of microorganisms, cells, and algae is negatively charged. Bioimaging of cationic CPs is driven through the employment of electrostatic interactions [68]. On the other hand, the interaction between the backbone of CPs and phospholipids or other hydrophobic components on the surface of biological targets is crucial for enhancing the binding between CPs and these targets. In recent years, many CPs have been reported for fluorescence imaging, NIR imaging, and PAI.

13.3.1 Fluorescence Imaging

Significant advancements in fluorescence imaging utilizing CPs have established this technique as an effective optical tool for disease diagnosis and treatment monitoring, owing to its high temporal resolution and sensitivity [69]. Fluorescence imaging of CPs mainly depends on the delocalized π-electron system within their conjugated structure. Light excitation leads to electron transitions, which ultimately release energy through fluorescence emission. Fluorescence properties of CPs are primarily influenced by several factors, including π–π^* transition, fluorescence quantum yields, and fluorescence lifetimes. These CPs can be used for imaging cells, bacteria, and algal microorganisms.

Current dependence on single antibody recognition modalities for targeted imaging of tumor cells is still unsatisfactory. The development of cellular multi-imaging techniques can enhance the detection capabilities for various tumor cells, which is beneficial for cell sorting and clinical diagnosis. Wang et al. developed a biologically functionalized multicolor CPNs for imaging and detection of targeted tumor cells [70]. Four types of CPs (P10–P13) with blue, green, yellow, and red emitting light were obtained via Suzuki cross-coupling polymerization of various monomers (Figure 13.7a). To prepare CPNs, four carboxyl-functionalized nanoparticles were obtained through co-precipitation, using hydrophobic interactions between CPs with different fluorescent colors and poly (styrene-co-maleic anhydride) (PSMA), respectively. These CPNs were further modified with antibodies (anti-EpCAM or anti-ErbB2) and successfully utilized to distinguish SK-BR-3 and MCF-7 breast cancer cells. In addition, multicolor CPNs prepared through mixing different colored CPs in a certain ratio and precipitated with PSMA could undergo FRET (Figure 13.7b). Fluorescence-emitting multicolor CPNs covering the entire visible spectral region could be acquired via excitation with short-wavelength excitation light. Utilizing this feature, antibody-conjugated CPNs enabled multi-channel cellular imaging on MCF-7 cells (Figure 13.7c). Compared with the single antibody recognition pattern of tumor cells, the binding of CPNs labeled with different antibodies achieved better targeted imaging and detection specificity.

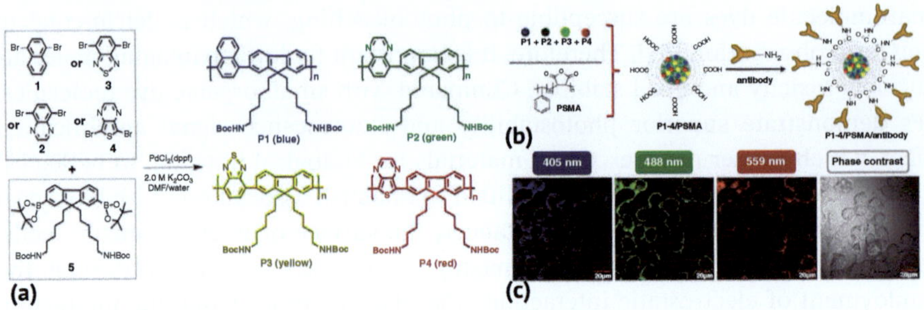

Figure 13.7 (a) Synthetic routes and chemical structures of CPs for cell imaging. (b) Schematic diagram of antibody-modified multi-color conjugated polymer nanoparticles. (c) Fluorescence imaging of multi-color nanoparticles excited via different excitation wavelengths. *Source*: [70]/with permission of John Wiley & Sons.

In clinical practice, infections resulting from a mixture of pathogenic bacteria are prevalent among patients. Rapid identification of pathogens through fluorescence imaging within a short period of time is more helpful for clinical applications. Yuan et al. investigated the distinguish between Gram-negative bacteria (*Escherichia coli*), Gram-positive bacteria (*Bacillus subtilis*), and fungi (*Candida albicans*) through fluorescence imaging in buffer solutions of different ionic strengths with a cationic CPs poly(phenylene vinylene) derivative (PPV-NMe_3^+, P14) [71]. Due to the different extracellular envelope compositions of various pathogenic bacteria, PPV-NMe_3^+ was designed with side chain with quaternary ammonium group, which could be made to interact specifically with different pathogenic bacteria by adjusting the concentration of phosphate buffer (Figure 13.8a). Differential imaging of PPV-NMe_3^+ with various bacteria at different concentrations of phosphate buffer was manifest (Figure 13.8b). The binding of PPV-NMe_3^+ to *C. albicans* was observed at a lower PBS concentration (5 mM), indicating that PPV-NMe_3^+ integrated into the cell wall of *C. albicans*, which comprised glucans, chitin, and mannoproteins. The binding ability of PPV-NMe_3^+ to *E. coli* was enhanced at a concentration of 20 mM PBS. Due to the outer membrane and thinner cell wall of *E. coli*, PPV-NMe_3^+ mainly adhered to the bacterial surface, resulting in a more positive ζ potential for *E. coli*. The binding between PPV-NMe_3^+ and *E. coli* was mainly realized via electrostatic interactions. The strongest binding capacity of PPV-NMe_3^+ to *B. subtilis* was observed at high PBS concentration (40 mM), which was attributed to the relatively thick cell wall of *B. subtilis* leading to the binding of PPV-NMe_3^+ to its cell wall through hydrophobic interactions. This method greatly reduces the analysis time compared to traditional cultivation methods. In addition, Wang et al. synthesized a CPs with intrinsic FRET backbone, and functional side chains poly[(9,9-bis{6′-[N-(triethylene glycol methyl ether)-di(1H-imidazolium) methane]hexyl}-2,7-fluorene)-co-4,7-di-2-thienyl-2,1,3-benzothiadiazole]tetrabromide (PFDBT-BIMEG, P22) was used for rapid differentiation and killing of microbial pathogens [72]. The functional side chain of PFDBT-BIMEG comprised bis-imidazolium (BIM) group and ethylene glycol (EG) moiety, which were capable of interacting with microorganisms through electrostatic and salt-bridging mechanisms. Binding of pathogenic bacteria with PFDBT-BIMEG caused changes in the polymer aggregation

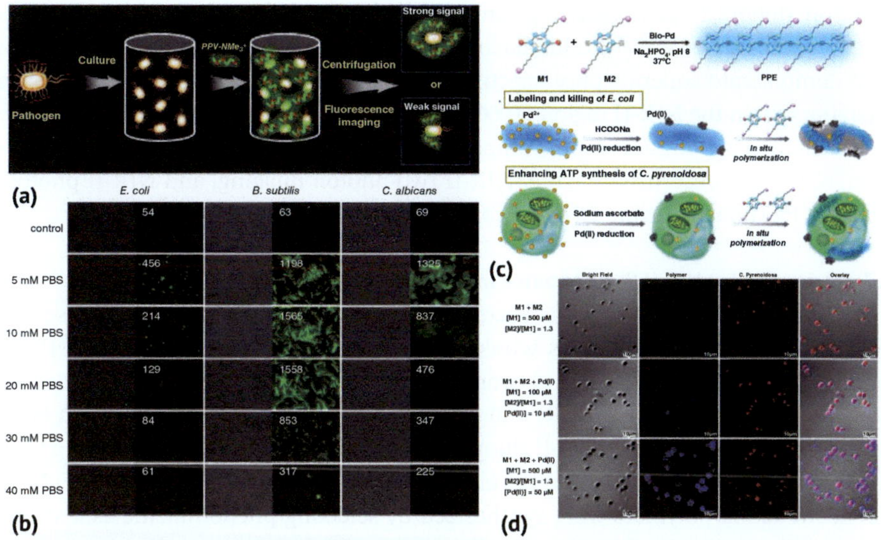

Figure 13.8 (a) Schematic representation of cationic CPs used to distinguish between different microbial pathogens. (b) Fluorescence imaging of *E. coli*, *B. subtilis*, and *C. albicans* mixed with cationic CPs in different concentrations of PBS buffers. *Source*: [71]/with permission of John Wiley & Sons. (c) Schematic diagram of in situ polymerization of cationic CPs polyphenyleneethynylene (PPE) on the surface of *E. coli* and *C. pyrenoidosa* catalyzed via Pd. (d) Fluorescence images of *C. pyrenoidosa* surface polymerization at different monomer concentrations. *Source*: [73]/with permission of John Wiley & Sons.

structure, which could be subsequently detected through alterations in fluorescence signals, resulting from intramolecular FRET.

The application of functional polymers in cell surface engineering not only modulates cell activity but also provides long-term monitoring. Qi et al. developed an in situ synthesis method using palladium catalysis to directly synthesize photoactive CPs polyphenyleneethynylene (PPE, P15) on the surface of *C. pyrenoidosa* for bioimaging and modulation of biological functions [73]. Pd^0 catalyst was first reduced in situ on the surface of *C. pyrenoidosa*, and then two cationic monomers were selected for Sonogashira polymerization on the cell surface (Figure 13.8c). Upon addition of monomers M1 and M2 polymerization, blue fluorescence was observed around *C. pyrenoidosa* (Figure 13.8d). The intensity of *C. pyrenoidosa* blue fluorescence increased with increasing monomer concentration. This in situ-formed PPE could be utilized not only to monitor microorganisms during polymerization but also to enhance ATP synthesis in *C. pyrenoidosa* by increasing light absorption and facilitating cyclic electron transfer in algal microorganisms.

13.3.2 NIR Imaging

CPs with long wavelength absorption (NIR region) have been increasingly designed and synthesized due to deep tissue permeability in organisms. The majority of donor–acceptor (D–A) structured CPs are synthesized via Stille or Suzuki polymerization [74]. D–A structured CPs are synthesized through the coupling of different electron donors (e.g. thiophene, benzodithiophene, etc.) and electron acceptors (e.g. benzothiadiazole, diflurothiadiazole, etc.) [75]. These CPs exhibit NIR absorption and demonstrate superior photoelectric conversion efficiency (PCE). Fluorescence imaging within the NIR-II region (1000–1700 nm) yields less tissue autofluorescence and increased penetration depth compared to other imaging regions [76]. In this section, NIR-II fluorescence imaging, NIR-II two-photon imaging, and NIR-II photoacoustic (PA) imaging using CPs as a contrast agent are presented.

13.3.2.1 CPs for NIR-II Fluorescence Imaging

NIR-II fluorescence imaging has become an important tool in biomedical research for disease diagnosis as well as whole-body imaging [77]. In recent years, NIR-II emitting CPs have emerged as a prominent research focus due to their excellent signal-to-background ratio (SBR) and enhanced tissue penetration depth. Wu et al. proposed a dual-enhanced NIR-II fluorescence of CPs for mouse brain imaging [78]. The CPs were designed based on the classical donor–acceptor (D–A) structure. Nine semiconducting polymers were synthesized by selecting phenothiazine as a strong donor and coupling it with acceptors that possess varying electron affinities, such as benzothiazole and its derivatives (Figure 13.9a). Semiconducting polymers reduced the non-radiative decay pathway of polymers in the aggregated state through the aggregation-induced luminescence properties of phenothiazines. Secondly, the semiconducting polymers minimized fluorescence quenching through the site resistance effect across different layers, which further enhanced the fluorescence. The bulky side groups of the synthesized NIR-II P3c semiconductor polymer dots (Pdots, P16) effectively inhibited π-stacking aggregation and excimer/exciplex formation, which led to a pronounced anti-aggregation-caused quenching (ACQ) effect in the densely packed polymer chains. Compared to the performance of the pristine polymers in tetrahydrofuran solution, these Pdots showed about 21-fold fluorescence enhancement in aqueous solution with a fluorescence quantum yield of about 1.7%. Imaging of P3c Pdots with AIE and anti-ACQ effects demonstrated clearer vascular structures, providing a higher signal to background ratio (Figure 13.9b). In addition, due to the large π-conjugated backbone, CPs are usually converted into CPNs for biological applications. Huang et al. prepared semiconducting polymers (PTQ, P18) with high NIR-II emission (~1057 nm) for efficient NIR fluorescence imaging and PTT [79]. Unlike strong electron acceptors such as BBTDs, PTQs utilized weaker electron acceptors that were not only more amenable to functional group modification but

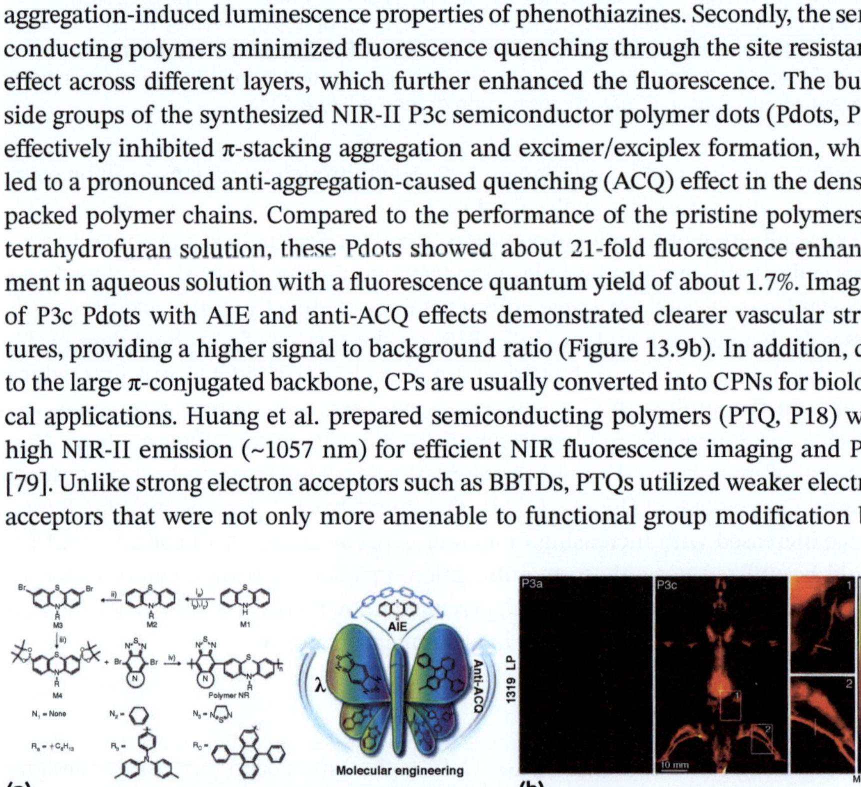

Figure 13.9 (a) Schematic diagram of the formation of semiconductor polymer dots. (b) NIR-II fluorescence image of the vasculature of the hind limb of mouse following intravenous injection of P3a Pdots. *Source*: [78]/with permission of John Wiley & Sons.

also exhibited red-shifted absorption properties, resulting in greater brightness compared to most organic NIR-II fluorophores. The nanoparticles L1057 NPs assembled with DSPE-PEG2000 and PTQ were able to visualize the whole body and cerebral vasculature in real time, demonstrating great potential in biomedical imaging.

13.3.2.2 CPs for NIR Two-photon Imaging

The excitation light used for two-photon imaging typically falls within the NIR range of 700–1100 nm. This wavelength range facilitates imaging in deeper tissue levels, particularly in brain tissue. Two-photon imaging has the advantages of low autofluorescence, high penetration depth, and low photodamage [80]. The backbone structure of CPs provides significant polarizability and a high two-photon absorption cross-section (TPA cross-section), allowing the generation of a strong fluorescent signal upon two-photon excitation. Liu et al. developed an NIR-II excitable single-chain conjugated polymer dots (CPdots) for two-photon nondestructive cranial penetration deep brain imaging [81]. To facilitate intramolecular charge transfer, CPs poly{[4,4,9,9-tetrakis(4-(octyloxy) phenyl-4,9-dihydro-s-indaceno[1,2-b:5,6-b′]dithiophene)]-alt-co-[4,7-di(thiophen-2-yl) -2,1,3-benzothiadiazole]} (PIDTDBT, P17) was synthesized to introduce strong electron donor and acceptor structures into the polymer backbone (Figure 13.10a). For good water dispersibility in biological applications, PIDTDBT was prepared as nanoparticles (PIDTDBT dots) by nanoprecipitation. PIDTDBT dots achieved a fluorescence quantum yield of 20% in water and TPA cross-section of 1.21×10^3 GM at 1200 nm. In the craniotomy model, a femtosecond 1200 nm laser was employed to excite PIDTDBT dots with a particle size of approximately 4 nm. PIDTDBT dots facilitated the acquisition of a three-dimensional (3D) structural map of mouse cerebral vasculature, achieving a maximum imaging depth of 1010 μm (Figure 13.10b). Furthermore, deep 2PF imaging of intact mouse skulls using NIR-II-excited PIDTDBT dots still rendered the vascular network in the brain with high clarity (Figure 13.10c). The imaging depth of

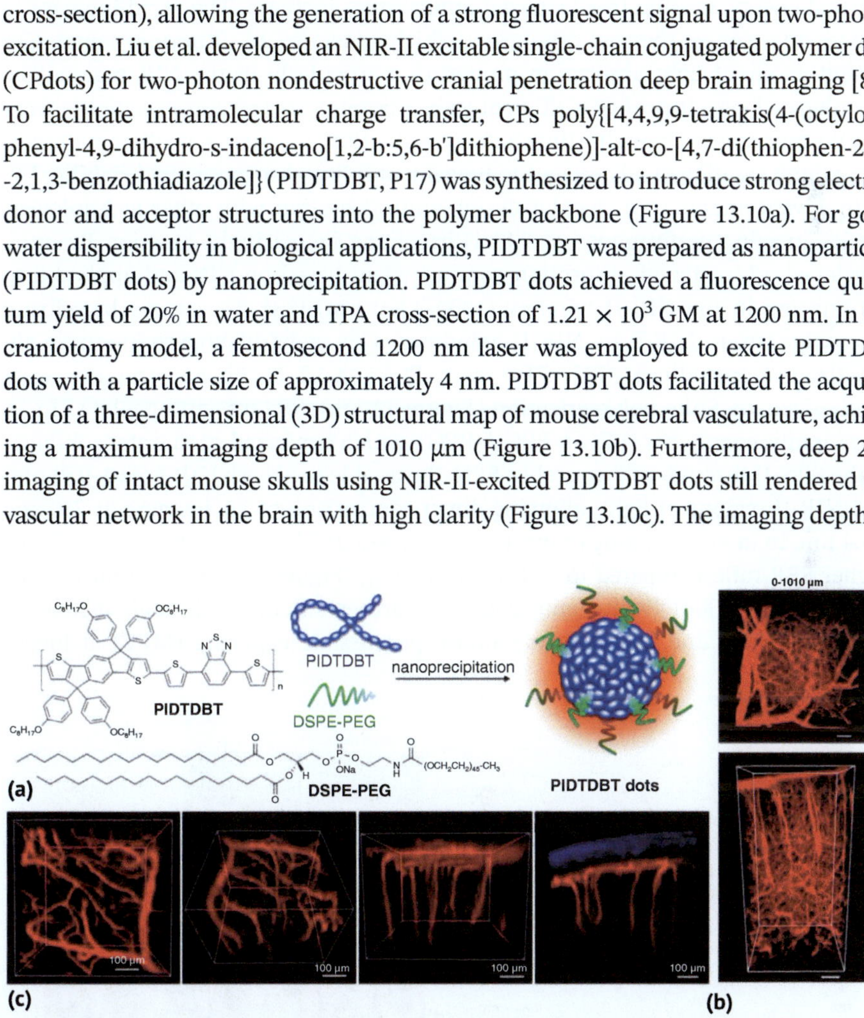

Figure 13.10 (a) Schematic representation of the synthesis of NIR-II photoexcited conjugated polymer dots (CPdots). (b) Three-dimensional (3D) reconstruction of cerebral vascular 2PF images at depth of 0–1010 μm. (c) Two-photon fluorescence 3D recomposition 2PF images of cerebral blood vessels under intact skull. *Source*: [81]/with permission of John Wiley & Sons.

the cerebral vascular network obtained through the intact skull could reach 400 μm. This result demonstrated that NIR two-region excitation light effectively penetrated the skull, achieving efficient two-photon excitation for high-contrast, non-destructive imaging of deep blood vessels in the cranial penetrating brain.

13.3.2.3 CPs for NIR PA Imaging

Compared with conventional optical imaging techniques, NIR PAI integrates optical excitation with ultrasound detection. This method is anticipated to serve as an effective contrast agent, enabling deep tissue penetration and providing excellent spatial resolution [82]. PAI involves the conversion of absorbed light energy into thermal energy, leading to a localized temperature increase. The rapid increase in temperature causes thermal expansion, which generates an ultrasonic signal. CPs exhibit excellent light absorption properties, particularly in the NIR region. The π-conjugated structure of CPs efficiently absorbs photons and converts their energy into thermal energy, inducing the photoacoustic effect. Compared to other materials, CPs exhibit significant advantages in biocompatibility, absorption, and stability [3]. With advancements in molecular design and integration with other technologies, the PA imaging performance of CPs is becoming increasingly superior. Pu et al. developed a broadband absorption contrast agent based on CPs and applied it to PA imaging in the NIR-II window (1000–1700 nm) [83]. To achieve broadband absorption from NIR-I to NIR-II windows, CPs poly(diketopyrrolopyrrole-alt-thiadiazoloquinoxaline) (SP2, P19) with D-A1-D-A2 backbone structure were synthesized via Stille polymerization (Figure 13.11a). SP2 was encapsulated in an amphiphilic copolymer (PEG-*b*-PPG-*b*-PEG), which conferred good water solubility to the SPN-II through self-assembly. SPN-II was able to absorb NIR-I (650–950 nm) and NIR-II regions of light, allowing direct comparison of PA imaging at 750 and 1064 nm. In in vivo imaging of live rat brains, NIR-II PA imaging showed a 1.5-fold higher SBR ratio compared to NIR-I PA imaging (Figure 13.11b). Technically, the SBR ratio of NIR-II PA imaging could be further enhanced through increasing the laser power due to higher maximum permissible exposure (MPE) at longer wavelengths.

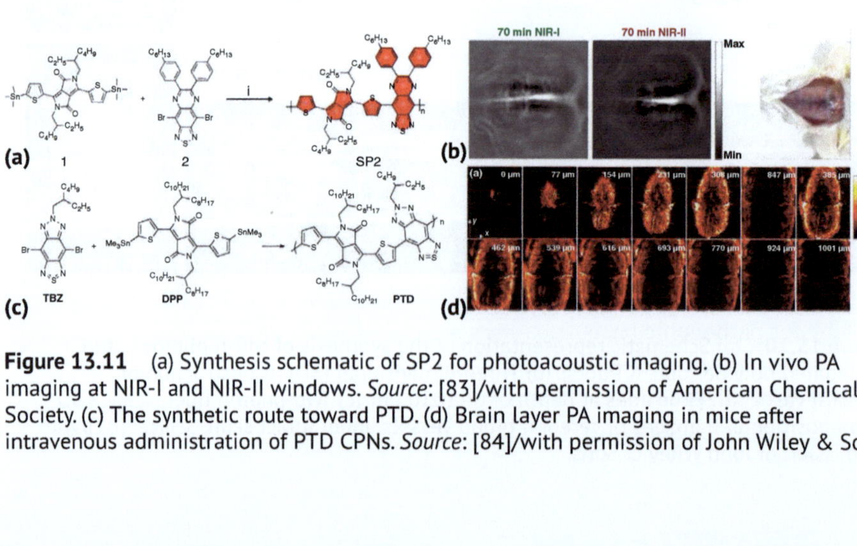

Figure 13.11 (a) Synthesis schematic of SP2 for photoacoustic imaging. (b) In vivo PA imaging at NIR-I and NIR-II windows. *Source*: [83]/with permission of American Chemical Society. (c) The synthetic route toward PTD. (d) Brain layer PA imaging in mice after intravenous administration of PTD CPNs. *Source*: [84]/with permission of John Wiley & Sons.

Moreover, exogenous contrast agent-assisted NIR-II high-resolution PA imaging offers deep penetration and high SBR ratio, which is expected to be used for the study of large-scale 3D biological structures. Liu et al. utilized NIR-II CPNs to assist optical-resolution photoacoustic microscopy imaging (ORPAMI) in accurately labeling brain and tumor vascular structures [84]. A new D-A1-D-A2-type CPs PTD (P20) was designed in this study, which involved the alternating copolymerization of thiadiazolobenzotriazole (TBZ) and thiophene-substituted diketopyrrolopyrrole (DPP) monomers through the Stille coupling reaction (Figure 13.11c). PTD was prepared into nanoparticles PTD NPs via microfluidics for biological applications. PTD NPs exhibited extremely high PA sensitivity under 1064 nm pulsed laser irradiation and were clearly visible even at concentration of 2 $\mu g\ mL^{-1}$. Under intact skull conditions, NIR-II ORPAMI imaging following the administration of PTD NPs clearly visualized cerebral blood vessels, with the imaging depth reaching 1001 μm (Figure 13.11d). This demonstrated that this imaging technique effectively penetrated biological tissues to obtain information on deep structures. These superior performances render the CPNs-assisted NIR-II ORPAMI technique an effective tool for biomedical applications.

13.4 Therapeutic Applications

13.4.1 Antimicrobial Therapy

In addition to biosensing and imaging diagnostics, CPs have demonstrated promising outcomes in the treatment of various diseases, including cancer, microbial/viral infections, and heart disease. In recent years, researchers have discovered that CPs can be sensitized under light conditions to generate singlet oxygen ($^1O_2$). This generated singlet oxygen can be exploited for the eradication of bacterial and tumor cells [53b, 85]. PDT is favored for its advantages, including minimal invasiveness and low toxicity [86]. In PDT, CPs act as photosensitizers to generate reactive oxygen species (ROS) with cell-killing ability, including single-linear oxygen ($^1O_2$), hydrogen peroxide (H_2O_2), and hydroxyl radicals (·OH). As depicted in Figure 13.12a, the absorption of light energy by the photosensitizer results in the excitation of electrons from the ground state ($^1PS^0$) to the short-lived singlet state ($^1PS^*$). Subsequently, these singlet electrons transition to a lower energy, yet longer-lived, triplet state ($^3PS^*$) [87]. The triplet state exciton (3PS*) can generate reactive radicals, such as superoxide radicals and hydroxyl radicals, through electron transfer to the surrounding substrate. Another method for generating ROS involves transferring energy to the photosensitizer through energy transfer, which converts triplet oxygen ($^3O_2$) to generate singlet oxygen ($^1O_2$). The two methods for generating free radicals are competitive and primarily depend on the solvent environment, the surrounding oxygen concentration, and the aggregation state of the photosensitizer. CPs-based PDT technology for pathogenic bacteria killing can solve the problem of bacterial multidrug resistance [88]. In addition to light-assisted methods, compounds with quaternary ammonium salt groups also have killing ability against many microorganisms [89]. Gram-negative bacteria are less susceptible to PDT antimicrobial therapies due to

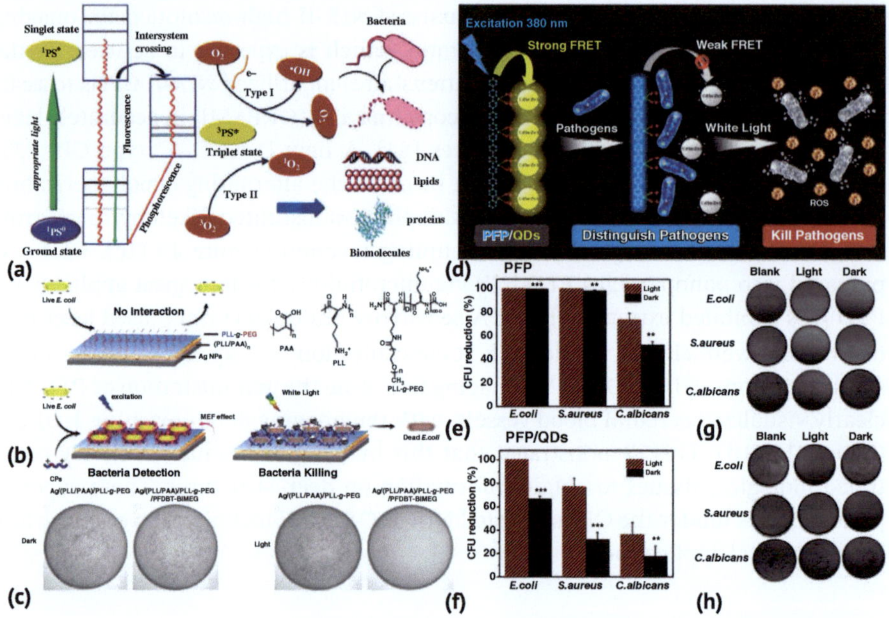

Figure 13.12 (a) Schematic diagram of ROS generation methods and antimicrobial mechanisms. *Source*: [87]/with permission of American Chemical Society. (b) Mechanism diagram of conjugated polyelectrolyte–silver nanostructures pair for bacterial detection and killing. (c) CFU for *E. coli* suspension incubated on conjugated polyelectrolyte–silver nanostructures pair in the dark and under white light irradiation. *Source*: [92]/with permission of John Wiley & Sons. (d) Schematic illustration of pathogen identification and antibacterial based on CPs and QDs. (e, f) CFU in various pathogens treated with (e) CPs and (f) CPs/QDs. (g, h) Antibacterial effect of (g) CPs and (h) CPs/QDs treated pathogen agar plates. *Source*: [93]/with permission of American Chemical Society.

the protection of the outer membrane and $^1O_2$ diffusion limitations. Bazan et al. designed a membrane-inserted conjugated oligoelectrolyte (COE) to improve the killing efficiency of PDT antimicrobial therapy against Gram-negative bacteria [90]. The structural design of the membrane-inserted conjugated oligoelectrolyte PTTP (P21) aimed to provide a red-shifted absorption spectrum compared to previously synthesized COE analogs. By incorporating electron donor (thieno[3,2-b]thiophene) and electron acceptor ([1,2,5]thiadiazolo[3,4-c]pyridine) subunits into the conjugated backbone to modulate the band gap, PTTP exhibited a maximum absorption at 507 nm and a maximum emission at 725 nm. PTTP could insert into lipid membranes and efficiently generate singlet oxygen species in situ, demonstrating efficient antimicrobial capacity.

Energy transfer with the aid of CPs to conventional photosensitizers is a sensitization strategy that has been developed in recent years. As this method capitalizes on the powerful light-trapping capacity of CPs, it significantly enhances the efficiency of $^1O_2$ production [91]. Wang et al. designed an optical detection system based on conjugated polyelectrolyte–silver nanostructure pairs for the detection and killing of bacteria [92]. Ag nanoparticles were initially adsorbed onto the functionalized quartz substrate, as shown in Figure 13.12b. Subsequently, the Ag nanostructures were enveloped with an isolation layer comprising PLL/PAA bilayer. Finally, additional layer of poly(l-lysine)-graft [3.5]-poly(ethylene glycol) (PLL-g-PEG) was applied to inhibit the adsorption of proteins and bacteria. Due to the negatively charged ether present on the bacterial surface and in the PEG, after the introduction of the cationic CPs PFDBT-BIMEG (P22), *E. coli* was assembled on the surface of the assay system via electrostatic interactions and produced fluorescent signals for bacterial detection. Subsequently, PFDBT-BIMEG transferred energy to oxygen under white light irradiation, generating ROS that can effectively kill bacteria. This energy transfer combined with the metal-enhanced fluorescence effect of Ag nanostructures amplified the fluorescence intensity of CPs and further improved the efficiency of ROS production. Colony counting showed that the platform containing PFDBT-BIMEG exhibited the highest killing efficiency of 99% under white light irradiation (Figure 13.12c). This bio-detection system provides a powerful method for bacteria capture, detection, and killing.

Subsequently, they proposed to utilize the energy transfer between CPs and quantum dots (QDs) to distinguish and kill pathogens. They constructed a new platform for pathogen differentiation and killing based on CPs-QD hybridized materials [93]. CPs-QD hybridized material integrated polyfluorene derivative PFP and CdSe/ZnS quantum dots, which improved FRET efficiency from PFP to QD (Figure 13.12d). Upon incorporation of different pathogenic strains, FRET of PFP to QDs was interrupted by competitive binding of the pathogen to PFP. The fluorescence intensity of CPs-QD hybridized materials varied from different pathogens due to differences in cell surface charge and structure. For different pathogens, such as *E. coli, S. aureus*, and *C. albicans*, treatment with CPs-QD was effective in reducing colony-forming units and decreasing the dark toxicity of CPs (Figure 13.12e–h). This work provides a promising therapeutic diagnostic platform for pathogen identification and killing.

13.4.2 Anti-tumor Therapy

CPs provide a variety of available tools, including modifiability, photothermal properties, and photodynamic properties, to address the challenges encountered in tumor therapy. The modifiable properties of CPs are widely used for drug delivery [94]. Yang et al. synthesized cationic pentathiol (5T) for targeting mitochondria and synergistic anticancer therapy [95]. 5T selectively accumulated in mitochondria and increased the anticancer activity by 2.9-fold. This multifunctional anticancer agent based on CPs combines optical monitoring capability and anticancer activity, providing a new avenue for cancer therapy.

Recent advances have focused on the design of photoactivated CPs and the development of phototherapeutic methods. PTT has received significant attention in cancer treatment due to its low side effects and induction of thermal ablation of cancer cells [96]. With efficient photothermal conversion and strong light absorption capabilities, CPs have found extensive applications in PTT for cancer treatment. The primary mechanism of CPs-based PTT involves the absorption of light energy, which is then converted into thermal energy, leading to localized hyperthermia that induces thermal damage and cell death in tumor tissues. Some classic CPs, including polypyrrole, polyaniline, and polythiophene, are used for PTT treatment of tumors [97]. Nowadays, CPs containing D–A structure have exhibited potent NIR absorption and impressive photothermal conversion efficiency [98]. D–A CPs can effectively promote the generation of excited state electrons due to the electron transfer between donor and acceptor units, increasing the release of heat and improving the therapeutic effect. Wang et al. developed an optogenetic system based on D–A-type CPs (P23) for activating cancer immunotherapy under NIR laser irradiation [99]. The system comprised photothermal CPNs and interferon-γ (IFN-γ) plasmids, which regulated by heat-shock promoter HSP70 (Figure 13.13a). CPs with photothermal effects were selected to copolymerize DPP electron-deficient units and electron-rich aromatic units, modulating the optical band gap and switching them to the NIR region. To enhance the stability of CPs in aqueous solution, D–A-type CPs were co-assembled with the amphiphilic lipid DSPE-PEG2000 to form photothermal CPNs. The temperature of the solution containing CPNs increased with irradiation time or nanoparticle concentration, indicating superior NIR photothermal performance (Figure 13.13b). When the photothermal CPNs absorbed NIR light and converted it into heat, HSP70 promoter was triggered, inducing gene expression of IFN-γ (Figure 13.13c). The secreted IFN-γ selectively bound to the IFN-γ receptor on tumor-associated macrophages (M2-type), inducing macrophage polarization. The activated macrophages secreted proinflammatory cytokines and necrosis factors, which exerted a cytotoxic effect on tumor cells.

In addition, the tuning of electron density in CPs contributes to the improvement of NIR-II fluorescence and PTT performance. Huang et al. proposed a strategy to mediate electron acceptor density, enhancing the radiative rate constant by decreasing the electron acceptor density in the main chain of CPs, which increases NIR-II fluorescence

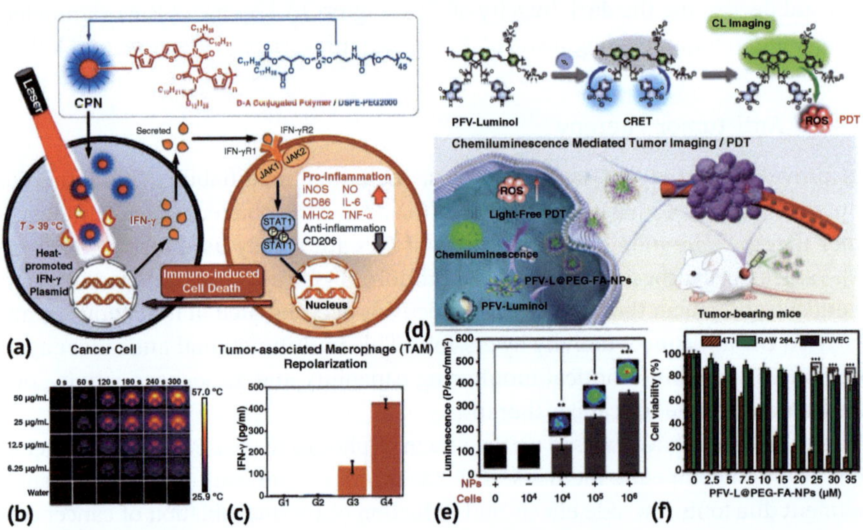

Figure 13.13 (a) Schematic illustration of NIR regulation of cancer immunotherapy based on PTT CPNs. (b) Infrared thermal imaging of PTT CPNs in solution under 808 nm laser irradiation. (c) The expression level of IFN-γ. G1: M2; G2: M2 4T1; G3: M2 4T1 (Plasmid); G4: M2 4T1 (Plasmid, Laser). *Source*: [99]/with permission of John Wiley & sons. (d) Schematic representation of PFV-Luminol for chemiluminescence and PDT therapy. (e) Chemiluminescence imaging of PFV-Luminol NPs in 4T1 cells. (f) Cytotoxicity of PFV-Luminol NPs to different cells. *Source*: [102]/with permission of American Chemical Society.

brightness [100]. Since fewer electron-withdrawing groups in the polymer backbone are highly beneficial for increasing NIR-II fluorescence intensity. Therefore, long alkyl side chain-modified bithiophene (2 TC) was added to the backbone of CPs as an electron donor to reduce the density of electron-withdrawing groups in the CPs backbone. The obtained CPs TTQ-2TC-4T (P24) not only exhibited increased solubility in organic solvents but also enhanced NIR fluorescence intensity (7.30-fold enhancement of NIR-II emission intensity) and triggered PTT. On the other hand, TTQ-2TC-4T and CO donor ($Mn_2(CO)_{10}$) were developed into nanoparticles for realizing NIR-II PTT/gas synergistic therapy. The nanoparticles exhibited high photothermal conversion efficiency (44.43%) at 1064 nm and NIR-II fluorescence imaging of blood vessels and tumors.

PDT typically employs light to stimulate the production of ROS from CPs to eliminate tumor cells. CPs can be activated using localized irradiation with light of a specific wavelength after reaching solid tumor, and then their energy is transferred to oxygen to generate ROS. The generated ROS can induce apoptosis and necrosis in tumor cells [101]. CPs exhibit better photostability and lower phototoxicity compared to conventional photosensitizers such as porphyrins. CPs can be engineered by optimizing their molecular structure to improve stability within organisms, improve biocompatibility, and reduce toxic side effects during treatment. Tang et al. synthesized a chemiluminescent CPs, PFV-Luminol (P25), for chemiluminescence and PDT treatment of cancer deep tissues [102]. PFV-Luminol was composed of conjugated polyfluorene vinylene (PFV) backbone and isoluminol-modified side chain (Figure 13.13d). The covalent attachment of chemiluminescent isoluminol groups to the main chain decreases the distance between them, enhancing chemiluminescence resonance energy transfer (CRET) efficiency. In the presence of pathologically high levels of various ROS, especially $^1O_2$, the energy generated by the reaction between isoluminol and ROS could be transferred to the PFV via CRET. Subsequently, excitation of the PFV receptor allowed spontaneous generation of green fluorescence and ROS. In addition, PFV-L@PEG-FA-NPs nanoparticles were constructed through self-assembly of PFV-Luminol with the amphiphilic copolymer DSPE-PEG-FA. The PFV-L@PEG-FA-NPs contained folic acid and exhibited superior targeting ability. Applying PFV-L@PEG-FA-NPs to in vitro tumor cell imaging, the luminescence signal was closely correlated with the number of tumor cells (Figure 13.13e). In addition, PFV-Luminol generated ROS during chemiluminescence, facilitating the destruction of tumor cells through PDT without the necessity for an external light source. The survival rate of 4T1 cells treated with PFV-L@PEG-FA-NPs (at a concentration of 30 μM) was only about 12% (Figure 13.13f). This study opens up new avenues for the design of chemiluminescent probes and their applications in disease diagnosis and therapy.

13.4.3 Antiviral Therapy

PDT therapy based on CPs can also be used to address the problems associated with antiviral drugs. PDT inactivates viruses by disrupting the viral structure through ROS generated using light irradiation of photosensitizers [103]. PDT demonstrates broad-spectrum antiviral activities without the development of viral resistance [104]. CPs exhibit light-harvesting ability and efficient ROS generation, allowing them excellent candidates for PDT-based antiviral therapy. Modification of CP side chains with positively charged groups enhances binding and photoinactivation of proteins via electrostatic interactions [105]. Nevertheless, the interactions between CPs and viruses remain underexplored. Wang et al. examined the interactions between CPs and viruses to elucidate the mechanism of photoinactivation of viruses [106]. SARS-CoV-2 pseudovirus was utilized as a model to further explore the photoinduced inactivation of RNA viruses using three cationic CPs with different π-conjugated backbone structures (fluorene-co-phenylene: PFP, thiophene: PMNT, phenylene vinylene: PPV). These three CPs preferentially bound electrostatically to negatively charged viral proteins due to positive charge. Additionally, the different backbone structures of CPs not only further enhanced their binding affinity to viruses through hydrophobic interactions but also conferred distinct ROS-generating abilities. As shown in Figure 13.14a, the pseudovirus consisted of SARS-CoV-2 envelope coated with S proteins and RNA genome-containing

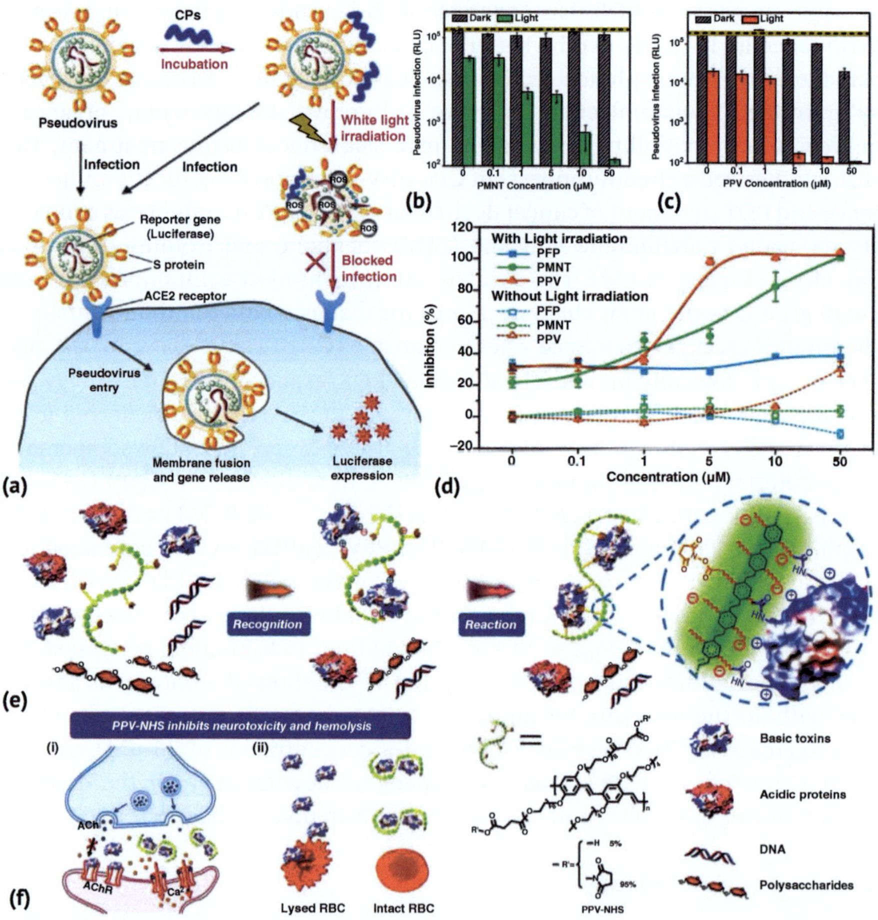

Figure 13.14 (a) Schematic representation of CPs photoinactivating virus. HEK293T/ACE2 cells were infected with (b) PMNT- and (c) PPV-treated pseudoviruses. (d) Pseudoviral inhibition efficiency with CPs concentration. *Source*: [106]/with permission of Springer Nature. (e) Schematic diagram of selective recognition of CPs and basic proteins. (f) Mechanism representation of anti-toxicity of CPs against basic protein. *Source*: [107]/with permission of Springer Nature.

luciferase as reporter gene. HEK293T cells overexpressing angiotensin-converting enzyme 2 receptors were used as host cells to mimic the infection process of SARS-CoV-2 and to analyze the infection through luciferase activity. PMNT and PPV demonstrated a dose-dependent inactivation effect on the pseudovirus under light exposure (Figure 13.14b–d). The relative light unit (RLU) decreased with increasing concentration. PPV showed superior efficacy compared to PMNT in photoinactivation. PPV could achieve complete photoinactivate viral nucleic acids following photoinactivation of proteins, thus exhibiting the highest photoinactivation efficiency. In contrast, the less pronounced photoinactivation effect of PFP on pseudoviruses was attributed to its promotion of gene transfection and delivery under dark conditions. This study is pivotal for the advancement of effective antiviral PDT reagents.

In addition, PPV polymers can be used as a broad-spectrum detoxifier for inhibiting the toxicity of basic proteins. Sun et al. developed a reactive CPs, PPV-NHS (P26), as a broad-spectrum antitoxin platform against basic toxins [107]. The PPV-NHS side chain was modified with carbocyclic acid and N-hydroxysuccinimide ester. PPV-NHS selectively recognized basic proteins through electrostatic interactions, subsequently facilitating the binding of basic proteins via amidation reactions (Figure 13.14e). Furthermore, PPV-NHS demonstrated the ability to inhibit cardiotoxin (CTX) toxicity both in vitro and in vivo (Figure 13.14f). In a mouse model, PPV-NHS administration was able to enhance the survival rate of toxin-attacked mice, demonstrating the effectiveness and potential application of CPs as an antitoxic platform.

13.5 Conclusion

In summary, this chapter reviews the advances and achievements in the design and application of CPs for biological detection, bioimaging, and disease treatment. The amplification of fluorescence signal/electrical signal based on CPs effectively improves the sensitivity of detection. Modulating the molecular structure and employing nano-assembly techniques can impart additional functionalities to CPs, including excellent photostability, biosafety, and high brightness. The versatility of CPNs plays an important role in biosensing, imaging, as well as in spatiotemporally modulated NIR phototherapy. These results indicate promising prospects for the application of CPs in the biological field, potentially facilitating further breakthroughs in clinical aspects.

References

1 Wang, J., Lv, F., Liu, L. et al. (2018). Strategies to design conjugated polymer based materials for biological sensing and imaging. *Coordination Chemistry Reviews* 354: 135.

2 a) Liu, B. and Bazan, G.C. (2004). Interpolyelectrolyte complexes of conjugated copolymers and DNA: platforms for multicolor biosensors. *Journal of the American Chemical Society* 126: 1942; b) Satrijo, A. and Swager, T.M. (2007). Anthryl-doped conjugated polyelectrolytes as aggregation-based sensors for nonquenching multicationic analytes. *Journal of the American Chemical Society* 129: 16020.

3 Wang, Y., Zhang, H., Wang, Z. et al. (2020). Photothermal conjugated polymers and their biological applications in imaging and therapy. *ACS Applied Polymer Materials* 2: 4222.

4 a) Gaylord, B.S., Heeger, A.J. and Bazan, G.C. (2002). DNA detection using water-soluble conjugated polymers and peptide nucleic acid probes. *Proceedings of the National Academy of Sciences of the United States of America* 99: 10954; b) Chen, X., Hussain, S., Abbas, A. et al. (2022). Conjugated polymer nanoparticles and their nanohybrids as smart photoluminescent and photoresponsive material for biosensing, imaging, and theranostics. *Microchimica Acta* 189: 83; c) Tao, Z., Hong, G., and Shinji, C. (2013). Biological imaging using nanoparticles of small organic molecules with fluorescence emission at wavelengths longer than 1000 nm. *Angewandte Chemie International Edition* 52: 13002; d) Lin, F., Duan, Q.-Y., and Wu, F.-G. (2020). Conjugated polymer-based photothermal therapy for killing microorganisms. *ACS Applied Polymer Materials* 2: 4331.

5 a) Feng, L., Zhu, C., Yuan, H. et al. (2013). Conjugated polymer nanoparticles: preparation, properties, functionalization and biological applications. *Chemical Society Reviews* 42: 6620; b) Kaeser, A. and Schenning, A.P. (2010). Fluorescent nanoparticles based on self-assembled π-conjugated systems. *Advanced Materials* 22: 2985.

6 Landfester, K., Montenegro, R., Scherf, U. et al. (2002). Semiconducting polymer nanospheres in aqueous dispersion prepared by a miniemulsion process. *Advanced Materials* 14: 651.

7 Kuehne, A.J.C. (2017). Conjugated polymer nanoparticles toward in vivo theranostics–focus on targeting, imaging, therapy, and the importance of clearance. *Advanced Biosystems* 1: e1700100.

8 a) Pu, K., Shuhendler, A.J., Jokerst, J.V. et al. (2014): Semiconducting polymer nanoparticles as photoacoustic molecular imaging probes in living mice. *Nature Nanotechnology* 9: 233; b) Zhang, D., Wu, M., Zeng, Y. et al. (2016). Lipid micelles packaged with semiconducting polymer dots as simultaneous MRI/photoacoustic imaging and photodynamic/photothermal dual-modal therapeutic agents for liver cancer. *Journal of Materials Chemistry B* 4: 589.

9 Gao, D., Zhang, P., Liu, Y. et al. (2018). Protein-modified conjugated polymer nanoparticles with strong near-infrared absorption: a novel nanoplatform to design multifunctional nanoprobes for dual-modal photoacoustic and fluorescence imaging. *Nanoscale* 10: 19742.

10 Zhao, Q., Zhao, H., Guo, Y. et al. (2018). Ultra-rapid detection of endogenous nitric oxide based on fluorescent conjugated polymers probe. *Analytical Chemistry* 90: 12663.

11 Thomas, S.W., 3rd, Joly, G.D., and Swager, T.M. (2007). Chemical sensors based on amplifying fluorescent conjugated polymers. *Chemical Reviews* 107: 1339.

12 Shen, Q., Song, G., Lin, H. et al. (2024). Sensing, Imaging, and Therapeutic Strategies Endowing by Conjugate Polymers for Precision Medicine. *Advanced Materials* 36: e2310032.

13 Förster, Th. (1948). Zwischenmolekulare Energiewanderung und Fluoreszenz. *Annalen der Physik* 437: 55–75.

14 a) Zhang, X., Hu, Y., Yang, X. et al. (2019). FÖrster resonance energy transfer (FRET)-based biosensors for biological applications. *Biosensors and Bioelectronics* 138: 111314; b) Chang, M.J., Kim, K., Park, K.S. et al. (2018). High-depth fluorescence imaging using a two-photon FRET system for mitochondrial pH in live cells and tissues. *Chemical Communications* 54: 13531.

15 a) Pu, K.Y. and Liu, B. (2009). Optimizing the cationic conjugated polymer-sensitized fluorescent signal of dye labeled oligonucleotide for biosensor applications. *Biosensors and Bioelectronics* 24: 1067; b) Brédas, J.L., Beljonne, D., Coropceanu, V. et al. (2004). Charge-transfer and energy-transfer processes in π-conjugated oligomers and polymers: a molecular picture. *Chemical Reviews* 104: 4971.

16 Wang, S., Liu, B., Gaylord, B.S. et al. (2003). Size-specific interactions between single-and double-stranded oligonucleotides and cationic water-soluble oligofluorenes. *Advanced Functional Materials* 13: 463.

17 Müllen, K. and Wegner. G. (1998). Electronic Materials: The Oligomer Approach. *Wiley-VCH, Weinheim.*

18 a) Duan, X., Yue, W., Liu, L. et al. (2009). Single-nucleotide polymorphism (SNP) genotyping using cationic conjugated polymers in homogeneous solution. *Nature Protocols* 4: 984; b) Biswas, D.S., Gaki, P., Cruz Da Silva, E. et al. (2023). Long-range energy transfer between dye-loaded nanoparticles: observation and amplified detection of nucleic acids. *Advanced Materials* 35: e2301402; c) Yang, Q., He, B., Chen, C. et al. (2021). A rapid, visible, and highly sensitive method for recognizing and distinguishing invasive fungal infections via CCP-FRET technology. *ACS Infectious Diseases* 7: 2816.

19 Nilsson, K.P. and Inganäs, O. (2003). Chip and solution detection of DNA hybridization using a luminescent zwitterionic polythiophene derivative. *Nature Materials* 2: 419.

20 Ge, Z., Gu, H., Li, Q. et al. (2018). Concept and development of framework nucleic acids. *Journal of the American Chemical Society* 140: 17808.

21 a) He, F., Tang, Y., Yu, M. et al. (2006). Quadruplex-to-duplex transition of G-rich oligonucleotides probed by cationic water-soluble conjugated polyelectrolytes. *Journal of the American Chemical Society* 128: 6764; b) Liu, L., Wang, X., Zhu, S. et al. (2021). Different surface interactions between fluorescent conjugated polymers and biological targets. *ACS Applied Bio Materials* 4: 1211.

22 Stender, H., Fiandaca, M., Hyldig-Nielsen, J.J. et al. (2002). PNA for rapid microbiology. *Journal of Microbiological Methods* 48: 1.

23 Ma, L., Huang, Y., Zhang, H. et al. (2020). Sensitive detection and conjoint analysis of promoter methylation by conjugated polymers for differential diagnosis and prognosis of glioma. *ACS Applied Materials & Interfaces* 13: 9291.

24 Shen, Q., Zhang, H., Huang, Y. et al. (2023). Sensitive detection of single-nucleotide polymorphisms by conjugated polymers for personalized treatment of hypertension. *Science Translational Medicine* 15: eabq5753.

25 Liu, Y., Meng, S., Qin, J. et al. (2022). A fluorescence biosensor based on double-stranded DNA and a cationic conjugated polymer coupled with exonuclease III for acrylamide detection. *International Journal of Biological Macromolecules* 219: 346.

26 a) Reik, W. (2007). Stability and flexibility of epigenetic gene regulation in mammalian development. *Nature* 447: 425; b) Robertson, K.D. and Jones, P.A. (2000). DNA methylation: past, present and future directions. *Carcinogenesis* 21: 461; c) Issa, J.P. (2004). CpG island methylator phenotype in cancer. *Nature Reviews Cancer* 4: 988.

27 Aganezov, S., Yan, S.M., Soto, D.C. et al. (2022). A complete reference genome improves analysis of human genetic variation. *Science* 376: eabl3533.

28 Song, J., Zhang, J., Lv, F. et al. (2013). Multiplex detection of DNA mutations by the fluorescence fingerprint spectrum technique. *Angewandte Chemie International Edition* 52: 13020.

29 Hu, T., Ma, C., Yan, Y. et al. (2022). Detection of DNA methyltransferase activity via fluorescence resonance energy transfer and exonuclease-mediated target recycling. *Biosensors* 12: 395.

30 Goodall, G.J. and Wickramasinghe, V.O. (2021). RNA in cancer. *Nature Reviews Cancer* 21: 22.

31 Buccitelli, C. and Selbach, M. (2020). mRNAs, proteins and the emerging principles of gene expression control. *Nature Reviews Genetics* 21: 630.

32 Dong, H., Lei, J., Ding, L. et al. (2013). MicroRNA: function, detection, and bioanalysis. *Chemical Reviews* 113: 6207.

33 Yang, Z., Qi, Y., Lai, N. et al. (2018). Notch1 signaling in melanoma cells promoted tumor-induced immunosuppression via upregulation of TGF-β1. *Journal of Experimental & Clinical Cancer Research* 37: 1.

34 Wark, A.W., Lee, H.J., and Corn, R.M. (2008). Multiplexed detection methods for profiling microRNA expression in biological samples. *Angewandte Chemie International Edition* 47: 644.

35 a) Volinia, S., Calin, G.A., Liu, C.G. et al. (2006). A microRNA expression signature of human solid tumors defines cancer gene targets. *Proceedings of the National Academy of Sciences* 103: 2257; b) Muhonen, P. and Holthofer, H. (2009). Epigenetic and microRNA-mediated regulation in diabetes. *Nephrology Dialysis Transplantation* 24: 1088; c) Ji, J., Lu, W., Zhu, Y. et al. (2019). Porous hydrogel-encapsulated photonic barcodes for multiplex detection of cardiovascular biomarkers. *ACS Sensors* 4: 1384l; d) Cerro-Herreros, E., Sabater-Arcis, M., Fernandez-Costa, J.M. et al. (2018). miR-23b and miR-218 silencing increase Muscleblind-like expression and alleviate myotonic dystrophy phenotypes in mammalian models. *Nature Communications* 9: 2482.

36 Cheng, Y., Dong, L., Zhang, J. et al. (2018). Recent advances in microRNA detection. *Analyst* 143: 1758.

37 Zhou, Y., Zhang, J., Zhao, L. et al. (2016). Visual detection of multiplex microRNAs using cationic conjugated polymer materials. *ACS Applied Materials & Interfaces* 8: 1520.

38 Tang, Y., He, X., Yuan, R. et al. (2020). Logic-signal-based multiplex detection of MiRNAs with high tension hybridization and multiple signal amplification. *Analyst* 145: 4314.

39 Xiao, F., Fang, X., Li, H. et al. (2022). Light-harvesting fluorescent spherical nucleic acids self-assembled from a DNA-grafted conjugated polymer for amplified detection of nucleic acids. Angewandte Chemie International Edition 61: e202115812.

40 a) Ohta, S., Glancy, D., and Chan, W.C. (2016). DNA-controlled dynamic colloidal nanoparticle systems for mediating cellular interaction. *Science* 351: 841; b) Maye, M.M., Kumara, M.T., Nykypanchuk, D. et al. (2010). Switching binary states of nanoparticle superlattices and dimer clusters by DNA strands. *Nature Nanotechnology* 5: 116.

41 a) Mirkin, C.A., Letsinger, R.L., Mucic, R.C. et al. (1996). A DNA-based method for rationally assembling nanoparticles into macroscopic materials *Nature* 382: 607; b) Storhoff, J.J., Elghanian, R., Mucic, R.C. et al. (1998). One-pot colorimetric differentiation of polynucleotides with single base imperfections using gold nanoparticle probes. *Journal of the American Chemical Society* 120: 1959.

42 Afshari, N., Al-Gazally, M.E., Rasulova, I. et al. (2022). Sensitive bioanalytical methods for telomerase activity detection: a cancer biomarker. *Analytical Methods* 14: 4174.

43 Hussain, S., Chen, X., Wang, C. et al. (2022). Aggregation and binding-directed FRET modulation of conjugated polymer materials for selective and point-of-care monitoring of serum albumins. *Analytical Chemistry* 94: 10685.

44 Huang, Y., Yao, X., Zhang, R. et al. (2014). Cationic conjugated polymer/fluoresceinamine-hyaluronan complex for sensitive fluorescence detection of CD44 and tumor-targeted cell imaging. *ACS Applied Materials & Interfaces* 6: 19144.

45 Bhalla, N., Jolly, P., Formisano, N. et al. (2016). Introduction to biosensors. *Essays Biochem* 60: 1.

46 Lakard, B. (2020). Electrochemical biosensors based on conducting polymers: a review. *Applied Sciences* 10: 6614.

47 Wang, Z.G., Zhan, P., and Ding, B. (2013). Self-assembled catalytic DNA nanostructures for synthesis of para-directed polyaniline. *ACS Nano* 7: 1591.

48 Bai, L., Chen, Y., Bai, Y. et al. (2017). Fullerene-doped polyaniline as new redox nanoprobe and catalyst in electrochemical aptasensor for ultrasensitive detection of Mycobacterium tuberculosis MPT64 antigen in human serum. *Biomaterials* 133: 11.

49 Su, X., Wu, L., Chen, G., Zheng, C. et al. (2024). Organic conjugated polymer nanoparticles enhanced tyrosinase electrochemical biosensor for selective, sensitive and rapid detection of bisphenol A. *Science of the Total Environment* 951: 175765.

50 Fan, G.C., Han, L., Zhang, J.R. et al. (2014). Enhanced photoelectrochemical strategy for ultrasensitive DNA detection based on two different sizes of CdTe quantum dots cosensitized TiO2/CdS: Mn hybrid structure. *Analytical Chemistry* 86: 10877.

51 Zhao, W.W., Xu, J.J., and Chen, H.Y. (2015). Photoelectrochemical bioanalysis: the state of the art. *Chemical Society Reviews* 44: 729.

52 a) Guo, L., Li, Z., Marcus, K. et al. (2017). Periodically patterned Au-TiO2 heterostructures for photoelectrochemical sensor. *ACS Sensors* 2: 621; b) Zhao, M., Fan, G.C., Chen, J.J. et al. (2015). Highly sensitive and selective photoelectrochemical biosensor for Hg2+ detection based on dual signal amplification by exciton energy transfer coupled with sensitization effect. *Analytical Chemistry* 87: 12340; c) Pang, X., Bian, H., Su, M. et al. (2017). Photoelectrochemical cytosensing of RAW264. 7 macrophage cells based on a TiO2 nanoneedls@ MoO3 array. *Analytical Chemistry* 89: 7950.

53 a) Marmiroli, M., Pagano, L., Savo Sardaro, M.L. et al. (2014). Genome-wide approach in Arabidopsis thaliana to assess the toxicity of cadmium sulfide quantum dots. *Environmental Science & Technology* 48: 5902; b) Zhu, C., Liu, L., Yang, Q. et al. (2012). Water-soluble conjugated polymers for imaging, diagnosis, and therapy. *Chemical Reviews* 112: 4687.

54 Lu, H., Hu, R., Bai, H. et al. (2017). Efficient conjugated polymer–methyl viologen electron transfer system for controlled photo-driven hydrogen evolution. *ACS Applied Materials & Interfaces* 9: 10355.

55 Liu, S., He, P., Hussain, S. et al. (2018). Conjugated polymer-based photoelectrochemical cytosensor with turn-on enable signal for sensitive cell detection. *ACS Applied Materials & Interfaces* 10: 6618.

56 Wu, T., Zhao, W., Gu, J. et al. (2019). Homoleptic cyclometalated iridium (III) complex nanowires electrogenerated chemiluminescence sensors for high-performance discrimination of proline enantiomers based on the difference of electron-transfer capability. *Talanta* 194: 98.

57 Chen, Y., He, Y., Zhao, J. et al. (2022). Hydrophobic localized enrichment of Co-reactants to enhance electrochemiluminescence of conjugated polymers for detecting SARS-CoV-2 nucleocapsid proteins. *Analytical Chemistry* 94: 4446.

58 Chen, H., Lu, Q., Liao, J. et al. (2016). Anodic electrogenerated chemiluminescence behavior and the choline biosensing application of blue emitting conjugated polymer dots. *Chemical Communications* 52: 7276.

59 Luo, J.H., Cheng, D., Li, P.X. et al. (2018). An electrochemiluminescent sensor based on functionalized conjugated polymer dots for the ultrasensitive detection of Cu 2+. *Chemical Communications* 54: 2777.

60 Zhao, J., Luo, J., Liu, D. et al. (2020). A coreactant-free electrochemiluminescence (ECL) biosensor based on in situ generating quencher for the ultrasensitive detection of microRNA. *Sensors and Actuators B: Chemical* 316: 128139.

61 Ho, H.A., Najari, A., and Leclerc, M. (2008). Optical detection of DNA and proteins with cationic polythiophenes. *Accounts of Chemical Research* 41: 168.

62 Tang, Y., Feng, F., He, F. et al. (2006). Direct visualization of enzymatic cleavage and oxidative damage by hydroxyl radicals of single-stranded DNA with a cationic polythiophene derivative. *Journal of the American Chemical Society* 128: 14972.

63 Zhang, P., Lu, C., Niu, C. et al. (2022). Binding studies of cationic conjugated polymers and DNA for label-free fluorescent biosensors. *ACS Applied Polymer Materials* 4: 6211.

64 Zhang, P., Qin, K., Lopez, A. et al. (2022). General label-free fluorescent aptamer binding assay using cationic conjugated polymers. *Analytical Chemistry* 94: 15456.

65 a) Park, C.H., Kim, J.P., Lee, S.W. et al. (2009). A direct, multiplex biosensor platform for pathogen detection based on cross-linked polydiacetylene (PDA) supramolecules. *Advanced Functional Materials* 19: 3703; b) Lee, K., Povlich, L.K., and Kim, J. (2010). Recent advances in fluorescent and colorimetric conjugated polymer-based biosensors. *Analyst* 135: 2179; c) Jung, Y.K., Kim, T.W., Kim, J. et al. (2008). Universal colorimetric detection of nucleic acids based on polydiacetylene (PDA) liposomes. *Advanced Functional Materials* 18: 701.

66 Song, S., Ha, K., Guk, K. et al. (2016). Colorimetric detection of influenza A (H1N1) virus by a peptide-functionalized polydiacetylene (PEP-PDA) nanosensor. *RSC Advances* 6: 48566.

67 Yang, C.S., Chang, C.H., Tsai, P.J. et al. (2004). Nanoparticle-based in vivo investigation on blood–brain barrier permeability following ischemia and reperfusion. *Analytical Chemistry* 76: 4465.

68 a) George, S., Hamblin, M.R., and Kishen, A. (2009). Uptake pathways of anionic and cationic photosensitizers into bacteria. *Photochemical & Photobiological Sciences* 8: 788; b) Liu, L., Duan, X., Liu, H. et al. (2008). Microorganism-based assemblies of luminescent conjugated polyelectrolytes. *Chemical Communications* 45: 5999.

69 Wei, J., Liu, Y., Yu, J. et al. (2021). Conjugated polymers: optical toolbox for bioimaging and cancer therapy. *Small* 17: e2103127.

70 Feng, L., Liu, L., Lv, F. et al. (2014). Preparation and biofunctionalization of multicolor conjugated polymer nanoparticles for imaging and detection of tumor cells. *Advanced Materials (Deerfield Beach, Fla.)* 26: 3926.

71 Yuan, H., Liu, Z., Liu, L. et al. (2014). Cationic conjugated polymers for discrimination of microbial pathogens. *Advanced Materials* 26: 4333.

72 Zhu, S., Wang, X., Yang, Y. et al. (2018). Conjugated polymer with aggregation-directed intramolecular Förster resonance energy transfer enabling efficient discrimination and killing of microbial pathogens. *Chemistry of Materials* 30: 3244.

73 Qi, R., Zhao, H., Zhou, X. et al. (2021). Pyridinium-activated alkyne-based spontaneous amino-yne click polymerization toward n-π conjugated polyelectrolytes. *Angewandte Chemie International Edition* 60: 5759.

74 Wang, Y., Feng, L., and Wang, S. (2019). Conjugated polymer nanoparticles for imaging, cell activity regulation, and therapy. *Advanced Functional Materials* 29: 1806818.

75 Zhang, W., Lin, W., Li, C. et al. (2019). Rational design of BODIPY-diketopyrrolopyrrole conjugated polymers for photothermal tumor ablation. *ACS Applied Materials & Interfaces* 11: 32720.

76 Cao, Y., Dou, J.H., Zhao, N.J. et al. (2017). Highly efficient NIR-II photothermal conversion based on an organic conjugated polymer. *Chemistry of Materials* 29: 718.

77 Hong, G., Antaris, A.L., and Dai, H. (2017). Near-infrared fluorophores for biomedical imaging. *Nature Biomedical Engineering* 1: 0010.

78 Zhang, Z., Fang, X., Liu, Z. et al. (2020). Semiconducting polymer dots with dual-enhanced NIR-IIa fluorescence for through-skull mouse-brain imaging. *Angewandte Chemie International Edition* 59: 3691.

79 Yang, Y., Fan, X., Li, L. et al. (2020). Semiconducting polymer nanoparticles as theranostic system for near-infrared-II fluorescence imaging and photothermal therapy under safe laser fluence. *Acs Nano* 14: 2509.

80 a) Lecoq, J., Parpaleix, A., Roussakis, E. et al. (2011). Simultaneous two-photon imaging of oxygen and blood flow in deep cerebral vessels. *Nature Medicine* 17: 893; b) Kim, H.M. and Cho, B.R. (2015). Small-molecule two-photon probes for bioimaging applications. *Chemical Reviews* 115: 5014.

81 Wang, S., Liu, J., Feng, G. et al. (2019). NIR-II excitable conjugated polymer dots with bright NIR-I emission for deep in vivo two-photon brain imaging through intact skull. *Advanced Functional Materials* 29: 1808365.

82 Wang, L.V. and Hu, S. (2012). Photoacoustic tomography: in vivo imaging from organelles to organs. *Science* 335: 1458.
83 Jiang, Y., Upputuri, P.K., Xie, C. et al. (2017). Broadband absorbing semiconducting polymer nanoparticles for photoacoustic imaging in second near-infrared window. *Nano Letters* 17: 4964.
84 Guo, B., Chen, J., Chen, N. et al. (2019). High-resolution 3D NIR-II photoacoustic imaging of cerebral and tumor vasculatures using conjugated polymer nanoparticles as contrast agent. *Advanced Materials* 31: e1808355.
85 Ji, E., Corbitt, T.S., Parthasarathy, A. et al. (2011). Light and dark-activated biocidal activity of conjugated polyelectrolytes. *ACS Applied Materials & Interfaces* 3: 2820.
86 Dougherty, T.J., Gomer, C.J., Henderson, B.W. et al. (1998). Photodynamic therapy. *Journal of National Cancer Institute* 90: 889.
87 Li, Z., Lu, W., Jia, S. et al. (2021). Design and application of conjugated polymer nanomaterials for detection and inactivation of pathogenic microbes. *ACS Applied Bio Materials* 4: 370.
88 Hamblin, M.R. and Hasan, T. (2004). Photodynamic therapy: a new antimicrobial approach to infectious disease? *Photochemical & Photobiological Sciences* 3: 436.
89 Li, P., Poon, Y.F., Li, W. et al. (2011). A polycationic antimicrobial and biocompatible hydrogel with microbe membrane suctioning ability. *Nature Materials* 10: 149.
90 Wang, B., Wang, M., Mikhailovsky, A. et al. (2017). A membrane-intercalating conjugated oligoelectrolyte with high-efficiency photodynamic antimicrobial activity. *Angewandte Chemie International Edition* 56: 5031.
91 Xing, C., Xu, Q., Tang, H. et al. (2009). Conjugated polymer/porphyrin complexes for efficient energy transfer and improving light-activated antibacterial activity. *Journal of the American Chemical Society* 131: 13117.
92 Wang, X., Cui, Q., Yao, C. et al. (2017). Conjugated polyelectrolyte–silver nanostructure pair for detection and killing of bacteria. *Advanced Materials Technologies* 2: 1700033.
93 Yuan, H., Zhao, H., Peng, K. et al. (2019). Conjugated polymer-quantum dot hybrid materials for pathogen discrimination and disinfection. *ACS Applied Materials & Interfaces* 12: 21263.
94 Jiang, L., Bai, H., Liu, L. et al. (2019). Luminescent, oxygen-supplying, hemoglobin-linked conjugated polymer nanoparticles for photodynamic therapy. *Angewandte Chemie International Edition* 58: 10660.
95 Yang, G., Liu, L., Yang, Q. et al. (2012). A multifunctional cationic pentathiophene: Synthesis, organelle-selective imaging, and anticancer activity. *Advanced Functional Materials* 22: 736.
96 a) Jung, H.S., Verwilst, P., Sharma, A. et al. (2018). Organic molecule-based photothermal agents: an expanding photothermal therapy universe. *Chemical Society Reviews* 47: 2280; b) Liu, Y., Xu, M., Chen, Q. et al. (2015). Gold nanorods/mesoporous silica-based nanocomposite as theranostic agents for targeting near-infrared imaging and photothermal therapy induced with laser. *International Journal of Nanomedicine* 10: 4747.

97 a) Feng, G., Mai, C.K., Zhan, R. et al. (2015). Narrow band gap conjugated polyelectrolytes for photothermal killing of bacteria. *Journal of Materials Chemistry B* 3: 7340; b) Yang, K., Xu, H., Cheng, L. et al. (2012). In vitro and in vivo near-infrared photothermal therapy of cancer using polypyrrole organic nanoparticles. *Advanced Materials* 24: 5586; c) Yang, J., Choi, J., Bang, D. et al. (2011). Convertible organic nanoparticles for near-infrared photothermal ablation of cancer cells. *Angewandte Chemie International Edition* 50: 441.

98 Li, S., Wang, X., Hu, R. et al. (2016). Near-infrared (NIR)-absorbing conjugated polymer dots as highly effective photothermal materials for in vivo cancer therapy. *Chemistry of Materials* 28: 8669.

99 Fu, X., Huang, Y., Zhao, H. et al. (2021). Near-infrared-light remote-controlled activation of cancer immunotherapy using photothermal conjugated polymer nanoparticles. *Advanced Materials* 33: e2102570.

100 Sun, P., Jiang, X., Sun, B. et al. (2022). Electron-acceptor density adjustments for preparation conjugated polymers with NIR-II absorption and brighter NIR-II fluorescence and 1064 nm active photothermal/gas therapy. *Biomaterials* 280: 121319.

101 Shen, Y., Shuhendler, A.J., Ye, D. et al. (2016). Two-photon excitation nanoparticles for photodynamic therapy. *Chemical Society Reviews* 45: 6725.

102 Li, L., Zhang, X., Ren, Y. et al. (2024). Chemiluminescent conjugated polymer nanoparticles for deep-tissue inflammation imaging and photodynamic therapy of cancer. *Journal of the American Chemical Society* 146: 5927.

103 Sobotta, L., Skupin-Mrugalska, P., Mielcarek, J. et al. (2015). Photosensitizers mediated photodynamic inactivation against virus particles. *Mini Reviews in Medicinal Chemistry* 15: 503.

104 Wiehe, A., O'Brien, J.M., and Senge, M.O. (2019). Trends and targets in antiviral phototherapy. *Photochemical & Photobiological Sciences* 18: 2565.

105 Duan, X., Liu, L., Feng, X. et al. (2010). Assemblies of conjugated polyelectrolytes with proteins for controlled protein photoinactivation. *Advanced Materials* 22: 1602.

106 Qi, R., Lv, F., Zeng, Y. et al. (2023). Photoactive conjugated polymer-based strategy to effectively inactivate RNA viruses. *NPG Asia Materials* 15: 17.

107 Sun, H., Liu, J., Lv, F. et al. (2021). Selective reaction of conjugated polymers with basic proteins for broad-spectrum antivirulence therapy. *NPG Asia Materials* 13: 21.

Index

Note: Page numbers in *italics* refers to figures and page numbers in **bold** refers to tables on the corresponding page.

Conjugated Polymer Synthesis and Materials Applications, First Edition. Edited by Yoshiki Chujo.

© 2026 Wiley-VCH GmbH. All rights reserved, including rights for text and data mining and training of artificial intelligence technologies or similar technologies. Published 2026 by Wiley-VCH GmbH.

p